S0-AIV-964

Rock Engineering

Rock Engineering

John A. Franklin

Maurice B. Dusseault

McGraw-Hill Publishing Company
New York   St. Louis   San Francisco   Auckland
Bogotá   Hamburg   London   Madrid   Mexico
Milan   Montreal   New Delhi
Paris   São Paulo   Singapore
Sydney   Tokyo   Toronto

Library of Congress Cataloging-in-Publication Data

Franklin, John A.
  Rock engineering.

  Includes bibliographies and index.

  1. Rock mechanics.   2. Rock excavation.
I. Dusseault, Maurice B.   II. Title.
TA706.F76   1989   624.1′5132   88-13621
ISBN 0-07-021888-9

Copyright © 1989 by McGraw-Hill, Inc. All rights reserved.
Printed in the United States of America. Except as permitted
under the United States Copyright Act of 1976, no part of this
publication may be reproduced or distributed in any form or by
any means, or stored in a database or retrieval system, without
the prior written permission of the publisher.

1234567890   DOC/DOC   89432109

ISBN 0-07-021888-9

*The editors for this book were Joel Stein and Ingeborg M. Stochmal,
the designer was Naomi Auerbach, and the production supervisor was
Richard A. Ausburn. This book was set in Century Schoolbook. It was
composed by the McGraw-Hill Publishing Company Professional and
Reference Division composition unit. Printed and bound by R. R.
Donnelley & Sons Company.*

Information contained in this work has been obtained by McGraw-
Hill, Inc., from sources believed to be reliable. However, neither
McGraw-Hill nor its authors guarantees the accuracy or completeness of any information published herein, and neither McGraw-
Hill nor its authors shall be responsible for any errors, omissions,
or damages arising out of use of this information. This work is
published with the understanding that McGraw-Hill and its authors
are supplying information but are not attempting to render
engineering or other professional services. If such services are
required, the assistance of an appropriate professional should be
sought.

For more information about other McGraw-Hill materials,
call 1-800-2-MCGRAW in the United States. In other
countries, call your nearest McGraw-Hill office.

Contents

Preface

The chapters of this book follow the rock engineering syllabus taught to fourth-year civil engineering and earth sciences undergraduates at the University of Waterloo in Ontario, Canada. The same students can take supplementary courses relating to surface works and underground construction, or on specialized themes such as laboratory and field testing. Other, more basic courses teach engineering to geologists and geology to engineers.

Having searched for a suitable course text, and finding none that covered the field of rock engineering in a way that would, in our opinion, prepare students adequately for a career in this field, we decided to try to write one ourselves. Our starting point was a compilation of course notes tempered by consulting and research experience. After preparing a first draft on this basis, we filled important gaps and added further depth by extensive referencing of extracts from recent journal publications and conference proceedings. The book is written not only for students, but also, and perhaps mainly, for the practicing civil or mining engineer. It is divided into three parts. The first describes rocks, their environment of water and stress, and how these are explored by drilling, geophysics, mapping, sampling, and index testing. Part 2 explains the various techniques available for design, the behavior of rock, testing for design purposes, and monitoring to check the predictions and assumptions. Part 3 covers field techniques for rock excavation and support.

A planned forthcoming book, *Rock Engineering Applications*, will further demonstrate how these techniques are applied to quarrying and the use of stone, to the design of foundations, slopes, dams, and reservoirs, and to mining and civil engineering works underground.

In many places we have sacrificed detail in favor of a broad and, we hope, balanced picture. Particular attention has been given to recommending supplementary reading wherever detail has been omitted for lack of space.

ACKNOWLEDGMENT

We gratefully acknowledge the following for sparing the time to review chapters of the manuscript, and for their detailed, constructive, and most helpful comments: Dr. W. Bawden, Queen's University and Centre de Recherche Noranda; Professor Z. T. Bieniawski, Pennsylvania State University; Dr. B. Brady and Dr. A. Siggins, CSIRO Division of Geomechanics, Melbourne, Australia; Mr. R. Clarke and Dr. G. Herget, CANMET, Ottawa, Ont.; Mr. D. L. Coursen, Explosives Technologies International, Inc., Delaware; Dr. P. Cundall, University of Minnesota; Mr. J. Dunnicliff, Instrumentation Consultant, Lexington, Mass.; Dr. K. Hadley, Exxon Company, New Orleans, La., Dr. O. Hungr, Thurber Consultants Ltd., British Columbia; Mr. E. Magni and Mr. C. A. Rogers, M. T. C. Downsview, Ontario; Mr. D. Martin, Piteau & Associates, British Columbia; Dr. A. H. Merritt, Gainesville, Fla.; Mr. L. Pascoli, Longyear Canada, Inc., North Bay, Ont.; Dr. J.-C. Roegiers, Dowell Schlumberger, Tulsa, Okla.; Professor B. Stimpson, University of Manitoba; and, from the University of Waterloo, Professors R. Farvolden and J. Greenhouse and graduate students C. Bennett, C. Fordham, and N. Maerz.

John A. Franklin
Maurice B. Dusseault

About Rock Engineering

1.1 Historical Perspective

Rock engineering can be traced back to the earliest days of mining and civil engineering. Greek and Egyptian stonemasons worked intricate patterns and constructed columns and interlocking mosaics of rock that for craftsmanship can be scarcely equaled. The pyramid of Cheops in Egypt was constructed 4700 years ago, of more than 2 million blocks of dressed limestone. The first dams were built in Egypt and Iraq around 2900 B.C. Even the earliest of miners faced problems of excavation and rock support on a par with those of present times, equipped with tools and techniques that were much inferior. Early military sappers developed the art of undermining fortifications, and these skills have evolved into present-day tunneling technology.

Recurring disasters throughout the world, notably landslides and rockbursts, kill many and damage much property each year. On average, at least one large dam ruptures each year and two more suffer serious accidents. In Ontario, Canada, alone, rockbursts have occurred at a rate of more than 20 per annum, and a similar number of fatalities from rockbursts have been reported each year in South Africa. All this underscores how little we understand of rock and how poorly equipped we are to control its behavior. Nevertheless, progress has been made. Innovations during the last 100 or so years include the invention of dynamite and AN/FO explosives (1867), electric detonators (1876), tunnel-boring machines (1881), sprayed mortar (1909), rockbolts (1918), shotcrete (1942), tungsten carbide drill bits (1940), the New Austrian Tunneling Method (1950), and oil-hydraulic percussive drills (1971). Reliable instruments are now available to monitor and warn of rock movements at the surface and in underground works. Computation techniques have evolved to the stage where rock conditions can be modeled and predicted with some degree of realism.

1.2   What Is Rock Engineering?

What is meant by *rock engineering,* and how does this differ from *rock mechanics;* what are the differences, if any, between a rock engineer, a geologist, and an engineering geologist?

According to one definition, engineers are people looking for the best available answer today, whereas scientists prefer to wait until tomorrow for the "right" answer. Rock engineers are similarly pressed for time and often have to make decisions based on judgment and experience rather than "truth." More specifically, the rock engineer's responsibility relates to materials of construction, mines, tunnels, foundations, dams, and landslides; in other words, his or her job is to predict and control the behavior of anything made of, in, or on rock.

The labels *rock mechanics* and *geomechanics* are narrowly defined. *Mechanics* is, according to Webster's dictionary, "a...science that deals with energy and forces and their effect on bodies." Certainly, one part of the rock engineer's job is to study the motion, creep, or bursting of rock in a foundation or a mine, but rock engineering is by no means just mechanics. It calls for an appreciation of the nature of geological materials, groundwater, and ground stresses, and the ability to judge conditions and predict how they will affect, and be affected by, construction work. The engineer is called upon not just to study nature passively, but to intervene, excavate, and stabilize, to extract minerals and to create caverns, highways, water reservoirs, and power projects, all the time attempting to preserve and protect the natural environment.

The term *geological engineer* encompasses the *rock engineer* and the *soil engineer,* who both have a different outlook from that of the classical geologist. The classical, or "pure," geologist tries to understand the history (genesis) of rocks, whereas the geological engineer, or engineering geologist, looks to the future, trying to predict the behavior of earth materials as aggregates or fill, and their stability in the walls of excavations. This forward-looking goal of "modeling" the behavior of rock in engineering works is fundamental to applied rock engineering.

Rock physicists (geophysicists) use the science of physics to study the behavior of rock materials. One of their tasks is to investigate geologic processes and how rocks behave under high temperatures and pressures deep in the earth's crust, mantle, and core. At great depths, exploration by drilling is very costly, so most of the available information on deep geological structures has come from geophysical exploration, using deep seismic reflection to map layers of differing sonic velocity and therefore density. Deep geological processes can sometimes also be inferred by observing, at surface, rocks that once were deeply buried but now are revealed through processes of tectonics and erosion.

Specialists in these related fields often work together to solve multidisciplinary problems. Many, during a career in geotechnology, apply their knowledge to several fields, and experience in one often conveys benefits in others. Overspecialization is a hazard in a world of complex technology, and a broadly based education can be a greater asset than a very detailed knowledge of one or two highly specialized topics.

1.3 Judgment and Approximation

Rock engineering is more a craft than either an art or a science. Craftsmanship depends on an understanding or "feel" for the materials and a dexterity in manipulating the tools of the trade. Appreciation of materials in this case requires an understanding of basic geology, groundwater, and ground stress regimes and of how these interact. Tools include the methods of site investigation and testing, modeling, computation, design, excavation and stabilization, instrumentation, and monitoring.

To be accomplished in this craft, perhaps the most important attribute is an ability to view a project as a whole and to temper a rigorous scientific method with common sense. In describing the rock engineering work at the Lagrande hydroelectric complex in Quebec, Canada, Murphy and Levay (1985)* remarked that "experienced judgment based on field observations is usually more valid in the assessment of rock treatment requirements than any strictly theoretical approach." Nevertheless, they made good use of mathematical models to dimension the rock pillars and caverns and to investigate rock stresses and seepage gradients around penstocks. The studies were always viewed as of qualitative rather than quantitative significance.

Disasters seem hardly ever to be caused by lack of precision or miscalculation, but much more often by neglect, ignorance, or overspecialization. Mother Nature has a habit of taking advantage of those who turn their backs on her. In geomechanics it pays to be wary, because subsurface conditions can hardly ever be known or predicted accurately in advance of construction. Decisions are made and cost estimates are based on the best information that can be obtained within a limited time and, more importantly, a limited project budget.

---

* D.K. Murphy and J. Levay, "Rock Engineering at the Lagrande Complex, Quebec," *Canadian Tunneling* (Tunneling Association of Canada, 1985), pp. 129–141.

Uncertainties are reduced by all available means to a level considered acceptable; but even then, construction often must accommodate the unexpected.

1.4   Rock Engineering and Related Publications

The First International Congress on Rock Mechanics was held in Portugal in 1966, and in 1987 the International Society for Rock Mechanics celebrated its silver jubilee. During those 25 years the body of literature has been expanding at an accelerating pace, with an ever-increasing number of conferences and technical journals devoted to this and closely related themes.

Therefore we have included with this introductory chapter the following brief guide to the literature of rock engineering and related disciplines, including abstract services, books, periodicals, newsletters, and conferences. There must surely be many omissions for which we apologize, but the list may perhaps help those looking for further background reading.

1.4.1   Abstract journals and bibliographies

Engineering Geology Abstracts (quarterly): American Geological Institute, Alexandria, Va.

GEODEX, abstracts relating to geotechnical subjects.

Geomechanics Abstracts, issued and bound with *International Journal of Rock Mechanics and Mining Sciences* (Pergamon, Oxford).

KWIC Index of Rock Mechanics Literature. Part 1 (to 1969), E. Hoek, Ed. (AIME, New York); Part 2 (1969–1976), J. P. Jenkins and E. T. Brown, Eds. (Pergamon, Oxford).

1.4.2   Books

Attewell, P. B., and I. W. Farmer: *Principles of Engineering Geology* (Chapman and Hall, London, 1976).

Bieniawski, Z. T.: *Rock Mechanics Design in Mining and Tunneling* (A. A. Balkema, Rotterdam and Boston, 1984), 272 pp.

Blyth, F. G. H., and M. H. DeFreitas: *A Geology for Engineers,* 6th ed. (Edward Arnold, London, 1974).

Brady, B. H. G., and E. T. Brown: *Rock Mechanics for Underground Mining* (Allen and Unwin, London, 1985), 527 pp.

Brown, E. T., Ed.: *Analytical and Computational Methods in Engineering Rock Mechanics* (Allen and Unwin, London, 1987), 259 pp.

Budavari, S. Ed.: "*Rock Mechanics in Mining Practice,*" Monograph ser. M5, S. Afr. Inst. Mining Mettal., Johannesburg, 282 pp. (1983).

Coates, D. E.: *Rock Mechanics Principles* (Mines Branch Monograph 874, CANMET, Ottawa, Ont., 1970).

Duncan, N.: *Engineering Geology and Rock Mechanics,* 2 vols. (Leonard Hill, London, 1969).

Finkl, C. W., Ed.: *The Encyclopedia of Applied Geology* (Van Nostrand Reinhold, New York, 1984), 644 pp.

Goodman, R. E.: *Methods of Geological Engineering in Discontinuous Rocks* (West Publ., St. Paul, Minn., 1976).

———: *Introduction to Rock Mechanics.* (Wiley, New York, 1980), 478 pp.

Hoek, E., and J. Bray: *Rock Slope Engineering,* 2d ed. (Inst. Min. Metall., London, 1978).

——— and E. T. Brown: *Underground Excavations in Rock.* (Inst. Min. Metall., London, 1980), 527 pp.

Jaeger, C.: *Rock Mechanics and Engineering,* 2d ed. (Cambridge University Press, Cambridge, 1972), 523 pp.

Jaeger, J. C., and N. G. W. Cook: *Fundamentals of Rock Mechanics,* 2d ed. (Chapman and Hall, London, 1979), 593 pp.

Krynine, D., and W. Judd: *Principles of Engineering Geology and Geotechnics* (McGraw-Hill, New York, 1959).

Lama, R. D., V. S. Vutukuri, and S. S. Saluja: *Handbook on Mechanical Properties of Rocks,* 4 vols. (Trans-Tech Publ., Rockport, Mass.). Series on Rock and Soil Mechanics: vol. 1, Vutukuri, Lama, and Saluja, 1974; vols. 2, 3, and 4, Lama and Vutukuri, 1978.

Legget, R. F.: *Geology and Engineering,* 2d ed. (McGraw-Hill, New York, 1962), 884 pp.

McLean, A. C., and C. D. Gribble: *Geology for Civil Engineers,* 2d ed. (Allen and Unwin, Boston, Mass., 1985), 314 pp.

Obert, L., and W. I. Duvall: *Rock Mechanics and the Design of Structures in Rock.* (Wiley, New York, 1967).

Roberts, A.: *Geotechnology* (Pergamon, Oxford, 1976).

Stacey, T. R., and C. H. Page: *Practical Handbok for Underground Rock Mechanics* (Trans-Tech Publ., Rockport, Mass.). Series on Rock and Soil Mechanics, vol. 12, 144 pp. (1986).

Zaruba, Q., and V. Mencl: *Engineering Geology* (Elsevier, New York, 1976).

1.4.3 Periodicals

Bulletin of the Association of Engineering Geologists (Association of Engineering Geologists, Dallas, Tex.).

Bulletin of the International Association of Engineering Geology (Paris, France).

Canadian Geotechnical Journal (Canadian National Research Council, Ottowa, Ont.).

CIM Bulletin (Canadian Institute of Mining and Metallurgy, Montreal, Que.).

Engineering Geology (Elsevier).

Engineering and Mining Journal.

Geotechnique (Institute of Civil Engineers, London).

International Journal of Mining and Geological Engineering (Chapman and Hall, London).

International Journal of Rock Mechanics and Mining Sciences and Geomechanics Abstracts (Pergamon, Oxford).

International Journal of Surface Mining (A. A. Balkema, Rotterdam and Boston).

Journal of the Geotechnical Engineering Division (American Society of Civil Engineers, New York).

Journal of the Soil Mechanics and Foundations Division (continued as *Journal of the Geotechnical Engineering Division,* American Society of Civil Engineers, New York).

Quarterly Journal of Engineering Geology (Scottish Academic Press, Edinburgh).

Rock Mechanics (Springer, Vienna).

Rock Mechanics and Rock Engineering (Springer, Vienna and New York).

Tunnelling and Underground Space Technology (Pergamon, Oxford).

1.4.4 Newsletters

News (International Society for Rock Mechanics; Secretariat: Lisbon, Portugal).

Geotechnical News (BiTech Publishers, Vancouver, B.C.).

1.4.5 Proceedings

Canadian Rock Mechanics Symposia (every 18 months approximately): Sponsored by
the Canadian Rock Mechanics Association (CARMA); various publishers.
Congresses of the International Association of Engineering Geologists (every 4 years).
Congresses of the ISRM (every 4 years): 1st—Lisbon (1966); 2d—Belgrade (1970); 3d—
Denver (1974); 4th—Montreux (1979); 5th—Melbourne (1983); 6th—Montreal
(1987).
International Congresses on Large Dams: Sponsored by the International Commission
on Large Dams (ICOLD).
Symposia of ISRM (annual): Various countries and sponsors.
U.S. Symposia on Rock Mechanics (annual): Sponsored by U.S. National Commission on
Rock Mechanics; various publishers.

Part 1

Ground Characterization

P1.1 The Four "Elements"

Earth, air, fire, and water, the four elements of the ancients, are replaced in modern geomechanics by soil, rock, water, and stress. Any description or classification of ground conditions has to address each of these four aspects.

The alchemical analogy can be carried further by considering transmutation of the elements. Soils and rocks are quite stable although eventually, in the extensive time frame of geology, soils change to rocks by lithification, and rocks to soils by processes of weathering. Water and stress are much more transient by nature. The water table rises and falls in the fissures and pores of the earth, depending on rainfall and the seasons. A tunnel excavated in saturated ground acts as a drain that removes water, and changes pressures and the directions of flow. Similarly it modifies the stress field, acting as a "stress raiser." Engineering works affect the environment in many ways, and, in turn, the environment governs the behavior of the ground.

P1.2 Characterization—Rock Mass and Rock Material

Each rock formation has a unique "character" that not only bears the imprint of its geological history, but also controls its future behavior in engineering works. Just as a human's character can be estimated in terms of attributes such as height, weight, and intelligence, so a rock can be characterized by a set of properties. Some of these, such as color and grain size, can be observed directly, whereas others, such as strength and durability, can be measured only by testing.

The geologist's traditional role has been to study the history or genesis of rocks, whereas the engineer's has been to predict the performance of these materials in works yet to be constructed. The engineer's observations, tests, descriptions, and classifications, although based on geological groundwork, are substantially different from those of the classical geologist (Deere, 1963; Merritt and Baecher, 1981).

In engineering applications, flaws and weaknesses govern the behavior of the rock and therefore assume a much greater importance than the solid "intact" (unbroken) material. Features such as joints, bedding planes, and surfaces of cleavage or schistosity, cut through the rock (Fig. P1.1). These are usually given the collective name *discontinuities,* although quite often a simpler term, such as joints, fissures, or fractures, is

Figure P1.1   Joints in slate; Les Méchins, Gaspé, Que., Canada.

used instead. Throughout this book, *joints* will be used in its broader sense, recognizing that a joint also has a narrower geological definition.

To illustrate the importance of joints, a freestanding vertical column of rock without them could in theory stand to a height of several kilometers before the weight became sufficient to cause crushing at the base (see example in Sec. 5.1.2). In contrast, a real rock column intersected by joints would behave as a stack of blocks, each relatively indestructible. The joints would give way long before the blocks themselves.

Joints are important not only because they limit the strength of the rock mass, as in the above example, but also because they control bulk deformation and flow of groundwater. Settlement in a foundation, for example, is caused more by a closure of joints than by the compression of solid blocks. Most flow of groundwater occurs along open joints rather than through pores in the rock material, unless the joints are widely spaced and tight, and the rock material is very porous.

At one extreme, the mechanical behavior of an intensely jointed rock mass approaches that of a granular soil, whereas at the other, when the rock is massive and the joints are poorly developed, it is closer to that of a laboratory specimen. The terms (*in situ*) *rock mass* and (*intact*) *rock material* are used to distinguish jointed from unjointed rock conditions (Fig. P1.2). Large-scale rock mass behavior must be taken into account in all real rock engineering problems, and a study of specimens in the laboratory is only one step toward gaining an understanding of in situ rock performance.

P1.3   Role of Testing and Investigation

Testing is used to measure properties that cannot be determined quantitatively by observations alone, such as to record the response of rocks to imposed stresses or strains. Tests can be grouped into four categories according to objectives as follows:

Index tests, whose purpose is to assist in rock classification and characterization.

Design tests, which are needed to provide data for design computations. They measure the material param-

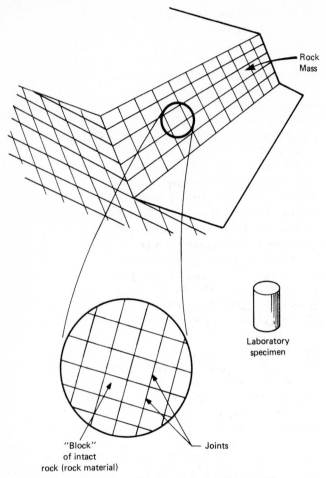

Figure P1.2 Definitions of *rock mass* and *rock material*.

eters for constitutive equations that describe rock behavior.

Quality control tests, which are used to check the quality of support or reinforcing materials, such as
rockbolts and shotcrete.

Research tests, which have no immediate practical application, but measure the fundamental behavior of
rocks and rock masses.

Index tests are described in Part 1 of the book, which is
concerned solely with characterization of the ground.

Design tests are described in Part 2 on rock behavior, and quality control tests in Part 3 on excavation and support techniques. Tests of a research nature are outside the scope of a textbook on rock engineering. Methods for measuring index characteristics are given in *Suggested Methods* (ISRM, 1981) and subsequent individual procedures published in the *International Journal of Rock Mechanics and Mining Sciences,* and also in publications of the American Society for Testing and Materials.

Index tests provide a pointer or indication of rock quality. They supplement and quantify visual observations, and thus become part of the process of rock description and classification. They must therefore be quick, simple, and inexpensive. Tens or even hundreds of specimens need to be tested, and very little time and effort can be spent precisely measuring the characteristics of just one or a few pieces of rock. For the same reasons, testing is done in the field if at all possible, using portable equipment, and as part of a program of field exploration, core logging, and outcrop mapping.

The selection of an appropriate suite of tests from the many available depends on the type of rock and also on the application. Some tests have quite broad application, whereas others may be appropriate for limestones as opposed to granites, or for the design of tunnel supports as opposed to the evaluation of rock for crushing as aggregate. Test requirements are determined to some extent by established classification systems or by contract specifications, but even in such cases, the design of a testing program requires great care to obtain the most information for the least work and expense.

Chapter 6, which brings to a close this first part of the book, also brings together the earlier chapters by focusing on how characteristics of the ground are explored and mapped in the field. The armory of techniques available for this purpose includes air photography, outcrop, trench and test pit mapping, exploratory drilling, and the excavation of exploratory shafts and adits. The end product is a site investigation report, which presents an interpretation of soil, rock, water, and stress conditions and a forecast of how the character of the ground will affect the feasibility, nature, and cost of construction.

References

Deere, D. U.: "Technical Description of Rock Cores for Engineering Purposes," *Rock Mech. Eng. Geol.*, **1**, 18–22 (1963).

ISRM: "Suggested Methods for Rock Characterization, Testing, and Monitoring," ISRM Commission on Testing Methods, E. T. Brown, Ed. (Pergamon, Oxford, 1981), 211 pp.

Merritt, A. H., and G. B. Baecher: "Site Characterization in Rock Engineering," *Proc. 22d U.S. Symp. Rock Mech.* (Cambridge, Mass., 1981), pp. 47–63.

2

Character of Rock Materials

2.1 Rock Descriptions

2.1.1 Sequence of description

Concise and consistent rock descriptions are needed when logging core, rock outcrops, or hand specimens in the laboratory. The following checklist can help ensure that all relevant characteristics are included in a consistent sequence.

1. Intact rock
   - Rock name
   - Color, mineral composition, alteration
   - Texture, grain size and shape
   - Porosity, density, water content
   - Strength, isotropy, hardness
   - Durability, plasticity, swelling potential

2. Jointing
   - Block size and shape
   - Number and types of joint sets
   - Characteristics of each set

This is an extension of the traditional system of the geologist, to which parameters of engineering significance have been added. It includes properties that can be observed directly, and others best quantified with the help of index testing. The results can be presented as a written description, a data matrix tabulation, or a classification.

The rock name is usually printed in capital letters, followed by a string of descriptors. Characteristics of the intact rock material are given first and then those of jointing. The following is an example of the intact rock part of the description only. (Jointing is covered in Chap. 3.)

> SANDSTONE, medium greenish grey; 60% quartz, 30% limestone rock fragments, 10% interstitial crystalline calcite cement; micro-cross-bedded, fine to medium sand sizes with occasional rounded quartz pebbles; porosity (10–20) 14%; easily broken by light hammer blow and slightly friable; block size (5–40) 20 cm, typical shape 1:3:3.

Note that quantitative data, when available, take preference to descriptors such as *large* or *abundant*. When such descriptors have to be used for brevity, they are defined in an appendix to the report. Quantities are expressed as a range (in parentheses) followed by a typical or representative value. Sometimes different types of descriptors can be combined, for example, "(5–15) 12-mm feldspar phenocrysts in a 1–2-mm quartz-feldspar matrix," which for clarity combines aspects of grain size and mineral composition.

The basic rock description can be augmented by any number of supplementary tests. In most practical applications the budget is limited, and only a few of the more relevant tests can be included. The testing program must be designed carefully and economically to suit the application.

2.1.2 Geological names and family characteristics

Classical rock names used without qualification can be misleading in an engineering context. For example granite, when weathered or faulted, can behave more like a crumbly sand than like the monolithic material implied by the name. Hence there is a need to qualify names with descriptors like *weathered, porous,* or *friable* or, preferably, to supplement them with quantitative index test data.

A further problem is the proliferation of names with obscure origins and meanings (Bates and Jackson, 1980). Terms such as *psammite* and *calcilutite,* used in an attempt at greater precision, can have the reverse effect and are not likely to be understood when they appear in engineering reports. For the igneous rocks which comprise only 25% of the earth's crust there are over 2000 names, reflecting subtle mineralogical changes, usually with little engineering significance. In contrast, sedimentary rocks such as shales, sandstones, and limestones have few names, yet are more abundant and can exhibit a much broader range of engineering properties.

Rather than omitting names, however, simple ones are needed. Information conveyed by generic rock names goes far beyond the scale of the hand specimen to imply global characteristics and patterns of great engineering significance, such as related to rock structure, jointing, weathering, and groundwater flow. Rock names are indispensable when attempting to infer rock mass characteristics by interpolation or extrapolation from drillhole or outcrop data. To assign names, the rocks are first identified, according to their genetic type, as igneous, sedimentary, or metamorphic, based both on large-scale structure and on features of the hand specimen; different criteria of nomenclature apply within each category (Travis, 1955; Folk, 1954, 1959; Boggs, 1967; Dearman, 1974).

For engineering reports, names should be selected from a short list of common and better known alternatives, such as those proposed by the ISRM Commission on Classification (ISRM, 1981a; see Fig. 2.1). Rocks should be assigned these "engineering names" unless more specific ones have been established by a study of mineral composition in thin section, and are likely to be familiar to the reader. Lesser known or local terms should be defined.

Modifiers play an important role in qualifying the names of rocks with intermediate grain size or mineral content. For example, *silty sandy SHALE* is used to describe a rock with at least 50% clay-sized particles, but with a substantial content of coarser sizes. The modifiers *carbonaceous, calcareous,* or *argillaceous* are used for sedimentary rocks containing a significant yet secondary content of organics, carbonates, or clay minerals, respectively.

Note that the modifier gives no information on where the secondary mineral is to be found. A calcareous sandstone may contain calcium carbonate either as grains or as intergranular cement, grain coatings, partings, or segregations along bedding planes. Information on the geometric interrelationship of constituent minerals should be conveyed as a separate part of the description, under the heading "texture," as discussed in Sec. 2.2.2.1.

Rock materials as a whole tend to fall into just a few "families," with the members of each family having broadly similar features from an engineering point of view. These and their key characteristics are now reviewed. Included in this discussion are not only the features of the intact material, but also those typical of the jointing to be found in these rock types. This to some extent preempts the discussion in Chap. 3, which covers the all-important jointing characteristics in much greater detail.

2.1.3 Igneous and metamorphic families

Igneous and metamorphic rocks are characterized by a crystalline or, more rarely, glassy texture with very low porosity, usually less than 2%, unless the rock has been weathered and leached. The strongest

Figure 2.1 Common rock names and their geological definitions. (*Based on Dearman, 1974; ISRM, 1981a.*)

Genetic Group		Sedimentary					Metamorphic		Igneous		
Structure		Bedded					Foliated	Massive	Massive—Jointed		
		Fragmental (detrital grains)		50% grains are of carbonate	50% grains are of fine-grained igneous rock	Chemical-organic rocks	Crystalline or glassy (cryptocrystalline)				
				Limestone	Volcanic ash		Quartz, feldspars, micas, acicular dark minerals	Depends on parent rock	Light-colored minerals are quartz, feldspar, mica, and feldsparlike minerals		
Grain size, mm	Texture								Acid	Intermediate	Basic
60	Very coarse-grained (Rudaceous)	Grains are of rock fragments	Rounded grains: conglomerate; Angular; grains: breccia	Calcirudite	Rounded grains: agglomerate; Angular grains: volcanic breccia				Pegmatite		
2	Coarse-grained	Grains are of rock, quartz, feldspar, and clay minerals							Granite	Diorite	Gabbro
	Medium-grained (Arenaceous)	Sandstone: grains are mainly mineral fragements. Quartz sandstone: 95% quartz, voids empty or cemented. Arkose: 75% quartz, up to 23% feldspar, voids empty or cemented. Graywacke: 75% quartz, 15% fine detrital matrix, rock and feldspar fragments		Calcarenite	Tuff	Saline rocks: halite, anhydrite, gypsum, Limestone, Dolomite, Peat, Lignite, Coal	Gneiss: alternate bands of granular and flaky minerals	Quartzite, Marble, Granulite, Hornfels, Amphibolite	Microgranite	Microdiorite	Dolerite
0.06	Fine-grained (Argillaceous or lutaceous)	Mudstone; Shale: fissile mudstone; Siltstone: 50% fine-grained particles		Calcilutite (chalk)					Rhyolite	Andesite	Basalt
0.002	Very fine-grained	Claystone: 50% very fine-grained particles				Chert flint					
	Glassy								Volcanic glasses: obsidian, pitchstone, tachylite		

16

rocks are found among this group, but there are also some very weak and easily split types.

2.1.3.1 Igneous family.

An igneous rock is one that has solidified from molten material (*magma*). It may be crystalline or glassy (vitreous) or a combination of both. Igneous rocks are further subdivided according to their grain size and their light or dark color. Light and dark varieties of the coarse-grained igneous rocks (grain size typically larger than 2 mm) are termed *granite* and *gabbro,* respectively. Medium-grained (0.06–2 mm) equivalents are *microgranite* and *diabase*. Fine-grained (smaller than 0.06 mm but still visible) equivalents are *rhyolite* and *basalt*. Glassy equivalents are *obsidian* and *tachylyte*.

Those geologists whose classification system is based on the inferred genesis (origin) of the rock, rather than on any single observable quantity such as grain size, would argue with the above definitions. They would classify gabbro and granite as intrusive (plutonic) rocks, and rhyolite and basalt as extrusive ones, based mainly on grain size, but also on the macroscopic shape of the rock unit and on several other factors.

Basalt, probably the most common extrusive rock, is characterized by flow banding, often with porous zones (pumice) and columnar jointing caused by cooling (see Fig. 3.2). Sometimes the lava flows are interbedded with horizons of volcanic ash. Basalts extruded under water often show a pillow structure (*pillow lavas*).

Granite, the most ubiquitous intrusive rock, is found in many states of decomposition. In northern regions it tends to show few or no signs of weathering, whereas in southern, more tropical climates, weathering penetrates commonly to depths of 30 m and sometimes 300 m. Granite contains feldspars and ferromagnesian minerals that are chemically unstable and tend to break down into clay minerals. Weathering is initiated by dissolved materials in rainwater and is accelerated at the high groundwater temperatures associated with tropical environments.

2.1.3.2 Metamorphic family.

A metamorphic rock is one derived from a preexisting igneous, sedimentary, or metamorphic rock as a result of a marked change in temperature or stress. *Dynamic* metamorphism generates intense stresses locally, which tend to deform, fracture, and pulverize the rock. *Regional* metamorphism affects an extensive area through an increase in pressures and temperatures. *Contact* metamorphism results from the heating of host rock in the vicinity of a body of intruded igneous magma.

Truly metamorphic rocks are by definition those in which the majority of minerals have been altered by metamorphism: heat, pressure, or a combination of both. Recrystallization has progressed to the stage where the original fabric has been obscured or destroyed. Rocks which have been only slightly metamorphosed and retain the principal features of the primary material, such as sedimentary bedding or little-altered igneous mineralogy, can be called by their primary names along with the prefix *meta,* as in metasandstone or metagranite.

When metamorphism is more advanced, new textures and minerals require the use of metamorphic rock names defined according to texture and mineral composition. A foliated metamorphic rock is called a *gneiss* when separated into bands of lighter and darker minerals. Often mica occurs as subordinate partings between bands of granular quartz and feldspar. Such rocks are termed *schist* when the content of mica, chlorite, and other platy minerals is sufficient to create foliation and fissility, usually when the content of platy minerals exceeds 50%. They are termed *slate* when mica is absent, yet the fine-grained rock is both hard and fissile. The nonfoliated metamorphic rocks are defined primarily according to mineral composition, the main types being *hornfels* (nonschistose and fine-grained), *marble* (crystalline carbonate minerals), and *quartzite* (crystalline quartz).

2.1.4 Sedimentary families

Sedimentary rocks come from many sources and include several distinctly different families. The detrital sedimentary rocks are composed mainly of broken fragments derived from a preexisting rock or from the weathering products of such rocks that have been transported by rivers, winds, or glaciers to their places of deposition. For example, about 25% of the sedimentary rock of the world is sandstone. The chemical sedimentary rocks (e.g., rock salt) are composed mainly of precipitates from solution. They usually have a crystalline texture.

Sedimentary rocks are first subdivided according to their predominant (greater than 50%) mineral composition. Those containing mainly carbonate minerals, identified if necessary by an acid test, are termed *limestones* or *dolomites* according to their magnesium content. Those consisting mainly of noncarbonate rock fragments and hard minerals such as quartz or feldspar are termed *conglomerate, sandstone,* or *siltstone* according to the predominant grain size. Those containing at least 50% of clay minerals are termed *claystone* if homogeneous, or *shale* if laminated and fissile.

2.1.4.1 Sandstone family. Grain size defines the difference between conglomerates, sandstones, and siltstones, but in engineering applications, porosity is more important. Sandstone porosities are greatest when the rock is poorly graded, that is, composed of grains within a narrow size range. They are least when small particles fill the voids

between larger ones, or when void space is filled by intergranular cement. The strongest sandstones are formed by crystallization of siliceous cements as overgrowths on the grains. The resultant rock can have a microtexture of interlocking crystals similar to the mosaic of an igneous rock, imparting similar mechanical characteristics. Weaker sandstones have little or no cement, or a cement of calcite or clay.

The geologist has traditionally distinguished also between ortho-quartzite for a sandstone composed mainly of quartz, arkose for one composed mainly of feldspar, and graywacke for one composed of rock fragments. The distinction has little significance in engineering because quartz, feldspar, and rock fragments are more or less equally competent as constituent grains. It is usually best to use the simpler generic term *sandstone*.

Characteristics related to the sedimentary depositional environment are the most important to note when characterizing the large-scale features of sandstone formations. Bedding joints are usually extensive and planar. Differences in rock mass behavior result from the beds being thin or thick and also from the presence or absence of parting materials different from, and often weaker than, the beds that they separate. Descriptions should include the typical spacings and thicknesses of such partings and planes of weakness. Joints other than bedding are often oriented perpendicular to the bedding planes, and are often discontinuous.

A rare type of flexible sandstone, termed *itacolumite,* is found in France, Brazil, the United States, and India (Dusseault, 1980). The thin beds flop from side to side when held at one end and shaken. Grain boundary cracks allow a limited amount of rotation. The behavior is similar to that of the human knee or elbow joint that is free to swing until the joint encounters its internal stop.

The pyroclastic rocks, volcanic breccia (coarser than 2 mm) and tuff (finer than 2 mm), can be regarded as a special class of sandstones, but are of igneous origin, being composed of debris ejected from volcanos. They are composed of ash fragments, primarily angular shards with a high glass and volcanic rock content, and are identified both by their microtexture and by their interbedded association with basalt or rhyolite lava flows. Often the grains become welded together by heat, in which case the appearance and engineering properties become similar to those of an igneous rock.

2.1.4.2 Limestone family. This family of *carbonate* rocks includes limestones, composed mainly of the mineral calcium carbonate, also the dolomites and dolomitic limestones that contain magnesium carbonate, and the argillaceous carbonates that contain some clay.

As with sandstones, porosity is the principal attribute distinguishing the mechanical character of the different family members. *Primary* porosity is present in most limestones, resulting from incomplete filling of voids between rounded grains. *Secondary* porosity can result from crystal lattice expansion caused by conversion of calcite to dolomite, the process of dolomitization. Further voids, up to the size of large caverns, are created by leaching and solution along grain boundaries or joints.

Limestones exhibit a great variety of textures, some porous and some dense. There are two principal types, fragmental and crystalline. The grains in a fragmental limestone are similar to those in a sandstone, but composed of calcium carbonate, a weaker mineral than quartz or feldspar. They may be in the form of shells and shell fragments (fossiliferous limestones), rounded pellets with an onionlike structure of concentric shells (oolitic limestones), or irregular grains derived from fragmented limestone rocks (lithic limestones). Other textural varieties are found in reef limestones and in recrystallized or partially recrystallized marbles and dolomitic rocks. The individual textures have only limited mechanical significance except as a result of the porosities associated with them. The more porous the rock, the lower its strength, and the greater its deformability and permeability.

The jointing patterns of limestones are again controlled by sedimentary bedding. The greater solubility of limestones, however, has at many locations resulted in *karstic* solution features that are unique to limestone terrain (Chap. 4). Groundwater travels at greater velocities along the bedding and orthogonal joint sets than through the rock matrix, so that beds are preferentially dissolved. Tubular pipes develop at the intersections of joints. The extent of solution varies greatly, from joints that are open by just a few millimeters to major potholes and caves. Some cavities remain open, whereas others are filled by sediments or by the collapse of overlying strata.

2.1.4.3 Shale family. This family of clay-bearing rocks includes materials as varied as weakly cemented claystones that break down readily in water, and much more rocklike materials approaching the character of slates. The clay content leads to characteristics that are common to the family as a whole, including a general weakness and susceptibility to softening and disintegration under the action of atmospheric weathering agencies. The shales are typically "shaley," in other words, they split readily along the planes of bedding. A distinction is often made between true shales which are flaky and easily split, and mudstones which, because of a random orientation of clay mineral platelets, are much less fissile.

Different qualities of shale within the family can be differentiated by appropriate index testing. The slake-durability test (Sec. 2.3.3.1) measures susceptibility to cycles of wetting and drying. Those materials with a slake-durability index of less than 80% may be regarded as soillike and should be further tested by measuring their plasticity index (Sec. 2.3.3.2). Those that are more rocklike may be characterized by measuring their strength and strength anisotropy using the point-load strength test (Sec. 2.3.1.3). A shale rating system based on these three index properties is described in Chap. 3.

The inherent weakness of shales means that they are often overstressed even at shallow depths in underground works, and they tend to creep and flow more readily than most other rock types (Chap. 10). They also have a tendency to shrink and swell in response to changes in moisture content. The younger shales from Cretaceous age onward are particularly likely to contain swelling clay minerals, such as montmorillonite (Secs. 2.2.1.3 and 10.4.2).

2.1.4.4 Salt family. Members of this family include rock salt (sodium chloride or halite), potash (potassium chloride or sylvite), anhydrite, and gypsum (sodium sulfate in anhydrous and hydrous forms). All rocks in the group are water-soluble and geologically are termed *evaporites,* having been formed by the evaporation of water from saline inland lakes and seas.

Members are named according to their predominant chemical and mineral composition. They are found both in massive deposits of economic significance and also thinly interbedded between strata of sandstone or shale.

As a group, these materials are weak and moderately brittle at low temperatures and pressures but become plastic and ductile at the moderate depths encountered in typical underground mining operations. Differences in strength and ductility occur between members of this family mainly as a result of differences in mineral composition. The rocks are therefore classified mainly by mineralogical and chemical testing.

2.2 Physical Characteristics

2.2.1 Color, mineral and chemical composition

2.2.1.1 Color. Whereas it has little direct bearing on engineering behavior, color should be included in a rock description because it helps one to visualize a rock and, perhaps more importantly, to correlate between *marker beds,* where they occur at different locations and depths. It also conveys information not only on primary mineral composition but also on patterns of weathering and alteration. Shales, for example, are often brown in the weathered zone, grading to gray at greater depth.

Like rock names, descriptions of color should be kept simple. The principal color names are the spectral *hues* in common usage: red, orange, yellow, green, and blue, plus white, gray, and black. Other common rock hues include pink, brown, and buff (yellowish brown).

Besides hue, two further color dimensions, *chroma* and *value,* are included in a precise color designation as needed in building-stone applications. The chroma of a color, its intensity or strength, can be designated by modifiers such as vivid, brilliant, or weak, and the value of a color by the terms light, medium, or dark. These terms can be combined in forms such as dark greenish gray, vivid purple, medium red-brown, or speckled brown and white. Color segregations that impart a characteristic texture to the rock should be noted, often in combination with mineralogical information, for example, "GNEISS, containing 10–30-mm-thick light pinkish gray quartz and feldspar bands, alternating with 2–5-mm-thick brownish black mica partings."

As in all aspects of rock description, discretion must be applied in selecting an appropriate level of detail for the project. Greater precision in color description can when necessary be obtained using color chips as comparators, such as the 40 chips in the Rock Color Chart (Geological Society of America, 1963) or the 248 in the similar but more extensive Soil Color Chart.

2.2.1.2 Common rock-forming minerals. Full and quantitative measurements of mineral composition are not often needed in rock engineering. Approximately 1700 mineral species are recognized, mostly "accessory" minerals found in very small quantities. Rarely do they have much influence on the mechanical behavior of a rock. Usually they can be identified only with the aid of a thin-section and polarizing microscope.

There are, however, some six common rock-forming mineral assemblages that control the mechanical properties of most rocks encountered in engineering projects (Franklin, 1970). These can usually be identified quite readily in the field, at least in terms of broad categories, either with the naked eye or with a hand lens or stereo microscope. Diagnostic characteristics include color, crystal form, cleavage, and hardness, which can be assessed by a simple scratch test using a pocket knife; for example, to distinguish calcite from quartz. The six categories are listed below in decreasing order of mechanical quality. As might be expected, the competence of a rock is largely governed by the hardness of the minerals it contains, and by other textural attributes such as platiness, which imparts fissility to the rock material.

Quartzofeldspathic: Acid igneous rocks, quartz and arkose sandstones, gneisses and granulites; usually strong and brittle. Important indexes: porosity, quartz/feldspar ratio, and feldspar freshness. Quartz is the most common rock-forming mineral and a major component of granites and most sandstones. It is usually transparent or white to gray, glassy, and not easily scratched. Feldspars are the principal constituents of most igneous rocks and arkose-type sandstones and include a spectrum of types intermediate between orthoclase and albite (the alkali feldspars) and between albite and anorthite (the plagioclase feldspars). Common diagnostic characteristics include an opaque, pink to white color and well-developed lines of cleavage often visible in the rectangular faces of exposed crystals. They are quite easily scratched with a pocket knife.

Lithic/basic: Basic igneous rocks (basalts and gabbros), lithic and graywacke sandstones, amphibolites; usually strong and brittle. Important indexes: porosity, texture, quartz content, freshness of dark minerals. They are characterized by mafic (dark-colored) grains such as the basic minerals (amphiboles and pyroxenes), and also rock fragments (hence lithic). Because these can seldom be identified individually without a microscope, and because their mechanical characteristics are quite similar, they may be grouped together as *mafic minerals* in a hand specimen description. When fresh, these minerals have a hardness only a little less than that of quartz. However, they are often at least partially altered or weathered into clays with a consequent reduction in hardness.

Micaceous: Schists which by definition contain more than 50% platy minerals, and those gneisses containing more than 20% mica; often fissile and weak. Important indexes: strength anisotropy, fissility, mica and quartz content, porosity. Micas and platy minerals, such as biotite, muscovite, and chlorite, occur as a minor but important component of some igneous rocks and as the major component of schistose metamorphic rocks. They are identified, as a group, by their hexagonal platiness and well-developed cleavage, and individually by their color. Biotite is typically brown to black, muscovite silver, and chlorite green. Their platiness and frequently their segregation into bands of high mica content impart weakness to the rocks in which they occur. The micas are often affected by alteration or weathering and occur as soft clayey inclusions.

Carbonate: Limestones, marbles, and dolomites; weaker than categories 2 and 3 and soluble over geological time spans; normally brittle, and viscous and plastic only at high temperatures and pressures. Important indexes: porosity, texture, calcite/dolomite ratio, quartz and clay content. Carbonate minerals such as dolomite and calcite, the principal components, are recognized mainly by the ease with which they are scratched, and by their effervescence in dilute hydrochloric acid. They occur as equidimensional crystals, grains, or fossil fragments and, because of their solubility, often as intergranular cements and pore fillings. They are usually translucent white to light buff, although sometimes darker or even black.

Saline: Rock salt, potash, and gypsum; usually weak and plastic, sometimes viscous, particularly in deep mines; soluble over engineering time spans. Important indexes: mineral composition and solubility. Salts such as halite occur either as significant mineral deposits or as subordinate beds, intergranular cements, or grains in sedimentary rock formations. These "evaporites" are soluble and are deposited from sea brine solutions. Diagnostic features common to the group include a characteristic softness and usually a translucent to pinkish white color. Individual members of the group can be identified by crystal form (for example, halite is characteristically cubic whereas gypsum tends to be fibrous), or by chemical tests.

Pelitic (clay-bearing): Mudstones, slates, and phyllites; often viscous, plastic, and weak. Important indexes: slake durability, quartz and clay content, porosity, and density. Clay minerals such as illite, kaolinite, and montmorillonite occur as the main constituents in shales and slates, and as secondary alteration products in many igneous and metamorphic rocks and limestones. They are fine-grained and therefore difficult to identify, except by inference from their softness and generally grey-green-brown coloration. Positive and sometimes quantitative identification of clay mineral types can be obtained by x-ray or differential thermal analysis. Such identification can be important because some of the clay minerals are not only soft, but also prone to swelling, whereas others, such as the more common illites and kaolinites, are dimensionally more stable. Further details of this important assemblage are given below.

The preceding is basically a petrographic classification determined by the presence and abundance of the characteristic minerals: quartz, dark grains, fresh feldspars, salts and carbonates, altered minerals, mica and platy minerals, clay minerals, pores, and cracks.

2.2.1.3 Clay minerals. Rocks containing clay minerals tend to be the most troublesome, the most abundant, the most diverse, and the most difficult to study because of their fine-grained nature. These factors justify a more detailed account of clays than of other mineral types.

Clay minerals are *phyllosilicates,* or platy silicate minerals. They are found not only in soils, but also in weathered igneous and metamorphic rocks, in sedimentary rocks such as shales, sandstones, and limestones, and as infillings in joints and faults.

The most troublesome is the smectite family of clay minerals, of which the most common is montmorillonite. This group is described further, in the context of swelling, in Chap. 10. In addition to their swelling behavior, smectites have an extremely low shear strength and a very low permeability. Even 2–4% of smectite clay in the pore space of a sandstone can reduce its permeability by several orders of magnitude.

Smectites often form from the degradation of volcanic ash, and whenever volcanic rocks are present, smectite must be suspected, either as discrete beds or as an alteration of grains. A bed of siliceous volcanic ash can convert to smectite over a period of a few million years, forming a unique rock called bentonite (after Fort Benton, Wyo.), which is composed almost entirely of smectite (usually montmorillonite).

At high pressures and temperatures, smectite changes first to illite, then to other clay minerals with much superior properties. Smectite seems to be destroyed and absent if the sedimentary rocks have been buried deeper than 3500 m or exposed to temperatures above 200°C for a long time. Sedimentary rocks older than Jurassic usually contain little smectite.

Illite is common in shales, but rare in igneous rocks. Pure illites do not swell; the platelets are stacked and tightly bound one to the other by cationic bonding, which is not easily ruptured by ordinary means. The surface area of illites is much lower than that of the smectite family.

Kaolinite, used to make fine-quality bone china, is formed by the weathering of feldspar in tropical climates, and is the most common mineral in fault gouge and joint fillings in igneous rocks. Samples of igneous rock taken from within a few hundred meters of the surface often appear "milky" because of the partial alteration of feldspars. "Rotten" granites, andesites, and other igneous rocks are weaker almost entirely because of their feldspar alteration. Kaolinite has the largest grain size found in the common clay minerals, and also displays the least plastic behavior and swelling potential, because it adsorbs little water. Pure intact kaolinite has a much higher permeability than other clays. The bonding between clay platelets is, however, not as strong as that of an illite.

Bauxite and brucite are found in tropical weathering regimes. Their presence in older rocks is evidence of intense leaching above the water table in tropical climates. Bauxite is a red, claylike mineral mined for its aluminum. It does not swell at all, and has a high shear strength compared to other clay minerals. Brucite is a magnesium mineral often found as a metamorphic or weathering product. Its properties are similar to those of bauxite.

Vermiculite is a clay mineral found in large crystals in metamorphic rocks and rarely in clay shales. It has a very high surface area and can be extremely weak, particularly if presheared. Chlorite is a metamorphic mineral often found in shales and slates, formed as the end product of low-temperature low-pressure metamorphic processes acting on clay-rich sedimentary rocks. It has no swelling potential and reasonable strength properties.

Other, much less common clay minerals can give rise to engineering problems in varying degrees when found on site. These include attapulgite, halloysite, and dickite (kaolinite family) and sepiolite, nacrite, saponite, and lepidolite (smectite family).

2.2.1.4 Nonclay platy minerals. Mica is a large-grained analogue of clay minerals, and in fact muscovite mica is chemically and crystallographically almost identical to the clay mineral illite.

Molybdenite (molybdenum sulfide) and graphite (elemental carbon) are not clay minerals, but they form fine platelets which can result in extremely low shear strengths. Both are valuable as dry lubricants in machinery. Molybdenite, formed by low-temperature hydrothermal emplacement, is often found as a coating on joints in igneous or metamorphic rocks. Graphite, the end product of high-temperature or high-pressure metamorphosis of organic materials, is found in the form of graphitic slates, schists, and gneisses. The major lead-zinc ore body at Mount Isa Mines in Queensland, Australia, is located between thick bands of graphitic slate.

2.2.1.5 Mineralogical descriptions. In the mineralogical description, the key constituents should be listed as estimated percentages in decreasing order of abundance. A granite, for example, may be described as containing "approximately 50% quartz, 30% white to buff feldspars, 10% mafic minerals, and 10% biotite mica." When part of the mineral content is obscured because of the fineness of grain, this may be stated in a description such as the following: "approximately 20% (1–8) 3-mm angular quartz in a black microcrystalline matrix."

Percentages of the principal rock-forming minerals provide key indexes to mechanical performance. Quartz percentage is an important characteristic of sedimentary rocks, particularly of siltstones and shales, and an indicator of their strength and abrasiveness. The total clay content and at least the approximate percentages of clay mineral types are useful indicators of the potentially plastic and swelling behavior of shales. The mica content is an essential index to the character of a schist.

2.2.1.6 Field identification. Minerals are distinguished optically, by eye or with a microscope. The most useful diagnostic characteristics

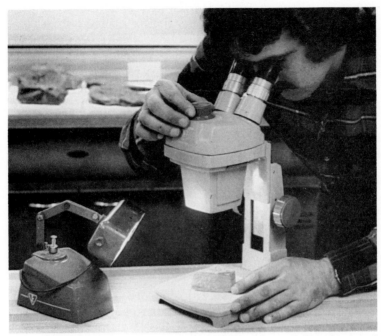

Figure 2.2 Stereo-zoom microscope for rapid examination and identification of rock samples.

are shape, color, transparency, cleavage, and hardness. A minimum of simple equipment, such as a hand lens, a pocket knife, and an acid bottle, is essential for rock characterization when logging outcrops and core. Often identification of the principal mineral constituents requires no greater degree of sophistication.

Larger projects justify an improvised field laboratory, with portable yet still quite rudimentary instruments and index testing machines. A zoom stereo-binocular microscope (Fig. 2.2) is useful for examining rocks. It is superior to a hand lens because it provides greater magnification while maintaining depth of focus and field of view. A good light source is essential. Specimens should be washed to remove dust and surface coatings. A knife or needle can be used under the microscope to assess mineral hardnesses.

2.2.1.7 Thin-section microscopy. Simple magnification of a broken rock surface, although sufficient for classification, is inadequate for a full mineralogical analysis. Rock-forming minerals and textures can be positively identified only by examining thin sections under a polarizing microscope (Fig. 2.3).

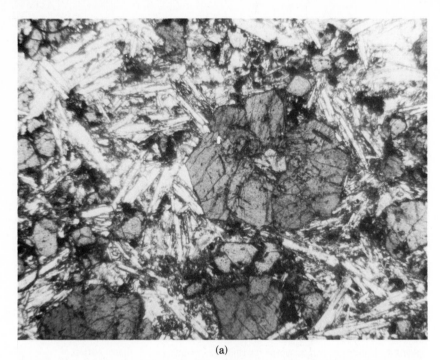

(a)

Figure 2.3 Thin-section micrographs showing typical textures for igneous sedimentary and metamorphic rock. All are viewed through crossed polarizing filters. (*a*) Igneous rock shows typical crystalline texture of a Scottish gabbro, with large dark crystals of olivine and pyroxene in a matrix of platy plagioclase feldspar. (*b*) Sedimentary rock shows typical fragmental (clastic) texture. In this wind-blown (eolian) Pennant Sandstone from England, well-rounded quartz grains are surface-stained with iron oxide and overgrown with silica, which forms a strong cement. (*c*) Metamorphic rock shows a typical foliated texture. In this schist from the Himalayas, Pakistan, the plates of mica and more equidimensional but still flattened quartz show a very strong preferred orientation, the result of recrystallization under high compressive stress. (*Courtesy of E. Appleyard and D. Lawson, University of Waterloo, Ont.*)

Montoto (1983) gives an excellent review of the petrophysical characteristics of intact rock materials and the petrographic procedures for their measurement.

To prepare a thin section, a chip of rock cemented to a microscope slide is ground to transparent thinness so that the rock fabric can be observed in transmitted light. Normal light reveals the texture of grain boundary contacts, grain sizes, and shapes. Polarized light reveals the identity of minerals by producing characteristic birefringent patterns and colors. For example, quartz gives a black-white extinction as the microscope slide is rotated in plane-polarized light, whereas feldspars, calcite, and other minerals are characterized by different colors and color banding. Staining techniques make certain types of mineral more readily identifiable under the microscope; cal-

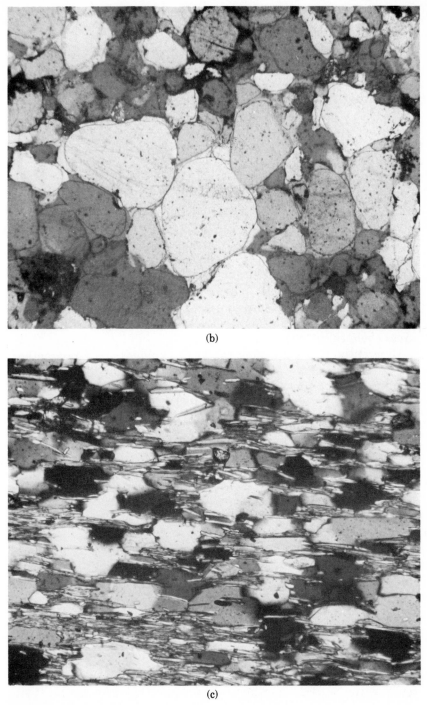

(b)

(c)

Figure 2.3. *(Continued)*

cite and dolomite can be distinguished from each other in this manner. The techniques are specialized and call for the expertise of a petrographer (Hamblin, 1962).

Quantitative measurements are more tedious and require a microscope with a measuring stage and a counting system. The thin section is moved step by step beneath the field of view, counting the number of times the cross hairs come to rest on a particular mineral type.

2.2.1.8 Other methods of microscopic and chemical analysis. Light-transmitted polarizing microscopy can be combined with light-reflected fluorescence microscopy. Supplementary studies using a scanning electron microscope (SEM) achieve much higher resolutions than are possible using an optical microscope, and are of great help in studying the pore structures and fabric of fine-grained rocks (Fig. 2.4). The SEM gives a topographical view of the rock surface. With a detector for back-scattered electrons, the different minerals can be discerned by the different emission levels corresponding with different atomic numbers of the mineral phases. In addition, an x-ray detector gives chemical information.

Differential thermal analysis (DTA) techniques and x-ray methods are needed to identify and, in some cases, to quantify the percentages of different clay minerals in an argillaceous rock (Grim, 1953). Clay minerals can be identified rigorously by x-ray diffraction, nuclear magnetic resonance, chemical analysis, or infrared spectroscopy; x-ray

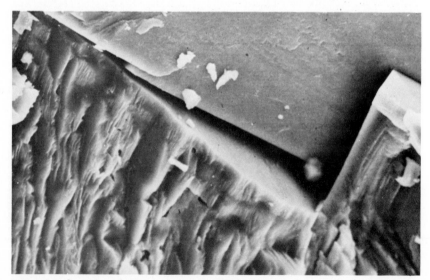

Figure 2.4 Scanning electron microscope (SEM) micrograph, revealing a triangular pore between calcite crystals in a crystalline limestone. (*Courtesy of P. Churcher.*)

diffraction is the most common and useful method. To identify clay minerals semiquantitatively, the size fraction less than 2 μm is removed, mounted on a ceramic or glass slide so that all the clay minerals are oriented parallel to the slide surface, treated with ethylene glycol, and x-rayed.

For rocks in general, chemical analysis techniques are seldom as valuable as mineralogical ones, because most rocks and rock-forming minerals are monotonously similar in their chemical composition. A typical granite, for example, may have an almost identical chemical composition to that of a sandstone or even a shale, whereas the mineralogical compositions and textures of these three rocks are completely different. Chemical analysis can, however, be useful in certain specific situations. For example, total carbonate content and calcite/dolomite ratios can be determined using the Chittick apparatus, which measures the volume of CO_2 produced when a sample is dissolved in HCl (Dreimanis, 1962).

2.2.2 Textures, grain sizes, and shapes

2.2.2.1 Textures. The *texture,* or *fabric,* of a rock is the size, the shape, and the arrangement of constituents on the scale of the hand specimen, one or a few centimeters in size. This contrasts with *structure,* which is the arrangement of rock mass components on a scale of several meters. Structure includes such features as macrobedding, faulting, and folding. Textural and structural differences are diagnostic in distinguishing between igneous, metamorphic, and sedimentary rocks, which have very similar mineral assemblages.

Before any description or testing can be accomplished, the rock mass or drill core is subdivided into geotechnical mapping units (GMU), within the boundaries of which the rock is regarded as homogeneous and therefore capable of being described by a single set of characteristic properties. The arrangement of constituents within each GMU is termed *texture,* whereas the arrangement and juxtaposition between the GMUs is termed *structure.*

The following are definitions of some of the more common and useful textural descriptors.

Crystalline rocks consist of an interlocking mosaic of crystals (Fig. 2.3*a*), whereas *fragmental* (detrital or clastic) rocks are made up of grains or fragments that are seldom in such close contact (Fig. 2.3*b*). The crystalline rocks, because of their interlocking texture, are typically stronger, less porous, and less deformable than fragmental varieties with similar mineral composition; a quartz-feldspar granite is usually much stronger than a quartz-feldspar (arkose) sandstone. All igneous and most metamorphic rocks are crystalline, whereas the fragmental rocks are exclusively sedimentary. Some sedimentary rocks, however, are crystalline, notably the evaporites deposited from solution. A partially crystalline texture can result from metamorphism, alteration, or secondary crystallization in the pore space of a sedimentary rock, leading to textural descriptions such as "subrounded 1–2-mm quartz grains in a microcrystalline calcite matrix."

A texture may be termed *homogeneous* or *heterogeneous,* depending on whether or not all parts of the sample have a near-identical texture and mineral composition. It may be termed *isotropic* or *anisotropic* depending on whether or not preferred orientations are visible (Fig. 2.3*c*). For example, a quartzite is homogeneous and anisotropic if it consists entirely of elongated quartz grains with their long axes parallel. Anisotropy can result either from segregation (bands or partings with different mineral compositions or grain sizes) or from preferred orientation of the long axes of platy or elongated grains such as micas or clays, or even occasionally from preferred orientation of crystallographic axes when the crystals themselves are equidimensional.

2.2.2.2 Grain sizes and shapes. Grain sizes and size distributions can be given either numerically as a range and a typical value, such as "(1–5) 3 mm," or with reference to the standard particle size nomenclature for soils, such as "silty sand with some gravel sizes." The commonly accepted size designations, chosen for easy spacing on a logarithmic scale, are clay (finer than 0.002 mm), silt (0.002–0.06 mm), sand (0.06–2 mm), gravel (2–60 mm), cobbles (60–200 mm), and boulders (coarser than 200 mm). The designations are more appropriate for sedimentary rocks than for igneous and metamorphic ones. For consistency, it may therefore be better to use numerical values rather than size names, irrespective of the rock type.

Grain sizes can be estimated visually with the aid of either a microscope graticule or a set of reference samples. Such samples can be prepared from a sand by sieving and mounting the sieved fractions on pieces of card labeled with the grain-size range.

The shape of fragmental or crystalline grains can be described in terms of the relative lengths of orthogonal grain axes, hence terms such as *equidimensional* (1:1:1); *platy* or *discoid* (two long axes and one short); and *fibrous* or *prolate* (two short axes and one long). Fragmental grains which have been subjected to sedimentary attrition may be further described by terms such as *angular, subangular, subrounded, rounded,* or *well rounded.* These descriptors are usually combined with each other and with mineralogic and grain-size information in a form such as: "discoid, well rounded quartzite pebbles in a silty sand matrix."

2.2.3   Porosity, density, and water content

2.2.3.1   Importance of pores and microfissures.   From an engineering standpoint, pores are by far the most important "mineral" in a rock because they are the weakest. They govern physical attributes such as strength (Fig. 2.5), deformability, and hydraulic conductivity.

Porosity, the ratio of pore volume to total volume, is characterized by descriptors such as *dense, porous,* or *very porous* or, when feasible, by numerical values obtained by testing. *Primary* porosity results from the presence of pores in the form of either subrounded intergranular space, or elongated microfissures at grain boundaries. Microfissured rocks usually are weaker than porous ones without microfissures but with the same total porosity (Kelsall et al., 1986). Joints introduce a *secondary* porosity.

2.2.3.2   Definitions.   Rock is composed of three phases: solid minerals, water, and air; the latter two, together, fill the pore space. Various parameters describe the relative percentages of these phases.

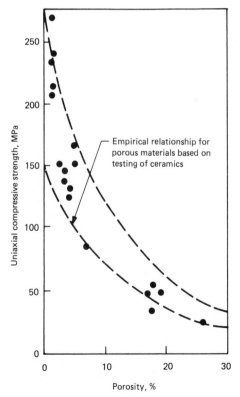

Figure 2.5   Effect of porosity in reducing the strength of rocks and other brittle materials. Data points relate to tests on a vesicular lava from California. (*After Kelsall, Watters, and Franzone, 1986.*)

Dry density or *unit weight* is defined as the weight of solids divided by the total volume of the specimen.

Porosity is pore volume as a percentage of total volume.

Degree of saturation is the ratio of water to pore space by volume.

Water content is the ratio of water to solids by weight.

These parameters are determined by measuring the weight and volume of solid, liquid, and gaseous phases (ISRM, 1981*b*). Knowing two or more parameters, the others can be obtained by simple equations, since all are interrelated.

2.2.3.3 Porosities of typical rocks. Sands have porosities of 30–50%, whereas the porosity of a sandstone, because of compaction and cementing, is often reduced to less than 5% and sometimes less than 1%. Muds can have porosities as high as 70–80%, whereas after compaction the equivalent shale has a primary porosity usually in the range of 5–20%.

Young carbonate rocks commonly have primary porosities in the range of 20% for coarse, blocky limestone, to more than 50% for poorly indurated chalk. Most dolomite is formed by geochemical alteration of calcite. This mineralogical transformation causes an increase in porosity because the crystal lattice of dolomite occupies about 13% less space than that of calcite. Secondary porosities in carbonate rocks can be very high as a result of solution and the formation of karstic cavities (Sec. 4.1.4.3).

Igneous and metamorphic rocks have a primary porosity rarely greater than 2%, although weathering, which progresses outward from the grain boundary cracks and particularly from joints, can occasionally result in porosities of 50% or more. Such is the case for granites in which the feldspars and micas have been altered and leached, leaving a fragile skeleton of loosely packed quartz crystals.

In sedimentary rocks, particularly sandstones, porosity decreases systematically with depth, at a rate of about 1.3% for every 300-m depth (Fig. 2.6). Similar effects are observed in igneous and metamorphic rocks, but these are less pronounced because the materials have a very low porosity even at shallow depth.

2.2.3.4 Measurement of porosity and water content. Porosity can be observed and measured under the microscope by preparing a thin or polished section of rock that has been impregnated with colored opaque plastic, usually fluorescine dye carried in a fluid monomer cementing agent (Franklin, 1969). This allows pore space to be distinguished and measured by the same counting techniques as used for mineral deter-

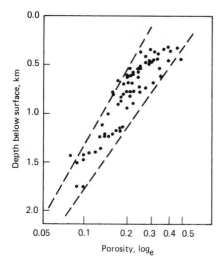

Figure 2.6 Typical reduction in the porosity of rock caused by depth and the increasing weight of overburden. (*Thompson and Schatz, 1986.*)

minations. The microscope is useful also for evaluating pore space configuration.

Physical methods for porosity determination rely on various alternatives for measuring the three phases, solids, water, and air (Manger, 1966). Weights of solids and pore water are usually measured simply by weighing the specimen at natural water content, drying at a standard 105°C, and then reweighing to determine dry weight and water loss.

Problems are encountered with rocks such as oil shales that contain volatile materials other than water, and with clay shales that contain pore water tightly bound to the rock-forming minerals. Rocks contain both "free" and "fixed" water (Chap. 4). When a specimen is removed from the ground, the free water drips out or evaporates. The remainder is retained within the rock fabric by capillary forces and by molecular attraction in some cases. Even heating to temperatures above the 100°C boiling point of free water is insufficient to remove the more tightly bound water molecules. In the case of shales, for example, temperatures of several hundred degrees, but well below those at which the clay is fired into a ceramic (1100°C), are needed to remove the last traces of interstitial water vapor. Therefore the usual drying temperature of 105°C is an arbitrary one. Convention dictates that the "natural" water content of rock materials is referred to this arbitrary datum.

Total volume is easily measured when the specimen has a regular geometry; a core specimen with flat ends, for example, has a volume equal to the area of its base times its height. The volume of an irregular specimen or sample of aggregate can be measured by saturating the rock in vacuum, or by coating pieces with a film of plastic or wax.

The saturated or coated specimens are then placed carefully into a measuring cylinder, and their volume is determined either from water displacement in a measuring cylinder, or as the weight or volume of water overflow. An alternative method employs Archimedes' principle. The saturated or coated rock is weighed first in air, then suspended in water. The difference in weights (grams) is equal to the volume of water displaced (cubic centimeters).

The most common method for measuring interconnected pore volume is to vacuum-saturate and surface-dry the specimens, weigh them, oven-dry them, and weigh them again. Pore volume is determined from the weight of water to completely fill the pores, which is the difference between saturated and dry weights. This technique works well when the rock minerals are water-insensitive, but not if they expand or shrink, in which case the pore volume and even the volume and weight of minerals are variables. Such problems can be overcome to some extent by using a fluid such as kerosene in place of water.

A mercury porosimeter is often used to measure the porosities of limestones and sandstones in oil reservoirs. The core specimen is dried, weighed, and placed in the porosimeter to measure first the total volume by mercury displacement. The pore air volume is then measured using the mercury screw pump to increase ambient pressure, measuring the pressure-volume relationship, and applying Boyle's law (Washburn and Bunting, 1922; Ramana and Venkatanarayana, 1971). The mercury technique can also give information concerning the pore throat size distribution.

The preceding methods measure only "interconnected" porosity; they treat any trapped pores, bubbles, or vesicles as part of the volume of solids. When, occasionally, such bubbles form an important part of the rock fabric, it becomes necessary to distinguish between *interconnected* and *total* porosities. Interconnected porosity is measured as above. Total porosity is measured by crushing the rock to a fine powder and measuring the volume of powder by fluid displacement in a pycnometer or density bottle. The total volume of pores is calculated as the difference between the volume of the specimen and that of the crushed particles.

2.3   Mechanical Characteristics

2.3.1   Strength

2.3.1.1   Strength and strength index tests. Even when the rock is closely jointed, intact strength is important because along with the joint wall roughness, it determines the shear strength of the rough joint surfaces. Information on strength is therefore essential in the rock description. Terms such as *strong* or *weak, fissile* (easily split) or *friable* (easily crumbled) when used should be defined (Ingram, 1953). In the absence of more sophisticated testing, simple hammer and pocket knife methods allow rock strengths to be described in such terms as "easily broken by a light hammer blow" or, more concisely, according to carefully defined criteria, as given in Fig. 2.7 (ISRM, 1981*a*; Kirsten, 1982).

2.3.1.2 The uniaxial compressive strength test. For many years the *uniaxial,* or *unconfined,* test was the main quantitative method for characterizing the strength of rock materials (Hawkes and Mellor, 1970). Mount Isa Mines, for example, have performed more than 20,000 uniaxial compressive strength tests during a 30-year period (J. V. Simmons, personal communication). The test is still the basis of many rock classifications, although for field characterization it has been largely superceded by the simpler point-load method described below.

In the uniaxial test, a rock cylinder with a length two to three times its diameter is cut from core using a diamond saw, and the ends are ground flat and perpendicular to the cylinder axis using a lapping machine. It is loaded by a compression testing machine, using spherical seatings to ensure that the load is applied axially. Compressive stress is calculated by dividing the applied force by the specimen cross-sectional area. The stress level is increased steadily until the specimen fails, when the applied stress is equal to the uniaxial compressive strength of the rock.

Drawbacks of the test include the time and effort needed to cut and

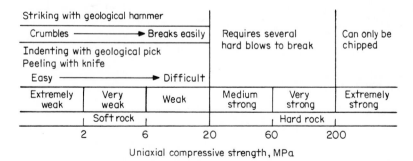

Figure 2.7   Rock strength designations and crude test criteria for estimating intact rock strength in the field. (*Modified from ISRM, 1981a; Kirsten, 1982; Canadian Geotechnical Society, 1985.*)

flatten specimen ends. The test can be used only when core is available in suitable lengths, and when the rock is sufficiently sound to allow machining.

2.3.1.3 The point-load strength test. In the point-load test, core specimens or irregular rock fragments are loaded to failure between the conical platens of a portable and lightweight tester (Fig. 2.8a). Failure load P and platen separation distance D are measured, and the point-load strength index is calculated by applying a correction to the uncorrected point-load strength P/D^2 to account for specimen size and shape, as discussed below (Broch and Franklin, 1972; ISRM, 1985).

The test has three variations, diametral, axial, or irregular lump, the choice of which depends on the available specimen geometry (Fig.

(a)

Figure 2.8 Point-load test. (a) Portable tester. (b) Appearance of core after diametral testing.

(b)

Figure 2.8   *(Continued)*

2.9). No specimen machining is needed. The *diametral* test loads a stick of rock core across its diameter. The platens should make contact with the core no closer than one core diameter from the nearest free end. If the core is anisotropic, it is rotated so that the platens make contact along a single plane of weakness. The *axial* test loads rock disks, such as those produced by diametral testing (Fig. 2.8b), in a direction perpendicular to the planes of weakness. The *irregular lump* test uses specimens of irregular shape such as of broken, crushed, or blasted rock.

The size-corrected point-load strength index I_{s50} is defined as the value of P/D^2 that would have been measured by a diametral test with $D = 50$ mm. No correction is needed when D is close to 50 mm, such as when testing NX core whose diameter is 54 mm. When D is substantially different from 50 mm, one of two methods of size correction is employed:

1. When possible, a range of sizes is tested in order to plot log P against log D^2, which is generally a straight-line graph. The value of P at $D^2 = 2500$ mm^2 ($D = 50$ mm) is obtained by interpolation or extrapolation.

2. When only single-sized material is available, a correction factor $F = (D/50)^{0.45}$ is calculated, from which I_{s50} is obtained as $F(P/D^2)$.

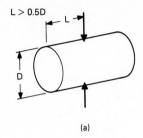

(a)

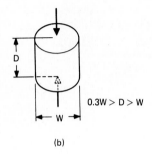

(b)

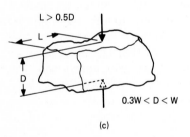

(c)

Figure 2.9   Three types of point-load strength test, showing limitations on the geometry of specimens. (*a*) Diametral test. (*b*) Axial test. (*c*) Irregular lump test. (*ISRM, 1985.*)

Shape errors are avoided for platy or discoid specimens by applying load along the shortest specimen dimension, and by measuring the minimum specimen width W perpendicular to the loading direction. The uncorrected point-load strength is then taken as P/DW (rather than P/D^2), which may be size-corrected in the manner described above.

This is a convenient method for measuring not only strength, but also strength anisotropy. The point-load strength anisotropy index I_{a50} is defined as the ratio of greatest to least strengths, measured perpendicular and parallel to the planes of weakness. When the rock core is drilled nearly perpendicular to the planes of weakness (bedding, for example), diametral testing gives the lowest strength values and separates the rock into disks with flat end faces. The disks are then rotated and subjected to axial point-load strength testing to give maximum point-load strength values.

Uniaxial compressive and point-load strengths are often closely correlated, the former usually being 20–25 times the latter. However, since the ratio can vary in the range of 15–50 in exceptional cases, the prediction of uniaxial strength from point-load strength is unreliable unless confirmed by running both types of test on the same rock type. Point-load strength is best used directly for rock classification rather than as a means of predicting uniaxial compressive strength.

2.3.1.4 The Brazilian strength test. Like the point-load method, the Brazilian test belongs to a family of "indirect tensile" strength tests in which compression applied at the platens generates combined tension and compression in the central part of the specimen. A crack starting in this region propagates parallel to the axis of loading (Fairhurst, 1964; Mellor and Hawkes, 1971). The specimens are disks with flat and parallel end faces. They are loaded diametrally, but along line contacts as opposed to the point contacts of the otherwise similar diametral point-load test.

As an index testing method, the Brazilian test is more convenient than the uniaxial compressive test, but less so than the point-load test which postdates it. This is because of the need for core and for machining of end faces. The method is sometimes used as a design test. Brazilian tensile strength, given by $0.64P/D^2$, is approximately equal to the uniaxial tensile strength of the rock. The test is also used for measuring anisotropy of strength and elastic modulus (Chaps. 8 and 9).

2.3.2 Hardness

2.3.2.1 Concept of hardness. The concept of hardness is applied loosely to rock as a synonym for strength (e.g., soft rock, hard rock; see Fig. 2.7), but hardness is more strictly defined as a property of metals and of the rock-forming minerals. Depending on the test method, we can define scratch, indentation, abrasion, impact, and rebound hardnesses. Only the rebound tests have been used to any extent for rock characterization. Abrasion hardness and impact toughness measurements are employed in special applications, such as for the evaluation of building stones and for the assessment of rock drilling and cutting equipment. An interesting sandblast test for abrasion resistance has been described in Verhoef (1987).

2.3.2.2 Scratch and indentation tests. The original scratch test was devised by Friedrich von Mohs in 1882. The Mohs scale of hardness uses

10 common minerals as standards, with talc ($H = 1$) as the softest and diamond ($H = 10$) as the hardest. On this scale, calcite has a hardness of 3 and quartz of 7. Indentation hardness tests measure the penetration of a sphere, cone, or pyramid indentor into the test surface. They include the Brinell, Vickers, Knoop, and Rockwell techniques, developed by metallurgists.

2.3.2.3 Rebound tests. The Schmidt rebound hammer (Fig. 2.10), initially used for concrete testing, can also be used to give a quick, approximate estimate of rock strength. The very portable type L hammer used in rock testing contains a spring-loaded plunger that is pressed against an outcrop surface or core specimen. When fully depressed, a weight is released that strikes the plunger and rebounds by an amount registered by a pointer and scale, giving the Schmidt rebound number. The height of rebound is proportional to the hardness (resilience) of the rock surface, zero for very soft materials and close to 60 when the surface is hard and unbroken (Poole and Farmer, 1980; Haramy and DeMarco, 1985).

The test is simple and can be useful in the field to confirm the quality of rock exposed in foundations, tunnels, and shafts. The surface to be tested must be clean, smooth, and unaffected by joints and fissures; otherwise unrealistically low values of the rebound number are recorded. Rock core specimens for testing in the laboratory must be se-

Figure 2.10 Schmidt hammer test for rock hardness. In situ testing at Noranda's Hemlo mine, Ontario, Canada.

Figure 2.11 Shore scleroscope. Like the Schmidt hammer, a rebound test for hardness.

curely mounted in a heavy steel V block, or the results will be influenced by energy losses in the mounting.

The Shore scleroscope (Fig. 2.11) also makes use of the rebound principle by recording the rebound height of a small diamond-tipped hammer which is released down a glass tube to strike the rock surface. The impact energy is much less than that of the Schmidt hammer, and the surface tested much smaller; the results reflect more the hardness of individual minerals than bulk characteristics. Several tests are needed at different locations on the rock surface to average out local variations. The test is, however, quick and convenient for rock characterization in the laboratory. Both the Schmidt rebound and the Shore scleroscope tests have the disadvantage of giving a relatively high scatter of results, and a small range of applicability. They give zero values when the rock is moderately soft and so can seldom be used, for example, on shales.

2.3.2.4 Abrasiveness tests. Abrasiveness tests are used to evaluate the rate of wear of tools used to cut rock (West, 1981). They measure the capacity of rock to abrade or grind a cutting tool such as a drill bit or the disk cutter of a tunnel-boring machine (Chap. 15).

In the Taber test, for example, rock abrasiveness is evaluated by measuring the weight loss of abrasion wheels with a standardized geometry and hardness that travel over the rock surface at a standardized velocity for a standardized distance (Tarkoy and Hendron, 1975). Taber abrasiveness A_r is the reciprocal of weight loss. A low value indicates that the rock is relatively abrasive.

2.3.3 Durability, plasticity, and swelling potential

2.3.3.1 Slake durability. All rocks are more or less affected by wetting and drying. Materials like unweathered granites and well-cemented quartzitic sandstones are "durable" because they can survive many cycles of wetting and drying without disintegration. They are nevertheless weaker when wet because of the action of water in microcracks. The results of strength tests should therefore state the water content of the rock at the time of testing.

In contrast, many clay-bearing rocks (shales and some weathered igneous rocks) and others containing minerals such as anhydrite will swell or disintegrate when exposed to atmospheric wetting and drying. Not only are they weaker when wet, but they are also permanently weakened.

Slake durability, defined as the resistance of a rock to wetting and drying cycles, can be assessed by immersing samples in water and noting their rates of disintegration. In the ISRM form of slake-durability test, the slaking process is standardized and the results are measured (Franklin and Chandra, 1972). The apparatus is shown in Fig. 2.12. Ten lumps of rock, each weighing between 40 and 60 g, are placed in a sieve mesh drum and oven-dried. The drum is then immersed in a water bath and slowly rotated; samples of moderate to low durability progressively disintegrate and fragments leave the drum through the sieve mesh. After 10 min the drum with the remaining fraction of the sample is removed, redried, and subjected to an identical second cycle

Figure 2.12 Slake-durability test, used to assess the resistance of rocks to cyclic wetting and drying.

of slaking. The second-cycle slake-durability index I_{d2}, used as the reference for classification purposes, is the weight of dry sample remaining in the drum after two cycles of slaking, expressed as a percentage of the initial dry sample weight. I_{d2} values approach zero for samples that are highly susceptible to slaking, and approach 100% for the more rocklike materials that slake very little.

2.3.3.2 Plasticity. The Atterberg limit tests that measure the liquid and plastic limits and the plasticity index are recommended to further characterize the claylike shales, those with I_{d2} less than 80%. Samples for plasticity testing should be taken from material that remains in the water bath after slake-durability testing. Crushing and chemical dispersion are avoided because although they are effective in producing a pulverized product, the artificial "soil" bears little resemblance to the rock. Plasticity tests are inappropriate for nonplastic rocklike materials.

The plastic limit is determined first by repeatedly rolling the sample into a thread, by hand on a glass plate. The sample dries with each successive rolling. The plastic limit w_p is defined as the water content of the sample when it can be rolled no thinner than 3.2 mm (⅛ in) without breaking. If the sample cannot be rolled into a thread as thin as this, it is reported as NP (nonplastic) and need not be tested for the liquid limit.

The liquid limit is measured by filling the cup of a special tester with the same sample, scribing a central groove with a grooving tool, and then lifting and dropping the cup until the groove closes. It is defined as the water content required for the groove to close in 25 drops. The water content is adjusted and the test repeated at least three times. The liquid limit w_L is found by interpolating a graph of water content versus drop count.

The plasticity index I_p, also in percentage of water content, is defined as the difference between liquid and plastic limits. This, together with second-cycle slake-durability index and the point-load strength, can be used for the classification of shales (see Sec. 3.3.3.3 and Franklin, 1983).

2.3.3.3 Swelling potential. Swelling tests can be used to further characterize the behavior of rock types that, as a result of an initial screening by slake-durability testing, show evidence of potential swelling problems. They are too slow to be practical as primary index tests. Simple tests are outlined below, and the more complex swelling tests used mainly for purposes of design are described in Chap. 10.

Katzir and David (1968) describe a simple free-swell test on powdered rock, crushed and placed in a test tube. Water is added and the amount of volume expansion is recorded. Another form of ad hoc free-swell test is to determine volumetric expansion before and after allowing the rock to swell in water, by immersing the specimen in a nonpenetrating fluid such as mercury.

A qualitative index of slaking, swelling, and dispersion potential is described in Dusseault et al. (1983). The method requires observation of volumetric expansion (swelling) and nonexpansive deterioration (water slaking), and the tendency for the clay mineral present to autodisperse (dispersion potential). The rock is immersed in water, and observations are made over a 24-h period at progressively increasing intervals (1, 2, 4, 8, 15 min, and so on).

2.3.3.4 Other durability tests. Some of the more resistant rock types that are unaffected by water content changes may still be weakened or disintegrate when exposed to freeze-thaw cycles or to saline or other aggressive environments. When describing or classifying these materials for potential use as aggregates, building stones, or armor stones that are to be exposed to frost or to chemical or salt attack, it is important to subject them to freeze-thaw and other forms of durability assessment. The term *durability* itself must be qualified as slake durability, and so on, according to the nature of the test.

Various abrasion and accelerated weathering tests are often specified for evaluating the durability of riprap. Resistance to freezing and exposure to saltwater can be evaluated by the sulfate soundness test (AASHTO test T104 for ledge rock using sodium sulfate). Stone should have a loss not exceeding 10% after five cycles. In the Los Angeles abrasion test (AASHTO test T96), the stone should have a percentage loss of not more than 40% after 500 revolutions. In the freezing and thawing test (AASHTO test T103 for ledge rock, procedure A) the stone should have a loss not exceeding 10% after 12 cycles of freezing and thawing.

References

Bates, R. L., and J. A. Jackson, Eds.: *Glossary of Geology* (Am. Geol. Inst., Falls Church, Va., 1980), 751 pp.

Boggs, S.: "A Numerical Method for Sandstone Classification," *J. Sed. Petrol.*, **37** (2), 548–555 (1967).

Broch, E., and J. A. Franklin: "The Point-Load Strength Test," *Int. J. Rock Mech. Min. Sci.*, **9**, 669–697 (1972).

Canadian Geotechnical Society: *Canadian Foundation Engineering Manual*, 2d ed. (BiTech Publ., Vancouver, B.C., 1985), 456 pp.

Dearman, W. R.: "The Characterization of Rock for Civil Engineering Practice in Britain," *Proc. Colloq. Géologie de l'Ingénieur* (Liège, Belgium, 1974), pp. 1–75.

Dreimanis, A.: "Quantitative Gasometric Determination of Calcite and Dolomite by Using Chittick Apparatus," *J. Sed. Petrol.*, **32** (3), 520–529 (1962).

Dusseault, M. B.: "Itacolumites: The Flexible Sandstones," *Quart. J. Eng. Geol.* (London), **13**, 119–128 (1980).

————, P. Cimolini, H. Soderberg, and D. W. Scafe: "Rapid Index Tests for Transitional Materials," *ASTM Geotech. Test. J.,* **6** (2), 64–72 (1983).

Fairhurst, C.: "On the Validity of the Brazilian Test for Brittle Materials," *Int. J. Rock Mech. Min. Sci.,* **1** (4), 535–546 (1964).

Folk, R. L.: "The Distinction between Grain Size and Mineral Composition in Sedimentary Rock Nomenclature," *J. Geol.,* **62**, 344–359 (1954).

————: "Petrographic Classification of Limestones," *Bull. Am. Soc. Petrol. Geol.,* **43** (1), 1–37 (1959).

Franklin, J. A.: "Rock Impregnation Trials Using Monomers, Epoxide and Unsaturated Polyester Resins," *J. Sed. Petrol.,* 1251–1253 (Sept. 1969).

————: "Observations and Tests for Engineering Description and Mapping of Rocks," *Proc. 2d Int. Cong. Rock Mech.* (Belgrade, Yugoslavia, 1970), vol. 1, paper 1-3, pp. 1–6.

————: "Evaluation of Shales for Construction Projects: An Ontario Shale Rating System," Res. Rep. RR229, Ontario Ministry Transp. and Commun., Toronto, 9 pp. (1983).

———— and R. Chandra: "The Slake-Durability Test," *Int. J. Rock Mech. Min. Sci.,* **9**, 325–341 (1972).

Geological Society of America: *Rock Color Chart* (Geol. Soc. Am., New York, 1963).

Grim, R. E.: *Clay Mineralogy* (McGraw-Hill, New York, 1953).

Hamblin, W. H.: "Staining and Etching Techniques for Studying Obscure Structures in Clastic Rock," *J. Sed. Petrol.,* **32** (3), 530–533 (1962).

Haramy, K. Y., and J. J. DeMarko: "Use of the Schmidt Hammer for Rock and Coal Testing," *Proc. 26th U.S. Symp. Rock Mech.* (Rapid City, S.D., 1985), pp. 549–555.

Hawkes, I., and M. Mellor: "Uniaxial Testing in Rock Mechanics Laboratories," *Eng. Geol.,* **4** (3), 177–285 (1970).

Ingram, R. L.: "Fissility of Mudrocks," *Bull. Geol. Soc. Am.,* **64**, 869–878 (1953).

ISRM: "Basic Geotechnical Description of Rock Masses," ISRM Commission on Classification of Rocks and Rock Masses, M. Rocha, Coordinator, *Int. J. Rock Mech. Min. Sci. Geomech. Abstr.,* **18** (1), 85–110 (1981a).

————: "Suggested Methods for Rock Characterization, Testing, and Monitoring," ISRM Commission on Testing Methods, E. T. Brown, Ed. (Pergamon, Oxford, 1981b), 211 pp.

————: "Suggested Method for Determining Point Load Strength," ISRM Commission on Testing Methods, *Int. J. Rock Mech. Min. Sci. Geomech. Abstr.,* **22** (2), 51–60 (1985).

Katzir, M., and P. David: "Foundations in Expansive Marls," *Proc. 2d Int. Research in Eng. Conf. Expansive Clay Soils* (Texas, 1968).

Kelsall, P. C., R. J. Watters, and G. Franzone: "Engineering Characterization of Fissured, Weathered Dolorite and Vesicular Basalt," *Proc. 27th U.S. Symp. Rock Mech.* (Tuscaloosa, Ala., 1986), pp. 77–84.

Kirsten, H. A. D.: "A Classification System for Excavation in Natural Materials," *Siviele Ing. in Suid Afrika,* 293–308 (July 1982).

Manger, G. E.: "Method-Dependent Values of Bulk, Grain and Pore Volume as Related to Observed Porosity," *U.S. Geol. Survey Bull.* 1203, 20 pp. (1966).

Mellor, M., and I. Hawkes: "Measurement of Tensile Strength by Diametral Compression of Discs and Annuli," *Eng. Geol.,* **5**, 173–225 (1971).

Montoto, M.: "Petrophysics: The Petrographic Interpretation of the Physical Properties of Rocks," *Proc. 5th Int. Cong. Rock Mech.* (Melbourne, Australia, 1983), vol. B, pp. 93–98.

Poole, R. W., and I. W. Farmer: "Consistency and Repeatability of Schmidt Hammer Rebound Data during Field Testing," *Int. J. Rock Mech. Min. Sci. Geomech. Abstr.,* **17**, 167–171 (1980).

Ramana, Y. V., and B. Venkatanarayana: "An Air Porosimeter for the Porosity of Rocks," *Int. J. Rock Mech. Min. Sci.,* **8** (1), 2953 (1971).

Tarkoy, P. J., and A. J. Hendron, Jr.: "Rock Hardness Index Properties and Geotechnical Parameters for Predicting Tunnel Machine Performance," Rep. for National Science Foundation Grant GI-36468, Dept. Civ. Eng., Univ. Illinois, 325 pp. (1975).

Thompson, T. W., and J. F. Schatz: "Wellbore Stability and Reservoir Compaction," *Proc. 27th U.S. Symp. Rock Mech.* (Tuscaloosa, Ala., 1986), pp. 539–551.

Travis, R. B.: "Classification of Rocks," *Quart. Colo. School Mines,* **50** (1), 98 pp. (1955).

Verhoef, P. N. W.: "Sandblast Testing of Rock," *Int. J. Rock Mech. Min. Sci. Geomech. Abstr.,* **24** (3), 185–192 (1987).

Washburn, E. Q., and E. N. Bunting: "Determination of Porosity by the Method of Gas Expansion," *J. Am. Ceram. Soc.,* **5** (48), 112 (1922).

West, G.: "A Review of Rock Abrasiveness Testing for Tunneling," *Proc. Int. Symp. Weak Rock* (Tokyo, Japan, 1981), pp. 585–594.

Character of
the Rock Mass

3.1 Characteristics of the Rock Mass as a Whole

3.1.1 Overview

3.1.1.1 Jointing data in rock mass descriptions. Properties discussed in Chap. 2 were those characteristic of intact rock materials that can be observed and measured on the scale of a hand specimen. Those considered in the present chapter pertain to the rock mass on a larger scale and take into account the important influence of jointing.

These properties can be subdivided into two categories; those that relate to the "brokenness" of the rock mass (the intensity of jointing in a general sense), and those that relate to the features of individual joints and joint sets. Parameters in the first category, such as block size and rock quality designation (RQD), describe nondirectional rock mass properties, whereas those in the second category, such as orientation, spacing, and roughness, have little meaning unless applied to a specific joint set.

The checklist presented in the previous chapter as an aid to describing characteristics of intact rock now can be extended to cover those of the joints:

1. For the geotechnical mapping unit (GMU) as a whole (see Sec. 6.1.2.1)
   - Block size, RQD, and block shape
   - Number and types of joint sets

2. For each joint set within the GMU
   - Orientation
   - Spacing
   - Persistence, surface textures, and roughness
   - Aperture, filling, and wall strength
   - Seepage

3.1.1.2 Obtaining and presenting data on joint patterns. Conventional methods of measuring joint patterns by climbing over a rock face with a tape measure and compass are slow, difficult, and often dangerous. New methods of *digital photoanalysis* are being developed that may prove simpler (Franklin, Maerz, and Bennett, 1987). Black and white photographs of rock faces or the faces themselves are converted to digital form by a television scanner that replaces the image by an array of black, white, or gray pixels. The image is enhanced to bring out the joints as black lines on a white background, and this network is analyzed to obtain statistics that characterize the pattern. Block sizes and shapes can be quantified, both in a rock face and in a photograph of broken rock produced by blasting.

A common difficulty for the engineer in analyzing rock mechanics problems is to visualize the three-dimensional blocky structure from one-dimensional data such as a drillhole log, or from two-dimensional data such as a number of outcrop logs. Computer-assisted graphic display methods (Sec. 7.1.6) promise to be helpful. Priest and Samaniego (1983) have developed a computer program, DICHA, that generates, in two dimensions, a random discontinuity fabric using scan-line data as input.

3.1.2 Intensity of jointing

3.1.2.1 Block size. The concept of block size is analogous to that of grain size on a microscopic scale. The rock mass is conceived as being made up of discrete intact blocks bounded by thin joints, and its behavior as being governed by a combination of block and joint characteristics.

Block size is defined as the average diameter of a typical rock block in the unit to be classified. It is measured simply by observing an exposed rock face at the surface or underground, or a rock core obtained by drilling, or a pile of broken rock such as a slope talus or a muck heap after blasting. The observer selects the "typical" block and either estimates or measures its average dimension. In the case of a rock core he or she selects the typical core stick and measures its length. Like other characteristics, the block size can be expressed as a range and a typical value, for example, "block size (5–30) 12 cm."

Imprecision is inevitable because natural joints are difficult to distinguish from planes of weakness and blast-induced fractures, and because block shapes are often slabby rather than equidimensional. However, estimates are usually quite repeatable and adequate to convey the required rock mass quality information. Block sizes, like strengths and other characterization parameters, vary over several orders of magnitude (from millimeters to several meters) and so are plotted on logarithmic axes in a classification diagram. An error of even 10 or 20% is usually insignificant. More important is not to waste time with attempts at precise measurement, which often makes observations slower and less useful as an aid to field mapping.

3.1.2.2   Rock quality designation (RQD).

The parameter *core recovery* is the ratio of recovered core length to total length drilled, expressed as a percentage. RQD is the sum of the lengths of rock core pieces longer than 10 cm expressed as a percentage of a given total length drilled, usually a core run (Deere et al., 1969). RQD is a modified form of core recovery, and was developed as an aid to core logging on the premise that while core recovery itself is a useful index to rock quality, it becomes much more useful if very broken materials that are recovered are not counted as "core."

RQD is 100% for strong and massive rocks which when diamond drilled give pieces all of which are longer than 10 cm, and is near zero for closely jointed rocks. Quality designations are given as *excellent* (RQD in the range of 90–100%), *good* (75–90%), *fair* (50–75%), *poor* (25–50%), and *very poor* (less than 25%). Although RQD is affected by rock strength, because weak rocks tend to be broken during drilling, it is more a function of joint spacing and is closely related to block size.

RQD, when introduced, was a great improvement over rock classifications that stressed intact rock strength and ignored jointing. Block size might be considered as an alternative and perhaps a better parameter, because it can be visualized and has direct physical significance. It can vary from millimeters to meters, whereas RQD has a limited range. Very broken rocks have zero RQD values, irrespective of whether the pieces are 9 cm or smaller. Similarly, rocks with an average block size greater than about 40 cm will have an RQD of 100%, even if the blocks are large. Another limitation of RQD is that it is defined only for rock core, and cannot easily be applied in rock outcrop mapping.

3.1.2.3   Volumetric joint count.

An areal or volumetric density of jointing can be expressed in terms of the average number of joints per unit area or unit volume of the rock mass. The volumetric joint count J_v

(Barton et al., 1974) is obtained by adding the jointing intensities (number of joints per meter) measured for each individual set along lines normal to each set, using 5- or 10-m sampling lengths. For example, in the case of four sets,

$$J_v = \frac{6}{10} + \frac{24}{10} + \frac{5}{5} + \frac{1}{10} = 4.1/\text{m}^3$$

Barton et al. propose block size designations as follows:

Designation	J_v, joints/m^3
Very large blocks	<1.0
Large blocks	1–3
Medium-sized blocks	3–10
Small blocks	10–30
Very small blocks	>30
Crushed rock	>60

RQD is related approximately to J_v by the equation

$$\text{RQD} = 115 - 3.3\,J_v \qquad (\text{RQD} = 100 \text{ for } J_v < 4.5)$$

More often, the density or intensity of jointing is measured linearly as the average number per unit length intersecting a scan line along a length of drill core or across a rock face, without identifying the set to which each joint belongs. The average discontinuity spacing is the reciprocal of the average frequency. Priest and Hudson (1976) analyzed 7000 joint spacing values measured in chalk at the Chinnor tunnel in England as part of a tunnel-boring machine study. They found RQD to be related to average spacing S by the equation

$$\text{RQD} = 100 \, \exp\left(-\frac{0.1}{S}\right)\left(\frac{0.1}{S} + 1\right)$$

From this expression, an RQD value accurate to within 5% could be obtained by counting the joints and calculating their average spacing or frequency. Measurements on freshly exposed rock faces may in fact be superior to measurements of drill core because the problem of core breakage by drilling is avoided.

A negative exponential distribution fitted the mean discontinuity spacing values very well, and was found to be equally applicable to highly fractured mudstone and less fractured sandstone at the Kielder experimental tunnel. Joints formed by a more systematic geological process, such as current bedding or columnar jointing, tend to be spaced according to a log-normal distribution (Priest and Hudson, 1976; Meints, 1986).

3.1.3 Block shape

Most rock masses have a characteristic shape of block that depends on the number of joint sets and their relative orientations and spacings. A description of the block shape adds to the rock mass characterization and is particularly relevant in some engineering applications. For example, rock that is slabby as a result of jointing is likely to remain so after blasting and may well be unsatisfactory for use as fill or armor stone.

Concepts of block shape are similar to those of grain shape, but the terminology is somewhat different (Fig. 3.1). *Cubic* blocks are produced by three orthogonal and equally spaced joint sets. *Slabby* ones result from a single closely spaced set and two at a wider spacing. *Prismatic* or *columnar* blocks, such as are characteristic of certain basalt lava flows, are created by two closely spaced sets and one at a wider spacing (Fig. 3.2).

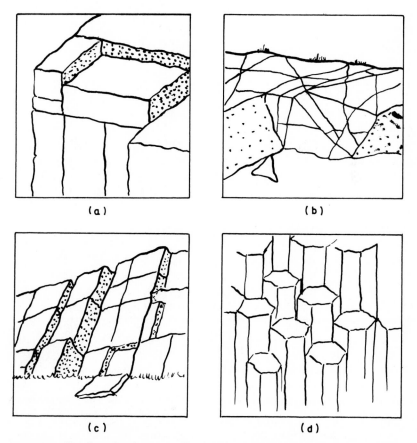

(a)   (b)   (c)   (d)

Figure 3.1 Sketches of rock masses illustrating (a) blocky, (b) irregular, (c) tabular, and (d) columnar block shapes. Compare with columnar basalt in Fig. 3.2. (*After ISRM, 1981.*)

(a)

(b)   (c)

Figure 3.2   Joints in Rocky Coulee, Wash., basalt flow. (*a*) Flow zones; entablature (above) and colonnade (below); (*b*) Closeup of columnar jointing, (*c*) Hexagonal column tops. (*Meints, 1986; courtesy of Joyce Meints.*)

These terms relate to a rock mass with exactly three joint sets, which is common but by no means universal. When just one or two sets are evident, others at a very wide spacing must be assumed if blocks are to be formed. Rocks containing more than three sets, or sets that are not orthogonal, produce wedge or multifaceted blocks that may be more or less equidimensional, slabby, or prismatic.

Terminology such as this is best supplemented by quantitative information on aspect ratios. The shape of a typical block is expressed in terms of its dimensions in two orthogonal directions as ratios of the shortest dimension, with or without a descriptor, in the form "slabby, 1:5:7," or "multifaceted, 1:1:2."

3.1.4   Identification of joint sets

A joint *set* consists of individual joints with similar physical and mechanical characteristics that occur in a nearly parallel array. The jointing in a region can usually be subdivided into two or more such sets, which together constitute the *jointing system*. Often each set is characteristically different from the others because of differences in geological origin and history.

Joint orientations are usually shown graphically as dots (poles) on a stereoplot. Sets, if present, are revealed as clusters of poles, and the stereoplot is often contoured to highlight concentrations of joints and, for each set, to quantify the statistical distribution of orientations about an average value (see Fig. 3.7).

Joint sets are distinct and easily defined when the clusters of poles are tight. This is usually the case in sedimentary rocks little affected by folding or faulting. Bedding joints, for example, are often near-horizontal. The reverse is true in some igneous and metamorphic rocks that have been affected by several stages of deformation. These may show a nearly "random" jointing pattern that can only be divided into sets by mapping large numbers of joints within small zones that are homogeneously jointed. Jointing is seldom truly random, having been formed as the result of well-ordered geologic processes. Patterns can be discerned given the right tools and resources.

3.2   Characteristics of Each Joint Set

3.2.1   Types of set

Just as intact rocks are characterized by rock names, so each discontinuity set should be characterized first by its generic type, if this can be identified. Identification of the generic type is not always easy, even for an experienced geologist.

3.2.1.1 Faults and shears. *Faults* and *shears* are joints along which there has been shearing movement. On a global scale, faulting results from relative movements of the continental plates and causes earthquakes. The Wegener hypothesis of continental drift, proposed in the 1920s, considered the earth's crust as a number of plates moving in relation to each other. Plate thicknesses are believed to be 70–100 km. Relative motion between adjacent plates amounts to a few centimeters per year, or to a few thousand kilometers of plate movement over 100 million years.

Surface movement between plates occurs in part along large faults. Perhaps the most famous, the San Andreas Fault, first came to the attention of Californian geologists in the 1890s and was brought to infamy overnight in 1906 by the San Francisco earthquake (Uemura and Mizutani, 1984). Displacements of up to 7 m appeared along a distance of approximately 430 km following the line of the fault. Careful research has shown that movements have amounted to more than 500 km over the last 70 million years (more than 7 mm per year).

On a smaller scale, faults tend to occur in sets, often clustered within a fault zone, but may be widely spaced. They have a greater displacement than shears, and are distinguished by an observable offset of beds or other marker horizons or by drag folding on either side of the plane of shearing. They are often accentuated by weathering and erosion, which removes the weaker fault breccia and gouge (Fig. 3.3). According to their directions of offset they may be classified as normal, reverse, thrust, or wrench faults. Faults and shears often have slip surfaces marked by slickensides (polished rock) or containing gouge (crushed rock, clay, or sandy fragments). They are continuous and weak unless healed and rewelded by hydrothermal injection of quartz or other strong vein materials. Fault gouge can form a water conduit or barrier, or a combination of the two, which when penetrated releases large quantities of water under pressure. It is probably true that failures of slopes and underground excavations are controlled more often by faults and shears than by any other geological feature.

Foliation shear zones originate from differential movement between adjacent layers of metamorphic rock, concentrated in the weaker lay-

Figure 3.3 Fault zone predicted by air photo interpretation, then exposed during construction of Science North museum, Sudbury, Ont., Canada. (*a*) Aerial view during construction. The Creighton fault zone was uncovered beneath the exhibit building. To the lower left of the photograph is the "principal shear," a weak zone gouged out by glacial ice. The resulting slot was incorporated into the architecture of the building (see *c*). (*b*) Cross section showing individual shears in the 100-m-wide fault zone, located by drilling an inclined exploratory hole. (*c*) Glacially polished surfaces and boulders wedged between the faces of the principal shear. Photo taken near the spiral glass-walled staircase in the completed building. (*Franklin and Pearson, 1985.*)

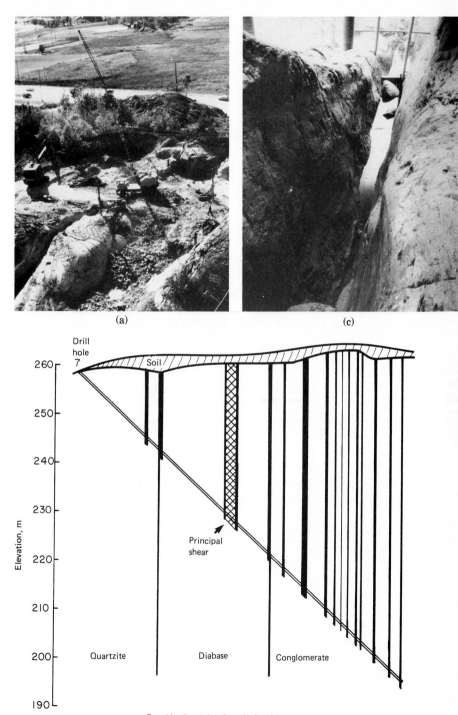

(a)

(c)

Drill
hole
7

Soil

260

250

240

230

Principal
shear

220

Elevation, m

210

200   Quartzite   Diabase   Conglomerate

190

Equal horizontal and vertical scales

(b)

ers, typically mica, chlorite, talc, or graphite schist, in a sequence of harder, massive rocks such as granite gneiss and quartzite. Although they extend over several hundred meters, offsets are typically small, from a few centimeters to 1 or 2 m. The thickness of gouge and crushed rock is typically just a few centimeters, but the adjacent rock is closely jointed, and may be altered or sheared for 1 or 2 m to either side. Foliation shear zones have been responsible for many design and construction problems on hydroelectric developments (Deere, 1974).

Shale mylonite seams (*clay mylonites*), responsible for various problems and particularly slope failures in shales, are discussed in Stimpson and Walton (1970) and also in Deere (1974). They consist of sheared shale gouge with a high natural water content and a low residual shearing resistance (10–20°). Clay mylonites are thin, having a typical width of only a few millimeters to 1 or 2 cm. The gouge is usually the same color as the wet host rock, so it is hard to recognize unless the rock core has been allowed to dry partially: the gouge retains water for longer than the intact shale, and appears darker. Soft clay mylonite seams can be identified when logging core by running a pocket knife along the core in the box, recognizing that the clay mylonite is usually much softer than the host shale.

3.2.1.2 Bedding joints. *Bedding joints* are associated primarily with sedimentary rocks, although flow bedding can occur also in volcanic materials. They are characteristically at or parallel to the boundaries of beds with different lithologies of different color, texture, and mineral composition. They are often marked by clay or shale partings that greatly reduce shear strength. Because of their depositional origin, they are often continuous and planar. Note that bedding is not always accompanied by jointing. A sedimentary rock may show color or mineral segregation banding with no corresponding planes of weakness. Bedding joints can be obscured and rewelded by metamorphism.

3.2.1.3 Cleavage joints. Cleavage is produced by deformation during metamorphism, and is marked by subparallel minerals that recrystallize approximately perpendicular to the maximum compressive stress. There may be more than one cleavage direction if metamorphism has occurred in several stages. Cleavage is often oblique to the bedding, and characterized by a uniform relationship to the axial planes of folds (Uemura and Mizutani, 1984). The cleavage planes are closely spaced but for the most part coherent, so only a few are joints, and a joint may grade into a plane of weakness, or vice versa. Hence the orientations of cleavage joints are easily defined, but not their spacings and persistences.

3.2.1.4 Other joints. *Sheet joints* run nearly parallel to the ground surface, curve to follow the topography, and are related to stress relief. For example, they occur in granite terrain as a closely spaced "onion skin" pattern near the surface, becoming tighter and more widely spaced at greater depth. Landslides are sometimes triggered by the buildup of groundwater pressure in sheet joints that lie parallel to steep valley walls.

Joints (without qualification) include the many sets that cannot be classified by genetic origin or by association with structures such as bedding or cleavage. They are sometimes subdivided by structural geologists into *shear* and *tension* joint categories according to their orientation with respect to an inferred tectonic stress field. Shear joints run parallel to faults and shears, and may be incipient shears along which there is no evidence of displacement, whereas tension (or "extension") joints run normal to the direction of minimum principal stress. The terms are not easily defined and can lead to confusion. As an alternative to genetic names, joint sets can be given arbitrary numbers or names, such as set 1, set 2,..., or vertical set.

3.2.2 Orientation

3.2.2.1 Definitions. In the well-known dip-strike notation used by geologists, *strike* is the direction of a horizontal line in the plane of the joint, and *dip* is the vertical angle measured downward from the horizontal to the *fall line* of greatest dip. Ambiguities arise because an east-west striking joint may dip either to the north or to the south. Further complexity can occur when directions are recorded from the cardinal points rather than from north, in a form such as "strike E 48° S, dip 30° SW."

The simpler and unambiguous form of dip magnitude and dip direction becomes mandatory when, in rock engineering, hundreds of jointing observations must be recorded and processed statistically. Dip magnitude B (Fig. 3.4) is the same as defined above, expressed as a two-digit number. Dip direction A is always expressed as a three-digit

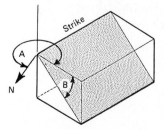

Figure 3.4 Dip direction A and dip magnitude B are used by rock engineers to define the orientation of joints, in place of the more traditional strike and dip.

azimuth between 0 and 360° measured clockwise from north to the horizontal projection of the fall line. The combined dip and dip direction thus appear in the rock description in the form "18° at 025°."

3.2.2.2 Measuring methods. Joint orientations are conventionally measured using a geological compass clinometer of the Brunton type. The measurements are slow and tedious and can be dangerous or impossible on steep, high rock cuts or in underground mine stopes. Improved metering devices have been developed that permit the recording of dip and dip direction at the touch of a button. They are easier to use overhead, such as in the crown of a tunnel.

Joint orientations and other characteristics can be determined directly from photographs, using digital photoanalysis techniques. One approach is to use stereophotogrammetry to measure the coordinates of three points in a joint plane, from which its orientation may be computed geometrically. This requires stereo pairs of photographs, an expensive stereocomparator, and much time and repetitious measurement, but can be partially automated. An alternative is to measure lineations on two nonparallel photographs and to combine the resulting rose diagrams by a variation of the geological three-point solution. This promises to be much quicker and more convenient.

3.2.2.3 Graphing and contouring. Data on joint orientations are easier to visualize and analyze if presented graphically. The simplest form is a *rose diagram,* which shows the frequency of joints as a function of their dip direction (or strike), but says nothing about their dip magnitude (Fig. 3.5).

For complete, three-dimensional information on joint orientations a polar stereonet (stereoplot) as shown in Fig. 3.6a is preferred. In this upper-hemisphere polar projection each joint plane is represented by its *pole,* a unique point on the diagram. Dip directions define the radii of the stereoplot. Dip magnitudes measured outward from the center of the graph define a series of concentric dip circles. A pole is plotted by locating the appropriate radius and dip circle, interpolating if necessary between lines in the grid.

The stereoplot used in this manner is simply a graph in polar coordinates, and joint poles can be plotted without knowing or understanding how a polar projection is obtained. However, both polar and equatorial projections need to be understood for follow-up work in applying the information to analyses of rock mass stability, hence the following brief explanation.

Imagine a line passing through and perpendicular to the joint, termed the *normal* to the joint, and also a reference upper hemisphere centered

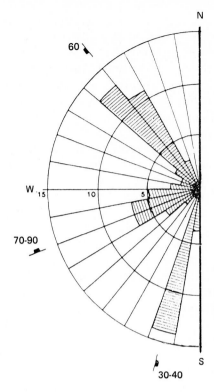

Figure 3.5 Rose diagram method for recording the directional frequencies of joint lineations. In the example, a third dimension is added by giving information on dip and dip direction and magnitude alongside each joint cluster. (*ISRM, 1981.*)

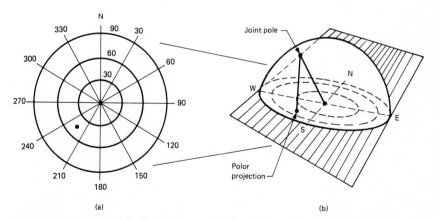

(a)   (b)

Figure 3.6 Stereographic method for recording joint directions. (*a*) Polar equal-area stereoplot, a graph with dip magnitude shown as concentric rings, and dip direction as radii. A horizontal joint plots at the center. (*b*) Upper-hemisphere projection, showing projection of joint pole onto equatorial plane.

where the normal meets the joint (Fig. 3.6b). The point where the normal intersects the hemisphere is termed the *joint pole*. To avoid three-dimensional graphs, the hemisphere is projected downward onto the equatorial plane to give a polar projection. Selection of the upper hemisphere is arbitrary and many authors use the lower hemisphere instead. The upper hemisphere is recommended for use in rock engineering because poles plot at their correct dip directions. For example, a joint dipping directly south plots along the 180° radius in the lower part of the stereoplot. The lower hemisphere convention results in an inversion in which southerly dipping joints plot to the north, easterly ones to the west, and so on. Note that a horizontal joint plots at the center of the stereoplot, whereas a vertical one plots on the equator.

Joint sets are defined by concentrations or "clusters" of poles. Well-developed sets are easily recognized as tight clusters (Fig. 3.7). Usually, however, the poles within a set are to some extent scattered, first by real differences from one joint to the next, even in the same set; second by errors of observation; and third by the roughnesses of individual joints measured locally rather than on a large scale. The orientation of each set can be expressed as a range and a typical value, for example, "orientation (35–45) 40° at (127–156) 140°," or it can be given more precisely in terms of statistical means and standard deviations.

Poorly developed sets show greater scatter and can be defined only by an objective contouring of the stereoplot. Microcomputer programs are available that take much of the work and personal bias out of plotting and contouring joint orientation data (e.g., Mahtab et al., 1972). The more usual manual contouring method uses a rectangular grid superimposed on the plot. A counting circle is placed with its center at each grid intersection in turn. The numbers of poles within the circle are noted for each intersection, and these values are contoured. Typically, a centimeter grid is used with a 20-cm-diameter stereonet and a 2-cm-diameter counting circle, giving overlap which leads to smooth contours.

Poles located on the equator at opposite ends of the same diameter are likely to belong to the same near-vertical set, plotting at one or the other end of the diameter only because of insignificant variations to either side of the vertical. The special peripheral counter used for such poles contains two counting circles separated by a center-to-center distance equal to the diameter of the stereonet. Counts in the two circles are added and the contours "wrapped round."

3.2.3 Spacing

3.2.3.1 Terminology. *Spacing* is defined as the average distance between adjacent joints in a set, measured normal to the joint plane. De-

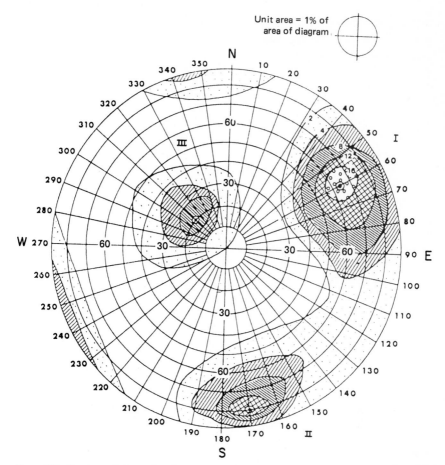

Figure 3.7   Contoured stereoplot showing the orientations of three sets of joints. Main sets I and II are approximately normal to each other; minor set III is nearly horizontal. (*ISRM, 1981.*)

scriptors can be used such as *widely* or *closely* spaced, *thickly* or *thinly* bedded. For engineering purposes, these should always be defined. The block size is related to joint spacing in that it is governed by, and approximately equal to, the average of spacings for all sets in the jointing system.

   Intensity of jointing is the inverse of spacing, the number of joints per meter. As an alternative parameter to spacing it has the practical disadvantage that large numbers must be counted when joints are closely spaced, say as close as 1 cm. The equivalent spacing can be estimated quickly and directly in the same way as the block size. It is also easier to visualize, and increases rather than decreases with improving rock mass quality.

3.2.3.2 Measuring methods. Spacing is measured by a tape traverse on the rock face or on a photograph. When possible, the direction of traverse is selected as normal to the joint set. Repeat traverses are made in different directions for each set. A correction is required if the traverse direction is other than normal to jointing.

This procedure is possible only for well-developed sets that can be identified in the field without resorting to stereoplot contouring. Otherwise the traverse must simultaneously record the spacings and orientations of all joints intersected, without any attempt at subjective separation into sets. Computer processing of data can then be used, first to identify the sets, then to compute spacings. A similar procedure can be used when logging joints in core.

3.2.3.3 Plotting and interpretation. Joint spacings, like all other rock mass properties, show a range rather than a unique value. Usually they are found to be log-normally or exponentially distributed so that geometric rather than arithmetic mean values should be used (Priest and Samaniego, 1983). For most practical purposes, however, spacings can be described in terms of a range and a typical value, the latter being obtained as the peak in the frequency histogram.

3.2.4 Persistence

An idealized plane containing a joint has three components: *joint segment* areas with no coherence, *rock bridge* areas as strong as the intact rock, and *weak segment* areas with reduced coherence, neither fully intact nor fully separated. Joint segments are often modeled as disk-shaped, or as terminating in straight edges where they meet other, nonparallel joints (Merritt and Baecher, 1981).

In sliding stability calculations, small intact rock bridges have a great effect on total shear strength, and the designer needs to know what percentage these represent of the total surface of sliding.

Persistence (or *continuity*) is defined as the percentage ratio of joint segment to total area measured in the plane of the joint. A 100% persistent joint can be followed without interruption for the full distance of exposure. Faults, because of their mode of formation, can be inferred as 100% persistent and often are plotted for many kilometers on geological maps. Most bedding joints are also highly persistent.

The geological term *development* relates to persistence. A well-developed joint set is one in which individual joints are closely parallel and persist without offset or interruption for considerable distances. Terms such as *master* joints, *major* joints, *minor* joints, or *micro* joints refer to persistence also.

Persistence becomes difficult to define and to measure when taken out of the context of an idealized model. For example, slightly offset "en echelon" joint segments might be taken as a single "wavy" joint with intermediate intact rock bridges, or as a collection of discrete joints. Also, persistence should be defined in terms of areas rather than lengths, but only lineations can be observed in practice. Probably easier to define and to measure are the *trace lengths* of joints on rock outcrops. Measurements of trace lengths, like spacings, appear to follow either an exponential or a log-normal distribution.

3.2.5 Surface characteristics

3.2.5.1 Surface textures. Textures form on joint surfaces at various stages in their geological history. They are diagnostic of genetic origin, and relevant also to mechanical behavior. Roughness, for example, affects both the shear strength and the permeability of a joint.

Primary textures, sedimentary in origin, are associated with bedding only, and include ripples and polygonal cracks. Secondary textures, resulting from metamorphism or joint formation, can be found in rocks of all types. They include rib and feather marks of extension-type jointing, created by rapid propagation of the joint at the time it was formed.

The textures of shears and faults are particularly important to define. *Fault zones* include many individual shears, some thick and some thin, usually with layers of brecciation and gouge. Slickensides (polished and striated surfaces, often coated with finely crushed rock) are frequent within the zone. A major fault found, for example, in a proposed dam foundation should be treated as a geotechnical mapping unit rather than as a single structural feature, and mapped and characterized in detail.

3.2.5.2 Roughness. A typical joint cross section is rough and wavy. As in electric or acoustic signals, the simplest waveforms are sinusoidal or sawtooth and have a unique wavelength and amplitude. A more random type includes many wavelengths, amplitudes, and frequencies, which can be determined by Fourier analysis.

To observe and measure roughness, a shadow can be cast on a rough rock joint facet by a straightedge held against the surface in bright sunlight or artificial lighting. Using photoanalysis, the edge of the shadow can be digitized and any required roughness parameter calculated (Franklin, Maerz, and Bennett, 1987). Undulations in the irregular shadow can be converted to true asperity amplitudes by a simple trigonometric factor: the roughness amplitude is the length of shadow multiplied by the cotangent of the angle of incidence of the lighting (Fig. 3.8).

Figure 3.8 Roughness measurement using the straightedge shadow method. (*Franklin, Maerz, and Bennett, 1987.*)

Other, usually less convenient methods of roughness measurement include using a compass clinometer with a variable-size base plate (Fecker and Rengers, 1971) or a profilometer clamped to the rock surface. A stylus or wheel traverses the surface along a line, while a transducer system reproduces the waveform on paper or as a magnetic record on tape or disk.

For routine rock description, consistent use should be made of descriptive terms like *rough, smooth,* and *polished,* which are qualitative but meaningful. They are usually applied to roughness on a scale of less than about 100 mm measured along the joint. Terms such as *undulating* or *planar* refer to "waviness" that can be observed only on a larger scale. Roughness may be quantified in terms of approximate amplitudes over a given length of observation. For example: "1-mm amplitude on a scale of 100 mm, with superimposed waviness of 80 mm in 10 m."

The rise angle i of the typical roughness asperity is an important parameter related to shear strength, measured in order to extrapolate shear strength from small to large scale (Chap. 11). Barton defines a joint roughness coefficient (JRC) on a scale of 1:20 (discussed further in Chap. 11). This is a parameter in an empirical shear strength criterion, and can be measured by a simple tilt test. Alternatively, JRC can be estimated by visual comparison with a set of 10 "typical roughness profiles" published in ISRM (1981).

For characterizing roughness, Reeves (1985) suggests the root-mean-square contact gradient Z_2, which is defined as

$$Z_2 = \frac{1}{L} \int_{x=0}^{x=L} \left(\frac{dy}{dx}\right)^2$$

where y is the amplitude at a particular point on the roughness profile about the centerline, x is the tangential distance along the profile, and L is the total length of the profile. The centerline divides the profile such that the sums of areas to either side of it are equal.

3.2.5.3 Wall strength.

The parameter *joint wall compressive strength* (JCS) was defined by Barton and Choubey (1977). Unaltered wall surfaces have a similar strength to that of the intact rock, which can be obtained directly by measuring uniaxial compressive strength, or estimated more quickly from the point-load strengths of cores or outcrop specimens.

When joint walls are thickly coated with the products of alteration or weathering, the "filling" can be sampled, described, and tested in the laboratory. Shear strengths of fillings are unlikely, however, to be the same as of the undisturbed, filled joint (Chap. 11).

Weakened wall rocks can be tested in the laboratory or the field using a Schmidt hammer, a simple and portable testing device (Sec. 2.3.2.3). JCS is then estimated as follows:

$$\log_{10}(\text{JCS}) = 0.00088\rho R + 1.01$$

where JCS is in megapascals, ρ is the dry rock density in kilonewtons per cubic meter, and R is the Schmidt rebound number of the joint surface.

When the altered layer is so thin that measurement becomes difficult, Barton found that taking JCS as one-quarter that of the unaltered rock gives a realistic approximation for obtaining a safe (lower-bound) shear strength. JCS is discussed further in Sec. 11.3.5.2.

3.2.6 Separation characteristics

3.2.6.1 Aperture.

The *aperture* of a joint (also known as its openness or separation) is the mean distance separating the two intact joint walls. Note that aperture includes the thickness of any filling that may be present. Joints may be termed *open* or *tight* according to whether their aperture is large or small. The aperture is usually greatest for near-surface joints as a result of rebound and stress release, and the joints become tighter as the depth increases.

Apertures are usually just a few micrometers wide, except where the rock has been loosened by near-surface weathering or blasting, or dissolved by water flowing through the joints. Measurements in surface outcrops, such as using a set of automobile "feeler gauges," can be misleading. Pyrak et al. (1985) measured apertures in the laboratory by making a cast using Wood's metal, an alloy that melts at about 70°C. After solidification, the rock can be broken or dissolved away, leaving a replica of the joint space.

Values more representative of conditions at depth can perhaps be back-calculated from measurements of rock mass permeability, assuming a simplified model of jointing with parallel-plate flow. Snow (1970) suggests from this type of observation that apertures are lognormally distributed, with many small apertures and few large ones.

Apertures sometimes can be measured using the downhole techniques of Chap. 6. A drillhole television camera or an impression packer reveal joints that are open and water-conducting and those that are not. In the rare cases where the apertures are very large, they can be measured directly using a scale or graticule. The integral core sampling method (Sec. 6.5.3) is potentially an excellent one for investigating apertures at depth, because rock core is recovered with a central reinforcing rod that retains the rock pieces at their original degree of separation.

3.2.6.2 Filling types. *Fillings* are defined as any materials within the joint whose properties differ from those of the rock to either side. The rock mass description should give both the nature and the thickness of fillings. When several types occur in layers, as is often the case in faults, each layer should be described.

Such materials have a variety of origins. Shale partings in bedding joints of a sandstone or limestone have a primary (sedimentary) origin. Veins of quartz or calcite in granites often originate shortly after emplacement of the igneous magma. Many types of filling materials, however, greatly postdate rock formation and result from alteration or weathering. Hydrothermal alteration by the injection of gases or liquids not only alters the wall rock but also deposits various additional materials as veins. Groundwater percolation can leach, dissolve, or deposit materials, depending on the chemistry and velocity of the water. True *infillings* can result from the transport of clayey and sandy sediments through an open system of joints, and larger solution cavities can even be filled by cobbles and boulders.

Filling materials vary greatly in their mechanical characteristics, from very soft to very hard and strong. Materials of extremely low strength, that call for precautions when encountered, include clayey, platy, and very soft minerals, such as montmorillonite, illite, chlorite, graphite, and talc. Clays are soft and soapy in texture and usually have a high water content. When they may have a bearing on rock mass stability they should be further characterized by x-ray, DTA, or optical mineralogical testing and by plasticity and water content measurements. The clay mylonites (Sec. 3.2.1.1) found in shales consist of finely ground and highly plastic shale created by shearing. Even when only 1 or 2 mm thick and dipping at a few degrees, they have been known to cause extensive slope failures.

Fillings of intermediate strength include those of sandy consistency, such as crushed or brecciated hard or moderately hard rock, lightly altered wall rock, or veins of calcite when weaker than the surrounding wall rock.

Most joints have no filling, and are neither strengthened nor weakened by filling materials. Their unaltered surfaces are in contact. Stained joints can be included in the "no filling" category. Staining is important as an indicator that a joint has conducted groundwater, but usually has little or no influence on strength or other mechanical properties.

Strong fillings such as vein quartz, calcite, and limonite can heal and recement the joint, which may become as strong as the surrounding rock.

3.2.6.3 Filling thickness.   Thicknesses of filling materials can be estimated visually and expressed in terms such as "thinner than 1 mm." Intermediate thicknesses should be expressed as a range and a typical value, for example, "(1–10) 3 mm," whereas substantial thicknesses generally require that the overall thickness value be accompanied by a sketch showing the types and layers of materials. Lindberg (unpublished communication) has measured a log-normal distribution for filling thicknesses in basalt at the Hanford, Wash., test site.

3.2.6.4 Joint seepage and moisture.   The water-conductive characteristics of joints may be measured using the permeability tests discussed in Chap. 4. In addition, amounts and locations of observed seepages should be noted in the rock mass description. They are indicative of the permeability of both the joint and the set to which it belongs, and of the elevation of the water table. The line of emergence of a water table separates joints that seep from those that do not.

Terms such as *dry, moist, slight,* or *moderate* seepage, *dripping, running,* and *flowing* may be used, preferably accompanied by some definition of terminology. Flow quantities from joints, if substantial, can be measured as the volume of a sample collected in a given period of time.

3.3 Rock Mass Quality Indexes

3.3.1 Weathering indexes

3.3.1.1 Definitions. *Weathering* is the process of weakening and/or disintegration of rocks under the influence of the atmosphere and hydrosphere (Fookes et al., 1971). Geologists distinguish between weathered and "fresh" (that is, unweathered) rocks by observing the degree of decomposition of minerals such as feldspars, amphiboles, and pyroxenes, which are among the first to decompose. Severe weathering is easily identified, but microscopic examination is needed to differentiate fresh and slightly weathered rocks.

A distinction is made between weathering and *alteration,* which results from deep-seated changes such as caused by hot hydrothermal fluids, high pressures, and high ambient rock temperatures. Weathering is a type of *diagenesis,* a term used for geological processes that occur under near-surface temperatures and pressures and result finally in lithification.

Note the important difference between the *state of weathering* of a rock and its *weatherability* (or its opposite, *durability*). Weatherability is a measure of susceptibility to weakening or disintegration during the time span of an engineering project (Chaps. 2 and 10).

3.3.1.2 Weathering processes and patterns. Weathering processes are classified as physical, chemical, and biological. *Physical* weathering dominates in arid or cold climates, producing an accumulation of rock fragments similar to the parent rock. *Chemical* weathering is more common in warm climates. It results in decomposition and the degradation of a large number of rock-forming minerals into just a few. Only the most resistant minerals, usually quartz and mica, remain unaffected. *Biological* weathering is generally less important, but chemical and physical weathering can be accelerated by biological processes, such as the production of humic or lignic acids.

A weathered rock mass often exhibits a *weathering profile,* grading from extremely weathered materials near the surface to less weathered materials and fresh, or unweathered, rock at greater depth (Fig. 3.9). Granites have been reported in Washington, D.C., as weathered to a depth of 24 m so badly that they could be removed with pick and shovel, but weathering may penetrate much deeper than this. Limestones have been found decomposed to depths of 60 m in Georgia, United States, and shales to 120 m in Brazil. During construction of the Keiwa hydroelectric project in Australia, the rocks were found weathered to depths in excess of 300 m. Weathering on the Russian platform has been reported at 1500 m.

Climate and time determine the state of weathering of the rock

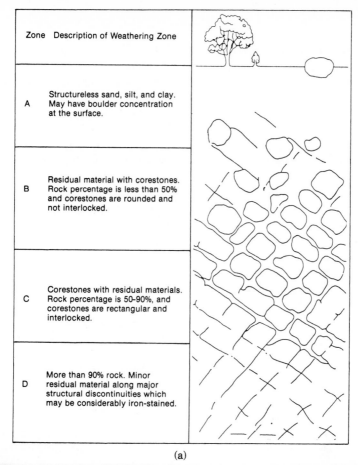

Zone	Description of Weathering Zone
A	Structureless sand, silt, and clay. May have boulder concentration at the surface.
B	Residual material with corestones. Rock percentage is less than 50% and corestones are rounded and not interlocked.
C	Corestones with residual materials. Rock percentage is 50–90%, and corestones are rectangular and interlocked.
D	More than 90% rock. Minor residual material along major structural discontinuities which may be considerably iron-stained.

(a)

(b)

Figure 3.9 Rock weathering. (a) Weathering profile, showing zones of increasing degree of weathering, and development of corestones and residual soil. (*Brand and Phillipson, 1984.*) (b) Corestones produced by spheroidal weathering of granite, Sonoran desert, near Tucson, Ariz.

mass. Weathering processes are more severe in tropical climates. Many areas affected by ice-age glaciation have been stripped of weathering products by glacial ice and meltwaters. In Canada, for example, the 10,000 years or so since the last glaciation have not been sufficient to allow significant further weathering, and the glaciated rock surfaces remain more or less fresh.

Weathering does not necessarily proceed uniformly downward. It occurs mainly along joints and spreads inward from the joints to increasingly affect the rock material. It can be particularly severe along faults and other major conduits for groundwater. Certain rock types or beds are preferentially weathered because they contain less stable minerals than other beds. Occasionally buried, or fossil, soils, which represent an earlier period of geological weathering at a time when they were exposed on surface, can be found beneath layers of competent, unweathered rocks.

Rocks that are permeable along joints but quite impermeable in the rock matrix typically develop a *corestone* pattern of weathering as shown in Fig. 3.9 (see also front cover). Blocks of fresh rock, usually rounded, with a strength similar to the parent material, are surrounded by residual soil. The corestones become larger and more rectangular and the soil component more minor with increasing depth. Most igneous rocks, particularly fine-grained ones, weather according to this pattern. Other rocks that are relatively permeable throughout the matrix, such as many sandstones, show a more uniform weathering in which the corestones are greatly weakened and may be quite irregular in shape. This is also typical for rocks that contain widely spaced joints and in which the rock materials contain metastable minerals.

3.3.1.3 Classifications of weathering grade.

Various weathering classifications have been proposed based on the extent of penetration from joints inward into the corestones, and on the degree of decomposition of the intact rock material itself. Dearman (1974) subdivided his classification according to criteria of physical disintegration and chemical decomposition. Brand and Phillipson (1984) describe the weathering classification used for granites and volcanics by the Geotechnical Control Office in Hong Kong, which classifies rock according to six grades and four zones (Table 3.1). Grade I to III materials are commonly referred to as rock and grades IV to VI as soil.

An alternative to a universal classification is to set up a scale of weathering grades for a particular project. This approach was applied to chalk bedrock by Wakeling (1970), who defined grades of weathering according to the size and content of corestones and the spacing of joints. The grades were found to be correlated with results from stan-

TABLE 3.1   Weathering Grade Designations Employed by Geotechnical Control Office in Hong Kong

Grade	Description	Typical distinctive characteristics
VI	Residual soil	Soil formed by weathering in place but with original texture of rock completely destroyed
V	Completely decomposed rock	Rock wholly decomposed but rock texture preserved No rebound from N Schmidt hammer Slakes readily in water Geological pick easily indents surface when pushed
IV	Highly decomposed rock	Rock weakened, large pieces can be broken by hand Positive N Schmidt rebound value up to 25 Does not slake readily in water Geological pick cannot be pushed into surface Hand penetrometer strength index >250 kPa Individual grains may be plucked from surface
III	Moderately decomposed rock	Completely discolored Considerably weathered but possessing strength such that pieces 55 mm in diameter cannot be broken by hand N Schmidt rebound value 25 to 45 Rock material not friable
II	Slightly decomposed rock	Discolored along discontinuities Strength approaches that of fresh rock N Schmidt rebound value greater than 45 More than one blow of hammer to break specimen
I	Fresh rock	No visible signs of weathering; not discolored

SOURCE: After Brand and Phillipson (1984).

dard penetration tests (SPT) and plate-bearing tests (Chaps. 6 and 9). Weathering grades were defined from grade I (unweathered, SPT greater than 35) to grade VI (extremely soft structureless chalk containing small lumps of intact chalk, SPT below 8).

3.3.2 Seismic indexes to rock quality

3.3.2.1 Overview. Geophysical methods, which measure the seismic, electrical, magnetic, gravimetric or radiometric properties of rocks, have two main applications: to map rock unit boundaries, faults, and folds, and to characterize the individual strata or rock units. Mapping applications of geophysics are covered in Chap. 6, and measurements of dynamic rock behavior in Chap. 9, whereas rock characterization aspects are reviewed in the present chapter.

For characterization, only the seismic methods are in common use. They measure the time taken for sound (low-frequency) or ultrasound (high-frequency) waves to travel through the rock from a transmitter to a receiver. Sonic velocity is obtained by dividing the travel distance by the travel time. It is greatly reduced by the presence of pores and joints. Measurements therefore give a useful indication of porosity and intensity of jointing (Young et al., 1985), hence of rock mass strength, deformability, and hydraulic conductivity.

3.3.2.2 P- and S-wave arrivals. The transmitted signal is composed of two principal components, P waves (compressional or longitudinal waves), which oscillate along the direction of wave propagation, and S waves (shear or transverse waves), which oscillate perpendicular to this direction. P waves, which travel faster than S waves and are the first to arrive at the receiver, are more commonly used to provide an index to rock quality. A P wave can travel through either a solid or a liquid, whereas an S wave can travel only through solid materials. Saturation with water increases the seismic P-wave velocity. Since shear waves can only pass through the mineral skeleton, the S-wave velocity is almost the same for both dry and saturated rocks.

S-wave velocity is useful in special applications and can be measured either by suppressing P-wave transmission using special shear wave generating transducers, or by carefully identifying the S-wave arrival on a record of the complete acoustic arrival waveform.

3.3.2.3 Methods of measurement. In the laboratory, a cylindrical specimen prepared for strength testing can be used first for ultrasonic velocity measurements, which are nondestructive. A piezoelectric transmitter and receiver are pressed against the flat end faces of the cylinder, which are usually greased to improve the acoustic coupling. Travel time of the ultrasonic wave generated by the piezoelectric transmitter is measured by a cathode ray oscilloscope, which displays the initiation pulse and the received signal on a time axis.

Field measurements make use of geophones, which detect sound in the audible frequency range. Ultrasound is difficult to use in the field because the high frequencies are rapidly attenuated and can be detected only over short distances. Measurements in the field can be made along the ground surface, between or along drillholes, or from drillholes or underground excavations to the surface. In the more common surface-to-surface method, a hammer is often used as a convenient energy source in place of the more conventional explosive charge. The most reliable method is one using an enhancement hammer seismograph, which has the capability of averaging the arrival waveforms from repeated sledgehammer blows and displaying the accumulated signal on an oscilloscope screen, thereby improving the signal-to-noise ratio.

Acoustic borehole logging tools commonly make no contact between the transducers and the rock. The seismic signal travels through a borehole fluid, either water or drilling mud, and shear waves are attenuated and arrivals difficult to discern. Travel times are affected by the distance between the transducers and the rock, and therefore depend on the radius of the hole and the nature of the fluid. Newer designs of logging tool overcome this problem by pressing the transducers against the walls of the hole. Even then, corrections must be made for the influences of the hole at high frequencies.

3.3.2.4 Indexes of rock quality. Sound velocity is a useful index because sound travels faster through stiff and dense rock than through a deformable rock containing pores or fissures. Porosity and intensity of jointing in turn control such physical characteristics as strength, deformability, and ease of excavation (Sec. 14.2.2.3). Laboratory P-wave velocities vary from less than 1 km/s in porous soft rocks to more than 6 km/s in dense hard rocks. Ranges of values for typical rock and soil types are given in Table 3.2. Velocities are significantly lower for microcracked rock materials than for porous rocks without microcracks but with the same total porosity (Kelsall et al., 1986).

TABLE 3.2 Typical P-Wave Velocities

	km/s
Clays	1.0–2.7
Shales	1.4–4.6
Slates	3.5–4.4
Sands	0.2–2.0
Sandstones	1.4–4.6
Coals	1.1–2.8
Salt rocks	3.5–5.5
Limestones	1.7–6.4
Marbles	5.0–6.0
Schists	3.5–7.7
Granites	4.0–6.1
Basalts	5.0–6.7
Water	1.46
Ice	1.0–4.0
Quartz	5.22
Glass	5.8–6.8
Steel	5.9–6.3
Aluminum	6.3–7.0

Whereas the laboratory velocity is governed almost exclusively by porosity of the intact rock, the field velocity is reduced as a function of the spacings, apertures, and fillings of joints. The *velocity ratio k,* given by the P-wave velocity measured in the field divided by that measured in the laboratory, is a useful guide to rock mass quality. It approaches 1.0 for rocks with widely spaced and tight jointing, and approaches zero as the rock becomes more intensely jointed. Deere et al. (1969) define a *velocity index,* which is 100 times the square of this ratio. The index varies from 0 to 20 for very poor rock, to 80–100 for excellent rock. A typical application is to dam foundations: sonic velocities are measured initially to determine the zones of rock requiring grouting, and then to confirm, by the increase in velocities, that the joints have been successfully filled with grout.

Tanimoto and Ikeda (1983) established a correlation, from underground measurements at 104 tunneling sites, between the velocity ratio *k* and fracture frequency *n,* defined by the number of joints per meter:

$$n = \frac{5.0}{k^2} - 4.0$$

Values of the *dynamic* Young's modulus, which can be computed from sonic velocities by applying equations of theoretical elasticity, have sometimes been used as an alternative to the "raw" velocity measurements for purposes of rock classification. Onodera (1963) expressed rock quality as a *soundness index,* defined as the ratio between the field and the laboratory dynamic Young's modulis, on a scale from 0.20 (bad) to 0.75 (excellent). He and also Knill (1970) used this index to measure the effectiveness of grouting in dam foundations. In this application, however, dynamic modulus is no better than velocity and can be confused with the *static* Young's modulus measured by loading tests. Static and dynamic moduli bear little relationship to each other because the rates and amplitudes of loading are quite different when the rock is stressed dynamically by a sonic pulse as opposed to mechanically by a jack. Dynamic moduli can be 10 times higher than static ones, particularly in saturated weak rock.

The *petite sismique* method, developed in France, makes use of S-wave velocity, wavelength, and attenuation measurements in combination, to characterize the ground. Quantitative correlations have been established between petite sismique parameters and other engineering properties, such as the static modulus of deformability (Hoek and Londe, 1974). Turk and Dearman (1986) have proposed a *seismic fissuration index,* which is a function of the difference in P- and S-wave velocities.

3.3.3 Classification systems

3.3.3.1 Concepts and historic development. Rock classifications can take into account one or several attributes and are termed *univariate, bivariate,* or *multivariate,* depending on whether they are based on one, two, or more parameters. The more observations, the more complete the picture of rock character that is obtained. There is a practical limit, however, to the amount of observation and testing that can be done. A compromise is reached by selecting the most relevant types of observation and test, and by simplifying the testing procedures as far as possible.

In the early days of engineering geology, rock names were used as the only indication of mechanical competence. This led to many cases of misinterpretation that were at first addressed by introducing "weathering classifications," which allowed a rock name to be modified by terms such as *fresh* or *highly weathered.* Although an improvement, a need for direct physical testing soon became apparent.

The first quantitative classifications were based on just a single parameter, compressive strength. Rocks were classified on a scale from weak to strong. Subsequent recognition of the importance of jointing led to a classification based on the concept of rock quality designation (RQD) (see Sec. 3.1.2.2).

Later it became evident that no single property could be expected to provide an adequate basis for rock mass evaluation. Bivariate systems were explored, such as one by Deere and Miller (1966) based on compressive strength and Young's modulus (Fig. 3.10). They introduced the concept of *modulus ratio,* the dimensionless ratio between Young's modulus and uniaxial compressive strength. The modulus ratio varied from high (500) to low (200). Bivariate classifications could easily be represented in graphic form and gave useful insights into rock characteristics. However, the early ones were limited in their practical application either because they were based on properties such as elastic moduli that were difficult and time-consuming to measure, or because the selected parameters were not those most relevant to rock mass performance.

There is no single accepted classification, either for rock materials or for rock masses. A choice may be made from several that have been published, or a system may be designed to suit the specific requirements of a particular project or suite of rocks.

3.3.3.2 Size-strength classification. A simple *size-strength* classification is given here as an example. It has been applied in a variety of projects and rock types and can be extended easily by further testing (Franklin, 1976, 1986). It is based on two parameters: block size (Sec.

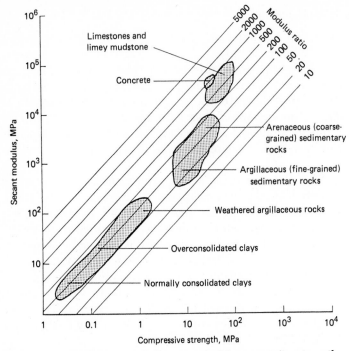

Figure 3.10 Classification according to uniaxial compressive strength and Young's modulus. (*After Deere and Miller, 1966; Hobbs, 1974.*)

3.1.2.1) and intact strength usually measured by point-load testing (Sec. 2.3.1.3).

The classification is shown graphically in Fig. 3.11. Broken and weak rock units plot toward the lower left of the diagram, whereas massive and strong rocks plot toward the upper right. The former are easy to excavate by ripping or simple mechanical excavation, but are difficult to support and stabilize. The latter require blasting but tend to be self-supporting.

The size-strength classification diagram can be contoured to combine block size and strength into a single degree of support number that can then be correlated with parameters of engineering performance, such as thicknesses of shotcrete and numbers of rockbolts and ribs required in a tunnel (Franklin, 1976).

3.3.3.3 Shale rating system. A shale rating system, as shown in Fig. 3.12, developed for the Ontario Ministry of Transportation (Franklin, 1983), is given as an example of how a classification can be purpose-designed for a particular suite of rocks. Shales are particularly variable materials and behave quite differently in engineering works, de-

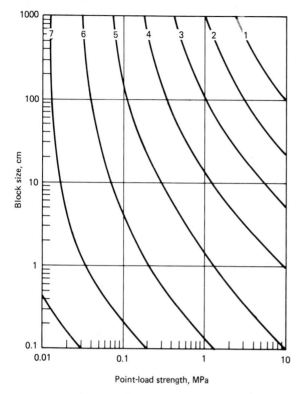

Figure 3.11 Size-strength classification for rocks. Contours show the *degree of support number D.*

pending on geological factors such as mineral composition and compaction.

The key performance variable for shales is their rate of breakdown during wetting and drying. The slake-durability test was devised to measure this property (Sec. 2.3.3.1). To assign a shale rating number, durability is measured first. The "soillike" shales with a slake-durability index of less than 80% are then further characterized by measuring the plasticity index of fragments passing through the slake drum. The "rocklike" shales with a slake-durability index greater than 80% are tested to determine their point-load strength. The tests are conducted on shale at natural water content and with load applied perpendicular to the bedding. The results are plotted on the rating chart to determine the corresponding rating values. Lines radiating at 2° intervals define rating values R_s in the range of 0.0–9.0.

Correlations have been explored between shale rating and performance parameters relating to the compaction and stability of shale embankments, slopes, and foundations.

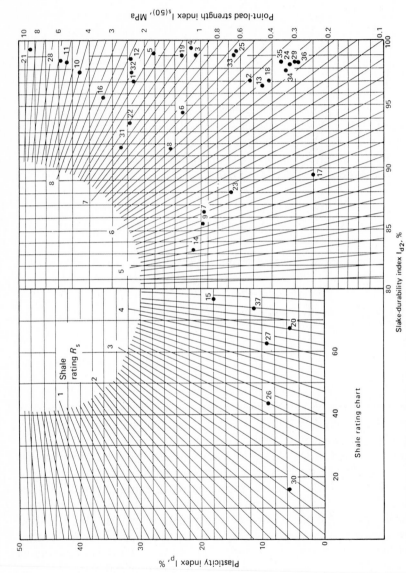

Figure 3.12 The R_s shale rating system. (*Franklin, 1983.*)

3.3.3.4 Rock mass rating (RMR) system. Other classification systems have been developed mainly for the empirical design of tunnels, among which the RMR and Q systems are perhaps the best known.

The RMR system was developed in South Africa in 1973, and is described in its latest form by Bieniawski (1984). RMR is the sum of six properties: uniaxial compressive strength, RQD, joint spacing, quality of the joints, groundwater conditions, and joint orientation. Tables allow determination of the parameters as a guide for the selection of excavation and support procedures, for openings of 4–12-m diameter.

3.3.3.5 Q system. The Q system was developed at the Norwegian Geotechnical Institute (NGI) by Barton et al. (1974) and considers six parameters: rock quality designation RQD, joint set number J_n, joint roughness number J_r, joint alteration number J_a, joint water reduction factor J_w, and stress reduction factor SRF. The tunneling quality Q is expressed as the product of ratios of pairs of these parameters as follows:

$$Q = \left(\frac{\text{RQD}}{J_n}\right)\left(\frac{J_r}{J_a}\right)\left(\frac{J_w}{\text{SRF}}\right)$$

Tables for determining the individual parameters are based on case histories of underground openings, mostly in igneous rocks. A problem arises since Q can vary by more than one order of magnitude depending on the assumed value of SRF. Water and stress conditions perhaps should be considered separately rather than included in a rock classification.

References

Barton, N., and V. Choubey: "The Shear Strength of Rock Joints in Theory and Practice," *Rock Mech.*, **10**, 1–54 (1977).
——, R. Lien, and J. Lunde: "Engineering Classification of Rock Masses for the Design of Tunnel Support," *Rock Mech.*, **6**, 189–236 (1974).
Bieniawski, Z. T.: *Rock Mechanics Design in Mining and Tunneling* (A.A. Balkema, Rotterdam and Boston, 1984).
Brand, W. E., and H. B. Phillipson: "Site Investigation and Geotechnical Engineering Practice in Hong Kong," *Geotech. Eng. (J. Southeast Asian Geotech. Soc.)*, **15**(2), 97–153 (1984).
Dearman, W. R.: "The Characterization of Rock for Civil Engineering Practice in Britain," *Proc. Colloq. Géologie de l'Ingénieur* (Liège, Belgium, 1974), pp. 1–75.
Deere, D. U.: "Engineering Geologists' Responsibilities in Dam Foundation Studies," *Proc. Eng. Fundamentals Conf. Foundations for Dams* (Pacific Grove, Calif., 1974), Am. Soc. Civ. Eng., New York, pp. 417–424.
——, A. H. Merritt, and R. F. Coon: "Engineering Classification of in situ Rock," Rep. AFWL-67-144, Air Force Systems Command, Kirtland Air Force Base, N.M. (1969).
—— and R. P. Miller: "Engineering Classification and Index Properties for Intact Rock," Rep. AFWL-TR-65-116, Air Force Weapons Laboratory (WLDC), Kirtland Air Force Base, N.M. (1966).

Fecker, E., and N. Rengers: "Measurement of Large Scale Roughnesses of Rock Planes by Means of Profilograph and Geological Compass," *Proc. Symp. Rock Fracture* (Nancy, France, 1971), paper I-18, 11 pp.

Fookes, P. G., W. R. Dearman, and J. A. Franklin: "Some Engineering Aspects of Rock Weathering with Examples from Dartmoor and Elsewhere," *Quart. J. Eng. Geol.* (London), **4** (3), 139–185 (1971).

Franklin J. A.: "An Observational Approach to the Selection and Control of Rock Tunnel Linings," *Proc. Conf. Shotcrete for Ground Support* (Easton, Md., Oct. 3–8, 1976), Am. Soc. Civ. Eng., New York, pp. 556–596.

———: "Evaluation of Shales for Construction Projects—An Ontario Shale Rating System," Res. Rep. RR229, Ontario Ministry Transp. and Commun., Toronto, 99 pp. (1983).

———: "Size-Strength System for Rock Characterization," *Proc. Symp. Application of Rock Characterization Techniques to Mine Design*, Am. Soc. Min. Eng. Ann. Mtg. (New Orleans, La., 1986), pp. 11–16.

———, N. H. Maerz, and C. P. Bennett: "Rock Mass Characterization Using Photoanalysis," *Proc. 28th U.S. Symp. Rock Mech.* (Tucson, Ariz., 1987); also *Int. J. Min. Geol. Eng.* (in press).

——— and D. Pearson: "Rock Engineering for Construction of Science North, Sudbury, Ontario," *Can. Geotech. J.*, **22**, 443–455 (1985).

Goodman, R. E.: *Methods of Geological Engineering in Discontinuous Rocks* (West Pub., St. Paul, Minn., 1976).

Hobbs, N. B.: "Settlement of Foundations on Rock," Gen. Rep., *Proc. Brit. Geotech. Soc. Conf. Settlement of Structures* (Cambridge, England, 1974), pp. 498–529.

Hoek, E., and P. Londe: "The Design of Rock Slopes and Foundations," Gen. Rep., *3d Int. Cong. Rock Mech.* (Denver, Colo., 1974), 40 pp.

ISRM: "Suggested Methods for Rock Characterization, Testing and Monitoring," ISRM Commission on Testing Methods, E.T. Brown, Ed. (Pergamon, Oxford, 1981), 211 pp.

Kelsall, P. C., R. J. Watters, and G. Franzone: "Engineering Characterization of Fissured, Weathered Dolorite and Vesicular Basalt," *Proc. 27th U.S. Symp. Rock Mech.* (Tuscaloosa, Ala., 1986), pp. 77–84.

Knill, J. L.: "The Application of Seismic Methods to the Prediction of Grout Take in Rock," *In Situ Investigations in Soils and Rocks* (Brit. Geotech. Soc., London, 1970), paper 8, pp. 93–100.

Mahtab, M. A., D. D. Bolstad, J. R. Alldredge, and R. J. Shanley: "Analysis of Fracture Orientations for Input to Structural Models of Discontinuous Rock," Rep. Invest. 7669, U.S. Bureau of Mines, 76 pp. (1972).

Meints, J. P.: "Statistical Characterization of Fractures in the Museum and Rocky Coulee Flows of the Grande Ronde Formation, Columbia River Basalts," M.S. thesis, Wash. State Univ., Pullman (1986).

Merritt, A. H., and G. B. Baecher: "Site Characterization in Rock Engineering," *Proc. 22d U.S. Symp. Rock Mech.* (Cambridge, Mass., 1981), pp. 47–63.

Onodera, T. F.: "Dynamic Investigation of Foundation Rocks in situ," *Proc. 5th U.S. Symp. Rock Mech.* (Minneapolis, Minn, 1963) (Pergamon, Oxford), pp. 517–533.

Priest, S. D., and J. A. Hudson: "Discontinuity Spacings in Rock," *Int. J. Rock Mech. Min. Sci. & Rock Mech. Abstr.*, **13**, 135–148 (1976).

——— and A. Samaniego: "A Model for the Analysis of Discontinuity Characteristics in Two Dimensions," *Proc. 5th Int. Cong. Rock Mech.* (Melbourne, Australia, 1983), vol. F, pp. 199–207.

Pyrak, L. J., L. R. Myer, and N. G. W. Cook: "Determination of Fracture Void Geometry and Contact Area at Different Effective Stresses," Fall 1985 Mtg., Am. Geophys. Union (San Francisco, Calif.).

Reeves, M. J.: "Rock Surface Roughness and Frictional Strength," *Int. J. Rock Mech. Min. Sci. Geomech. Abstr.*, **22**, 429–442 (1985).

Snow, D. T.: "The Frequency and Apertures of Fractures in Rock," *Int. J. Rock Mech. Min. Sci. Geomech. Abst.*, **7**, 23–40 (1970).

Stimpson, B., and G. Walton: "Clay Mylonites in English Coal Measures. Their Significance in Open Cast Slope Stability," *Proc. 1st Int. Cong. Int. Assoc. Eng. Geol.* (Paris, France, 1970), vol. 2, pp. 1388–1393.

Tanimoto, C., and K. Ikeda: "Acoustic and Mechanical Properties of Jointed Rock," *Proc. 5th Int. Cong. Rock Mech.* (Melbourne, Australia, 1983), vol. A, pp. 15–18 (1983).

Turk, N., and W. R. Dearman: "A Suggested Approach to Rock Characterization in Terms of Seismic Velocities," *Proc. 27th U.S. Symp. Rock Mech.* (Tuscaloosa, Ala., 1986), pp. 168–175.

Uemura, T., and S. Mizutani: *Geological Structures* (Wiley, New York, 1984), 309 pp.

Wakeling, T. R.: "Comparison of Results of Standard Site Investigation Methods Against the Results of a Detailed Geotechnical Investigation in Middle Chalk at Mundford, Norfolk," *Proc. Conf. in situ Investigations in Soils and Rocks* (May 1969), Brit. Geotech. Soc., London, pp. 17–22.

Young, R. P., T. T. Hill, I. R. Bryan, and R. Middleton: "Seismic Spectroscopy in Fracture Characterization," *Quart. J. Eng. Geol.* (London), **18**, 459–479 (1985).

4

Groundwater

4.1 Water in Rock

4.1.1 Benefits, hazards, and costs

Groundwater is a valuable natural resource, but sometimes also a major hazard and expense. On the positive side, groundwater pumped from wells is put to domestic and industrial use. It is employed to leach minerals from the ground, in solution mining, and in the production of geothermal steam. It provides a convenient barrier to contain hydrocarbons and compressed gases in underground storage chambers.

On the negative side, groundwater has the capacity to dissolve and transport toxic contaminants. Seepage into an open excavation or tunnel makes blasting difficult and unsafe. The engineer often is obliged to dispose of unwanted inflows, which are expensive to remove by pumping, or to seal by grouting. Water pressures trigger landslides, and outflows erode and carry away soils and weathered or closely jointed rocks. Acid drainage from sulfide-containing ores or waste heaps is a serious environmental problem. Lowering of the water table, intentional or accidental, can have many harmful effects, including interference with water-well supplies and agriculture, and can occasionally induce subsidence and even earthquakes (Evans, 1966).

In situations where water has a major destabilizing influence, drainage and grouting can have correspondingly beneficial effects. Often the high costs of installing groundwater control systems are more than offset by the benefits (Chap. 18).

4.1.2 Water in pores and fissures

4.1.2.1 Hydrologic cycle. Groundwater reaches the earth by precipitation, seeps into the ground, and returns to the surface by spring flow

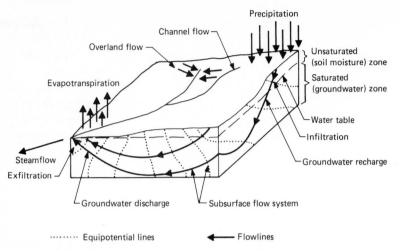

Figure 4.1   Hydrologic cycle. (*Freeze and Cherry, 1979.*)

and to the atmosphere by evapotranspiration (Fig. 4.1). The total quantity of water remains constant in this hydrologic cycle. Depending on the subterranean flow paths, the residence time of water in the earth can range from just a few weeks to thousands of years in some cases (Freeze and Cherry, 1979).

4.1.2.2   Primary and secondary porosity.   Intact rock contains pores and cracks between and within grains and crystals. This void space is termed *primary* porosity. Other voids in the form of joints, faults, and blast-induced fractures together form *secondary* porosity or *fracture* porosity. Voids are filled with water, air, and sometimes other gases and liquids. The proportionate volumes and weights of these constituents determine porosity, density, water content, and degree of saturation (Chap. 2).

4.1.2.3   Fixed and mobile water.   Some water clings to intergranular surfaces (usually clay mineral surfaces) so tightly that it can be regarded as a "constituent" of the rock (Chap. 2), whereas other water is free to move through the voids. The size, shape, and interconnection of the voids determine the *permeability* or *hydraulic conductivity* of the rock. The great majority of flow nearly always occurs through joints because although the joint porosity is often smaller than the primary (pore) porosity, the joints are more directly connected and offer a less tortuous path for water flow (Fig. 4.2). The groundwater in rocks can be visualized as consisting of a small volume of relatively mobile joint water, accompanied by a usually much greater volume of relatively stagnant pore water.

Figure 4.2 Seepage through rock joints prior to foundation grouting at Big Eddy Dam, Nairn, Ont., Canada.

4.1.2.4 Water table. The *water table* is a surface more or less parallel to the ground surface that separates saturated from unsaturated ground (Fig. 4.3). It is the locus of points where water pressure equals atmospheric pressure. Above the water table, a saturated *capillary fringe* often exists, in which pressures are maintained negative (less

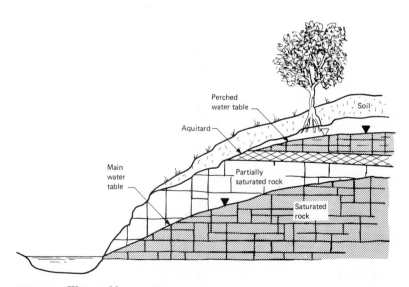

Figure 4.3 Water table concepts.

than atmospheric) by capillary (surface tension) forces in joints or between the fine grains of a soil. Above the capillary fringe, the rocks
and soils are not dry, but partially saturated. Below the water table,
the pressures in pore and joint water are positive.

In many situations, a single water table exists, below which the
groundwater pressure increases linearly with depth in a *hydrostatic*
pressure distribution identical to that beneath the surface of a body of
open water. However, multiple groundwater tables (nonhydrostatic
conditions) are common, for example, in an alternating sequence of
sandstones and shales in steep terrain, where each sandstone layer
acts as a more or less independent hydrogeologic regime, with its own
water pressure distribution.

A low-permeability stratum in an otherwise permeable deposit can
lead to perched water tables, as shown in Fig. 4.3. This illustration
shows a saturated lens with unsaturated conditions, both above and
below. Such lenses expand and contract or even disappear seasonally,
along with fluctuations in the elevation of the main groundwater table at greater depth.

4.1.2.5 Aquifers, aquitards, and aquicludes. An *aquifer* is a saturated,
permeable geologic unit that can transmit significant quantities of
water under ordinary hydraulic gradients. An *aquiclude* is a geologic
unit that is incapable of transmitting significant quantities of water.
An *aquitard* is intermediate in its hydraulic conductivity. Faults may
contain clayey layers that act as aquicludes, with or without broken
layers that perform as highly conductive aquifers.

A confined aquifer is contained between two aquitards: the water
level in a well rises above the top of the aquifer and, in artesian conditions, even above the ground surface as a flowing well. In an
unconfined aquifer, or water table aquifer, the water table forms the
upper boundary.

**4.1.3 Water supplies and problems of
contamination**

Growth of population and urbanization has been accompanied by an
increase in the demand for groundwater, which in North America satisfies about 20% of the total demand (Freeze and Cherry, 1979). Consumption in 1968–1970 totaled approximately 260 million cubic
meters per day in the United States and 1.71 million cubic meters per
day in Canada.

At the same time, urbanization has brought about an increase in
the production of waste materials which, without precautions, can
contaminate the groundwater. Deep well injection of liquid wastes

and also sanitary landfill for solid wastes both can lead to pollution. Further contamination occurs by seepage from sewage ponds and lagoons, and leaching of animal wastes, fertilizers, and pesticides. Special problems of contamination control arise with the need to dispose of mine tailings and radioactive wastes.

This book on rock engineering is concerned with the effects of water on engineering works in rock, and how to cope with associated water problems. It does not attempt to cover the extraction of water as a commodity. Water exploration and the yield, quality, and contamination of groundwater supplies are the focal points of textbooks on hydrogeology, such as Freeze and Cherry (1979).

4.1.4 Effects on engineering works

Water-related problems and countermeasures keep reappearing throughout this book. Table 4.1 and the following preview show the diverse ways in which water can affect construction.

4.1.4.1 Wetting effects. The mere presence of water, even in the absence of a flow or a pressure problem, is a hazard and an expense, particularly in the confined spaces of underground construction. Wet equipment is difficult to handle and control; shales turn to mud and construction equipment gets bogged down; geothermally heated and gas-saturated waters require extraordinary measures in ventilation to keep working conditions tolerable. During tests at a dam site in Greece, hot, sulfur-saturated waters were encountered in an adit. The atmosphere underground became so bad that silver coins pocketed in

TABLE 4.1   Adverse Effects of Water on Engineering Works

Wetting	Equipment becomes slippery and unsafe
	Mud obstructs construction traffic
	Blasting becomes more difficult
Weathering	Shales slake and disintegrate
	Clays and anhydrite swell
	Hard rocks break apart by ice jacking
	Salt rocks quickly dissolve
	Limestones contain a legacy of karsts
Flow	Pumping can be expensive
	Inrushes cause unexpected flooding
	Erosion removes surface deposits and fillings
	Contaminants are transported long distances
Pressure	Shear strength reduced along joints
	Water pressure in fissures disrupts rock
	Earthquakes caused by pumping and injection
	Subsidence induced by underdrainage

the morning had turned black by nightfall. The waters had to be dammed and diverted through pipes.

Blastholes fill with water. Blasting costs for wet holes are about double those for dry ones, because of difficulties in loading and detonation, and sometimes a requirement for alternative explosives. Water saturation affects the costs of shipping mined ores. Brawner (1968) estimated that even an extra 2% moisture in the iron ore at Knob Lake, Que., Canada, increased shipping costs by 12 cents per ton.

4.1.4.2 Slaking, swelling, and freezing. Changes in water content cause weakening, slaking, or swelling. Most rocks are weaker when wet, but regain their original strength when dried. Some rocks, however, slake (disintegrate) when wetted and dried repeatedly. Others contain minerals such as clays, pyrite, or anhydrite that swell in contact with water, air, or bacteria, and these rocks deteriorate irreversibly when subjected to wetting and drying cycles. Swelling mechanisms and problems are further discussed in Chap. 10.

In cold climates, frost and groundwater combine to create a further hazard. Closely jointed rock masses in which the blocks themselves are durable can loosen and disintegrate because of freezing and thawing of water in the joints.

4.1.4.3 Solution problems and karst. Most minerals are soluble to some extent. Solubility is greatest in salts (halite and gypsum), moderate in carbonate rocks (limestones and dolostones), and slight in silicates. Davis (1969) estimates that solution of silica can increase joint apertures in a quartzite rock by about 0.4 mm in 100,000 years. In the United States, nearly 15% of the land is underlain by "soluble" rock formations, mostly limestones (Davies, 1984).

In salt and potash mines, ingress of water that is less than fully saturated with salt means solution of salt, usually accompanied by serious problems of ground instability. An entire potash mine in Saskatchewan, Canada, was flooded in the late 1960s when high-pressure water behind the shaft several hundred meters above the potash zone could not be contained in time. Other flooding problems in Saskatchewan have arisen through weakening and cracking of the roof rock, allowing high-pressure water from overlying formations to enter the mine.

Except where salt rocks are found locally, active solution is rarely a problem during the life span of an engineering project. However, problems do occur because of the accumulated effects of "geological" solution over many thousands of years, which form karst (pipes and caves) by enlargement along joint surfaces and intersections.

Karst-related problems include foundation settlement and collapse, a high potential for leakage from reservoirs and beneath dams, severe difficulties in mapping and evaluating the groundwater flow regime, an increased potential for localized contamination of groundwater, and difficulties in selecting areas suitable for waste disposal. Because of exceptionally good drainage, most of the central parts of karst areas are impoverished in water and can be almost deserted, despite rainfall that may exceed 800 mm per year (Liszkowski, 1975).

Karst is named after the Karst region of Yugoslavia where such features abound. Types include sinkholes, pipes, caverns, caves, shafts, and other solution features (Fig. 4.4). Karst is found usually in limestone but occasionally in dolomite, gypsum, or rock salt. In major karst regions, thousands of kilometers of caves extend in places to depths of more than a kilometer. Channels develop vertically in the unsaturated zone, and horizontally or subhorizontally in the upper parts of the saturated zone. They usually follow the rock jointing system along and immediately beneath an existing or historic water table. Karst is not necessarily associated with the present-day groundwater regime and can have been formed under quite different hydrogeologic conditions in the past.

To dissolve, a rock formation must be brought into contact with water that is unsaturated in the minerals of which it is composed. Stagnant water soon becomes saturated, so moving water and rapid infiltration from the surface are essential for solution to continue.

Solution occurs preferentially along joints and particularly at intersections. This can be seen most clearly in the form of *clints* that follow the jointing in a limestone (Fig. 4.4b). Usually just a few of the joints become enlarged. Real joints are rough, giving rise to water flow velocities that vary in proportion to the square of the aperture. Therefore even on a microscale, water tends to flow along the widest channels (Sabarly et al., 1970). Because solution rates are greatest at higher velocities, apertures increase along these preferred flow paths. This leads to increased flow quantities through the enlarged channels, which gradually become the main flow routes at the expense of countless narrow fissures. Liszkowski (1975) calls this, appropriately, "the law of underground piracy."

4.1.4.4 Flow problems. High rates of inflow lead to high pumping costs during construction and, sometimes worse, to erosion of joint infillings, dissolution of soluble minerals, or external erosion and raveling of broken or low-durability rock materials. The water leakage beneath a dam has been known to exceed the flow entering its reservoir from the catchment area, leaving the reservoir dry.

Marulanda and Brekke (1981) describe a tunnel in Colombia bored through a sequence of alternating sandstones and shales. The rock

(a)

(b)

Figure 4.4 Solution in limestones. (a), (b) Earth-filled pipes in chalk; Welwyn Garden City, England. (c) Similar pipes caused extensive foundation settlement and structural damage at a nearby housing estate. (*Courtesy of Rock Mechanics Ltd., U.K.*) (d) Clints in limestone; Malham Cove, Yorks., England. (*Courtesy of P. Russell, University of Waterloo, Ont.*)

face erupted during the mucking cycle. An inflow of up to 3 m³/s carried with it 7000 m³ of sand from a bed of very friable sandstone, filling the tunnel, and covering all equipment and installations. A bulkhead was constructed 10 m behind the face and a drainage gallery excavated to intercept and divert the flow.

(c)

(d)

Figure 4.4 *(Continued)*

4.1.4.5 Pressure problems. High groundwater pressures counteract the contact forces between the faces of rock joints according to the principle of effective stress (Sec. 4.2.1.3), thereby reducing frictional strength and the overall stability of the rock mass. Pressures can develop behind surface layers of weathered, clayey, or frozen rock on a slope face. This "ice damming" and "clay damming" prevents natural drainage, and often is followed by bursting of the skin and by mudflows and avalanches. High water pressures also encourage buckling and heave of rock beds in foundation excavations.

Huder and Amberg (1983) describe a problem that developed in the Isla Bella tunnel in Switzerland, where a boring machine had to be removed under difficult conditions. Sheared rock was being squeezed like a paste through gaps in the lining, propelled by high water pressures. Because of the low permeability of the ground, the water inflow was small. Drain holes were fanned from the top heading, 5 m long radially and 15 m long in the direction of advance. After relieving the water pressure, excavation of the top heading proceeded quickly and conditions were noticeably improved, even though the total water removed amounted to no more than 6–8 L/s.

4.1.4.6 Drainage-induced subsidence and seismicity.

Subsidence can be caused either by caving, following the progressive collapse of an underground excavation, or by the removal of pore fluids from compressible soils. The latter mechanism, termed *drainage-induced subsidence,* can be caused by tunneling through the soils themselves or through the underlying rocks, or by pumping from water wells or oil fields over prolonged periods. The settlements usually extend more or less uniformly over broad areas, although differential settlements and surface tilting can result from nonuniform soil thicknesses or types.

Soils shrink and consolidate when drained. The silty soils of the Mexico City basin, for example, have been pumped for groundwater supply for many decades. During the period of 1938–1970, parts of the city subsided by as much as 8.5 m.

Similar but less severe problems were experienced in Stockholm, Sweden. Driving a subway tunnel through Precambrian bedrock caused a lowering of the water table by up to 6 m and drained the fine-grained clayey silt in glacial channels. Underdrainage was followed by an increase in effective stress, and by settlements, cracks in pavements and cellar floors, and damage to cables and water and sewer pipes. Morfeldt (1969) has estimated that an underdrainage of only 1.25 L/s could have been sufficient to account for these effects. The early problems led to official requests for careful pregrouting before tunneling through jointed and faulted bedrock.

Underdrainage can be avoided by taking precautions to minimize inflows. Grouting is the main method in civil works, although water-tight lining systems may be of benefit. Underground mining excavations are more difficult to control, being more extensive, and also because the economics of mining precludes costly grouting.

Groundwater recharge can be effective as a remedial measure. A fluid as valuable as oil or fresh water can be replaced with one of less value and in abundant supply, such as saline water in a coastal region. Injection through recharge wells keeps pace with extraction to maintain the water table at an approximately constant elevation.

Earthquake swarms may be triggered by pumping water or oil, by large-scale injection of fluids, or by the impounding of a reservoir. This phenomenon of induced seismicity is caused by changing the water pressures within faults that are in a state of quasi-equilibrium under the action of shear and normal stresses (Healy, 1975).

4.1.4.7 Gas inflows. Flow of oil and gas through rock is of considerable interest for the design of oil reservoir extraction strategy. Study of the flow of gases (methane, ethane, carbon dioxide) and of combinations of gases with other fluids is of critical interest not only for the development of gas wells but also for the drainage of methane and other potentially toxic or explosive gases from coal seams in underground mines (e.g., Harpalani and McPherson, 1986).

The release of methane from coal has had an increasingly important effect on the ventilation and safety of underground mines over the last 30 years, as mines have achieved greater depths and higher productivities. This trend, and especially the potential for commercially recovering methane in advance of mining from unminable coal, has prompted the development of sophisticated methods to predict methane desorption and migration through coal and the surrounding strata (SPE, 1986). For these studies, one must determine the permeability of the coal and its variation with factors such as stress, gas pressure, and time. Methane can form a special adsorbed layer on the free surfaces of coal, associated with the cleat and the natural porosity, which explains the very large quantities of gas per unit of coal volume.

4.2 Concepts of Pressure and Flow

We now turn to techniques for predicting when and where problems of excessive flows or pressures will occur. Armed with a basic knowledge of the laws of fluid flow through porous and fissured media, the engineer can predict pressure distributions and inflows likely to be generated by excavating an open-pit mine or a tunnel, and can design pumping, drainage, and grouting measures to keep these problems under control. Permeability tests and pressure and flow instrumentation provide the data needed for design, and the same instruments are used to monitor the behavior of groundwater during the construction and operating phases of the project.

4.2.1 Pressure

4.2.1.1 Hydraulic head, potential, and gradient. The *hydraulic head* (or simply head) at a given point in the ground is defined as the height

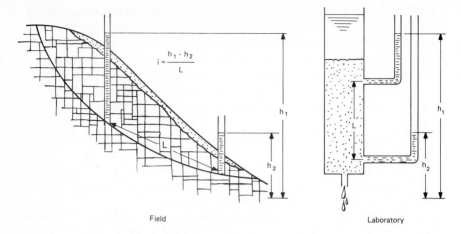

$$i = \frac{h_1 - h_2}{L}$$

Field

Laboratory

Figure 4.5 Definition of hydraulic gradient.

above an arbitrary datum to which water would rise in a standpipe. Head is expressed in units of meters of water (or simply meters). *Hydraulic potential* is defined as the product of hydraulic head and gravitational acceleration. Head differences are responsible for the flow of water between two adjacent points (Fig. 4.5). Flow always occurs from regions where the head is greatest to where it is least. *Hydraulic gradient* is defined as the difference in head between two points in the ground divided by the distance separating these points, in other words, *dh/dl*, the rate of hydraulic head loss per unit length of flow path. It is therefore dimensionless.

4.2.1.2 Potentiometric and equipotential surfaces. A *piezometer* is a device for measuring the hydraulic head at a point. If water levels in piezometers or wells tapping a confined aquifer are plotted and contoured, the resulting surface is a map of the hydraulic head in the aquifer, called a *potentiometric surface*. Groundwater within the aquifer flows down the slopes of this surface, in directions of greatest hydraulic gradient.

Similarly, a large number of piezometers distributed in three dimensions throughout an unconfined aquifer allow one to contour the *equipotential surfaces* of equal hydraulic head. The traces of these surfaces on a two-dimensional cross section are known as *equipotential lines*. Once the pattern of equipotential lines is known, flow lines can be constructed perpendicular to them, forming a *flow net*. Construction of flow nets is a powerful analytical tool and is further discussed in Sec. 4.4.1.

4.2.1.3 Principle of effective stress. The total stress across any arbitrary surface in the ground, such as a joint plane, is carried in part by the effective solid contacts, and in part by the water pressure between the faces of the joint. In 1925 this very important principle of effective stress was proposed by Karl Terzaghi. It can be formulated as follows:

$$\sigma_n = \sigma_n' + u$$

where σ_n is the total stess, σ_n' is the effective stress, and u is the pore or joint water pressure.

Take the case of a horizontal bedding joint in the rock foundation of a dam. The total vertical stress across the joint is constant and determined by the combined weight of the overlying rock and the dam. Suppose the water pressure u within this joint is now increased by filling the reservoir. The effective stress between the solid contacts has to decrease by a corresponding amount, with a consequent reduction in the shear strength of the joint. Unless the pressure is relieved by drilling drain holes or reduced by grouting, the dam may slide.

4.2.1.4 Consolidation. The theory of consolidation follows directly from the principle of effective stress as outlined above. *Consolidation* occurs when the grains of a saturated soil or the surfaces of a water-filled rock joint are pressed together by an externally applied load. At first the external load is carried entirely by the joint water, which cannot escape instantaneously (Fig. 4.6). The effective stress and therefore the shear strength of the joint at this instant may be quite low. The joint water pressure dissipates (decreases) as water migrates from the joint under the newly imposed hydraulic gradient. Flow finally

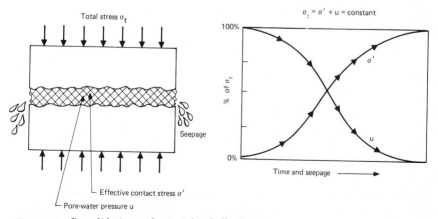

Figure 4.6 Consolidation and principle of effective stress.

stops when the total normal stress is carried by the asperity contacts. The equilibrium joint water pressure is governed by the head of water and not of overlying rock.

At any instant during consolidation, the principle of effective stress holds true. When the total stress across the rock joint is constant, it is equal to the sum of an (increasing) effective stress and a (decreasing) joint water pressure. Consolidation occurs so rapidly in most rocks, because of the high conductivity of the joints, that the transfer of load to the joint asperities is for all practical purposes instantaneous. Only in clay-filled joints is consolidation likely to be of any consequence. Design calculations must then take into account the reductions of shear strength that may occur because of excess pore pressures during dynamic loading.

4.2.2 Flow

4.2.2.1 Flow velocities.
Three different velocities of water flow are defined as follows. Perhaps the most common is the average flow rate through the bulk rock, termed the *specific discharge*. This has dimensions of velocity, and is defined as the volumetric flux Q (cubic meters) divided by the full cross-sectional area A of the specimen or flow channel (square meters) including both voids and solids.

The *average linear water velocity* is a measure of the speed with which water moves through the pores or joints, and is obtained by dividing the specific discharge by the areal or volumetric (averaged) porosity. It is faster than the specific discharge. Solutes are carried at the average linear velocity of the water.

Finally, the *average velocity of water particles,* which is self-explanatory, is greater than either the specific discharge or the average linear velocity. Water traveling through pores follows a tortuous path rather than a straight line. Actual particle velocities are difficult to measure.

4.2.2.2 Seepage forces.
Seepage forces are developed in overcoming friction on the joint surfaces. They are body forces and proportional to the potential gradient. For accurate modeling of rock mass behavior they have to be added to the forces generated by buoyancy (Sharp, Hoek, and Brawner, 1972). The forces generated by seepage have preferred directions, sometimes directions that are detrimental to stability.

4.2.2.3 Darcy's law, hydraulic conductivity, and permeability.
The French hydraulics engineer Henry Darcy in 1856, while conducting infiltration tests on fine-grained sands, noted a linear proportionality

between water flow and hydraulic gradient. His observation, known as Darcy's law, can be expressed as

$$Q = - KiA$$

or

$$v = - Ki$$

where Q is the flow rate (dx/dt), i is the hydraulic gradient (dh/dx), A is the cross-sectional area of the flow path, and $v = Q/A$ is the specific discharge.

In this equation K is known as *hydraulic conductivity*. It has the dimensions of velocity (L, T) and is a function not only of the nature of the soil or rock but also of the nature and temperature of the fluid. Darcy's law is empirical in that it rests only on experimental evidence. It describes the flow not only of water but also of oil and gas in deep geological formations.

The hydraulic conductivity K can be related to the permeability k, a measure that is independent of the fluid and characteristic of the soil or rock only,

$$K = krg/\mu$$

or

$$k = K\mu/rg$$

The parameter k is known as the specific or intrinsic permeability, or just *permeability; r* and μ are the density and the dynamic viscosity of the fluid, respectively; and g is the gravitational acceleration. K and k are actually tensors (similar to stress) requiring six independent values, such as the three principal directions plus the three principal magnitudes, to stipulate them fully.

Older texts often confuse hydraulic conductivity and permeability. However, hydraulic conductivity is the accepted term to relate flow to hydraulic gradient on the basis of Darcy's equation. Confusion can also arise because of the common use of several different units; conversion factors are given in Table 4.2. Permeability k has units of area; however, it is very small when measured in square meters, so petroleum engineers have adopted a separate unit called the *darcy,* which is equal to approximately 10^{-8} cm^2. In the water-well industry, hydraulic conductivity is normally expressed in gal/day/ft^2, whereas in rock engineering it is given in m/s. Typical values for rocks and soils are shown in Fig. 4.7.

4.2.2.4 Exceptions to Darcy's law. If Darcy's law were universally valid, a plot of specific discharge versus hydraulic gradient would give a straight line for all gradients. In practice, it proves to give a close approximation to conditions for both steady-state and transient flow, whether in saturated or unsaturated media, also in aquifers and aqui-

TABLE 4.2 Conversion Factors for Permeability and Hydraulic Conductivity

Unit	Permeability k			Hydraulic conductivity K		
	cm^2	ft^2	darcy	m/s	ft/s	gal/day/ft^2
cm^2	1	1.08×10^{-3}	1.01×10^8	9.80×10^2	3.22×10^3	1.85×10^9
ft^2	9.29×10^2	1	9.42×10^{10}	9.11×10^5	2.99×10^6	1.71×10^{12}
darcy	9.87×10^{-9}	1.06×10^{-11}	1	9.66×10^{-6}	3.17×10^{-5}	1.82×10^1
m/s	1.02×10^{-3}	1.10×10^{-6}	1.04×10^5	1	3.28	2.12×10^6
ft/s	3.11×10^{-4}	3.35×10^{-7}	3.15×10^4	3.05×10^{-1}	1	5.74×10^5
gal/day/ft^2	5.42×10^{-10}	5.83×10^{-13}	5.49×10^{-2}	4.72×10^{-7}	1.74×10^{-6}	1

SOURCE: Freeze and Cherry, 1979.

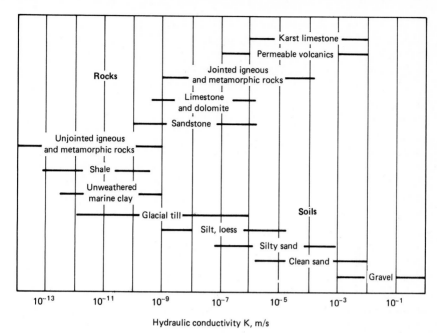

Figure 4.7 Typical hydraulic conductivities for rocks and soils. (*Freeze and Cherry, 1979.*)

tards, in homogeneous and heterogeneous, isotropic and anisotropic media, and for flow in jointed rocks as well as in granular materials.

Nonlinearity is found to occur, however, for flow through low-permeability sediments under very low hydraulic gradients (Reynold's number less than about 1), and also at high flow rates through high-porosity sediments and rocks with open fissures. Cornwell and Murphy (1985) describe investigations of flow through sawtooth joints in which the aperture was varied. They found that Darcy's law became invalid for Reynold's numbers greater than about 100. Flow rates that exceed the upper limit of Darcy's law are encountered in karstic limestones and extremely permeable volcanics, and in the blast-fractured rock that surrounds a tunnel (Sharp and Maini, 1972). Wittke (1973) has suggested that separate flow laws can be specified for the linear-laminar range (Darcy range), for a nonlinear laminar range, and for a turbulent range.

4.2.2.5 Storativity and transmissivity. These terms are used mainly in studies of water as a resource. The *specific storage* of a saturated aquifer is defined as the volume of water that a unit volume of aquifer releases from storage under a unit decrease of hydraulic head, such as caused by pumping. The *transmissivity* of a confined aquifer is defined as the product of its thickness and its hydraulic conductivity. The *storativity* of this aquifer is the volume of water that it releases from storage per unit surface area per unit decline in the component of hydraulic head normal to the aquifer's surface. If the stress compressibility of the aquifer or reservoir is a constant, then storativity is a constant. However, aquifers and reservoirs become stiffer with increased drawdown, and storativity therefore also declines.

4.2.2.6 Hydraulic conductivities of rocks. Typical hydraulic conductivity values for a range of rock conditions are shown in Fig. 4.7. Because most of the water flow occurs along joints rather than through the pore space, in situ tests nearly always give much higher conductivity values than laboratory tests on small specimens. Conductivity measured by laboratory testing is irrelevant in most engineering situations. It becomes important only when jointing is widely spaced or extremely tight, as at substantial depth, or when the rock material is extremely permeable. Louis (1974) gives the directional hydraulic conductivity of a set of continuous joints for either laminar or turbulent flow in terms of the following expression:

$$K = \left(\frac{e}{b}\right)k_f + k_m$$

where e is the mean aperture, b is the mean spacing, k_f is the hydraulic conductivity of the joint (itself a function of the square of the aperture, for laminar flow), and k_m is the primary conductivity of the rock matrix, often negligible. One joint per meter with an aperture of 0.1 mm gives a rock mass hydraulic conductivity of about 10^{-6} m/s, similar to the primary conductivity of a porous sandstone (Fig. 4.8). With a 1-mm aperture and the same spacing, the corresponding conductivity is 10^{-3} m/s, similar to that of a loose clean sand.

Primary conductivities of shales range from 10^{-9} to 10^{-12} m/s, so groundwater cannot move faster than a few centimeters per century through intact shale. Closer to the surface, however, joints impart a significant secondary conductivity. The primary conductivity of carbonate rocks varies greatly but is commonly less than 10^{-7} m/s. Many carbonate strata have appreciable secondary conductivity because of solution and enlargement of joints, particularly at the crests of anticlines and troughs of synclines.

Near-surface deposits of coal of Tertiary or Cretaceous age, even though only 1 m or so thick, are commonly a source of water for farms and small towns. Conductivities are typically in the range of 10^{-4}–10^{-6} m/s. Below about 100-m depth the hydraulic conductivities are much less, as the cleat is tight because of the higher stress.

Primary conductivities of igneous and metamorphic rocks typically range from 10^{-11} to 10^{-12} m/s or even much lower. However, local

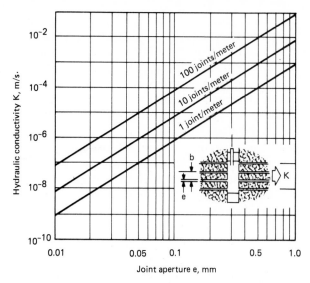

Figure 4.8 Effect of joint spacing and aperture on hydraulic conductivity. (*Hoek and Bray, 1974.*)

zones can be highly conductive, usually because of fracturing and joints. Water has been known to flow profusely into the shafts and drifts of hard rock mines even at depths below 1 km.

Bedded lava flows that solidify by rapid cooling often contain alternating layers of scoria (froth), closely jointed glassy rock, and coarse rubble zones between flows of more dense rock. Soils can form on top of the flows and subsequently be buried. Sowers (1976) computed an effective hydraulic conductivity for a scoria layer about 3 m thick in Iceland to be 1 m/s. The columnar basalt below this was by comparison an aquiclude, although its effective vertical conductivity ranged from 10^{-3} to 10^{-5} m/s.

Often the conductivity of the rock mass is anisotropic. In bedded sedimentary rocks, conductivity is typically about 1.5 times greater horizontally than vertically. Rock mass conductivity also decreases with depth, reflecting decreasing joint apertures and increasing spacings. Porosity and hydraulic conductivity are log-linearly related, so a small reduction in porosity leads to a large decrease in conductivity. At depth, therefore, well yields in oil-bearing rock have to be enhanced by hydraulic fracturing. The fractures are propped open by injecting sand.

4.2.3 Flow in joint networks

4.2.3.1 Assumption of an equivalent continuum. To this point we have assumed that the rock mass can be regarded as a continuum, much

like a soil, that flow can be modeled by Darcy's law, and that hydraulic conductivity can be expressed as an average value for an element of the rock mass. In most cases, the assumption gives quite reliable predictions, at least if one takes into account anisotropy of the conductivity tensor, and the important effects of stress and joint closure. The element in testing or modeling must contain sufficient joints and intersections to ensure that the average effects of jointing are reflected in the measured conductivity values.

However, the flow of water through individual fissures does require modeling in three special cases: first, when considering flow through a major conduit such as a fault; second, when attempting to predict conductivity of an equivalent continuum from characteristics of jointing rather than by testing directly; and third, when modeling hydraulically induced fractures and flow into an oil well, especially where the hydraulic conductivity of the surrounding matrix is low. Single-fracture models have been developed to analyze the behavior of hydraulically fractured wells (Witherspoon, 1986).

4.2.3.2 Flow through individual joints and joint networks. The flow volume between smooth parallel plates theoretically is proportional to the third power of the aperture. This cubic law also holds good for flow through real joints unless the aperture is smaller than 10 μm, when seepage is much greater than predicted by the cubic law, and is hardly reduced at all by further closure of the joint (Pyrak et al., 1985).

The equivalent hydraulic conductivity of networks of joints can be calculated from observations on individual joints. This was pioneered by Snow (1968, 1969), who determined the orientations and apertures of joints intersected by a borehole, assumed them to be infinite in length, and computed the conductivity of an equivalent continuum as an accumulation of individual joint conductivities. Snow gives relationships between porosity and the anisotropic conductivity tensor for three-dimensional joint geometries in which joint spacings or apertures differ with direction. Franciss (1985) gives similar information.

However, most real joints have a finite length. Characterization of a joint network cannot be considered complete until each joint has been described in terms of its effective aperture, orientation, and spacing, and also of its continuity (size or persistence), possibly even its position. Witherspoon and colleagues, also Oda and Hatsuyama (1985), have derived a conductivity tensor for the rock mass that takes into account each of these attributes. Long and Witherspoon (1985) have investigated the magnitude and nature of conductivity in joint networks with various degrees of interconnection, using a numerical code to generate sample joint networks in two dimensions.

4.2.3.3 Interdependence between stress, aperture, and conductivity. Hydraulic conductivity of the rock mass varies greatly when joints open and close as the result of small stress changes (Bawden et al., 1980; Gale, 1982). Pyrak et al. (1985) measured joint apertures by making a cast using Wood's metal, which melts below 100°C. They found that the contact area between granite surfaces reached values of 30% as the effective stress was raised to 85 MPa, above which aperture and flow rate became more or less constant.

Failure to take such changes into account probably was a prime cause of the Malpasset dam disaster in France, where closely jointed foundations exploded under the action of high water pressures. Since that time, rock engineers have become far more conscious of the need to consider stress when modeling groundwater flow and pressure distributions, and also of the need to relieve high pressures by drainage (Wittke and Leonards, 1985).

Voss et al. (1986) reviewed the many recent stress-conductivity experiments, including studies of the effects of scale and roughness and of varying shear and normal stresses and displacements. A shear displacement of 1 or 2 mm can be enough to augment the hydraulic conductivity of a joint by nearly two orders of magnitude. Barton et al. (1985) have proposed a constitutive model of joint behavior to simulate shear-displacement-dilation-conductivity coupling and normal-stress-closure-conductivity coupling.

Kafritsas et al. (1984) used the rigid-block distinct-element method developed by Cundall (Sec. 7.2.5.1) to investigate the coupling between groundwater flow and joint deformation, again using a cubic relationship between flow and aperture. Applying the model to a rock slope, the water pressures that developed along the potential surface of sliding were found to be greater than normally assumed, because of compression of the rock mass at the slope toe.

4.3 Pressure and Flow Measurements

4.3.1 Pressure measurements

Piezometers measure the water pressure or head at a point in a drillhole. They provide the information on groundwater needed for design, and record changes in groundwater pressures during construction (Huder and Amberg, 1983). The data thus obtained are used to check the predictions of design and the effectiveness of drainage and grouting.

These methods perhaps belong with the collection of monitoring techniques described in Chap. 12, but are described here for completeness in a groundwater context. Details of operating principles, such as for pneumatic and vibrating wire types of instruments, are given in Chap. 12 along with further information on the planning, design, and operation of monitoring systems.

4.3.1.1 Open-standpipe piezometers. The open-standpipe piezometer consists of a tube or pipe, slotted or perforated at its lower end, or fitted with a porous plastic or ceramic element, called a *piezometer tip*. In this and most other types of piezometer, the tip should be thoroughly saturated with de-aired water before installation, to prevent clogging and to ensure correct readings.

A hole is drilled, into which the standpipe is inserted (Fig. 4.9). Coarse-grained sand or gravel is placed as a filter immediately around the tip, usually with fine-grained sand above and below. The filter is hydraulically isolated by a bentonite clay plug above and, if necessary, below, and the remainder of the hole is backfilled with cement or cement-bentonite grout. Isolation of the tip is essential: open-hole

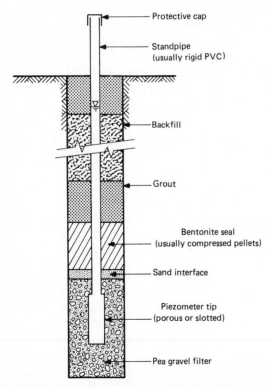

Figure 4.9 Open-standpipe piezometer installation.

measurements are unreliable because water pressures can vary considerably from one horizon to the next.

Water rises in the standpipe to a height sufficient to balance the groundwater pressure near the tip. This pressure can then be monitored using a simple water level probe, which consists of a twin-conductor cable with the ends connected to two coaxial brass cylinders separated by insulating material. Entry of the dipmeter tip into the water completes an electric circuit and activates a buzzer or lamp indicator in the cable reel. The cable is graduated along its length.

The large diameter of the standpipe tube means that water can take a long time to rise or fall, and no meaningful reading can be obtained until this level has stabilized. The response time of a piezometer may be critical when rapid fluctuations in groundwater pressure are expected and when the ground around the tip is relatively impermeable. Open-standpipe piezometers are inappropriate in such cases, which call for the use of electrical or pneumatic alternatives.

4.3.1.2 Hydraulic twin-tube piezometers. The hydraulic twin-tube piezometer consists of a porous plastic or ceramic tip connected by two flexible, small-diameter plastic tubes to a Bourdon gauge, or to a manometer in the gauge house. In a manometer the groundwater pressure is balanced by a head of mercury, which is read against a graduated scale.

Hydraulic twin-tube piezometers can be monitored at a central instrument house, and can be flushed periodically with de-aired water to prevent the formation of air bubbles that lead to reading inaccuracies. They are most often employed in earth dam applications, but only when the minimum head to be recorded is less than 5 m below any part of the piezometer tubing.

4.3.1.3 Electric piezometers. The electric types of piezometer consist of a porous plastic or ceramic tip connected to an electric transducer commonly of the vibrating wire or electric resistance strain gauge type (Fig. 4.10). They have a very rapid response to groundwater pressure fluctuations and thus may be used in materials of low hydraulic conductivity such as clays and in rocks with widely spaced and tight joints. Electric instruments are essential if one requires a continuous chart record of pressure fluctuations. They should be thoroughly insulated and waterproofed if they are to remain in service for some time.

4.3.1.4 Pneumatic piezometers. These consist of a porous plastic or ceramic tip connected to a pneumatic transducer (Fig. 4.11), which in turn is connected by flexible plastic tubes to a terminal panel or gauge

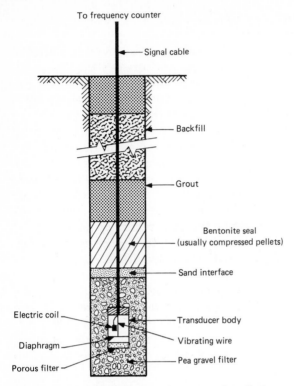

To frequency counter

Signal cable

Backfill

Grout

Bentonite seal
(usually compressed pellets)

Sand interface

Electric coil

Transducer body

Vibrating wire

Diaphragm

Pea gravel filter

Porous filter

Figure 4.10 Vibrating wire piezometer installation.
(*After Dunnicliff, 1988, Fig. 9.16.*)

house at the ground surface. Only a very small diaphragm deflection occurs when readings are taken, so volume change and response time are minimal. A further advantage is that the pneumatic piezometer may be installed at any elevation below or above the gauge house location; any magnitude of pressure can be measured. Pneumatic instruments tend to be less expensive than electric ones, and more reliable if precautions are taken to prevent kinking of tubes. They are commonly used for long-term monitoring at inaccessible locations, such as behind tunnel liners and beneath dams.

4.3.1.5 Multiple-port piezometers. If several piezometers are installed side by side in a single drillhole to form a piezometer "nest," great care is needed to ensure hydraulic isolation of each tip from the next. A more reliable method, although more expensive in terms of drilling, is to install only one piezometer per hole and to drill several holes side by side. A third, much superior method is to use a *multiple-port* (MP) piezometer system in a single hole, as described below.

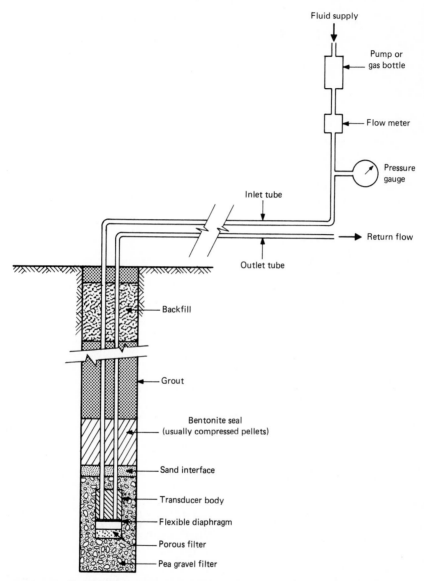

Fluid supply

Pump or gas bottle

Flow meter

Pressure gauge

Inlet tube

Return flow

Outlet tube

Backfill

Grout

Bentonite seal (usually compressed pellets)

Sand interface

Transducer body

Flexible diaphragm

Porous filter

Pea gravel filter

Figure 4.11 Pneumatic piezometer installation. (*After Dunnicliff, 1988, Fig. 9.14.*)

The MP system by Westbay Instruments of Vancouver, B.C., Canada, uses valved measurement ports inserted as couplings in a drillhole liner tube or casing. The complete assembly is installed in 76–115-mm-diameter drillholes. Packers are inflated with water, injected from within the casing, to seal the sections of hole between ports (Fig. 4.12). Larger and shallower drillholes may be sealed in-

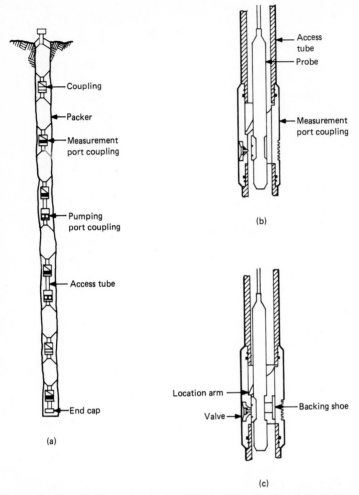

(a)

(b)

(c)

Figure 4.12 Multiple-port piezometer (MP system, Westbay Instruments Ltd., Vancouver, B.C., Canada). (*a*) String of downhole equipment, including packers and port couplings for testing and water sampling. (*b*) Probe in position. (*c*) Probe pressed against valve to monitor water pressure outside tube.

stead with sand and bentonite grout. To take measurements, the pressure-measuring probe is drawn up the casing until it locates the required port (Fig. 4.12*b*). A backing shoe is then activated to push the probe against the port valve, opening the valve, and connecting the transducer to groundwater outside the casing (Fig. 4.12*c*). A reading is taken. When the backing shoe is retracted, the valve reseals. The same system, fitted with supplementary pumping ports, can be used to take groundwater samples, and to give access for permeability testing.

A virtually unlimited number of measurement ports can be installed, spaced at 1.5 m or greater intervals according to the selectedlength of packer. Multiple-port systems have a number of important advantages, including reliable sealing and isolation of the individual strata, and the ability to verify that isolation has been achieved. Multiple groundwater pressure readings provide a very useful "log" of nonhydrostatic variations in the piezometric head.

4.3.2 Flow measurements

Water flow is monitored to assist in studies of reservoir leakage or contaminant migration, where one often needs to identify particular fissures carrying the majority of flow, and the rate of flow in these channels. Another application is to locate horizons of fissured rock in which to install piezometers or to carry out conductivity tests.

The most common method makes use of one of three types of tracer: a radioactive material, a dye, or a granular material such as a pollen that can be detected under the microscope. The time is measured for the concentrated tracer injected into one drillhole to appear in one or more nearby holes.

Flow can be measured in a single drillhole, as opposed to between holes, by observing the rate of dilution of a tracer or saline solution; or by using a small propeller (micromoulinette) with a device for counting the speed of rotation; or by the cooling effect of water flowing over an electrically heated wire.

4.3.3 Measurements of hydraulic conductivity

4.3.3.1 Laboratory measurements. Flow velocity (specific discharge) is measured at various known values of hydraulic gradient. First the core specimen is saturated with de-aired water. Its curved outer surface is sealed to prevent leakage, usually by a triaxial cell containing a pressurized rubber membrane (Sec. 8.2.4.1). A known water pressure is applied to one end of the core, and water emerging at the other end is collected in a graduated measuring cylinder. Flow in a given time interval is determined by measuring this volume or by weighing.

There are two basic types of laboratory conductivity test, constant-head and falling-head. In the constant-head test, the head differential H is maintained constant. From Darcy's law, the hydraulic conductivity of the specimen is calculated as

$$K = \frac{QL}{AH}$$

where Q is the steady volumetric discharge through the system, L and A are the length and the cross-sectional area of the core, and H is the head differential.

In the falling-head method, water is supplied to the lower face of the specimen through a calibrated tube of cross-sectional area a. The head is allowed to fall from H_1 to H_2 during time t. Hydraulic conductivity is in this case calculated from

$$K = \frac{aL}{At} \ln \left(\frac{H_1}{H_2}\right)$$

where ln is the base of natural logarithms.

The falling-head method is better for specimens with conductivity less than about 10^{-6} m/s, which includes most rocks.

A quicker laboratory method makes use of a gas permeameter in which air or an inert gas, such as dry nitrogen, is used instead of water. The gas flows more easily through small pores and fissures, so the coefficient of gas conductivity is much higher than that of water conductivity. The two values are, however, related, so water conductivities can be predicted from the cleaner, faster, and more convenient measurements using gas (Klinkenberg, 1941).

Gas flow rates are usually measured by an elutrometer, which consists of a glass tube containing a loosely fitting "float." The tube is mounted vertically. Gas flowing up the tube carries the float to a height proportional to the flow velocity. Different diameters of tube can be used for different flow rate ranges. Very small flow rates can be measured instead using a capillary tube connected to the downstream end of the specimen. Air escaping from the specimen displaces a small droplet of liquid or a soap bubble in the tube, allowing the flow rate to be measured directly against a scale. Yet a further alternative is to use a hot-wire flowmeter in which gas flow is measured from its cooling effect on an electrically heated wire.

To measure the permeability of materials with hydraulic conductivities less than 10^{-10} m/s, transient techniques may be used. These require pressure transducers to measure the decay of a pressure pulse, which is rapidly applied to one end of a specimen. Only the initial part of the pressure decay curve is required to determine the conductivity, as the shape of the curve is mathematically well-defined.

4.3.3.2 Tests in piezometers. Although the main purpose of a standpipe piezometer is to monitor the hydraulic head, it also provides a convenient means of measuring hydraulic conductivity. In a rising-head, or *bail,* test, a known quantity of water is removed, whereas in a falling-head, or *slug,* test, a known quantity is added. Hydraulic conductivity is then determined from a graph of water level recovery versus time, using the following expression:

$$K = M \frac{d^2}{Lt} \ln \left(\frac{H_1}{H_2} \right)$$

where d is the diameter of the standpipe, L is the length of the drillhole test section (whose diameter is D), and H_1 and H_2 are heads at the start and the end of time interval t. M is a coefficient that for $L/D = 4$ assumes values in the range of approximately 0.25–1.5, depending on the ratio of hydraulic conductivities parallel and perpendicular to the test hole (Hoek and Bray, 1974). For isotropic conductivity, such as in many igneous rocks, a value of 0.25 might be assumed, whereas when testing in a drillhole perpendicular to pronounced layering, a value of between 0.8 and 1.4 might be more realistic. Other forms of equations have been derived for use when testing under a variety of conditions of geology and geometry of the piezometer system (such as given in Freeze and Cherry, 1979). Constant-head permeability tests in piezometers also give quite consistent results, but require a somewhat more complicated control apparatus than bail or slug tests. The equation for interpretation is given in Sec. 4.3.3.3.

4.3.3.3 Packer tests. In this method of measuring hydraulic conductivity, the most common one in rock engineering applications, an exploratory drillhole is internally sealed or "packed" to isolate a test section (Fig. 4.13). When using a single-packer system, one packer is inflated or expanded mechanically to seal from that point to the base of the hole. Alternatively, two packers isolate a test section typically 3–6 m long at some intermediate location (Bennett and Anderson, 1982). Tests are conducted along the length of the hole and sometimes to depths of up to 300 m (Brawner, 1968). Louis and Maini (1970) suggested the refinement of using four packers in place of two. The outer packed sections act as transition zones to ensure that radial flow occurs in the central test section.

The single-packer method is best for tests during drilling, on completion of each drill run, and is essential when the rock mass is weak or intensely jointed and the hole likely to collapse. The drill rig remains set up over the hole, and problems of inadequate sealing or hole collapse can be rectified by redrilling. The single-packer method is more expensive because of standby payments for the drill rig and crew. The double-packer alternative is best when the hole is likely to be stable. In short holes, the tests can be completed after the drill has left the site and is no longer an expense to the project.

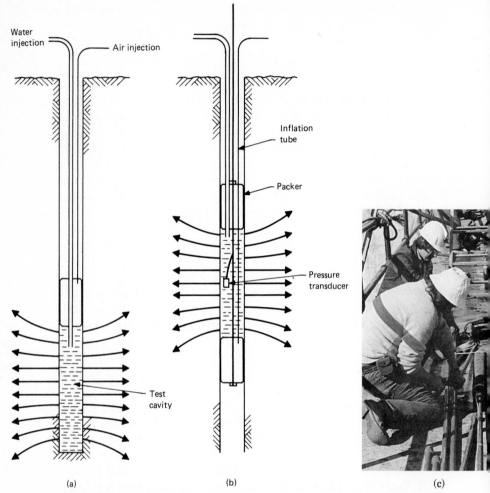

Figure 4.13 Packer permeability test method. (*a*) Single-packer test. (*b*) Double-packer test. (*c*) Assembly of double-packer system for testing the Big Eddy dam and its foundation, Nairn, Ont., Canada.

Hydraulic conductivity for a constant-pressure test using packers is given by the following equation:

$$K = \frac{Q}{2\pi L(H_1 - H_2)} \ln \left(\frac{2R}{D}\right)$$

where Q is the rate of injection, usually measured by a flow volume meter similar to the ones used to monitor household water supplies, H_1 is the head in the test section, H_2 is the total head measured at a distance R from the drillhole, L is the length of the test section, and D

is the diameter of the drillhole. H_1 can be measured directly using a downhole pressure transducer or, less reliably, it can be computed from the pressure measured at surface:

$$H_1 = \frac{P}{\gamma} + h_c - h_l$$

P being the gauge pressure, γ the unit weight of water, h_c the static head from the gauge to the test section, and h_l the loss of head in the delivery line.

This is similar to the well pump test in that it assumes that a piezometer is installed to monitor H_2 at a distance R from the hole. A more common and usually more convenient approach is to take all measurements in a single hole, in which case the hydraulic conductivity can be obtained from the simplified equation

$$K = N\left(\frac{Q}{H_1 - H_w}\right)$$

where H_w is the head of water and N is a coefficient that for $L/D = 4$ assumes values in the range of approximately 0.3–2.0, depending on the ratio of hydraulic conductivities parallel and perpendicular to the test hole (Hoek and Bray, 1974). For isotropic conductivity, such as in many igneous rocks, Hoek and Bray suggest a value of 0.3, whereas when testing in a drillhole perpendicular to pronounced layering, they suggest a value of between 1 and 1.5.

4.3.3.4 Lugeon test. The Lugeon method of water pressure testing was introduced by Maurice Lugeon, a French engineer, as a criterion for whether or not a rock required grouting. It is essentially a standardized form of packer test. Water is injected under an excess pressure of 1 MPa into a packed off section of drillhole. The flow rate is reported in *lugeon* units, where 1 lugeon is 1 liter per meter of packed off hole per minute, approximately equal to a hydraulic conductivity of 10^{-7} m/s. An often used (but by no means universal) grouting criterion is to grout if the water loss exceeds 1 lugeon.

Testing at an excess pressure as high as 1 MPa can be questioned not only in grouting, but more particularly in hydrogeological applications, because the pressure is likely to be high enough to open joints and to increase greatly the hydraulic conductivity being measured. It may even induce new fractures. As a general rule, tests for hydraulic conductivity should be carried out with the least possible disturbance to the existing groundwater regime.

4.3.3.5 Well tests. Hydraulic conductivity can also be evaluated using the *well test* or *drawdown test* common in ground water resource

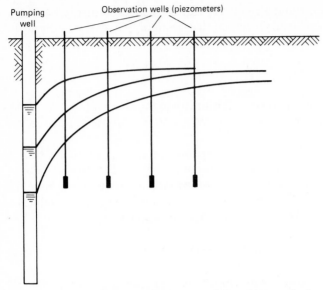

Figure 4.14   Permeability testing by the well drawdown method.

evaluations. A 20-cm or larger test well is bored and a submersible pump is lowered to the base of the hole (Fig. 4.14). Observation wells or piezometers are drilled and installed at various distances from the test well. Groundwater is pumped for an extended period while monitoring drawdown. Hydraulic conductivity is computed from the measured "cone of depression" in the groundwater table as a function of pumping rate.

These full-scale pumping tests are very expensive at depths greater than 30 m, although at shallower depths they are practical in some rock engineering applications. Packer tests and falling-head tests in piezometers are cheaper and can provide adequate data in many cases where pumping tests are not justified. Pumping tests are used mainly when exploitation of the aquifer by wells is contemplated, and when the test well can in all likelihood be converted to a producing well on completion of the testing. They provide reliable in situ measurements that are averaged over a large aquifer volume, and allow the measurement of horizontally anisotropic conductivity if observation wells are available in different directions.

4.4   Analysis of Groundwater Pressure and Flow

The prediction of pressure distributions and seepages and the design of drainage systems require a comparison of water pressures and flow paths before and after construction. Steady-state pressures and flow paths are expressed in terms of the Laplace equations with appropriate boundary conditions. Solutions are obtained using graphic methods (flow-net sketching), analog methods, or numerical techniques. The graphic methods are particularly suitable for an initial examination of the problem, and can be quite accurate given some practice in flow-net construction and reliable data on rock conditions.

4.4.1   Flow-net construction

A groundwater flow system can be represented by a three-dimensional set of equipotential surfaces and a corresponding set of orthogonal flow surfaces. A two-dimensional cross section through this flow system is called a *flow net* (Fig. 4.15). The sketching of flow nets by trial and error is one of the most powerful analytical tools for the investigation of groundwater flow. It requires a lot of practice but no sophisticated mathematics.

A flow line is an imaginary impermeable boundary in that there is no flow across it. All impermeable boundaries and aquicludes are therefore flow lines. A plane of symmetry, if one exists, can be taken as a flow line so that only half the problem need be solved.

An equipotential line is a contour of constant hydraulic head. A free water surface is therefore an equipotential line. The water table is neither a flow line nor an equipotential line, simply a line of variable but known hydraulic head. Equipotential lines must meet impermeable boundaries at right angles, and if the rock is isotropically permeable, they must intersect flow lines at right angles throughout the net. Symmetry may be used to reduce the complexity of the problem.

The area between two adjacent flow lines is known as a *stream tube*. If the flow lines are equally spaced, the discharge through each stream tube is the same. For isotropic ground, the flow net is usually constructed in squares, in which case the flow along any given stream

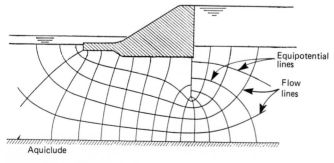

Figure 4.15 Typical flow net.

tube (per unit thickness of slice) can be calculated as the product of hydraulic conductivity and head drop.

In heterogeneous geological systems (those with layers of differing hydraulic conductivity), a tangent law of flow refraction must be satisfied at geological boundaries. In such cases, a flow net that is square in some formations must become rectangular when flow passes into others of different conductivity. Anisotropic conductivity conditions can also be analyzed by a geometric transformation, that is, by shrinking the dimensions of a cross section in the direction of greater permeability. Such a flow net is difficult to construct accurately, but qualitative flow nets can still be of assistance.

4.4.2 Electric analog simulation

The mathematics and physics of electrical flow and groundwater flow are similar (analogous). Darcy's law for the flow of water is analogous to Ohm's law for the flow of electric current. A continuity relationship applies in both cases; therefore they are both described by the Laplace differential equation. A specific discharge of water is analogous to current density; hydraulic conductivity is analogous to electric conductivity; and hydraulic head is analogous to electric potential.

This means that the flow of electric current can be used to solve steady-state groundwater flow problems by constructing an electric analog and tracing flow nets. There are two main types: the continuous (conductive paper) analog and the discrete (resistance or resistance-capacitance network) analog (Franciss, 1985).

Using the first, continuous approach, a sheet of conductive paper is cut in the same geometrical shape as the groundwater flow field. A voltage differential is established from one head boundary to the other, and a sensing probe and voltmeter are used to measure the potential distribution. Constant head boundaries are created with highly conductive silver paint, and impermeable boundaries are simulated by the unconnected edges of the paper model. Equipotential lines can be traced quickly so that a full equipotential net can be generated. The method can handle complex region shapes and boundary conditions but is generally restricted to isotropic, homogeneous systems, although, by placing many small cutout slots in the paper, anisotropic conditions may be approximated. Impermeable beds within the flow regime are made by cutouts, whereas extremely permeable beds, such as karst conduits, can be simulated with silver paint.

Resistance network analogs can be used in a similar manner. The conductive paper is replaced by a network of resistors connected to one another at the nodal points of a grid. A sensing probe measures the voltage at each of the nodal points in the network. By varying the resistances one can analyze heterogeneous and anisotropic systems. The resistance network method is more accurate and versatile than a conducting paper analog, but slower to construct and more expensive. It is not as powerful as the numerical methods that are nowadays used more often.

Resistance-capacitance networks (RC analogs) can be used to simulate storage as well as flow. The resistors are energy dissipators, similar in function to transmissivity, and the capacitors are reservoirs of electric charge, similar in function to storativity.

4.4.3 Numerical modeling

4.4.3.1 Continuum solutions. The hydraulic conductivities of most rocks are highly heterogeneous and anisotropic. Pressure distributions are nonhydrostatic, flow paths are frequently tortuous, and boundary and initial conditions are often complex. Under these circumstances, the simpler methods of analysis are inadequate, and such problems are most often solved using a numerical model. To construct a numerical model, one needs to know the size and shape of the region of flow, the initial and boundary conditions for head and flow, and the spatial distribution of conductivity. Several mathematical methods can be used.

In the *boundary-element method,* the boundary of the flow domain is discretized and the unknown boundary conditions in each subboundary are approximated.

Using the *finite-difference technique,* the flow domain is divided into a nodal grid and analyzed. A finite-difference equation is developed for every node in the grid. The resulting linear algebraic equations can be solved by many methods, one of the most common being relaxation, which involves repeated sweeps through the nodal net, applying the pertinent finite-difference equation at each node where the head is unknown. A set of starting values is assumed and is successively refined with each sweep until the change between steps is acceptably small. Also, the set of finite-difference equations can be solved directly by matrix methods.

The *finite-element method* requires subdivision of the flow domain into discrete elements, each of which can be assigned different and anisotropic properties. The elements in the finite-element method are designed by the modeler to suit the geological boundaries in each specific application, in contrast to the regular rectangular mesh employed in the finite-difference method. Deformable meshes can be designed which replicate drawdown of water tables in dynamic conditions.

Finite-element formulations are much more powerful than other methods for complex problems, and, by also using boundary elements to handle mixed boundary conditions, practically all continuum problems become solvable. The practical limitation is to obtain a sufficiently precise model of the various strata, and to quantify the material parameters adequately.

4.4.3.2 Discontinuum solutions. The alternative *discontinuum approach* is based on analyzing the hydraulics of flow through individual rock discontinuities. It is only feasible when the geometry of such channels can be defined. Methods of analysis based on fluid mechanics principles are discussed in Wittke (1973).

Cundall's distinct-element analytical method UDEC (Cundall, 1980) is capable of predicting the interaction of rock blocks undergoing large absolute and relative displacements, rotations, and fracturing into new, discrete bodies. The code permits stresses and displacements along the element boundaries (joints) to be calculated. UDEC contains a flow model for calculating flow along joints based on Darcy's law. It simulates flow through joints in a discontinuous, saturated rock mass where the motions of blocks and changes in aperture influence the conductivity along the joints.

References

Barton, N., S. Bandis, and K. Bakhtar: "Strength, Deformation and Conductivity Coupling of Rock Joints," *Int. J. Rock Mech. Min. Sci. Geomech. Abstr.,* **22** (3), 121–140 (1985).

Bawden, W. F., J. H. Curran, and J.-C. Roegiers: "Importance of Joint Behaviour on Potential Water Inflow into Underground Structures: An Analytical Approach," *Proc. 13th Can. Rock Mech. Symp.* (Toronto, Ont., 1980), Can. Inst. Min. Met. Spec. Publ., vol. 22, pp. 211–218.

Bennett, R. D., and R. F. Anderson: "New Pressure Test for Determining Coefficient of Permeability of Rock Masses," Tech. Rep. GL-82-3, U.S. Army Waterways Experiment Station, Vicksburg, Miss., 57 pp. (1982).

Brawner, C. O.: "The Influence and Control of Groundwater in Open Pit Mining," *Proc. 5th Can. Symp. Rock Mech.* (Toronto, Ont., Dec. 1968); also *Western Miner,* **42** (4), 42–55 (1969).

Cornwell, D. K., and H. D. Murphy: "Experiments with non-Darcy Flow in Joints with Large-Scale Roughness," *Proc. Int. Symp. Fundamentals of Rock Joints* (Bjorkliden, Sweden, 1985), pp. 323–332.

Cundall, P.: "UDEC—A Generalized Distinct Element Program for Modelling Jointed Rock," U.S. Army Contract DAJA 37-79-C-0548, European Research Office (1980).

Davies, W. E.: "Distribution and Characteristics of the Karst of the Northeastern United States," *Proc. Symp. Geol. Geotech. Problems in Karstic Limestone of the Northeastern U.S.* (Fredericksburg, Md., 1984), AEG/ASCE, pp. 1–14.

Davis, S. N.: "Porosity and Permeability of Natural Materials," in R. J. M. De Wiest, Ed., *Flow Through Porous Media* (Academic Press, New York, 1969), pp. 54–89.

Dunnicliff, J.: *Geotechnical Instrumentation for Monitoring Field Performance* (Wiley, New York, 1988, 577 pp.).

Evans, D. M.: "The Denver Area Earthquakes and the Rocky Mountain Arsenal Disposal Well," *Mountain Geol.*, **3**, 23–26 (1966).

Franciss, F. O.: *Soil and Rock Hydraulics: Fundamentals, Numerical Methods and Techniques of Electrical Analogs* (A. A. Balkema, Rotterdam and Boston, 1985), 170 pp.

Freeze, R. A., and J. A. Cherry: *Groundwater* (Prentice-Hall, Englewood Cliffs, N.J., 1979), 604 pp.

Gale, J.: "The Effects of Fracture Type (Induced versus Natural) on Stress-Fracture Closure-Fracture Permeability Relationships," *Proc. 23d U.S. Symp. Rock Mech.* (Berkeley, Calif., 1982), pp. 290–298.

Harpalani, S., and M. J. McPherson: "Mechanism of Methane Flow through Solid Coal," *Proc. 27th U.S. Symp. Rock Mech.* (Tuscaloosa, Ala., 1986), pp. 690–695.

Healy, J. H.: "Recent Highlights and Future Trends in Research on Earthquake Prediction and Control," *Rev. Geophys. Space Phys.*, **13**, 361–364 (1975).

Hoek, E., and J. W. Bray: *Rock Slope Engineering* (Inst. Mining and Metall., London, 1974), 309 pp.

Huder, J., and G. Amberg: "The Significance of the Piezometer in Rock Engineering," *Proc. 5th Int. Cong. Rock Mech.* (Melbourne, Australia, 1983), vol. B, pp. 99–104.

Kafritsas, J., M. Gencer, and H. H. Einstein: "Coupled Deformation/Flow Analysis with the Distinct Element Method," *Proc. 25th U.S. Symp. Rock Mech.* (Evanston, Ill., 1984), pp. 239–247.

Klinkenberg, L. J.: "The Permeability of Porous Media to Liquids and Gases," in *Drilling and Production Practices* (Am. Petrol. Inst., 1941), pp. 200–213.

Liszkowski, J.: "The Influence of Karst on Geological Environment in Regional and Urban Planning," *Bull. Int. Assoc. Eng. Geol.* 12, pp. 49–51 (1975).

Long, J. C. S., and P. A. Witherspoon: "The Relationship of the Degree of Interconnection to Permeability in Fracture Networks," *J. Geophys. Res.,* **90** (B4), 3087–3098 (1985).

Louis, C.: "Rock Hydraulics," Rep. 74SG035AME, Bureau Geol. Min. Research (BRGM), Orléans, France, 107 pp. (1974).

—— and Y. N. T. Maini: "Determination of in-situ Hydraulic Parameters in Jointed Rock," *Proc. 2d Int. Cong. Rock Mech.* (Belgrad, Yugoslavia, 1970), Rep. 1/32.

Marulanda, A. P., and T. L. Brekke: "Hazardous Water Inflows in Some Tunnels in Sedimentary Rocks," *Proc. Rapid Excavation and Tunneling Conf.* (San Francisco, Calif., 1981), pp. 741–752.

Morfeldt, C.: "Significance of Groundwater at Rock Constructions of Different Types," *Proc. Int. Symp. Large Permanent Underground Openings* (Oslo, Norway, 1969), pp. 311–323.

Oda, M., and Y. Hatsuyama: "Permeability Tensor for Jointed Rock Masses," *Proc. Int. Symp. Fundamentals of Rock Joints* (Bjorkliden, Sweden, 1985), pp. 303–312.

Pyrak, L. J., L. R. Myer, and N. G. W. Cook: "Determination of Fracture Void Geometry and Contact Area at Different Effective Stress," Fall 1985 Mtg. Am. Geophys. Union (San Francisco, Calif.).

Sabarly, F., A. Pautre, and P. Londe: "Quelques réflexions sur la drainabilité des massifs rocheux," *Proc. 2d Int. Cong. Rock Mech.* (Belgrad, Yugoslavia, 1970), Rep. 6-12.

Sharp, J. C., E. Hoek, and C. O. Brawner: "Influence of Groundwater on the Stability of Rock Masses," *Trans. Inst. Mining and Metall.* (London), **81**, 113–120 (1972).

—— and Y. N. T. Maini: "Fundamental Considerations on the Hydraulic Characteristics of Joints in Rock," in W. Wittke, Ed., *Percolation Through Fissured Rock* (Int. Soc. Rock Mech., Stuttgart, Germany, 1972).

Snow, D. T.: "Rock Fracture Spacings, Openings, and Porosities," *J. Am. Soc. Civ. Eng.,* **SM94**, 73–91 (1968).

————: "Anisotropic Permeability of Fractured Media," *Water Resources Research,* **5** (6), 1273–1289 (1969).

Sowers, G. F.: "Dewatering Rock for Construction," in *Rock Engineering for Foundations and Slopes, Proc. Specialty Conf. ASCE* (Boulder, Colo., Aug. 1976), pp. 200–216.

SPE: *Proc. Soc. Petroleum Eng. Unconventional Gas Techn. Symp.* (Louisville, Ky., 1986), 710 pp., 60 papers.

Voss, C. F., R. J. Bastian, and L. R. Shotwell: "A Coupled Mechanical-Hydrological Methodology for Modelling Flow in Jointed Rock Masses Using Laboratory Data for the Joint Flow Model," *Proc. 27th U.S. Symp. Rock Mech.* (Tuscaloosa, Ala., 1986), pp. 906–909.

Witherspoon, P. A.: "Flow of Groundwater in Fractured Rocks," *Bull. Int. Assoc. Eng. Geol.* 34, pp. 103–115 (1986).

Wittke, W.: "General Report on the Symposium 'Percolation Through Fissured Rock,' " *Bull. Int. Assoc. Eng. Geol.,* pp. 3–28 (1973).

———— and G. A. Leonards: "Modified Hypothesis for Failure of Malpasset Dam," *Proc. Int. Workshop on Dam Failures* (Purdue Univ., Lafayette, Ind., Aug. 6–8, 1985), 74 pp.

5.1 Forces, Stresses, and Their Effects

5.1.1 Concepts of force and stress

Forces and stresses are invisible but can be observed and defined in terms of their recognizable effects. Newton defined "force" as the cause of motion, and expressed it quantitatively as the product of a body's mass times its acceleration. About 100 years after Newton, "stress" was introduced as the intensity of force: the ratio of force per unit area. The visible effects of forces and stresses in rock engineering include landslides at surface and squeezing or rockbursting underground.

5.1.2 Worked example

The following simple example shows how stress can be calculated and compared to rock strength. It introduces the concepts of density, weight, force, stress, strength, and strain, and serves as an introduction to more complex geomechanics problems. The example demonstrates that a monumental column constructed by stacking 1-m cubes of granite (Fig. 5.1) can, in theory, reach a height of over 7 km before collapse. We will discuss why this could not happen in reality and, after introducing the concept of elasticity, will use the same "model" to determine by how much the column shortens during construction, as a result of its own weight.

5.1.2.1 Mass and weight.
A gram is defined as the *mass* of a cubic centimeter of water; 1 m³ of water therefore has a mass of 1000 kg, de-

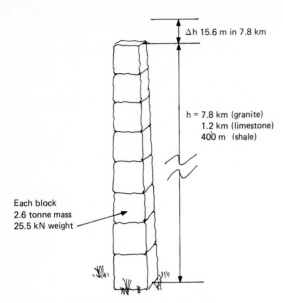

Δh 15.6 m in 7.8 km

h = 7.8 km (granite)
   1.2 km (limestone)
   400 m (shale)

Each block
2.6 tonne mass
25.5 kN weight

Figure 5.1 Granite column example to illustrate concepts of stress and strain.

fined as 1 t (tonne). Mass density is the mass per unit volume expressed in appropriate units.

The *relative density* of a rock (its unit mass compared with that of water) depends on its porosity and mineral composition. Granite is composed mainly of quartz and feldspar whose relative densities are 2.65, and if the porosity of the granite were zero, its relative density would also be about 2.65. However, granites usually have a porosity of 1–2%. Consequently the relative density of dry granite is about 2.60. 1 m^3 of granite has a mass of about 2.6 t.

Weight is defined as the force exerted by gravitational acceleration acting on a mass and is calculated as mass times this acceleration. *Standard gravity* at the surface of the earth is 9.80665 m/s^2. A 1-kg mass, in SI units, therefore weighs 9.80665 N. (1 newton corresponds to the weight of a small apple, about 0.1-kg mass.) The granite block with a mass of 2.6 t (2600 kg) weighs 25,497 N, or about 25.5 kN.

5.1.2.2 Contact stresses and strengths. The granite cube has a ground contact area of 1 m^2. During construction, each additional block or meter of column height h adds 25.5 kN to the weight of the column. The combined weight at the column base is $h \times 25.5$ kN. The contact stress, or *foundation bearing pressure,* obtained by dividing this weight by the contact area, is $25.5h$ kN/m^2, written as $25.5h$ kilopascals (kPa) (1 Pa = 1 N/m^2). The usual convention in rock mechanics is to write a compressive stress as positive, a tensile stress as negative.

This result has an importance beyond the present example because it can be inverted and used to estimate the vertical stress at a depth below the earth's surface, which within a granite (see Sec. 5.3.1.2) increases by about 26 kPa/m (or 1.13 lb/in^2 per foot in U.S. customary units).

To determine how high the column can be built, we must compare the stress at its base with the strength of the blocks. The correct strength value to use is the uniaxial compressive strength obtained by laboratory testing (Chap. 8).

Typical granites classify as very high strength rocks with uniaxial compressive strengths of about 200 MPa. To reach a foundation stress of this magnitude, the column would have to reach a height of 200,000/25.5 = 7.8 km (nearly 5 mi). The maximum height for a column constructed of concrete or a medium-strength (30-MPa) rock, such as a porous limestone, would be about 1.2 km, and that for a 10-MPa shale, about 400 m.

Evidently this 7-km column could never be built. Idealized models can be misleading, and the assumptions used in generating any model should always be studied to see if they are reasonable.

First, a vertical column would be impossible to construct. Nonverticality, caused by slight nonparallelism of the block faces, would lead to toppling. External forces, such as wind loadings and earthquakes (ground movement), would greatly aggravate the stability problem.

Second, random strength variations have been ignored. Mechanical properties are not unique and in over 7000 blocks, even with strict quality control, there would be some substantially weaker than the assumed 200 MPa. If one of these weaker blocks were found near the base of the column, it would limit the strength of the column as a whole. Furthermore, to ensure freedom from long-term creep, stresses should not exceed one-half the short-term rock strength.

5.1.2.3 Introduction to linear elastic behavior. Strain (distortion) is inevitable when applying a force or stress to any material. The strains are much smaller in rock than in a material such as rubber, so small that they usually pass unnoticed. Therefore strains are measured with the help of sensitive strain-gauging instruments.

Strain is defined, as in the simple spring model of Fig. 5.2, as the dimensionless ratio of length change ΔL to original length L:

$$\epsilon = \frac{\Delta L}{L} = \frac{dL}{L}$$

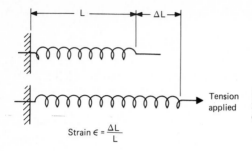

Strain $\epsilon = \dfrac{\Delta L}{L}$

Tension applied

Figure 5.2 Spring model. Strain is defined as the ratio of length change to original length.

In rock mechanics, strain is usually expressed in *microstrain,* where 1 microstrain = 10^{-6} m/m. A typical range is 100–20,000 microstrain. Each of the many types of rock has its own characteristic *stress-strain behavior,* often approximated by an idealized stress-strain model, as described in Chaps. 9 and 10. In 1678, Sir Robert Hooke noted that for many materials (later found to include rubber, metals, and also rocks at low stresses and temperatures), small strains are not only reversible but also linearly proportional to stress. This gave rise to Hooke's law:

$$\sigma = E\epsilon$$

where Young's modulus E is the constant of proportionality or slope of the linear stress-strain curve. Strain is dimensionless, being the ratio of two lengths, so that Young's modulus has units of stress. E values are measured by applying stresses to laboratory specimens and noting the corresponding strains (Chap. 9). Values for various rock types are given in Table 9.1.

This equation allows us to calculate the shortening of the granite column on the assumption that the blocks are behaving in a linear elastic manner according to Hooke's law.

5.1.2.4 Strain and column shortening. First the strain (and change in height) of the lowermost block can be found knowing that this block is loaded close to its strength of 200 MPa when the height of the superimposed column reaches about 7.8 km.

From Hooke's law, $\epsilon = \sigma/E$, where the stress σ in the lowest block may be taken as 200 MPa, equal to the strength of granite, and E for typical solid granite as about 50,000 MPa (50 GPa). Hence $\epsilon = 200/50,000$, or 4000 microstrain.

The corresponding reduction in height of this block is obtained as 4 mm from the definition $\epsilon = dL/L$, by multiplying the computed strain by the original block height of 1 m. Note that in granite, a relatively brittle material, as little as 4 mm of elastic strain is experienced by a 1-m block up to the instant of failure.

To compute total column shortening, note that both the stress and the strain levels increase linearly from zero to maximum, top to bottom, and that an "average" condition exists halfway up. Hence the 7800-m-high column of blocks experiences an average of 2 mm strain per meter, for a total shortening of 15.6 m.

5.1.2.5 Introduction to inelastic behavior. If stress levels always remained in the elastic range, the resulting small strains would give little cause for concern. However, real rocks often creep, swell, rupture, and in general deform much more than predicted by an idealized elastic model.

The *elastic limit* of rock can be exceeded as a result of either geological disturbances or activities of humans. The effects are no longer subtle and unimportant. Earthquakes are generated by slippage along faults or the boundaries of moving continental plates. Foundations settle or slide when the applied loading exceeds the rock's bearing capacity. The rock in tunnels and underground mines bursts, falls, or squeezes when natural stresses are aggravated by the presence of excavations. Deep drillholes spall and close when the stresses exceed the strength of the rock.

The responsibility of the rock engineer is to estimate or measure the levels of stress in the ground, to predict increases of stress during the project, and to forecast and guard against the consequences.

5.1.2.6 Vector representation of forces. Forces and stresses are vector quantities. In other words, they have not only magnitude, but also direction. Solution of everyday rock mechanics problems requires the adding, subtracting, and resolving of force vectors using simple graphic methods, as outlined below. Further detail is given in textbooks on mathematics and rock mechanics.

The sum or resultant of vectors **A** and **B** is a vector **C** formed by placing the initial point of **B** on the terminal point of **A** and then joining the initial point of **A** to the terminal point of **B** (Fig. 5.3a). This graphic method can be extended to any number of force vectors by constructing a *polygon of forces*, as shown in Fig. 5.3b. The only precaution is to ensure that the directions of forces to be added "flow" around the perimeter of the polygon. The resultant, a single force equal to the sum of all others, is a vector joining the start to the end point of the polygon.

The components of vector **A** are defined as vectors **B**, **C** (and **D** if in three dimensions), which are aligned along parallel rectangular axes (cartesian coordinates), and which when added give vector **A** (Fig.

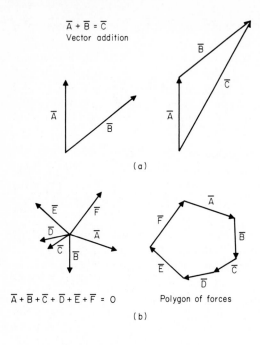

$\overline{A} + \overline{B} = \overline{C}$
Vector addition

(a)

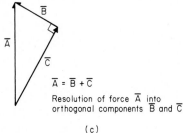

$\overline{A} + \overline{B} + \overline{C} + \overline{D} + \overline{E} + \overline{F} = 0$ Polygon of forces

(b)

$\overline{A} = \overline{B} + \overline{C}$
Resolution of force $\overline{A}$ into
orthogonal components $\overline{B}$ and $\overline{C}$

(c)

Figure 5.3 Vector addition and resolution of forces. (*a*) Addition of two forces. (*b*) Any system of forces in equilibrium can be represented as a closed polygon of forces. (*c*) A force can be resolved into two orthogonal components, three when in three dimensions.

5.3*c*). To facilitate computations, force vectors are "resolved" into components parallel and perpendicular to a line or plane of interest, such as a joint plane or potential surface of sliding.

Engineers need to have at their fingertips the technique of resolving forces. The concept of equilibrium, expressed by Newton as the equality of action and reaction, leads to the solution of many types of engineering design problems. A worked example relating to slope stability is given in Sec. 7.2.6.2.

5.1.3   Behavior of stressed rock

Rock behavior under experimental conditions is discussed in Chaps. 8 through 11. The following is an introduction to the various ways in which stresses manifest themselves in nature and in engineering works.

5.1.3.1 Earthquakes. The most obvious evidence of ground stress is an earthquake, which is caused by stick/slip shear displacement along a fault plane. Some active faults display a pattern of continuous steady movement accompanied by "swarms" of small earth tremors. Others remain dormant for extended periods during which they progressively accumulate energy, which is stored as strain in the surrounding rocks. This stored energy can be very large if the stressed volume of rock is large, and is released suddenly and catastrophically when it reaches a critical level. Most earthquakes are confined to narrow belts that encircle the earth along the margins of moving continental plates (Fig. 5.4). The belts are marked by concentrations of active faults and volcanoes.

5.1.3.2 Rockfalls and rockbursts. It is not so much ground stresses themselves that cause a problem in engineering works, as the release of these stresses, whether violent or gradual and pervasive. We can differentiate five principal classes of rock behavior: stable, swelling, falling, squeezing, and bursting. Only the latter three processes are stress-induced.

Rock falls occur under the action of gravity alone, whereas rockbursts are propelled by the sudden and violent release of energy in much the same manner as earthquakes. In fact, many of the larger rockbursts can be detected by regional seismic monitoring stations, where they have a recorded magnitude of up to Richter 5.5. Outbursts of rock and gas occur catastrophically in some coal, trona, and salt

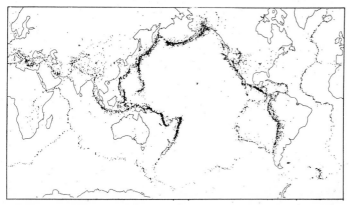

Figure 5.4 Earthquake epicenters follow clearly defined trends when mapped on a global scale. (*Eiby, 1980.*)

mines under the combined action of high stresses and substantial volumes of compressed methane or carbon dioxide.

Even near the surface, explosive bursting is experienced as buckling and heaving of rock floors in quarries and basement excavations (Franklin and Hungr, 1978). Heave problems occur most often in highly stressed rock that is also horizontally bedded, and where groundwater pressures are also high. The Chippewa Canal in Ontario, Canada, for example, when drained for cleaning was found to have a buckling type of heave along its invert, caused by stress release. Quarry floors at many locations in North America have experienced "pop-ups" that appear suddenly, sometimes violently. These have the appearance of sinuous ridges extending sometimes for several hundred meters, with a ridge height of about 1 m and a similar width (Fig. 5.5). They are caused by buckling that follows quarry excavation and the relief of vertical stresses.

5.1.3.3 Squeeze. Another mode of stress release is by a more gradual squeezing. Although not as hazardous, this can cause very serious and extensive damage to mine openings and tunnel linings, and to civil engineering works at surface.

Figure 5.5 Pop-up; Dufferin Aggregates quarry, Milton, Ont., Canada. (*Franklin and Hungr, 1978.*)

In one form of squeezing, the rock is either very soft or intensely jointed, and deforms like toothpaste in a ductile manner. In another, the excavation of a cavern or basement releases locked-in stresses and allows the horizontal beds to slide inward over one another like a deck of cards (Fig. 5.6). The inward movement cannot be resisted by the weak bedding planes, and shear displacement can occur for distances as great as several hundred meters from the excavation. Wall displacements can reach magnitudes of 200 mm or more. Displacements of 50 mm are commonplace (Franklin and Hungr, 1978; Quigley et al., 1978).

Stress-release displacements cannot be prevented, except by application of a restraining stress of similar magnitude to the original stress, which is impractical in nearly every case. Nor is the rate of displacement easy to predict. Displacements have been known to continue for more than 80 years. For example, in the "wheel pit" excavations for hydroelectric turbines in Niagara Falls, N.Y., wall convergence has been going on since 1905 when the wheel pits were excavated. The movement has been monitored and fluctuates annually but continues nevertheless. Horizontal convergences of about 50 mm during construction have been followed by an additional 1 mm each year, amounting to 70 mm during the period of 1905–1970 (Fig.

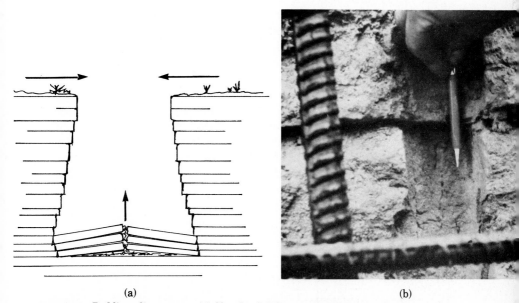

(a)   (b)

Figure 5.6 Bedding slip squeeze. (a) Sketch of sliding mechanism in excavation walls and tendency for upward heave and buckling in invert. (b) Evidence of bedding plane slip manifested as offset along a blasthole in an open cut; Hamilton, Ont., Canada. (*Quigley et al., 1978.*)

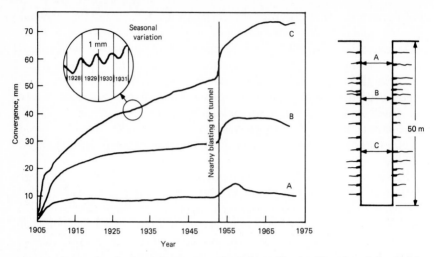

Figure 5.7 Long-term horizontal convergences of the walls of a 50-m-deep "wheel pit" excavation housing a hydroelectric turbine; Niagara, Ont., Canada. (*Lo et al., 1975.*)

5.7). "Rigid" concrete and masonry members spanning horizontally between the wheel pit walls have buckled. The turbines have been closed down because of consequent misalignment (Lo et al., 1975).

The only course of action is to accommodate the movements rather than resist them: to assume that movements will occur, will continue, and for open excavations in horizontally bedded, stressed rock, could amount to as much as 100 mm. An allowance of 50 mm has proven insufficient in some cases.

Underground rock excavations may be left unlined, or the liner may be flexible with wire mesh reinforcement to prevent fallout in the event of cracking. In open-cut excavations a gap is needed between newly excavated rock walls and any building to be constructed near them. The gap may be left open or filled with a deformable buffer material such as expanded polystyrene or polyurethane, placed either as sheeting or by spraying onto the exposed rock. The buffer must be sufficiently deformable to prevent excessive stress transfer from the rock to the structure. When constructing against heavily blasted rock excavations, the rock itself may have loosened sufficiently to absorb some of the potential ongoing displacement.

5.2 Stresses in Three Dimensions

5.2.1 Stress components

5.2.1.1 Stress on a surface.
Forces and therefore stresses are transmitted from place to place in the rock mass, through solid material

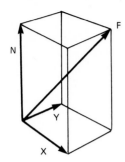

Figure 5.8 Vector resolution of forces on plane.

and across joints, from block to block. Consider the forces transmitted across a rough planar feature, such as a joint. However many individual forces may be acting on the plane, at contacts between asperities, they can be added vectorially to give a single *resultant force* at some oblique angle to the rock surface.

The reverse is also true. It is possible and often convenient to "resolve" the resultant force vectorially into components parallel to each of three orthogonal axes. The system of axes is often chosen, as in Fig. 5.8, so that force N is normal (perpendicular) to the plane, and the other two (X, Y) lie in it. The two components of force acting within the plane are termed *shear* forces, and the orthogonal component is termed the *normal* force. Similarly, shear and normal stresses are defined in terms of these force components per unit of surface area.

5.2.1.2 Stresses on an elemental cube. Consider now a very small cubic element of the rock mass, an approximation to a "point" within the mass. On each of the six faces of the cube, the resultant stress can be resolved into three components, a normal stress that acts perpendicular to the face, and two shear stresses that act at right angles to each other and parallel to the face.

In total there are 18 stress components, three on each of the six faces (Fig. 5.9a). However, only six are independent: for the cube to be in translational equilibrium, the normal stress components on opposite faces must be equal and opposite. For rotational equilibrium, the shear stress components on adjacent faces must be equal and opposite. Therefore the complete stress condition on the cubic element can be represented by a tensor with three normal stress components along the diagonal, and three independent and symmetrical shear stresses to either side:

$$[\mathbf{S}] = \begin{bmatrix} \sigma_x & \tau_{xy} & \tau_{xz} \\ \tau_{yx} & \sigma_y & \tau_{yz} \\ \tau_{zx} & \tau_{zy} & \sigma_z \end{bmatrix}$$

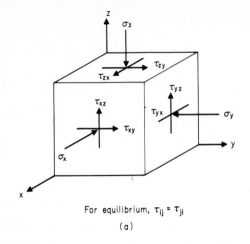

For equilibrium, $\tau_{ij} = \tau_{ji}$

(a)

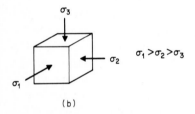

(b)

Figure 5.9 Stress components: (a) On faces of cubic element. For equilibrium, $\tau_{ij} = \tau_{ji}$. (b) Cube rotated into principal stress direction; shear stresses on faces become zero.

Note that for equilibrium, $\tau_{xy} = \tau_{yx}$, $\tau_{yz} = \tau_{zy}$, and $\tau_{zx} = \tau_{xz}$.

If the elemental cube is now rotated within the rock mass while keeping its centroid at the same location, the magnitudes of the stress components on its faces will be found to vary continuously. They depend on the directions of the arbitrary system of axes to which they are referred. *Transformation equations* are available to allow the calculation of stress components in one axis system from those in another (e.g., Jaeger and Cook, 1969).

It can be shown that for every general stress state as defined above, there is a unique orientation of the cube, reached by simple rotations, where all shear stresses disappear. At this orientation, all three components of normal stress reach their maximum amplitudes, termed *principal stresses* (Fig. 5.9b). The directions in which they act are termed principal stress directions. The principal stress tensor at the point now appears as follows:

$$[\mathbf{S}]_{\text{ppl}} = \begin{bmatrix} \sigma_1 & - & - \\ - & \sigma_2 & - \\ - & - & \sigma_3 \end{bmatrix}$$

This leads to the conclusion that, as an alternative to reporting six independent elements of the stress tensor, we can fully and conveniently represent the state of stress at any point in a rock mass by three principal stress magnitudes and their three orthogonal directions. There are many situations in rock engineering where the directions can be estimated accurately, leaving only the magnitudes to be measured. For example, in sedimentary basins unaffected by intense tectonism, the vertical stress is a principal stress, and the other two directions can often be estimated from structural features of the basin.

5.2.2 Simple stress states

The three principal stresses are not in general equal. Conventionally they are termed *major, intermediate,* and *minor* principal stresses, according to their relative magnitudes. Rarely one or two out of the three are zero in magnitude. The following are some simple stress states that can be found in the ideal situation of laboratory testing, and in special circumstances in the field.

5.2.2.1 Uniaxial stress. The simplest of all cases is when there is only one component of stress. In the uniaxial compressive strength test (Chap. 8), the major principal stress is applied by loading a cylinder of core along its axis. Its magnitude is given by the ratio of applied load to specimen cross-sectional area. The intermediate and minor principal stresses are zero because no loads are applied radially to the cylinder.

There are few rock engineering situations in which stresses are close to uniaxial. Slender pillars in a room-and-pillar mine approach such a condition. Short and squat pillars are uniaxially stressed only where exposed at mid-height. They are triaxially constrained by roof and floor rock at top and bottom and in the center of the pillar.

5.2.2.2 Biaxial stress. A *biaxial* stress state is one in which there are two principal stresses and the third is zero, as for a square-plate model loaded along its edges in the laboratory. Provided that the plate is thin, all stresses are zero perpendicular to the plate's surface.

Very importantly, a biaxial stress state exists at any free surface in the rock mass. For example, stresses in the walls of an underground excavation can only exist in directions parallel to the wall. Unless rockbolts or liners are installed, there can be no reaction at the wall, and there can be no radial stress without such reaction, else the rock would behave as a mechanism and pieces would fly off the wall. During excavation, therefore, all preexisting stresses in the radial direction are released.

5.2.2.3 Axisymmetric triaxial stress. In conventional triaxial strength tests, a cylindrical specimen is loaded both along its axis and also by fluid pressure around its curved surface. The axial stress, which is greater than the cell pressure, is the major principal stress. The two orthogonal components of intermediate and minor principal stress in this case are equal because they are applied by a single pressurized fluid.

A test like this, in which intermediate and minor principal stresses are equal, is called an *axisymmetric* triaxial test. The alternative "true" (multiaxial or polyaxial) triaxial test requires a rock cube with opposing pairs of faces stressed to different levels independently, and is seldom carried out because of mechanical difficulties. Stress configurations for this and other types of laboratory tests are described in Chap. 8.

5.2.2.4 Hydrostatic stress.

A *hydrostatic* state of stress corresponds to the pressure that acts on the surface of a small solid object immersed in a liquid. Because a liquid cannot resist shearing, the three principal stresses are equal and there are no shear stresses. If a difference were to exist temporarily between, say, horizontal and vertical stress components, the liquid would flow until a hydrostatic state were reestablished. This stress condition can be replicated in the laboratory in a hydrostatic compression cell in which the fluid pressure acts equally on all faces of the specimen.

Hydrostatic stress conditions may be assumed to exist in the groundwater in a rock mass, provided that the water is more or less at rest, and that the pore and fissure spaces are interconnected. Under these conditions, the pressure at any given depth is equal in all directions and can be calculated by multiplying the unit weight of water by the depth below the water surface or groundwater table. When these conditions do not apply, for example, when groundwater bodies are separated by aquicludes or when there is substantial flow, the water pressure distribution may be very different from a hydrostatic one.

Hydrostatic stress conditions can also exist in the rock itself, but only at great depths, temperatures, and pressures, sufficient to cause the rock to lose its shear strength and behave as a fluid. Stronger rock types lose their shear strength at depths of several tens of kilometers, whereas the weaker and more plastic ones can behave as viscous fluids much closer to the surface. Near-hydrostatic stress conditions exist in the earth's core and lower mantle, in salt or potash deposits at depths of hundreds of meters, and in soft clay shale even at depths of tens of meters. The term hydrostatic is often assumed to refer to the weight of a water column. To avoid confusion, the term *isotropic lithostatic* stress condition can be used to differentiate hydrostatic rock stresses from fluid pressure.

5.2.3 Stress distributions

The preceding are examples of *homogeneous* stresses, which are constant from place to place in the rock element or specimen. At all points the principal stresses are equal in magnitude and direction. We try to achieve this state in a test specimen, but it rarely occurs in nature, and then only at locations far removed from man-made excavations and at a constant given depth below the ground surface.

Vertical stresses in reality increase with depth, and stress concentrations occur near excavations and soft or hard inclusions. Figure 5.10 shows one example, the case of a circular tunnel of radius a, where the far-field stress has a value p in all directions (a hydrostatic stress field, as discussed). The radial stress increases from zero at the exposed wall, and approaches the far-field value p at a distance of three or four radii from the tunnel wall. The tangential component of stress, because of the concentrating effect of the tunnel, is $2p$ all around the tunnel wall, and approaches p deeper into the rock mass. The stress components at any given radius r can be calculated as follows:

$$\sigma_r = p\left(1 - \frac{a^2}{r^2}\right) = \text{radial stress}$$

$$\sigma_t = p\left(1 + \frac{a^2}{r^2}\right) = \text{tangential stress}$$

For a circular tunnel in a uniaxial vertical far-field stress regime (vertical stress = p, horizontal stress = zero), compressive stress magnitudes at mid-height in the tunnel walls are $3p$, whereas at the crown

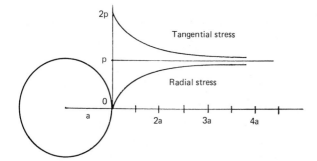

Figure 5.10 Stresses around a circular tunnel in an elastic rock under a hydrostatic far-field stress of magnitude p. In the walls of the tunnel, the radial stress is zero, and the tangential stress is twice the far-field stress. The tunnel has little effect on surrounding rock stresses further than two diameters from the walls of the excavation.

and invert, the stress is tensile and equal to $-p$. Elsewhere around the tunnel, the stresses vary systematically in magnitude and direction.

In general, excavations have the highest stress concentration at sharp corners, so when loaded with identical far-field stresses, a tunnel of square cross section has more pronounced stress concentrations than a circular one.

The distributions of stresses around excavations can be calculated as discussed in Chap. 7, provided one knows the nature of the rock (e.g., elastic, plastic, or viscous) and the magnitudes and directions of *virgin* stresses existing in the rock mass before the start of excavation work. In the remainder of this chapter we discuss the nature and causes of these virgin stresses, and how they can be estimated or measured.

5.3 Nature and Causes of Rock Stress

5.3.1 Predicted and measured stresses

5.3.1.1 Gravitational stresses. Ground stresses result from several different natural, geological, and artificial phenomena. The easiest to predict and the most universal are those caused by gravity.

The gravity-stress component increases with depth because of the increasing thickness and weight of overburden. It is vertical (directed toward the center of the earth) and is one of the three principal stresses except in situations where tectonic processes are active, or where the topography is varying so as to rotate the stress field. In most engineering situations the vertical stress is purely gravitational and may be calculated directly and reliably by multiplying the unit weight of the rock by the depth below surface. For example, as discussed in Sec. 5.1.2.2, the relationship between vertical gravitational stress and depth in granitic rocks is a linear one:

$$\sigma_v = 0.026z$$

where σ_v is the vertical stress in megapascals, and z is the depth below surface in meters.

In flat-lying sedimentary rocks, the vertical stress is taken to be the weight of the individual overburden beds. It can be calculated for a number of discrete strata:

$$\sigma_v = g(t_1 d_1 + t_2 d_2 + \cdots + t_i d_i + \cdots + t_n d_n)$$

where t is the thickness, d is the mass density of each bed (usually from a geophysical borehole density log), and g is gravitational acceleration. Major deviations from this value are not to be expected in nature: it can be assumed to be correct to within a few percent in almost all circumstances.

Horizontal stresses are also developed by gravitational loading. If rock were a freestanding column, as in the earlier example, the horizontal gravity-stress components would be zero. In the ground, however, rock is confined by more rock and is unable to expand laterally. Confining stresses have to develop to resist this expansion. Their magnitude can be calculated, using the theory of elasticity, as a function of Poisson's ratio, the ratio of horizontal expansion to vertical compression in a uniaxially compressed cylinder. For most rocks, with Poisson's ratio in the range of 0.1–0.25, the horizontal stress, if caused by gravity alone, would be about one-quarter of that acting vertically at any given depth. The horizontal gravity stresses would be equal in all directions and proportional to depth below surface.

Granular sediments develop horizontal stresses not so much because of the Poisson effect, but as a result of grain rearrangement and frictional behavior. According to theory, the lowest ratio of horizontal to vertical effective stress found in nature should be about 0.3. For shallow sediments of significant cohesive strength, the horizontal component could therefore be significantly less than the vertical. At depths of 5 km or more in sedimentary basins, where pressures and temperatures are high, diagenesis, shock loading caused by earthquakes, and other factors might be expected to return the stress more closely to a hydrostatic condition.

However, horizontal stress components calculated by the preceding assumptions (elastic or frictional granular responses) seldom bear any resemblance to those measured in situ. The actual horizontal stresses are usually much greater than those predicted, for reasons discussed below.

5.3.1.2 Trends in actual, measured rock stress values.

How do actual ground stresses, measured using the methods described later in this chapter, compare with those that would be predicted if caused by gravity alone? Figure 5.11a, a compilation of such measurements, shows that the vertical stress component is usually quite close to that predicted from the weight of overburden. The density of granitic rocks rich in quartz and feldspar is about 2650 kg/m^3, and of basic and ultrabasic rocks about 3300 kg/m^3. This results in a vertical stress gradient in the range of 26.0–32.4 kPa/m, depending on the density of the rock formations at the site (Herget, 1986).

The average horizontal stress in hard rocks, however, is much greater than gravitational. Compiling data for granitic rocks of the Canadian Shield, Herget (1986) has shown that the ratio of average horizontal to vertical stress approaches 1.5 at depths of 1.0–2.5 km

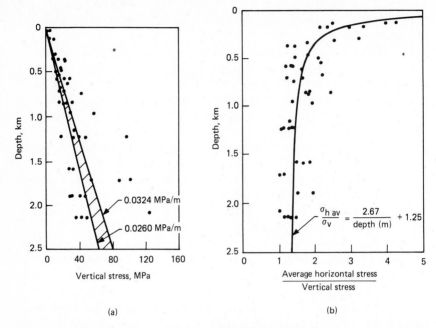

(a)   (b)

Figure 5.11 Measured stress variations with depth. (*a*) Vertical stress component can usually be predicted from weight of overburden. (*b*) Ratio of horizontal to vertical stress is about 1.5 at depths below 1 km, but rapidly increases to values greater than 5 in the 100 m nearest to surface. (*Data from Herget, 1986.*)

(Fig. 5.11*b*; see also Palmer and Lo, 1976; Brown and Hoek, 1978). However, at shallower depths the ratio of horizontal to vertical stress is much greater and increases rapidly toward the ground surface. At a depth of 100 m, where the vertical stress is about 2.6 MPa, horizontal stresses of about 14 MPa are commonly measured in southern Ontario, Canada, giving a ratio of 5. Very much higher ratios even than this occur within tens of meters of surface, where significant horizontal compressive stresses combine with a vertical stress close to zero.

This pattern of stress distribution is quite different from the pattern in "normally consolidated" soils in which the ratio of horizontal to vertical stress, called the *coefficient of earth pressure at rest,* is more or less constant at shallow depths. Horizontal rock stresses are not simply a fraction of the vertical stresses caused by overburden loading, and a simple ratio is meaningless, particularly close to surface where it can become nearly infinite in hard rocks. Stresses in rocks are best expressed as individual components and not as ratios. Horizontal components cannot be calculated: they should be measured for a particular site, or estimated from the trends of nearby measurements.

Softer sedimentary rocks show a somewhat different trend, intermediate between the soils and the hard rocks. Many younger sediments are found in flat-lying basins, neither faulted or folded, nor eroded to any significant degree. The major principal stress can then be vertical.

Fossil stresses are not necessarily preserved in the rock forever. One would not expect the more closely jointed and weaker rock materials, which can easily yield, to withstand extreme differences between principal stress components over extended periods of time. This is borne out by stress measurements in the weaker shales, where horizontal stresses seldom exceed the vertical. The theoretical maximum ratio is controlled by the strength of the rock, which may be exceeded during erosional unloading, particularly in weak clayey rocks, leading to horizontal bedding plane slickensides and clay mylonites (Sec. 3.2.1.1).

5.3.2 Sources of ground stress

Stress measurements in rock have been obtained at many locations by a variety of methods, and they show a consistent trend. Why then are actual horizontal stresses so different from those predicted by gravitational theory? Various plausible explanations are outlined below, one or several of which may apply. This is an area of active geomechanics research, and engineers and scientists do not yet fully understand the ways in which stresses are affected by such processes as diagenesis, rebound, plastic straining, and bedding plane slip during the deposition and denudation of rocks. Clearly, different processes and therefore stresses may apply in igneous, sedimentary, and metamorphic environments.

5.3.2.1 Effects of rock anisotropy. Calculations suggest that an explanation as simple as rock anisotropy may account for at least a part of the "abnormally" high horizontal stresses observed. Amadei and Savage (1985) and subsequently Savage, Amadei, and Swolfs (1986) conclude that, for stratified rock, the gravity-induced stress field depends on the thicknesses of the layers and on how rock stiffnesses vary with depth.

Most stratified rocks are more rigid in the bedding direction than across stratification, and also become stiffer with depth. The stress distributions predicted for these conditions appear similar to those obtained by measurement: the calculated horizontal stresses are larger than the vertical, particularly at shallow depth where the spacing of subhorizontal joints is small. However, to obtain a quantitative match between predicted and measured trends, somewhat unrealistic values must be assumed for rock properties. Anisotropy therefore may not be the whole answer to explaining the observed trends in ground stress data.

5.3.2.2 "Fossilized" stresses resulting from denudation.

The high horizontal stress components present today at the earth's surface may at least in part represent the "fossilized" effects of overlying earth, rock, and ice that has since been removed. Rock appears to have a "memory" of previous loading, resulting from plastic straining.

Much the earth's surface was covered by 1 km or more of ice during the Pleistocene era, which applied a vertical stress at the base of the ice sheet equivalent to the weight of ice (9 MPa/km). This generated a corresponding horizontal gravity stress equal to about one-quarter of the vertical component. When the ice melted, vertical stress was relieved (disappeared), there being nothing to impede upward rebound of the rock surface. Horizontal stresses would not be relieved in the same manner. Under some conditions, the horizontal stress today may therefore be close to that at the time of full ice loading (e.g., Herget, 1973).

This hypothesis is supported semiquantitatively by the similarity of measured horizontal stress values with those calculated from the geologically determined thickness of the ice sheet. Confirmation is obtained also from tests on overconsolidated clay soils, those that have been compressed by glacial ice. These exhibit a pattern of high horizontal stresses, much as do rocks. Ice thicknesses can be estimated from consolidation tests on clay samples, by observing the *overconsolidation ratio* from the break point in a stress-strain curve obtained by one-dimensional loading.

Ice melting is not the only form of geological denudation. At many if not most places the present ground surface was once covered by several kilometers of soil and rock strata that have since been removed by weathering and erosion. High stresses measured in southern latitudes where there has been little or no glaciation may perhaps be explained by this erosion, on the basis of an unloading hypothesis similar to that for ice melt described earlier. When a sedimentary basin has been uplifted and eroded, for example, the major principal stress is often reported as horizontal within a few hundred meters of the surface, at least four times as high as vertical, and much higher in well-cemented strata.

5.3.2.3 The effects of topography.

As the continents are raised by the buoyancy forces of isostacy, their upper surfaces are denuded by river systems and marine and glacial erosion. We noted how generalized denudation might relieve vertical stresses in a given stratum while allowing the horizontal stresses to remain much as before. A further effect is that the ground surface rarely remains flat. It is cut by ravines and valleys and sculpted into hills and mountains. This results in considerable local modification to the ground stress regime.

In mountainous terrain, near-surface ground stress trajectories tend to follow the topographical contours. For example, in Norway, near-surface tunnels excavated alongside steep valley walls have experienced rockbursting confined to the side closest to the valley wall, midway between springline and crown. This can be attributed to the influence of topography on ground stress distribution.

Another example of topographic effects is the heave of valley bottoms and the floors of quarries and basement excavations. Beds in valley floors when exposed by excavations for structures such as dams are often found to have slipped or buckled upward under the compressive action of high horizontal stresses. Similar buckling has been observed in quarries and other excavations. Glacial unloading and postglacial valley erosion in the clay shales of the plains of western Canada and the United States have generated slickensided bedding planes, upturned valley lips, and upwarped valley base strata, all evidence of high horizontal stresses and differential stresses in beds of different stiffnesses.

5.3.2.4 Tectonic stresses. While ground stresses reflect present-day gravity and overburden and also probably the overburden conditions that existed long ago, they must also be affected at least at some locations by continental drift and associated tectonic processes.

Relatively rigid plates that form the continents are being moved by convection currents in the earth's mantle, at rates of up to 12 cm per year, and in different directions depending on the location. In places the continents are moving apart, and in others they are being thrust one beneath the other. Earthquake belts marking the lines of interplate movements are well defined (Fig. 5.4). Near them, new faults appear and old ones are reactivated; crustal blocks rise, fall, or shear horizontally. The jostling of blocks can hardly occur without major redistributions of stresses along and near their boundaries, stress changes that for the most part go unmonitored and unrecorded.

Rummel (1986) reviews our present knowledge of tectonic stresses from the geophysical standpoint. Seismic records show that earthquakes between the contacts of continental blocks occur to crustal depths down to several hundred kilometers, below which the crust is essentially aseismic, presumably because the rocks are fluid and creep rather than rupture at these high temperatures and pressures. We might expect a maximum shear stress at a depth of between 5 and 10 km, depending on the fluid pressure. At greater depths, shear stresses are likely to diminish gradually.

Ample direct evidence shows that tectonic stresses are active. In western Europe, Klein and Barr (1986) found a consistent northwest-southeast trend in the major principal stress direction, dominated largely by plate tectonic boundary forces. In North America, Sbar and Sykes (1973) point to the relationship between stress patterns and plate tectonics. Along the Rocky Mountain front and for several hundred kilometers eastward, the largest of the two horizontal stress components is always directed perpendicular to the mountain front (Bell and Gough, 1980). In California, sedimentary rocks and even soils of recent age have been upturned and contorted by displacements along the San Andreas Fault.

In the vicinity of faulting or folding, the direction of the tectonic fabric either controls or reflects the principal horizontal stress directions. Far from active tectonics, stress effects are still felt, and usually the orientation and the relative magnitudes of the two horizontal stresses can be estimated with considerable confidence.

5.4   Methods of Stress Measurement

Rock stresses are sometimes measured during site investigation to provide design data, sometimes during mining or construction to check design predictions and to warn of any overstressing caused by excavation.

Rough estimates based on experience and global or regional data (e.g., Fig. 5.11) may suffice for the majority of engineering designs of rock excavations. Stresses may have to be measured rather than estimated when the cost of erroneous assumptions could outweigh that of measurements, for example, when planning a major mining operation in unexplored terrain, or designing a large hydropower cavern. Several alternative techniques are available for this purpose, some of which are described in detail in ISRM (1987).

5.4.1   Overcoring methods

5.4.1.1   The overcoring principle.   The most common overcoring methods rely on the fact that rock core is relieved of most of its in situ stress when drilled, and expands to nearly its initial unstressed configuration.

A simple analogy serves to explain the overcoring principle. A rubber band is stretched and then marked with lines, say 10 mm apart. The band is cut, allowing the original 10-mm length to shrink to some shorter value. The original tension in the band can be calculated from the known initial and final lengths of the segment, knowing the coefficient of elasticity of the rubber. The overcoring technique works the same way but in reverse because rock is in compression rather than tension.

Stress relief is usually measured by inserting a strain measuring device or diameter gauge into a small "pilot" drillhole, recording initial readings, and then drilling a concentric hole of larger diameter to remove an annulus of core with the stress meter inside (Fig. 5.13). The stress before overcoring is estimated either by applying external stress to the core until the stress meter again registers its initial values, or by computation from initial and final strain measurements, assuming linear elastic isotropic properties (or some other known behavior) for the rock. Gonano and Sharp (1983) provide a numerical solution for cross-anisotropic rock.

The overcoring type of stress measuring equipment has the advantage of permitting the determination of stresses moderately deep in the rock, away from disturbing influences of near-surface loosening and nearby underground excavations. The probes can be installed and left in place without overcoring if the object is to determine stress changes rather than virgin stress (Chap. 12). Like most alternatives, they need considerable expertise to provide reliable readings, and analysis of the results must account for the stress disturbance caused by the drillhole itself.

5.4.1.2 The USBM stress gauge. The U.S. Bureau of Mines (USBM) stress gauge (also called *deformation meter*) is perhaps the best known example of this type of instrument (Fig. 5.12). Its measuring probe contains three beryllium-copper or stainless-steel cantilever sensors: Each is instrumented with resistance strain gauges to measure cantilever bending and hence drillhole diameter changes in three directions at 120°. The electric cable from the probe is threaded through the drill rods and out through a water swivel to a strain gauge read-out unit (Hooker et al., 1974).

The drillhole is advanced at full overcore diameter to the first required depth of measurement. A small pilot hole of EX (37-mm) size is drilled beyond this depth (Fig. 5.13a). The USBM probe is inserted using a string of installing rods, pressing the cantilevers against the walls of the pilot hole. Initial readings are taken. A thin-walled overcoring bit is then used to extend the full-diameter drillhole past the measuring instrument, taking continuous readings of diameter changes as the bit advances (Fig. 5.13b). The overcore is then removed and its modulus is measured in a biaxial compression chamber, reapplying a stress level similar to that anticipated at the depth of

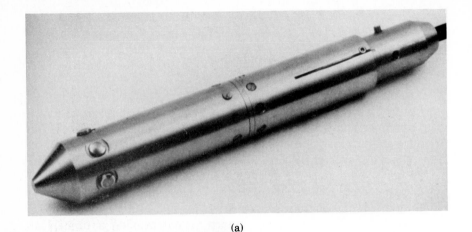

(a)

Figure 5.12 U.S. Bureau of Mines (USBM) stress-measuring "borehole deformation" gauge. (*a*) "Torpedo" probe with three pairs of measuring buttons attached to cantilevers within the probe body. (*b*) Probe in its calibrating device. Three strain gauge bridges can be seen in the background.

(b)

measurement, and using the same USBM probe in the laboratory as was used down the drillhole. The thick-walled cylinder formula gives Young's modulus from applied hydraulic pressure and internal diameter change.

The USBM probe is waterproof, robust, and reliable, and operates

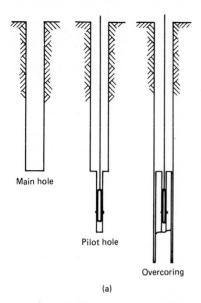 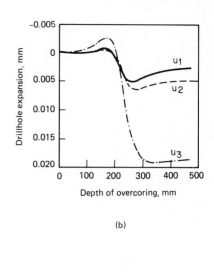

Main hole

Pilot hole

Overcoring

(a)

(b)

Figure 5.13 USBM gauge installation and measuring sequence. (a) Main hole is drilled, followed by a pilot hole, by insertion of the gauge, and by overcoring. (b) Continuous measurements across three diameters during overcoring show when readings have stabilized, and reveal any unreliable measurements.

very well in water-filled holes. It is therefore well suited to measurements in downholes drilled from the surface. It does, however, require a large-diameter hole and therefore bulky drilling equipment of the sort normally used only for geotechnical investigations at the ground surface. Other overcoring methods are more suitable for underground measurements because they can make do with smaller drills, such as the exploratory type of drill found more often in underground mines.

5.4.1.3 The South African "doorstopper." The doorstopper gauge (Fig. 5.14), developed in South Africa at the Council for Scientific and In-

Figure 5.14 South African "doorstopper" gauge. (*Courtesy of Roctest, Montreal, Que.*).

dustrial Research (CSIR), consists of a four-element 45° strain gauge rosette bonded to a rubber casting in a molded plastic cell (Leeman, 1964; Stickney et al., 1984). The end of a standard BX (58-mm) drillhole is flattened using a special facing bit. An adhesive is applied, and a "doorstopper" is pressed against the flat rock face. Initial gauge readings are taken after the adhesive has set, the electric leads are removed, and the drillhole is cored at its original diameter to relieve stresses on the central core. The core of rock with the attached doorstopper is removed, and further strain readings are taken. Stresses are calculated from the strain changes, knowing the Young's modulus of the rock and assuming elastic rock behavior.

This was one of the earliest stress determination techniques and is still widely used. It is particularly suitable for underground measurements because the method is very quick, and the overcoring size is much smaller than in alternative techniques. Coring can be done using light, exploratory drills. The technique works better than most when the rock is closely fractured or highly stressed and liable to disking, because only a small plug and not a larger overcore need be recovered.

Measurements are affected, however, by the geometry of the end of the hole, so the results need correcting. The technique is susceptible to inaccuracies if the rock is anisotropic, and like many of the alternative methods that rely on resistance strain gauges, it is difficult or impossible to use in a water-filled or very wet hole.

5.4.1.4 Lisbon, CSIR, and CSIRO overcoring methods. Other types of rock stress measuring instruments apply resistance strain gauges to the curved walls rather than to the flat end of the drillhole. These quite similar devices include the Portuguese (LNEC) stress tensor tube (Pinto and Cunha, 1986), the South African (CSIR) triaxial gauge (Leeman and Hayes, 1966), and the Australian (CSIRO) triaxial gauge (Fig. 5.15) (Worotnicki and Walton, 1976). They allow stresses to be resolved in three dimensions rather than in a single plane perpendicular to the hole, which is a limitation when using either the USBM or the doorstopper technique. They require a larger drillhole and a larger drill than for the doorstopper method, because the original pilot hole containing the resistance strain gauges must be overcored rather than advanced.

In all but the CSIR device, the resistance strain gauges are preencapsulated into an epoxy resin tube to maintain their alignments and to protect against mechanical damage and groundwater.

Jenkins and McKibbon (1986) compare CSIR, CSIRO, and USBM overcoring methods of stress determination at two mines in the United States. Various practical problems came to light, such as the

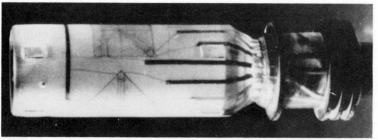

Figure 5.15 CSIRO (Australia) hollow inclusion stress cell, a fully encapsulated array of nine strain gauges for grouting into an EX sized drillhole. (*Courtesy of Geokon Inc., Lebanon, N.H.*)

need to cure CSIRO gauges at 60°C for 24 h before use, so as to relieve accumulated stress in the plastic cylinder of the gauge. Problems occurred as a result of temperature changes and cable flexure, with improper gluing of the gauges to the rock, and with fractures in the rock. Nevertheless, the CSIRO gauge has yielded many excellent results, even in soft sedimentary rocks, such as in the coal fields of eastern Australia.

These techniques are among the most popular for rock stress determination. Three-dimensional measurement at a single position and in a single hole is easier and much less expensive than combining two-dimensional measurements from differently oriented holes. The methods give reliable results provided that redundancy and cross-checking are built into the program. This is done by measuring more than the minimum of strains required in theory to resolve the stress tensor.

5.4.1.5 The Swedish SSPB-Hiltscher method. Most overcoring techniques are restricted to shallow drillholes, usually no longer than 70 m, so underground access is needed if measurements are to be made at greater depth. This can be a problem when exploring for a powerhouse cavern or waste repository, before tunnels and shafts have been excavated.

The Swedish State Power Board's SSPB-Hiltscher equipment developed in 1975 allows overcoring in near-vertical drillholes several hundred meters deep (Hiltscher et al., 1979; Hallbjorn, 1986). Stresses have been measured in water-filled holes down to 500 m, and also in a 45° inclined hole 90 m long.

The probe is a triaxial device, with three strain gauge rosettes at 120°, each of which contains longitudinal, transverse, and 45° gauges. The strain gauge adhesive is an acrylic resin with a pot life of about 20 min. The gauges are submerged in a protective "glue pot" of adhesive, and the probe, suspended on the end of a cable, is dropped down the hole. On striking the bottom, the glue pot sinks and the gauges are forced outward into contact with the rock walls. To record the probe azimuth, a preheated liquid around a compass needle solidifies during a period of 2 h while the strain gauge adhesive is setting. This method has widespread application in situations where deep measurements from the ground surface are essential, and where hydrofracturing is regarded as insufficiently reliable.

5.4.1.6 The photoelastic glass plug. The photoelastic glass stress plug is a glass cylinder with a centrally drilled hole. Photoelastic fringe patterns are observed before and after overcoring to give stress magnitudes and directions (Sec. 7.3.1.4). The technique, which is based on the *rigid inclusion* principle, is a biaxial one and has some merit in that the cell is robust, water-insensitive, and gives the directions of stress components in two dimensions very effectively. However, it is a rather insensitive way of measuring stress magnitudes and appears not to be widely used.

5.4.2 Slotting stress meter

Most overcoring devices are expendable, expensive, and slow, which is unfortunate in view of the variability of ground stresses and the need for large numbers of measurements if variations are to be adequately sampled and treated statistically. Bock (1986) describes an experimental device that is reusable, does not rely on overcoring, and permits many densely spaced, rapid measurements at reasonable cost. He found that the borehole slotting technique gave the in situ stress state with comparable accuracy faster, cheaper, and more reliably than overcoring and hydraulic fracturing tests.

The slotting stress meter is completely self-contained in its stress release operations and strain measuring capabilities. A half-moon-shaped slot is cut axially into the drillhole wall, using a thin diamond-impregnated blade. The original version is designed for use in HQ drillholes of 96-mm diameter, and uses a pneumatically driven saw to cut a slot 0.8 mm wide and a maximum of 32 mm deep. Slot closure is monitored by a sensitive strain sensor. At any given location, at least three slotting tests, with cuts in different radial directions, are made. In one experiment, 12 two-dimensional stress measurements, each based on six slotting tests, were carried out within a single day. Many measurements in a short hole length allowed delineation of structurally controlled stress domains, including unstressed and stressed layers within a zone 5 m thick.

5.4.3 Near-surface undercoring

Near-surface measurements of ground stresses can be made using an undercoring rather than an overcoring technique. In one Italian method (R. Ribacchi, personal communication) a flat rock surface is prepared, for example, in the walls of an exploratory tunnel. Measuring points are fixed to this surface in a circular pattern, and diametral distances are measured between each pair of points, using a sensitive portable displacement gauge. A central hole is drilled between the measuring points, and further readings are taken to determine the convergence of the drillhole. The preexisting ground stresses in the plane of the prepared surface may then be calculated.

5.4.4 Pressurization methods (strain nulling)

5.4.4.1 Flat-jack methods.
Flat-jack techniques offer a sometimes less expensive alternative to overcoring and are often used for the measurement of near-surface rock stresses in the exposed roof, walls, or floor of an excavation at the surface or underground. A flat jack is an inflatable "cushion" made of two thin steel sheets welded around their perimeter and filled with oil or a similar fluid. It is inserted into a slot cut in the rock, then inflated to expand the slot (Rocha et al., 1966; Pinto and Cunha, 1986; ISRM, 1987).

A pattern of displacement measuring points is first bonded to the rock surface, and initial readings are taken of the precise distances between points. A slot is then cut, preferably with a circular diamond-impregnated blade (Fig. 5.16a). The flat jack is inserted into the slot and inflated until the initial displacement readings have been reinstated (Fig. 5.16b and c). The pressure at which this is achieved gives a measure of the rock stress that existed perpendicular to the plane of the slot before it was cut.

Circular sawing limits the depth of the slot to a little less than the radius of the blade. Deeper slots can be prepared by careful line drilling using overlapping drillholes, but the slot is then irregular and of much greater thickness than the flat jack. Grout needed to fill the gap may adversely affect the results, so that sawing is much preferred.

A single flat jack measures stress in one direction only. Three-dimensional stress evaluation requires an exploratory chamber with slots at various directions in the walls and roof. The measured stresses apply to rock close to the surface of the test chamber and are therefore affected by the presence of the chamber itself, and by any blast damage or loosening. Interpretation of the results in terms of remote, virgin ground stresses requires the use of numerical solutions for the stress distribution around the chamber in which the measurements are made.

(a)   (b)

(c)

Figure 5.16   Flat-jack stress determination. (*a*) Sawing slot. (*b*) Inserting flat jack. (*c*) Reading slot width.

5.4.4.2 Hydrofracturing. As noted above, deep stress measurements are difficult using overcoring methods because of the length of cables and because of waterproofing problems. Hydraulic fracturing is often the only method available for measuring stresses at several hundred meters to several kilometers, for example, for studies of oil and gas reservoirs and for the design of deep underground excavations before access becomes available via mine shafts or exploratory galleries. It yields directly the magnitude of the least compressive principal stress regardless of the mechanical properties of the rock mass, and other stress components may be estimated (Haimson, 1974; Zoback and Haimson, 1982; ISRM, 1987).

A single or double *straddle* packer system is "set" (inflated) at the required depth so as to isolate a test cavity. For best results, a relatively impermeable section of the hole without joints is chosen. A liquid is injected into the test cavity and its pressure raised while monitoring the quantity injected (Fig. 5.17). A sudden surge of fluid accompanied by a sudden drop in pressure indicates that formation hydrofracture (fracture initiation, or "breakdown") has occurred. The hydrofracture continues to propagate away from the hole as fluid is injected, and is oriented normal to the least principal stress direction.

Once the hydrofracture has traveled about 10 drillhole diameters, injection is stopped by shutting a valve, and the *instantaneous shut-in pressure* is measured. The process is repeated several times to ensure a consistent measurement of this pressure, which is equal to the minor principal stress. The difference in the breakdown pressure between the first and second trials gives the tensile strength. The direction of the fracture plane can be determined by using a television camera, an ultrasonic scanner, or an impression packer, as described in Chap. 6.

When the plane of fracturing is nearly parallel to the drillhole, the following expressions, first developed by Hubbert and Willis (1957), may be used to calculate the principal effective stress components:

$$\sigma'_{min} = P_s - P_0$$

$$\sigma'_{max} = T + 3P_s - P_f - P_0 \text{ (1st cycle)}$$

$$\sigma'_{max} = 3P_s - P_r - P_0 \text{ (subsequent cycles)}$$

where T is the tensile strength of the rock, P_s is the instantaneous shut-in pressure, P_f is the fracture initiation pressure, P_r is the fracture reopening pressure, and P_0 is the initial pore-water pressure.

5.4.4.3 Hydraulic tests on preexisting fissures (HTPF method). Cornet (1986) suggests a method similar to hydrofracturing but without frac-

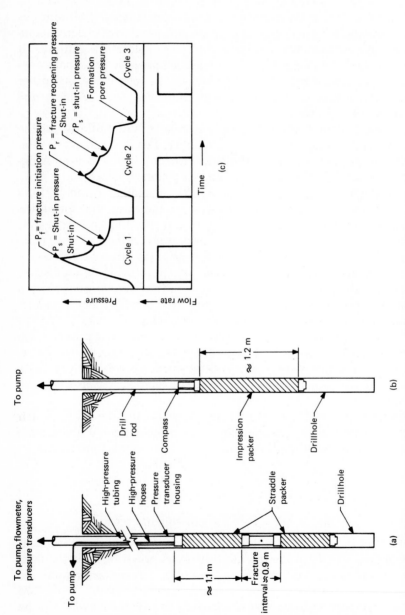

Figure 5.17 Hydrofracture stress determination. (*a*) Straddle packer used to generate hydrofracture. (*b*) Impression packer used to investigate direction of fracture. (*c*) Appearance of pressure-time hydrofracture record. (*From ISRM, 1987.*)

turing. Individual hydraulic pressure tests are conducted on preexisting rock joints at various inclinations in order to determine the complete regional stress tensor. A portion of the drillhole where a single joint has been identified, such as by televiewer, is sealed off with a straddle packer and then pressurized at a large enough flow rate to reopen the joint. The total normal stress supported by the joint can be determined by one of two methods, using either the instantaneous shut-in pressure or constant-pressure steps.

5.4.4.4 Dilatometer stress determination (sleeve fracturing). The *sleeve fracturing* method described by Stephansson (1983) is very similar to hydrofracturing except that a membrane is interposed between the rock and the fluid. Axial drillhole fractures can thus be generated without allowing fluids to penetrate the joints.

The device is essentially the same as the drillhole dilatometer used for measuring rock mass deformability, described in Chap. 9, and consists of a screw pump and an expandable drillhole probe. Rock stresses and rock deformability can be determined in one and the same test. The dilatometer is first calibrated by inflation in a metal cylinder, then expanded in the drillhole, and the elastic parameters of the rock are determined from the relationship between the pressure and the volume of injection. In situ stress is then measured by increasing the pressure until the rock fractures. Orientation of the fracture, which propagates as a plane normal to the direction of minor principal stress, is determined from an impression on a plastic adhesive tape wrapped around the sleeve. Using the hydrofracture equation given above, the critical fracture initiation pressure and the known tensile strength of the rock provide a value for the principal total stress difference $(3P_2 - P_1)$. The principal stresses may then be separated by repressurizing the drillhole to determine the shut-in pressure, which is the inflection point of the pressure-volume curve at which the fracture is reopened.

Hydrofractures can propagate long distances, whereas sleeve-induced ones are limited to within one or two radii from the drillhole wall. Also, in a sleeve fracture test, no information can be obtained from the pressure-volume behavior of fluid in the fracture. In spite of these apparent limitations, the method is attractive. Stephansson (1983) reports that stress values determined by sleeve fracturing at the Hanford, Wash., proposed radioactive waste repository site compared closely with those obtained by hydrofracturing and USBM overcoring methods. Sleeve fracturing is cleaner, quicker, and more convenient than hydrofracturing, mainly because it eliminates the need to transport and pump large quantities of pressurizing fluid.

5.4.5 Soft inclusion methods in viscoplastic rocks

A thin metallic wafer (flat jack) can be used to measure stress in a highly stressed viscoplastic rock such as potash or halite. The wafer is encapsulated between two halves of a core of the same rock, which is then reinserted in a drillhole, aligning the wafer perpendicular to the required direction of measurement. Cell pressure, read off a gauge or transducer, is raised above the estimated normal stress, and the cell is sealed. Pressure in the cell approaches the field stress at a rate controlled by the viscosity of the rock. Usually several months are required to obtain an accurate value. Once installed, the same cells can be used to monitor changes in stress.

To determine a full stress state, several wafer cells need to be installed at different orientations. Usually the principal stress directions can be estimated accurately because the original stress state was hydrostatic, and the excavation geometry is known.

Other methods in salt rock include long-term installation of an inflated cylindrical dilatometer, or creation of an oil-filled cavity in the salt rock which is sealed off and monitored. Cylindrical cells, being nondirectional, measure some combination of stresses, and the results are difficult to interpret in terms of a nonhydrostatic stress field.

5.4.6 Methods under development

5.4.6.1 Problems at depth. Knowledge of the orientation and magnitude of in situ stresses is critical to the design of hydraulic fractures for oil and gas well stimulation. The only stress measuring technique currently in use for deep-level measurements in oil and gas wells is hydraulic fracturing. The method is expensive, and the hydrofractures can themselves lead to future well control problems (Teufel and Warpinski, 1984). At great depth, because of drillhole closure or spalling, the conventional methods of stress determination become inoperative, and information must be obtained from novel and often less reliable sources. Various alternative methods are under investigation. If they work, they are likely also to prove useful in more conventional, shallower applications.

5.4.6.2 Sonic velocity methods. Various researchers over the years have tried using sound velocity to measure stress in the rock mass. High stresses close pores, and sound travels faster through the denser rock material (Sec. 3.3.2.4). To date, the results have been at best imprecise, because different rock types start off with different porosities and respond differently to stress changes.

Mao and coworkers (1984) appear to have achieved some measure of success in determining the stress-induced velocity anisotropy around a drillhole. In a downhole probe, eight transducers measured velocities in radial and tangential directions. Relative travel times were determined by a pulsed loop technique and also by cross correlation of digitized waveforms. Stress changes of about 0.6 MPa could be detected, corresponding to 8% of the absolute stress level.

5.4.6.3 Wellbore breakout and borehole ellipticity. Wellbore breakout occurs in the form of spalling along lines parallel to the wellbore axis, in the direction of minimum radial stress (Fig. 5.18). Wellbore breakouts therefore give an indication of principal stress directions (Bell and Gough, 1980), particularly useful when the packers needed for hydrofracturing are ineffective in providing a seal.

Klein and Barr (1986) used a four-arm caliper to log near-verticaloil and gas wells drilled in the North Sea, the Atlantic Ocean, and onshore Britain. Good agreement was found between maximum and minimum horizontal stress directions inferred from breakout analyses and those inferred from various types of stress measurement, where available.

5.4.6.4 Stress relaxation (DSCA). In the method of differential strain curve analysis (DSCA), a core specimen is restressed after its recovery from the base of a deep drillhole. Proponents of the method claim that the complete stress tensor can be determined from oriented core samples extracted from a single drillhole, regardless of depth (Teufel and Warpinski, 1984; Dey and Brown, 1986). If so, the method will find wide application, since measurements by reloading in the laboratory are much more convenient than those with strain gauges at depths of hundreds of meters.

The principle of the method is that on being cored, the specimen expands, and this expansion includes both an instantaneous elastic compo-

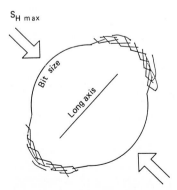

Figure 5.18 Wellbore breakout.

nent and an inelastic component. The elastic component is lost, but the inelastic component can be measured by reloading. The method relies on the majority of microcracks present in the sample having been created by the relief of in situ stresses during and after cutting of the core.

Specimens are prepared with resistance strain gauges, and are loaded hydrostatically to a pressure that is sufficient to close essentially all the crack porosity. Projecting the asymptotic slope of the strain-pressure curves back to zero pressure gives the contribution of crack closure to the strain recorded by each gauge. With a minimum of six appropriately oriented gauges, one can arrive at a tensor of three principal crack strains and their directions, which can be interpreted to give the in situ stress state.

Teufel and Warpinski (1984) conducted experiments on sandstone cores using spring-loaded clip-on gauges. Monitoring began 7 h after the core was cut. Strains amounting to 80–200 microstrain, depending on the direction of measurement, were recorded over 40 h, after which equilibrium was reached. A viscoelastic solution was used to back-calculate the principal horizontal stress magnitudes from the differential strain recovery magnitudes. The calculated principal stresses were within 12% of those determined by hydraulic fracturing.

5.4.6.5 Kaiser effect. When rock is loaded by a testing machine in the laboratory, it emits subaudible noises (acoustic emission) at a rate that increases with the level of applied stress. The *Kaiser effect* is a sudden increase in the rate of acoustic emission that appears to occur when the applied stress has reached a level greater than any previously experienced by the rock specimen (Kaiser, 1953; Hughson and Crawford, 1986).

The technique for using this phenomenon as a measure of preexisting stress is therefore quite similar to the DSCA method described before, in that specimens are recovered from any depth in a drillhole and are reloaded in a laboratory testing machine. There may be some correspondence between the stress required for crack closure in a DSCA test and that required to trigger the Kaiser effect. If it can be shown to give sufficiently precise and reliable results, the Kaiser method has all the advantages of DSCA, particularly in that it requires no downhole instrumentation.

Specimens are cored in three orthogonal directions, or recored if necessary from the initial core, and then loaded in a uniaxial compression testing machine until the *recalled maximum stress* can be identified by the sudden increase in acoustic emission rate.

The Kaiser effect has been demonstrated in the laboratory, although the same experiments have shown that some rocks lose their "recollection" of previous loading history after a period of a few days or weeks. The ability to retain and recall this memory accurately after millions of years remains in some doubt.

References

Amadei, B., and W. Z. Savage: "Gravitational Stresses in Regularly Jointed Rock Masses—A Keynote Lecture," *Proc. Int. Symp. Fundamentals of Rock Joints* (Björkliden, Sweden, 1985), pp. 463–473.

Bell, J. S., and D. I. Gough: "Intraplate Stress Orientations from Alberta Oil Wells," in *Evolution of the Earth*, Geodynamics ser., vol. 5 (Am. Geophys. Union, Washington, D.C., 1980), pp. 96–104.

Bock, H.: "In situ Validation of the Borehole Slotting Stressmeter," *Proc. Int. Symp. Rock Stress and Rock Stress Measurements* (Stockholm, Sweden, 1986), pp. 261–270.

Brown, E. T., and E. Hoek: "Trends in Relationships between Measured in situ Stresses and Depth," *Int. J. Rock Mech. Min. Sci. Geomech. Abstr.*, 15, 211–215 (1978).

Cornet, F. H.: "Stress Determination from Hydraulic Tests on Pre-existing Fractures—The HTPF Method," *Proc. Int. Symp. Rock Stress and Rock Stress Measurements* (Stockholm, Sweden, 1986), pp. 301–312.

Dey, T. N., and D. W. Brown: "Stress Measurements in a Deep Granitic Rock Mass Using Hydraulic Fracturing and Differential Strain Curve Analysis," *Proc. Int. Symp. Rock Stress and Rock Stress Measurements* (Stockholm, Sweden, 1986), pp. 351–357.

Eiby, G. A.: *Earthquakes* (Heinemann, Exeter, N.H., 1980), 198 pp.

Franklin, J. A., and O. Hungr: "Rock Stresses in Canada, Their Relevance to Engineering Projects," *Rock Mech.* (Springer), suppl. 6, pp. 25–46 (1978).

Gonano, L. T., and J. C. Sharp: "Critical Evaluation of Rock Behavior for in situ Stress Determination Using Overcoring Methods," *Proc. 5th Int. Cong. Rock Mech.* (Melbourne, Australia, 1983), vol. A, pp. 241–250.

Haimson, B. C.: "Determination of Stresses in Deep Holes and around Tunnels by Hydraulic Fracturing," *Proc. U.S. Rapid Excavation and Tunneling Conf.* (1974), vol. 2, pp. 1539–1560.

Hallbjorn, L.: "Rock Stress Measurements Performed by Swedish State Power Board," *Proc. Int. Symp. Rock Stress and Rock Stress Measurements* (Stockholm, Sweden, 1986), pp. 197–205.

Herget, G.: "Variations of Rock Stresses with Depth at a Canadian Iron Ore Mine," *Int. J. Rock Mech. Min. Sci.*, 10, 37–51 (1973).

———: "Changes of Ground Stress with Depth in the Canadian Shield," *Proc. Int. Symp. Rock Stress and Rock Stress Measurements* (Stockholm, Sweden, 1986), pp. 61–68.

———: *Stresses in Rock* (A. A. Balkema, Rotterdam, 1988), 179 pp.

Hiltscher, R., J. Martna, and L. Strindell: "The Measurement of Triaxial Rock Stresses in Deep Boreholes and the Use of the Rock Stress Measurements in the Design of the Construction of Rock Openings," *Proc. 4th Int. Cong. Rock Mech.* (Montreux, Switzerland, 1979), vol. 2, pp. 227–234.

Hooker, V. E., J. R. Aggson, and D. L. Bickel: "Improvements in the Three Component Borehole Deformation Gage and Overcoring Technique," Rep. Invest. RI 7894, U.S. Bureau of Mines (1974).

Hubbert, M. K., and D. G. Willis: "Mechanics of Hydraulic Fracturing," *Trans. Am. Inst. Min. Eng.*, 210, 153–168 (1957).

Hughson, D. R., and A. M. Crawford: "Kaiser Effect Gauging: A New Method for Determining the Pre-existing in-situ Stress from an Extracted Core by Acoustic Emissions," *Proc. Int. Symp. Rock Stress and Rock Stress Measurements* (Stockholm, Sweden, 1986), pp. 359–368.

ISRM: "Suggested Methods for Rock Stress Determination," *Int. J. Rock Mech. Min. Sci. Geomech. Abstr.*, 24(1), 53–73 (1987).

Jaeger, J. C., and N. G. W. Cook: *Fundamentals of Rock Mechanics* (Methuen, London, 1969), 513 pp.

Jenkins, F. M., and R. W. McKibbon: "Practical Considerations of in situ Stress Determination," *Proc. Int. Symp. Application of Rock Characterization Techniques in Mine Design* (1986), Am. Inst. Min. Eng., New Orleans, La., pp. 33–39.

Kaiser, J.: "Erkenntnisse und Folgerungen aus der Messung von Geräuschen bei Zugbeanspruchung von metallischen Werkstoffen," *Arch. f. Eisenhüttenw.*, **24**, 43–45 (1953).

Klein, R. J., and M. V. Barr: "State of Stress in Western Europe," *Proc. Int. Symp. Rock Stress and Rock Stress Measurements* (Stockholm, Sweden, 1986), pp. 33–44.

Leeman, E. R.: "The Measurement of Stress in Rock, Part 2," *J. South Afr. Inst. Min. Metal.*, **65**(2), 102–108 (1964).

——— and D. J. Hayes: "A Technique for Determining the Complete State of Stress in Rock Using a Single Borehole," *Proc. 1st Int. Cong. Rock Mech.* (Lisbon, Portugal, 1966), vol. 2, pp. 17–24.

Lo, K. Y., C. F. Lee, J. H. L. Palmer, and R. M. Quigley: "Report on Stress Relief and Time-Dependent Deformation of Rocks," Can. Nat. Res. Council Spec. Project 7307, Univ. Western Ontario, London, Ont., (1975) 320 pp.

Mao, N., J. Sweeney, J. Hanson, and M. Costantino: "Using a Sonic Technique to Estimate in situ Stresses," *Proc. 25th U.S. Symp. Rock Mech.* (Evanston, Ill., 1984), pp. 167–175.

Palmer, J. H. L., and K. Y. Lo: "In situ Stress Measurement in Some Near Surface Rock Formations—Thorold, Ontario," *Can. Geotech. J.*, **13**(1), 1–7 (1976).

Pinto, J. L., and A. P. Cunha: "Rock Stress Determinations with the STT and SFJ Techniques," *Proc. Int. Symp. Rock Stress and Rock Stress Measurements* (Stockholm, Sweden, 1986), pp. 253–260.

Quigley, R. M., C. D. Thompson, and J. P. Fedorkiw: "A Pictorial Case History of Lateral Rock Creep in an Open Cut into the Niagara Escarpment Rocks at Hamilton, Ontario," *Can. Geotech. J.*, **15**(1), 128–133 (1978).

Rocha, M., J. Baptista Lopes, and J. DaSilva: "A New Technique for Applying the Method of the Flat Jack in Determination of Stresses inside Rock Masses," *Proc. 1st Int. Cong. Rock Mech.* (Lisbon, Portugal, 1966), vol. 2, pp. 57–65.

Rummel, F.: "Stresses and Tectonics of the Upper Continental Crust—A Review," *Proc. Int. Symp. Rock Stress and Rock Stress Measurement* (Stockholm, Sweden, 1986), pp. 177–186.

Savage, W. Z., B. P. Amadei, and H. S. Swolfs: "Influence of Rock Fabric on Gravity-Induced Stresses," *Proc. Int. Symp. Rock Stress and Rock Stress Measurements* (Stockholm, Sweden, 1986), pp. 99–110.

Sbar, M. L., and L. R. Sykes: "Contemporary Compressive Stress and Seismicity in North America, an Example of Intra-plate Tectonics," *Geol. Survey of Am.*, **84**(6), 1861–1882 (1973).

Stephansson, O.: "Rock Stress Measurement by Sleeve Fracturing," *Proc. 5th Int. Cong. Rock Mech.* (Melbourne, Australia, 1983), vol. F, pp. 129–137.

Stickney, R. G., P. E. Senseny, and E. C. Gregory: "Performance Testing of the Doorstopper Biaxial Strain Cell," *Proc. 25th U.S. Symp. Rock Mech.* (Evanston, Ill., 1984), pp. 437–444.

Teufel, L. W., and N. R. Warpinski: "Determination of in situ Stress from Anelastic Strain Recovery Measurements of Oriented Core: Comparison to Hydraulic Fracture Stress Measurements in the Rollins Sandstone, Piceance Basin, Colorado," *Proc. 25th U.S. Symp. Rock Mech.* (Evanston, Ill., 1984), pp. 176–185.

Worotnicki, G., and R. J. Walton: "Triaxial Hollow Inclusion Gauges (CSIRO) for Determination of Rock Stresses in situ," *Proc. ISRM Symp. Investigation of Stress in Rock—Advances in Stress Measurement* (Sydney, Australia, 1976), pp. 1–8.

Zoback, M. D., and B. C. Haimson: "Status of the Hydraulic Fracturing Method for in situ Stress Measurements," *Proc. 23d U.S. Symp. Rock Mech.* (Berkeley, Calif., 1982), pp. 143–156.

6

Site Investigation

6.1 Planning

6.1.1 The key to design and construction

Successful design and construction require a reliable forecast of soil, rock, groundwater, and ground stress conditions. Without such a forecast, even the most up-to-date methods of design are of little or no use. Unexpected problems arise during construction and lead to last-minute changes and escalating costs, construction specifications become obsolete, and contractors take designers and owners to court because their cost estimates were based on false assumptions.

> Designers and contractors are able to cope with very major adverse geological conditions as long as they know about them in advance so that the design, the specifications and the construction planning can take them into account sufficiently. If not, extensive delays, cost overruns and controversy will certainly develop. If a latent adverse geological feature remains undetected during both the design and construction phases, the potential of failure during operation in the following years remains (Deere, 1974).

6.1.2 Objectives, scope, and cost of investigations

6.1.2.1 Production of maps and cross sections. The main aim of a site investigation is to determine soil, rock, groundwater, and ground stress conditions as they affect construction, and to produce plans and cross sections to show as clearly and reliably as possible the following information:

Contours of topography, pre- and postconstruction

Contours of top of bedrock (bedrock profiles)

Contours of base of weathered rock, if distinct

Contours of major stratum boundaries in both soil and rock

Contours of the groundwater table

Maps are prepared to define the boundaries of geotechnical mapping units (GMU), each of which is assigned its own set of descriptive and index properties (Sec. 3.1.1.1). Computers permit rapid storage, processing, and editing of geostatistics, as well as experimentation with alternative forms of presentation. Digital terrain models allow contouring not only of the ground surface, but also of the surfaces that separate soil from rock, and one geotechnical unit from another. Scales can be altered at will, overlays prepared, and cross sections extracted to examine the site from all angles.

The text of a site investigation report is almost supplementary to the maps and sections it contains. It is good practice to separate the report into one part covering facts and another covering interpretations. The factual section contains the terms of reference, describes the scope and nature of investigations, and presents the results. The interpretive section draws inferences regarding conditions between outcrops, test pits, and drillholes; suggests the degree of confidence to be placed on these interpolations and extrapolations; identifies key construction problems; and discusses the implications on various aspects of the work, including excavating and support requirements and costs.

6.1.2.2 Scope and cost of investigations. Investigations vary, depending on the size of the project, the depth of investigation, the complexity of the ground, and the amount of information already available. An investigation costs less for a highway at surface than for a highway tunnel because of the shallower depths to be investigated. A tunnel investigation in turn costs less than an investigation for an underground hydroelectric chamber which, although more localized, requires a more detailed knowledge of ground conditions so that large spans can be properly designed and stabilized (West et al., 1981).

A typical investigation costs between 0.25 and 1% of the capital cost of the project for an easily accessible site with simple geology, whereas more complex and remote sites may require a site investigation costing 5% or more of the total.

6.1.3 Optimization and phasing of work

Site investigations must make the best use of the limited time and budget available. Numbers, depths, and types of exploratory drillholes and tests are selected depending on the thickness and nature of the soil cover and the underlying rock. This poses a chicken-and-egg problem, in that the investigation can only be designed with some foreknowledge of ground conditions, and ground conditions can only be determined after some investigation.

The best compromise is to phase the investigations, starting with a reconnaissance and continuing with more detailed investigations as the character of the site unfolds (Table 6.1). Each successive phase is planned, then revised with the benefit of earlier results. The phases start with the least expensive and most general. Before even setting foot in the field, full use is made of existing maps and air photographs. This desk study is followed by a site reconnaissance to examine landforms, outcrops, and patterns of vegetation and drainage, followed as necessary by geophysical surveys, shallow test pitting, and deeper drilling and sampling.

Phasing allows development of a picture of the ground that comes more clearly into focus with the step-by-step addition of new data. The geotechnical engineer, like the surveyor, should work "from the general to the particular." The main features of the regional geological setting must be established before the details. This leads to a more reliable interpretation of local data and to better interpolation between points of observation.

Phasing of the investigations is often formalized, for major projects, into prefeasibility, feasibility, and design stages, with each stage taking months or even years. On smaller projects, the stages tend to blend into each other with some degree of overlap.

6.2   Desk Studies

6.2.1   Research of maps and reports

Making the most of published reports, maps, and air photographs depends on knowing what to look for and where to find it. Careful background research will bear fruit even for projects in remote and unpopulated areas.

Topographic maps are essential to plan field investigations and as base maps for recording geotechnical data. In remote areas the reliability of topographic mapping should be checked, as errors of tens of meters have been known to result from contouring of treetops instead of the ground surface.

Geological maps of various types and scales are usually available. Bedrock maps show the distribution of pre-Pleistocene (rock) materials beneath soil and in some countries also the locations and boundaries of bedrock outcrops. Quaternary maps show the nature of the unconsolidated surficial deposits, often with detailed infor-

TABLE 6.1 Phasing of Site Investigation

Primary (Research) Phase

Research of topographic and geological maps and reports

Research of regional seismic and ground stress data

Walking reconnaissance of the site and of regional outcrops, exposures of the same rocks in nearby tunnels and mines

Study of nearby water-well records

Air photo study using existing photography

Discussions with local residents and specialists

Preparation of base maps

Preliminary report on site conditions: plan next phase

Secondary Phase

Detailed logging of rock outcrops, statistics on jointing

Exploratory test pits and trenches: sampling

Index testing in situ and in the laboratory

Reconnaissance by helicopter or light airplane

Probe hole drilling using auger or air track

Seismic or electromagnetic geophysical traverses

Special air photography and supplementary interpretation

Preliminary, limited core drilling, logging, and testing

Definition of soil-rock interface topography

Rock mass classification

Progress report: plan next phase

Tertiary Phase

Further core drilling, vertical, inclined, or horizontal

Jointing information from drillhole television, impression packers, oriented core, integral-core sampling

Downhole geophysical logging and tomography between holes to define rock mass quality and individual faults, etc.

Downhole testing using a dilatometer to measure deformability

Packer testing for hydraulic conductivity

Installation of piezometers to study the groundwater regime and to monitor piezometric pressures

In situ stress determinations

Excavation and logging of exploratory trenches, adits, shafts

Large-scale in situ strength and deformability tests

Full evaluation of soil, rock, groundwater, and stress regimes

Final site investigation report

mation as to their genesis, grain size, and characteristic land form. Pedologic or agronomic soil maps are produced for agricultural purposes and can provide further information on soil thicknesses and types. Speciality maps may include engineering terrain evaluations; aggregate resource inventories; mineral potential, earthquake hazard, and land use maps; and aeromagnetic, gravity, and other geophysical surveys.

Previous groundwater investigations are a valuable source of data not only on groundwater, but also on soil and rock aquifers. Environmental agencies collect well logs and compile them into regional maps of the water table, water quality, and groundwater yields from various depths. The remarks made by well drillers on their boring logs, although not always "scientific," can give useful information on soil types, drilling conditions, and bedrock depths.

Sites in urban or suburban areas are usually surrounded by others that have been investigated in the past. Geotechnical consulting firms accumulate many site investigation reports within their home territory. Naturally they are reluctant to release such proprietary information which, however, may be legitimately obtained from the clients who commissioned the work. For example, highway authorities maintain libraries of investigations for all major roads and bridges. Many of the more important investigations have also been published as case studies in the technical literature.

When working in an unfamiliar area, the geotechnical engineer should consult with local specialists from universities, colleges, government, and municipal agencies who during many years have become familiar with local ground conditions and problems. Consulting local specialists at an early stage helps to direct site investigation efforts and funding and to avoid blunders that result from a lack of familiarity with local geology and such mundane matters as the availability and cost of drill rigs.

6.2.2 Air photos and remote sensing

6.2.2.1 Overview. A wide variety of aerial exploration techniques commonly termed *remote sensing* are available. They all measure energy emitted and reflected by the land surface. Large areas can be surveyed and interpreted geologically, rapidly, and economically by aircraft or earth orbital satellite. Digital or printed photographs from earth satellites are becoming increasingly useful in geotechnical applications because of improved resolution and enhancement. They can now be enlarged to give a resolution comparable to conventional air photographs, and are available for several different wavelengths covering the visible and invisible spectra. Printed images or a computer tape containing the information as an array of pixels can be purchased.

Remote sensing methods are useful for detecting regional features such as folds, faults, and surface drainage patterns. They provide the images needed for photogrammetric production of topographic maps. McEldowney and Pascucci (1979) present an excellent review of the various techniques and give examples of their use in fault studies for nuclear power plant sites in the United States.

6.2.2.2 Multispectral imagery. Black and white or color photography makes use of just one small part of the electromagnetic spectrum. Other types of imagery are available of radiations reflected from the earth's surface at other wavelengths within and outside the visible range. These include infrared and ultraviolet photography and radar images. Multispectral optical-mechanical scanning permits data acquisition through a broad portion of the spectrum at many discrete frequencies. The various wavelength alternatives can be studied singly or in combination.

Electromagnetic sensors may be classified as imaging or nonimaging. Imaging sensors include cameras, optical-electromechanical scanners, passive-microwave scanners, and radar. The most commonly used photographic methods are panchromatic black and white, color, color infrared, and black and white infrared. Nonimaging sensors include infrared and microwave radiometers, laser and microwave profilers, radio frequency sounding instruments, gamma ray spectrometers, and the gravimetric and electromagnetic induction devices reviewed in Sec. 6.4.

6.2.2.3 Black and white photographs. Conventional black and white photography of a 40-km^2 area should cost about $2000 at a 1:6000 scale. The cost of color photography of a similar area and to a smilar scale may be about 25% greater. Libraries contain air photos of even the most remote areas since nowadays they provide the basis for both topographic surveying and geological interpretation (Fig. 6.1). These can usually be obtained quickly and at nominal cost, except in countries where they are withheld for political or military reasons (Mollard, 1962).

Overlapping stereo pairs of photographs give a three-dimensional image of the ground surface. Viewing devices range from a small pocket stereoscope that can be carried on a field reconnaissance, to a larger mirror stereoscope more suited for office work, to a very large and expensive stereocomparator used for topographic surveying. The study and interpretation of air photographs is a skill that can only be

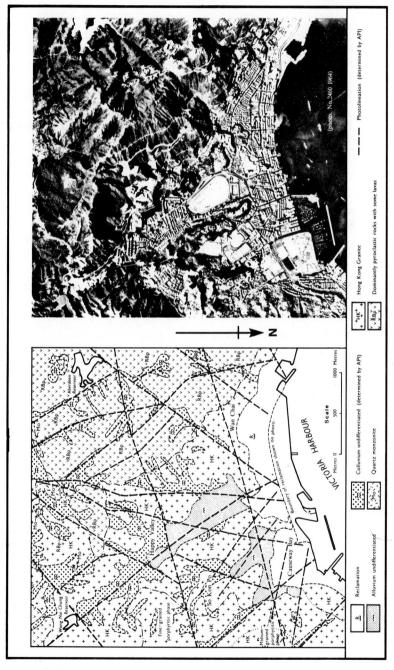

Figure 6.1 Air photo mapping of lineations and zones of rock, soil, and fill; Victoria Harbour, Hong Kong. *(Geotechnical Control Office, 1984.)*

167

developed with practice. Patience is needed, because many features become apparent only after several viewings.

A combination of photographs taken vertically and obliquely from the air, and horizontally from the ground, give the best possible coverage by allowing the site to be viewed from all angles, which is particularly important when viewing steep rock slopes.

6.2.2.4 Sequential photographs. Sequential photographs taken at intervals of years are often available and are useful for time-related problems. For example, meander patterns can be examined to determine whether river banks are stable. A comparison of old and new air photos in an urban environment helps determine whether depressions contain man-made fill. Old photographs can sometimes prove better than new ones because of scale differences, or because recent construction activity has disturbed the ground surface. Light and climatic conditions also vary; air photos taken during periods of drought, for example, can reveal different features from those taken after prolonged heavy rain.

6.2.2.5 Color, infrared, and false-color photography. Color photography, which registers broad-spectrum radiation within the visible range, is of great assistance in differentiating patterns of vegetation, which in turn may be used to interpret underlying soil, rock, and groundwater conditions.

Infrared (long-wavelength, low-frequency) photography is a heat measuring method that records radiation in the wavelengths of 8–13 μm. It is useful for detecting zones of saturated ground that are conspicuously warmer or colder than the surrounding dry terrain, such as fault zones or saturated soils subject to landsliding. Close-up infrared photographs might reveal blocks of loose rock that have become isolated from the surrounding warmer rock mass.

Photographs should be taken at times of greatest temperature and moisture contrast, such as early in the day after the surface has been warmed by the sun, and during dry spells following wet weather. False-color infrared photography during late summer can sometimes also reveal hidden wet areas.

6.2.2.6 Ultraviolet and radar imagery. Ultraviolet and radar photography detect radiations of short wavelength and high frequency which have greater penetration and can "see through" cloud cover and, in the case of ground-transmitted radar (Sec. 6.4.4.2), even through soil and rock. Airborne radar can be of the pure reflectance type, which reflects off soil or rock but not vegetation or clouds and is particularly useful in heavily vegetated or perennially overcast regions. A common technique is side-looking airborne radar (SLAR), a scanning technique using a narrow emitted beam and a collection antenna or similar system. At another wavelength (microwaves), radar penetrates the ground to a depth of several meters, but is strongly reflected by water, and this is used to rapidly explore the state of saturation of soil or rock. A space-shuttle-borne sensor of the latter type was responsible for discovering major underground flow systems in the eastern Sahara desert, despite a total lack of surface evidence.

6.3   Natural and Man-Made Rock Exposures

6.3.1   Near-surface reconnaissance

6.3.1.1   Relative merits of outcrop and drillhole observations. The purpose of surface and shallow subsurface reconnaissance is to obtain the maximum of information on overburden and rock conditions without the expense of drilling, and also to establish a program for later drilling operations. Drillholes are no substitute for observations of an extensive outcrop when one is available, and much drilling is done unnecessarily through ignorance or neglect of nearby rock exposures. The very existence of mapped or observed outcrops gives important and often overlooked information. Outcrops and patches of soil cover can usefully be surveyed and plotted on the site maps, and the pattern can generally be interpreted in terms of overall variations of bedrock depth (soil thickness).

Only limited information on rock structures and jointing can be obtained from drill-core logging and downhole viewing, because only small-scale features can be seen. The core sticks rotate, and weak seams, often more important than the solid rock, are softened or washed out by drill water. In contrast, an outcrop, trench, or test pit, gives an extensive view of rock structures and jointing, and little disturbance except that caused by weathering. Rock exposed in this manner readily provides samples for testing.

However, near-surface rocks are usually more weathered than those at depth, and more closely jointed, with closer joint spacings and larger apertures. The intact rock often tends to be weaker. Fault gouge and breccia can rarely be seen at surface, because these soft materials are either heavily vegetated or removed by erosion. Exposed surface rock can give a false impression of strength if the weaker materials are missing. Only the erosion gulleys or vegetated zones, clearer on air photographs than on the ground, are present to bear witness to the missing layers.

Drillholes have the advantage of being cheaper than shafts and adits to explore depths greater than can be trenched. They are convenient for investigating not only soil and rock, but also groundwater and ground stress conditions. They provide samples for examination and testing, and also a hole that can be studied with periscopes, cameras, and downhole testing and logging devices, and later used to house monitoring instruments.

6.3.1.2 On-site and off-site observations.   Outcrops need not be on site to be of value. Off-site outcrops give even more useful information than on-site ones when the strata of interest are too deep to explore on site, but are brought to the surface at some distance by faulting, folding, or topographic variation. Often the most pertinent observations are those made at a distance from the site in quarries, road cuts, or natural outcrops, or in mines or tunnels. Remote observations are valid if lateral variations are either negligible or predictable, which is often the case. Many rock formations continue for hundreds of kilometers with little variation. Uniformity, or the lack of it, can usually be determined from published information.

6.3.1.3 Test pits and trenches.   Rock outcrop exposures can be supplemented by trenches and test pits that penetrate through shallow overburden soils to expose the bedrock surface, or even deeper to investigate the weathered upper layer of rock (Fig. 6.2). In many foundations this is the main method of determining the depth to a satisfactory bearing stratum. When exploring a dam foundation, a trench is commonly excavated along the full length of the dam axis to expose the abutments and the valley floor. Trenches provide excellent access for in situ testing and sampling, provided that they can be excavated deep enough to reach rock of the required quality for testing.

The depth limitation of a normal backhoe is usually about 4 m. Trenches excavated by bulldozer have no such depth limitation, provided that the walls remain stable or are shored or benched to allow safe inspection and logging, and there is sufficient space laterally for access. Most municipalities have rigorous safety codes for excavation practice, and these must be studied before undertaking field work.

Hatheway and Leighton (1979) describe trenching as the most definitive of all subsurface exploratory methods, revealing both obvious and subtle geological features. Many hazards that would otherwise remain undetected, such as active or potentially active faults (*capable* faults), can be mapped. The authors give a detailed field procedure for trenching and trench logging. In summary, they recommend that trenches must be judiciously located, survey-controlled, safely excavated, adequately shored, logged in detail, and properly diagnosed.

6.3.1.4 Adits and shafts. When justifiable, excellent rock exposures can be made available by driving exploratory shafts and adits, which penetrate deeper than trenches into unweathered and strong rock formations. They are most often used in conjunction with investigations for major projects, such as dams and underground power chambers. They not only provide access for the mapping of rock structures and rock quality variations, but can be expanded locally into test chambers for large-scale in situ testing, for the installation of monitoring instruments, and for rock improvement by grouting and anchoring. Often exploratory adits and shafts serve an important long-term purpose, such as for drainage or as safety exits.

Adits are excavated by blasting. Exploratory shafts can be blasted or bored by a large-diameter raise boring or blind shaft boring rig, or an auger rig as used for sinking caissons (Chaps. 14 and 15). Alternatively, they can be cored using a calyx-type drill at diameters typically from 1.0 to 1.8 m. The method is similar to that used for core drilling at smaller diameters, except that special wedging devices or blasting are needed to break the core from the bedrock. Exploratory shafts are

(a)

Figure 6.2 Exploration by test pit. (a) Backhoe excavation gives rapid penetration through soil overburden. (b) Example of a test pit log in soil and weathered rock. (*Courtesy of Geotechnical Control Office, Hong Kong Government.*)

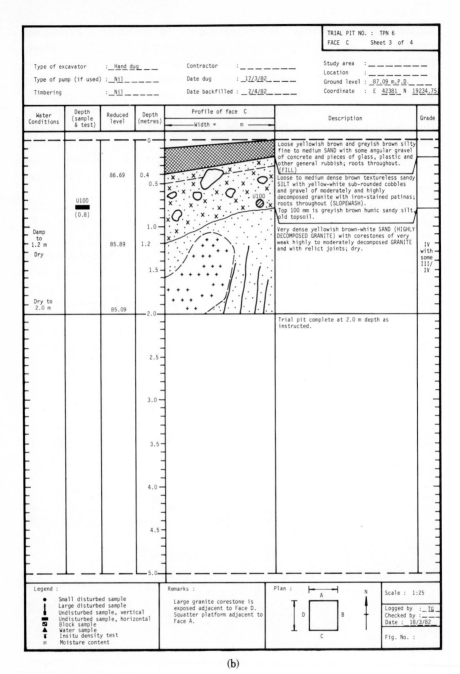

		TRIAL PIT NO. : TPN 6		
		FACE C Sheet 3 of 4		

Type of excavator   : _Hand dug_ _ _

Type of pump (if used) : _Nil_ _ _ _ _

Timbering   : _Nil_ _ _ _ _

Contractor   : _ _ _ _ _ _ _ _

Date dug   : _17/3/82_ _ _ _

Date backfilled : _2/4/82_ _ _ _ _

Study area   : _ _ _ _ _ _ _ _
Location   : _ _ _ _ _ _ _ _
Ground level : _87.09 m.P.D._ _ _ _
Coordinate : E _42381_ N _19234.75_

Water Conditions	Depth (sample & test)	Reduced level	Depth (metres)	Profile of face C —Width = m—	Description	Grade
		86.69	0.4 0.5		Loose yellowish brown and greyish brown silty fine to medium SAND with some angular gravel of concrete and pieces of glass, plastic and other general rubbish; roots throughout. (FILL)	
	U100 (0.8)		1.0		Loose to medium dense brown textureless sandy SILT with yellow-white sub-rounded cobbles and gravel of moderately and highly decomposed granite with iron-stained patinas; roots throughout (SLOPEWASH). Top 100 mm is greyish brown humic sandy silt, old topsoil.	
Damp to 1.2 m Dry		85.89	1.2 1.5		Very dense yellowish brown-white SAND (HIGHLY DECOMPOSED GRANITE) with corestones of very weak highly to moderately decomposed GRANITE and with relict joints; dry.	IV with some III/ IV
Dry to 2.0 m		85.09	2.0		Trial pit complete at 2.0 m depth as instructed.	
			2.5			
			3.0			
			3.5			
			4.0			
			4.5			
			5.0			

Legend :
• Small disturbed sample
| Large disturbed sample
| Undisturbed sample, vertical
▬ Undisturbed sample, horizontal
▣ Block sample
▲ Water sample
Ⓧ Insitu density test
m Moisture content

Remarks :
Large granite corestone is exposed adjacent to Face D. Squatter platform adjacent to Face A.

Plan :

Scale : 1:25

Logged by : _TG_
Checked by : _ _ _
Date : _18/3/82_
Fig. No. :

(b)

Figure 6.2   (*Continued*)

usable only if groundwater inflow can be controlled, and if the shaft walls remain stable and safe enough to inspect. They are often cased where they pass through soil and weathered, weak bedrock. Windows can be opened in the casing to permit inspection, testing, and sampling of the weaker zones.

6.3.2 Observations at rock exposures

6.3.2.1 Geotechnical mapping of exposures.
The characteristics of rock exposed in outcrops and trenches are recorded in the form of logs, such as shown in Fig. 6.2. Underground rock conditions are recorded on an expanded sketch that shows the crown, haunches, and invert of a tunnel or drift (Proctor, 1971), and the north point in the case of a shaft (Fig. 6.3).

The exposure is made safe and clean, usually by scaling and jetting. Jointing patterns are sketched or photographed when the face has partially dried, when the cracks are wetter and darker. The boundaries of GMUs are marked on the maps, and each is given a name or number. Characteristics of each GMU are then tabulated in a systematic sequence, as described in Chaps. 2 and 3. The intact rock description gives first the rock name (in capitals) and then its color, texture, grain size, and mineral composition, followed by mechanical index data such as strength and durability. The jointing description gives

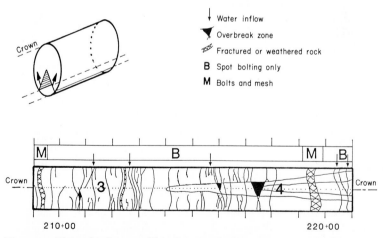

Figure 6.3 Example of a tunnel log. Jointing, bedding planes, and features such as overbreak and seepage are plotted on an opened out view of the tunnel. Conventionally, the view looks upward, with the direction of drive to the right. Top and bottom limits of the mapped strip correspond to 7 o'clock and 5 o'clock lines. The invert cannot be mapped, being always obscured by debris. Geotechnical mapping units (GMUs) are here given a number designation. (*After Nelson and O'Rourke, 1983.*)

first the bulk characteristics of block size and the number and types of joint sets, and then, for each set, its orientation, spacing, roughness, persistence, filling, and water seepage characteristics. When time permits, index measurements are made to supplement the visual record.

A GMU is defined as a volume of rock, or area on the log, that is sufficiently uniform in both its material and jointing characteristics to be given a single description (Franklin, 1986). Depending on the scale of mapping, a single *mixed* GMU sometimes contains more than one rock type, such as a thinly interbedded sequence of sandstones and shales. If so, its description starts with a statement naming the rock members, their percentages, and typical thicknesses. Each rock type is then described separately.

The log is completed with further information on water inflows (locations, rates, and, when applicable, stabilized levels) and on the stability of the excavation walls. It should also state the ease or difficulty with which a trench or pit was excavated.

A pocket dictating machine and a Polaroid or video camera are recommended for rapid recording of conditions. Bulldozers and backhoes uncover a lot of ground very quickly, and under the usual constraints of an investigation, need to be kept moving from place to place. Exposures often must be shotcreted or backfilled soon after excavating to prevent them from collapsing.

6.3.2.2 Sampling of jointed rock. Samples are usually taken for laboratory measurements of mechanical properties, if not from each exposure, at least from each GMU on site. Disturbance can be minimized by special techniques to take and preserve large samples that contain the joints and soft seams of interest. Block samples of hard, intact rocks can usually be obtained without any special precautions by prying the rocks loose from exposed faces. They can be shipped to the laboratory where they are cored to provide test specimens.

More fragile, jointed, or weak samples are seldom found loose in sufficient size, and have to be removed from the rock mass using one of several alternative methods. Soft rocks such as shale can be cut by a tungsten carbide tipped chain saw (Fig. 6.4), whereas block samples of harder rocks are prepared by either line drilling around the block perimeter, or sawing with a circular diamond saw blade. Line drilling tends to damage the sample, and a circular saw can cut only as deep as the radius of the blade. Wire sawing, in which an abrasive powder is fed in a water slurry along a continuous wire loop, is an alternative in rocks of moderate strength such as slate or limestone and requires several predrilled holes to give the wire saw access.

Large-diameter calyx coring is an excellent sampling technique in rocks of any strength. A rockbolt or dowel is usually installed central

(a)

(b)

Figure 6.4 A tungsten carbide tipped chain saw can be used to cut a sample of soft rock for testing, in this case a shale.

to the core or block before the start of drilling or cutting, to allow the block to be lifted and to prevent it from separating along joints. Particular care is needed when packing and shipping samples to minimize disturbance caused by vibrations, drying, or exposure to frost. The samples are often waxed, then wrapped in aluminum foil, then packaged in chips of plastic foam to separate them from the container walls. Stimpson et al. (1970) describe a method for encapsulating samples in polyurethane foam to give them extra protection (see also Sec. 11.2.2).

6.3.3   Fault activity and seismic risk

6.3.3.1   Assessing fault activity.   Faults and shears, when identified beneath structures such as concrete dams and nuclear power stations, require a special assessment to evaluate their potential for future activity (shear displacement), their effect on foundation stiffness, and the need for remedial treatment. Potential activity may also need to be assessed for faults that intersect tunnels or caverns.

According to the U.S. Bureau of Reclamation documentation on fault activity and assessment criteria for dam foundations, a fault is "inactive" only if it can be proven not to have suffered a shear displacement during the preceding 10,000-year period. Faults are classified according to evidence which may include marker beds of known age that traverse the fault plane, offsets, scarps, and historical or recorded seismic activity. However, strong earthquakes along a fault tend to recur in cycles of several hundred years with periods of quiescence in between. Hence the absence of earthquakes for several hundred years does not provide assurance that the fault is inactive.

Murphy et al. (1979) describe techniques for dating faults to determine the most recent fault movement. These include stratigraphy, paleontology, and structural analysis, as well as laboratory radiometric methods such as carbon-14, potassium-argon, rubidium-strontium, and uranium-thorium methods of age dating.

In the nuclear power industry, the potential for movement to recur along a fault has been based on the history of faulting within the past 500,000 years. In regulatory jargon this is defined as the *capability* of the fault; U.S. Nuclear Regulatory Commission (NRC) (1978) criteria define a capable fault as one that has exhibited one or more of the following characteristics.

Movement at or near the ground surface at least once within the past 35,000 years or movement of a recurring nature within the past 500,000 years

Macroseismicity instrumentally determined with records of sufficient precision to demonstrate a direct relationship with the fault

A structural relationship between a fault and a capable fault according to the above characteristics such that movement on the capable fault could be reasonably expected to be accompanied by movement on the other fault

6.3.3.2 Seismic risk. Different categories of engineering structures present different levels of risk and consequences of failure when subjected to earthquake motions. Engineering designs must often take into account the potential hazard of an earthquake occurring during the life of a structure.

Seismic zones within the United States and Canada have been mapped and are reported in the literature (e.g., U.S. Department of the Army, 1977; Canadian Geotechnical Society, 1985). Seismic zone designations in the United States, which differ somewhat from those in Canada, are tabulated as follows together with associated damage designation and seismic coefficients:

Zone	Damage	Coefficient
0	None	0
1	Minor	0.025
2	Moderate	0.05
3	Major	0.10
4	Great	0.15

The selection of earthquake motions for use in engineering designs depends on the type of engineering analysis to be performed (Krinitzsky and Marcuson, 1983). An *equivalent static load analysis* treats the earthquake loading as an inertial force applied statically to the center of mass. The magnitude of this force is the product of the structural mass and the seismic coefficient. Historically, seismic coefficients have been chosen by structural engineers on the basis of experience and judgment. They can be obtained from maps of the seismic activity of an area, and require no separate geological or seismologic investigations.

In the equivalent static load method, the weight, usually of a dam, is multiplied by the seismic coefficient, and this is applied as a horizontal force at the center of gravity of the section or element. In *dynamic analysis*, accelerograms are assigned to a site from typical response spectra. Peak amplitudes and ground velocities are obtained from several methods that involve magnitude of earthquake, distance from source, and corresponding motions; or they may be assigned on the basis of recorded earthquake intensity, usually using the Mercalli scale.

The abridged Mercalli intensity scale is shown in Table 6.2. For most of the world, the historic data are available only as intensities. Relationships between modified Mercalli intensity and earthquake acceleration, velocity, and duration are given in Krinitzsky and Marcuson (1983).

TABLE 6.2   Modified Mercalli Intensity Scale of 1931

Intensity	Damage
I	Not felt except by a very few under especially favorable circumstances.
II	Felt only by a few persons at rest, especially on upper floors of buildings. Delicately suspended objects may swing.
III	Felt quite noticeably indoors, especially on upper floors of buildings, but many people do not recognize it as an earthquake. Standing motor cars may rock slightly. Vibration like passing of truck. Duration estimated.
IV	During the day felt indoors by many, outdoors by few. At night some awakened. Dishes, windows, doors disturbed; walls made cracking sound. Sensation like heavy truck striking building. Standing motor cars rocked noticeably.
V	Felt by nearly everyone, many awakened. Some dishes, windows, etc., broken; a few instances of cracked plaster; unstable objects overturned. Disturbance of trees, poles, and other tall objects sometimes noticed. Pendulum clocks may stop.
VI	Felt by all, many frightened and run outdoors. Some heavy furniture moved; a few instances of fallen plaster or damaged chimneys. Damage slight.
VII	Everybody runs outdoors. Damage negligible in buildings of good design and construction; slight to moderate in well-built ordinary structures; considerable in poorly built or badly designed structures; some chimneys broken. Noticed by persons driving motor cars.
VIII	Damage slight in specially designed structures; considerable in ordinary substantial buildings with partial collapse: great in poorly built structures. Panel walls thrown out of frame structures. Fall of chimneys, factory stacks, columns, monuments, walls. Heavy furniture overturned. Sand and mud ejected in small amounts; changes in well water. Disturbed persons driving motor cars.
IX	Damage considerable in specially designed structures; well-designed frame structures thrown out of plumb; great in substantial buildings, with partial collapse. Buildings shifted off foundations. Ground cracked conspicuously. Underground pipes broken.
X	Some well-built wooden structures destroyed; most masonry and frame structures destroyed with foundations; ground badly cracked. Rails bent; landslides considerable from river banks and steep slopes; shifted sand and mud; water splashed (slopped) over banks.
XI	Few if any (masonry) structures remain standing; bridges destroyed. Broad fissures in ground; underground pipe lines completely out of service. Earth slumps and land slips in soft ground. Rails bent greatly.
XII	Damage total. Waves seen on ground surfaces; lines of sight and level distorted. Objects thrown upward into the air.

SOURCE: Abridged from Krinitzsky (1986).

6.4   Use of Geophysics in Site Investigation

6.4.1   Overview of methods

The range of geophysical techniques available is extensive and growing (Telford et al., 1976; U.S. Department of the Army, 1979; Griffiths and King, 1981; Keary and Brooks, 1984) . Several of the seismic, electric, magnetic, and gravimetric methods originally developed for mineral and oil prospecting have been successfully adapted for geotechnical engineering applications and are now being used increasingly.

Geophysical techniques have two applications, first to map structural or lithological boundaries, and second to give index property measurements related to the mechanical character of materials within each rock or soil unit. The application to rock characterization has been outlined in Chap. 3, and in the present chapter we focus on mapping applications only. Rock quality information is, however, obtained as a by-product, even when the main purpose is mapping.

Geophysical methods have a special application in the detection of underground cavities such as abandoned mining excavations and solution caverns in karstic limestones. Belesky and Hardy (1986), for example, find that gravity, electric resistivity, radar (Fig. 6.5), and seismic techniques are quite reliable in karst, whereas the alternative of drilling a large number of holes is costly and time-consuming. Extrapolation between borings is difficult because of the complex configuration of interconnecting cavities.

Many surface and downhole geophysical methods are similar, although the equipment has been miniaturized and modified for downhole use. Downhole tools (Sec. 6.6.2) transmit and detect over shorter distances, so they can measure ground characteristics with greater resolution but on a smaller scale.

6.4.2   Seismic methods

6.4.2.1   Basis for seismic measurements. Geophysical seismic measurements take advantage of two phenomena, that sound waves travel at different velocities in different rocks, and that they are reflected and refracted by beds of contrasting acoustic impedance. The source of sound (transmitter) can be a sledgehammer, a special shotgun, or a small explosive charge. The receiver can be a piezoelectric transducer or an electromagnetic geophone. Piezoelectric devices transmit and detect high frequencies over short distances, usually 1 m at most, and are used mainly in the laboratory and occasionally in borehole logging tools. An electronic signal proportional to the particle velocity or ac-

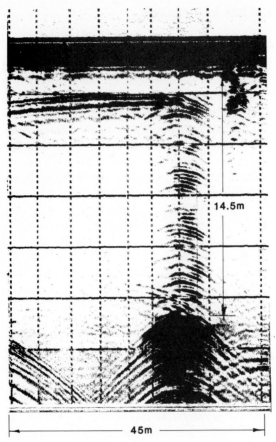

14.5m

45m

Figure 6.5 Image produced by ground-probing radar, revealing at a depth of 14.5 m the top of a limestone cave; Yarangobilly cave system, southeastern New South Wales, Australia. (*Courtesy of Dr. A. F. Siggins, CSIRO Division of Geomechanics.*)

celeration of the rock is displayed on an oscilloscope, recorded on paper, or stored digitally on computer tape.

Wave velocity is the distance the sound wave travels divided by the time of travel. High velocities (3000–6000 m/s) are associated with competent, saturated, and sparsely jointed rocks. Unsaturated porous rocks and shattered or deeply weathered rocks have low velocities (less than 3000 m/s). Usually the compressional sound-wave (P-wave) travel time is measured, but the shear-wave (S-wave) travel time can also be helpful when attempting to correlate with other soil and rock properties.

Reflected and refracted sound waves can be used to map interfaces between materials of different sonic properties, such as the soil-bedrock interface, and the boundaries between weathered and fresh rock and underlying strata. The amount of energy reflected and the angle of refraction depend on the *acoustic impedance contrast* across the boundary. Attenuation, the decay of sonic wave amplitude and an associated change of frequency content, can be a serious problem for shallow surface surveys using low-energy sources. However, more recent and advanced analytic techniques such as tomography (Sec. 6.4.2.3) are beginning to use the attenuation characteristics of strata as another source of valuable information.

6.4.2.2 Refraction and reflection. Refraction seismic surveys, the most common in geotechnical applications, measure the velocity of sound that travels by direct or critically refracted paths from transmitter to receiver. Reflection surveys, which are more costly and require more complicated data processing, rely on signals being reflected off a layer and then detected at surface.

The reflection method is used conventionally to explore deep geologic structures such as voids at depths of up to 300 m, and fault zones and stratum interfaces at much greater depth. Hunter et al. (1984), however, report that seismic reflection mapping can work well at much shallower depths, and have used seismic reflection to map the soil/rock contact as well as structures within 20-m-thick overburden. They suggest using a 12-channel enhancement seismograph and a hammer source. Research at the University of Waterloo, Ont., Canada, using seismic reflection has given spectacular detail below depths of 8 m under the right conditions, and in the future could replace seismic refraction methods in many engineering applications (J. Greenhouse, personal communication).

6.4.2.3 Seismic tomography. Seismic tomography (a concept similar to x-ray tomography in the medical field), offers a means for developing a map of the distribution of seismic velocity and attenuation values below surface (Gordon et al., 1970; Wong et al., 1983). Seismic transmitters and receivers are arranged to give a large number of path directions (*rays*) passing through the zone of interest. Combinations of transmitter-receiver layouts can be used: underground to surface, surface to underground, and between underground locations. Data are converted into a three-dimensional map of the ground using an algorithm that reconstructs the spatial distribution and boundaries of soils and rocks with different velocities or attenuations.

6.4.3   Seismic instruments

6.4.3.1   Single-channel seismographs.   Hammer seismographs use a sound wave generated by a sledgehammer blow to the rock surface or to a steel plate pressed into soil (Fig. 6.6a). Travel time is measured by a counter in the recording instrument, which is triggered by a switch fixed to the hammer shaft, and stopped when the sound wave reaches the receiver. The method is quick and convenient for small-scale engineering surveys, avoiding the problems associated with the use of explosives.

Some instruments display the complete waveform of arrivals on a cathode ray tube which may be photographed, digitized, or recorded on paper tape for a permanent record (Fig. 6.6c), whereas others give direct digital readings of first arrival time only. Waveform display is more expensive but superior. It has the advantage of allowing the user to choose the instant of first arrival, useful when the waveform is less than sharp, and also to determine the arrival time for the S-wave component. *Signal enhancement* models allow the waveform to be sharpened by averaging the signals from repeated hammer blows, a definite advantage when taking measurements through soil or weathered rocks that tend to produce a noisy waveform with a poorly defined first arrival.

6.4.3.2   Multichannel seismographs.   The portable multichannel equipment most often used in refraction surveys for engineering applications consists of an amplifying and a recording unit powered by batteries, with an array of 12 or 24 geophones spaced at between 5 and 15 m and connected to a multiple-channel magnetic tape or chart recorder (Fig. 6.6b). The seismic signal is usually generated by a small explosive charge or an impactor. Multiple-channel systems with an explosive source are the best for high-quality mapping; they are faster and provide superior resolution to single-channel methods. Single waveforms can be compared with each other to differentiate P, S, and other arrivals. Sound attenuation can also be examined as an aid to interpretation. Usually the array of waveforms is plotted so that reflecting beds stand out sharply in approximately the correct geometrical positions. Their true positions can be obtained by further processing of the seismic data.

6.4.4   Other geophysical techniques

In geotechnical applications, resistivity, magnetometry, and gravimetry are used less often than seismic surveys, but are helpful either individually or in combination to assist in certain types of mapping.

(a)

(b)

Figure 6.6 Seismic exploration methods. (*a*) Hammer seismic method. (*b*) Multichannel seismograph (Oyo Corp. instrument). (*c*) Example of seismic refraction trace using the Oyo seismograph with a 1-m geophone spacing. Travel time for arrivals at the nearest four geophones was 15 ms, giving a velocity in the upper (soil) layer of 0.27 km/s. Velocity in the lower (rock) layer was 1.32 km/s. The interface is at a depth of 1.8 m.

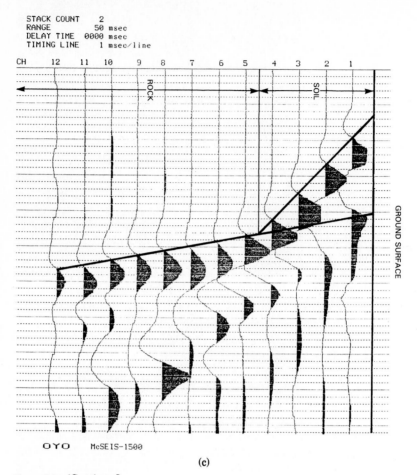

```
STACK COUNT    2
RANGE         50 msec
DELAY TIME  0000 msec
TIMING LINE    1 msec/line
```

(c)

Figure 6.6 *(Continued)*

6.4.4.1 Resistivity and conductivity. Conductivity of the ground, its ability to transmit electric currents, depends largely on the water content; thus soils are nearly always more conductive than rocks, and saturated clays are more conductive than dry sands. Changes in pore-water salinity also have great impact on ground conductivity. Measurements of ground conductivity or its inverse, resistivity, can be used to delineate the boundaries between soil and rock, and between soil or rock strata of different water content. The earth's resistivity varies over 12 orders of magnitude, from $1 \times 10^7 \, \Omega \cdot$ m in dry gabbro to $1 \times 10^{-5} \, \Omega \cdot$ m in metallic deposits. The easiest strata to define are those having electrical properties, such as resistivity, natural potential, polarization potential, and magnetic permeability, that are distinctly different from those of surrounding materials (i.e., high contrast).

The conventional methods of resistivity measurement use four steel or brass stakes (electrodes) driven into the ground. An electric current is passed between the outer pair of electrodes, and the voltage drop across the other pair is a measure of the electric resistance of the ground. By progressively increasing the separation between electrodes, it becomes possible to penetrate to increasing depths below surface until, for example, the depth of bedrock below the soil cover is detected by an increase in the resistivity values.

Recently developed electromagnetic (EM) methods permit a more rapid measurement of conductivity without electrodes and using portable equipment. One such instrument, the Geonics EM-34, consists of two wire-wound coils, one of which acts as a transmitter and the other as a receiver of electromagnetic radiation (Fig. 6.7). Transmission of the electromagnetic waves through the ground depends on the ground's conductivity, so conductivity may be measured as a function of attenuation between transmitted and received signals. The portability of this instrument makes it particularly useful for contouring the bedrock beneath the soil in remote and extensive areas.

Another form of EM instrument, the very low frequency (VLF) device, because it consists of a receiver only, is even more portable. It detects the local polarization of the 10–30-kHz signals provided by submarine guidance stations around the globe.

6.4.4.2 Ground-penetrating radar. Ground-penetrating radar is like conventional impulse radar in that it operates at frequencies of 100–500 MHz, except that the electromagnetic energy is transmitted through soil and rock instead of air. Energy is backscattered from the boundaries of strata that have a marked contrast in dielectric constant. The relative dielectric constant of air is unity, and that of most minerals is less than 10, whereas the dielectric constant of fresh water is about 100. Therefore water is extremely "opaque" to radar. A groundwater table can be detected with great precision, but a high water table may prevent exploration at greater depth.

This is a quick, high-resolution method for locating cavities such as karsts (Church and Webb, 1986; Belesky and Hardy, 1986). Figure 6.5 shows successful use of the method even as deep as 14.5 m in limestone, in spite of a water table at 3 m. Other applications include the mapping of salt domes, because salt is very dry and therefore transparent to radar, whereas the surrounding rocks contain water. Radar can detect and locate individual water-filled or clay-filled joints and faults, weathering fronts, and lithological contacts such as between igneous and sedimentary rock. It was used successfully for site investigation at a Washington, D.C., subway station to determine a three-dimensional structural picture (Rubin and Fowler, 1978).

(a)

(b)

Figure 6.7 Electromagnetic exploration to determine depths to bedrock for town site development in northern Ontario. Instrument is the Geonics EM34-2. (*a*) Calibration on a bedrock outcrop. (*b*) Running a conductivity traverse along a proposed road alignment to assess the rock to be blasted.

6.4.4.3 Gravimetry and magnetometry. Measurements of anomalies in the earth's gravitational and natural magnetic fields are most useful in mineral exploration, where gravitational and magnetic anomalies may result from concentrations of heavy or conducting minerals such as nickel ore or lead-zinc ore. Used on a smaller scale, however, the techniques have some application in site investigation.

Gravimeters make relative measurements of the vertical component of the earth's gravitational field. Given the right combination of conditions, they can detect soil-filled channels in bedrock, shallow mine openings, potholes, and other forms of near-surface cavity. A gravity anomaly, identified by contouring the gravity values, is detectable only when the cavity is empty or loosely filled and has a diameter nearly as large as its depth below surface. Cavities deeper than about 30 m, particularly when filled, are difficult to detect even with more precise microgravity methods.

6.5 Exploratory Drilling

Drilling and coring methods, and percussive and rotary techniques used in well drilling and rock blasting, are discussed in detail in Chap. 14. Here the methods are summarized, with further detail on how core is logged and on how drilling, probing, and sampling are used as part of a site investigation.

6.5.1 Boring and probing

6.5.1.1 Soil boring and penetration testing methods. Soil boring, testing, and sampling methods are used in the upper part of an exploratory rock drillhole, which usually passes through soil, and also in some soft or weathered rock formations, which are best bored and sampled as though they were stiff clays or dense sands.

The *standard penetration test* (SPT) used in routine soil investigations also has applications for field evaluation of soft or weathered rocks. A 51-mm O.D. split spoon sampler is driven into the ground by a 63.5-kg hollow-centered cylindrical falling weight or "hammer" that slides along the drill rods for a free-fall height of 760 mm. By counting the blows needed to drive the sampler into the ground, one obtains an estimate of the driving resistance and hence of in situ strength. In the standard test, the sampler is driven through three successive intervals of 150 mm. The first 150-mm blow count is ignored because the sampler is usually passing through soft soil debris fallen from higher up the hole. The remaining two 150-mm counts are added to give a blow count for the lower 300 mm, which is called the SPT N index. A loose sand has $N = 4$–10, whereas a very dense sand has $N > 50$. SPT N values have been correlated with parameters such as foundation bearing capacity. They are usually plotted in the form of a log showing the variations that occur in strata penetrated by the sampling tool.

The SPT log is intermittent because blow count readings can be taken only while driving the split spoon, which when full must be recovered to remove the sample. A quicker alternative *dynamic cone-penetrometer test* is available for situations where the blow counts are more important than the samples. A solid cone of 10-cm^2 cross section is driven into the ground in place of the split spoon. Blow counts are recorded without stopping to take samples. A continuous profile of penetration resistance is thus obtained.

6.5.1.2 Open-hole percussive probing. *Open-hole* drilling produces a hole without samples. Water-well rotary drills can be used for this purpose, or the air-track equipment whose main use is for the drilling of blastholes. Air-track drills can penetrate much more rapidly through soil and rock than core drills. In site investigation they are useful to "probe" through soil to detect and map the contours of the bedrock surface, and to test the bedrock for hidden soft seams and cavities.

For bedrock contour mapping, large numbers of probes are often driven on a regular grid pattern, whereas for foundation probing, only one or two probes per footing may be needed. Cavities and soft layers become evident from sudden increases in the penetration rate, which is monitored continuously. The engineer or technician supervising the probing uses a stopwatch and marks the drill rod with a crayon at the required intervals of depth.

A layer of boulders or loose, slabby rock commonly separates soil from the more competent rock horizons. These loose blocks can be mistaken for solid bedrock unless precautions are taken to drill through them and carefully detect underlying soft zones. In such situations, the technique of percussive probing can be improved by the acoustic sounding method (Stimpson, 1976). A geophone is installed in rock in a hole central to the pattern of probing, to pick up the sound generated at the drill bit. The amplitude of transmitted sound increases suddenly when the probe drill penetrates through the soil and hits the bedrock surface.

6.5.1.3 Horizontal probing. Horizontal drilling is particularly useful for tunnel exploration beneath mountainous terrain, where the cost of drilling from the surface can be very high. The exploratory probes can be either predrilled before the start of tunneling, or driven ahead of the tunnel face during construction. Open-hole water-flush rotary drilling with a noncoring bit is normally used. In the absence of core, the rock quality is inferred from characteristics of the cuttings in the return water, from the drillers' observations of the behavior of the drill, and from video inspection and geophysical logging of the drillhole walls. Parameters that can be measured by instrumenting the drill include thrust, penetration rate, rotary speed, and water pressure (Majtenyi, 1976).

6.5.2 Core drilling

6.5.2.1 Rock coring. The most common and convenient method for obtaining rock samples for examination and laboratory testing is by core drilling using a coring bit impregnated with diamonds. Various arrangements are possible (Fig. 6.8). Full-sized drill rigs can be mounted on trucks or tracked all-terrain vehicles, or on skids for towing or winching into position. Lightweight drills can be transported by backpack or air-lifted to explore terrain that would otherwise be inaccessible. Exploration over water is accomplished by drilling from platforms or spud-mounted barges, or from the winter ice in northern latitudes.

The core diameter should be as large as possible, BX (42 mm) or preferably NX (54.7 mm), unless the rock is particularly strong and massive. Small diameters, for example, the AX (30.1 mm) and EX (21.5 mm) sizes that are standard in mining, are more susceptible to fracturing as a result of core barrel vibrations and should be discouraged in geotechnical engineering applications. Large-diameter drilling usually produces core sticks in greater lengths, with greater recovery and with a reduced likelihood of mechanical damage.

Special precautions are needed when the rock is fragile, to protect the core from mechanical disturbance by the rotating drill barrel and from the erosive action of drilling water. For routine sampling, a double-tube core barrel is used in which the inner tube retains the core and remains stationary while the outer tube rotates around it. Even with this precaution, disturbance and sometimes damage to the core is unavoidable because of the need to remove it from the bit end of the barrel by hammering the barrel when the core gets stuck or, preferably, by extruding the core using water pressure.

Triple-tube core barrels overcome this problem and are used as an added precaution when the rock is very soft or fractured, and when the core must be reconstructed to record the in situ pattern of jointing. These contain an inner split cylinder or PVC liner that is removed together with the rock core sample (Chap. 14).

Equipment must be selected carefully, but the skill of the drill operators is particularly important in obtaining good-quality core for

(a)   (b)

(c)

Figure 6.8 Core drilling. (*a*) Typical drill rig. Site investigation, La Paz, Bolivia. (*b*) Lightweight rig mobilized by helicopter. (*c*) Small drilling platform for offshore exploration. (*Courtesy of Longyear Canada, Inc.*)

logging and testing. The variables of drilling, applied thrust, rotational speed, and flush water pressure must be selected and monitored and adjusted continuously to give optimum results depending on the ground through which the drill is penetrating.

6.5.2.2 Planning the pattern of drillholes. Because they represent a substantial proportion of the cost of site investigation, coreholes should be placed where they are likely to be most useful, and only to the depth needed. Regular grids of holes are inappropriate except where the geology is completely unknown, which is seldom the case by the time preliminary studies have been completed. The rock engineer must carefully consider the geology and groundwater conditions of the site, and the pattern of stresses to be imposed on the ground by the structure or excavation.

The total length of coring that can be done is usually known, within limits, in terms of the available site investigation budget and precedents for similar work in ground of similar complexity. The overall budget is first divided between surface and subsurface components of the investigation, using judgment to decide which elements are likely to bring the greatest benefits. Drill rig mobilization costs are subtracted from the drilling budget, and the balance is divided by an estimate of unit price per meter of drilling to obtain a first estimate of the total length that can be cored. Comparison with requirements leads to an optimization of drilling locations, depths, spacings, and numbers of holes. Readjustments are common as data are collected, and to accommodate sampling, downhole testing, and the installation of piezometers and other instruments.

6.5.2.3 Inclined and horizontal coring. Vertical drilling is best when the boundaries and discontinuities are near-horizontal, but gives very poor results when structures are steeply dipping or vertical. In the extreme, a single vertical drillhole can pass entirely through one bed of an upturned sedimentary sequence without exploring the others, or may pass alongside and parallel to a fault that it was meant to investigate. At least some holes should be aligned near-perpendicular to the planar features of interest, even though this will usually mean increasing the total cored length.

Horizontal coring is an excellent way to explore along the line of a proposed tunnel. Particularly in rugged topography, drilling from the surface is expensive, and much drill footage is wasted before arriving at the depth of interest. Barr and Brown (1983) used a fully instrumented, lightweight, portable hydraulic diamond drill experimentally in an underground limestone quarry. They report that this Atlas Copco Diamec 250 rig can core horizontal holes 56 mm in diameter and up to 150 m long using aluminum rods. Geotechnical data obtained in this research included rock strength indexes and the detection of open, clay, or gouge-filled and water-bearing discontinuities.

Much greater lengths of horizontal coring and probing are possible using heavier, specially instrumented equipment (Chap. 14). Horizontal drillholes up to 1.1 km long were cored for the U.S. Atomic Energy Commission at Mercury, Nev., using conventional wire-line methods, a 6-m-long NX core barrel, and an air-powered hydraulic core drilling rig with a 610-mm feed stroke. Holes were surveyed with a Sperry-Sun single-shot and multishot survey instrument. Directional control varied from fair to very good. Several 800-m-long holes terminated within 1–2 m of a 3-m-wide target. Similar distances have been achieved in the Athabasca Oil Sands, drilling in from a river valley wall.

6.5.3 Core orientation

Core tends to rotate during drilling, and only with special precautions can the in situ directional characteristics of joints be determined from the core. These aids to *core orientation* are distinct from the techniques for surveying the alignment and curvature of the drillhole itself, discussed in Chap. 14.

The simplest method is to try to reconstruct the core in the core box. Provided that the recovery is 100%, and with care and sufficient time, the geotechnical technician can piece together the core sticks like a jigsaw puzzle. This becomes difficult when one or more joints are perpendicular to the direction of drilling, as with horizontal bedding in vertical core, because it then becomes difficult to determine whether the core has been rotated.

This problem may be overcome by using a special nonrotating scribing core barrel, such as the *Craelius core barrel,* which scratches a straight line along one side of the core close to the bit, and before the core has had a chance to rotate. The scribe marks are aligned when reassembling the core, and the azimuth of the scribe line is noted.

Another method, the *integral coring technique,* was developed by the National Civil Engineering Laboratory (LNEC) in Portugal (Rocha and Barroso, 1971). A small-diameter hole is drilled first, into which a reinforcing rod is grouted to hold the broken core together during large-core drilling. The main drillhole is then advanced to overcore the reinforced rock. The integral sampling method works well even in sheared rocks containing substantial amounts of clay and weathering products.

6.5.4 Supervision and logging

6.5.4.1 Drill supervision. The drilling contractor usually provides the drilling equipment together with a two-person crew consisting of driller and helper. The driller's responsibility is to obtain high-quality samples and maximum recovery. The driller is not a soil or rock engineer, and is rarely familiar with the geology of the site or the requirements of the project. Therefore a third member of the team, a drilling inspector, directs the drilling program, maintains the quality of sampling, records drilling progress, and logs the core. This person is usually a qualified geotechnical technician or junior engineer.

6.5.4.2 Drilling log. The drilling log, not to be confused with the core log, records details of the drilling operations required for an adequate interpretation of the ground characteristics. An accurate record of depths is a top priority. Drill rod lengths are measured and counted as they are added to the drill string. The inspector keeps track of the depth of the drill bit at all times, and records on the log the depths at the start and end of each core run. Sometimes not all the drill core is recovered when the core barrel is withdrawn from the hole. The remnant core has to be redrilled at the start of the next run, a potential source of confusion in the recording of depths.

As core is transferred from the barrel to the box, core recovery is measured and recorded as a percentage of the length drilled in each run. Recovery is affected by both the quality of the rock and the quality of drilling. Remarks in the drilling log should note zones where core may have been lost or excessively broken because of drilling problems such as vibration or loss of water circulation.

Stabilized groundwater levels can only be recorded several hours or days after the completion of the drillhole, because the injection of drill flushing water results in temporary mounding of the water table. Nevertheless, water levels in the hole are recorded at the start and end of each day of drilling as a general guide to rock mass drainage characteristics. Changes in the color of the drill water are indicative of changes in geology. Elevations where water circulation is suddenly lost are recorded as an indication of the presence of fractures or cavities. If the hole "makes water," the drill bit has intercepted an aquifer under artesian pressure; the elevation of the top of the artesian aquifer is noted on the drilling log.

6.5.4.3 Core handling, preservation, and storage. Core can look different and acquire different mechanical characteristics if it is mishandled, poorly packaged, or subjected to a prolonged period of unprotected storage. It should be removed as gently as possible from the barrel, then reassembled, matching and fitting together the pieces (Fig. 6.9). A sturdy core box is placed on trestles at a convenient working height. Core is loaded into the box from left to right, top to bottom, so that it "reads like a book," with higher elevations toward the top left. Marker blocks, clearly labeled to show the depth, are placed to separate each core run from the next and at the start and end of the box. Labeled spacer blocks are inserted to fill gaps and to replace samples removed for testing. The completed box is labeled to show the project and drillhole numbers, depth interval, and date of drilling. Identical labels are needed on the inside and outside of the lid and on the box end, depending on the method of storage. A label should be visible without having to lift or move boxes that are stored in racks.

Core deteriorates when it is dried and rewetted, and water-sensitive core to be sampled later for test specimens must be kept moist. During logging, it should be protected by intermittent spraying with water. Sections of core not being currently examined should be covered with a damp cloth. Sections of core about to be examined should be carefully washed to reveal the rock fabric. Color photographs are usually taken at this time, while the core is still moist, and before sampling or strength testing. Photography requires good lighting, clear labeling, and depth marks or a scale. Several boxes from a single drillhole may be photographed alongside each other to fill the field of view. The photograph should, if possible, be taken at right angles to the boxes, by standing them on edge or by shooting down on them from an elevated position.

6.5.4.4 Logging of drill core. The approach to logging is similar, whether describing a drill core or a rock exposure (Sec. 6.3.2.1), except that in a core log some properties such as persistence cannot be recorded at all, whereas for others, such as spacing, only partial information can usually be obtained.

GMU boundaries are marked, and each GMU is given a description in terms of intact rock and jointing (Chaps. 2 and 3). The log is completed by transferring information obtained during drilling and from downhole measurements when performed, including the depths at which water was first encountered and any locations of lost circulation. When available, stabilized groundwater levels are transferred from the drilling log, together with notes concerning drilling problems such as unusually rapid or slow rates of drill bit penetration. A typical core log is shown in Fig. 6.10.

6.5.4.5 Sampling from core box. When logging is complete, samples for testing are selected and a record is taken of depths, sample numbers, and descriptions. The samples themselves are marked clearly

(a)

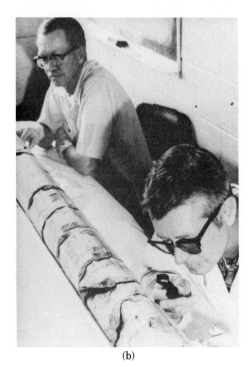

(b)

Figure 6.9 Core logging. (a) Core box photography. The core pieces have been carefully rearranged, the spacer blocks labeled, and the core washed and allowed to partially dry to reveal shaley layers. (b) Inspecting and logging of oil-field core.

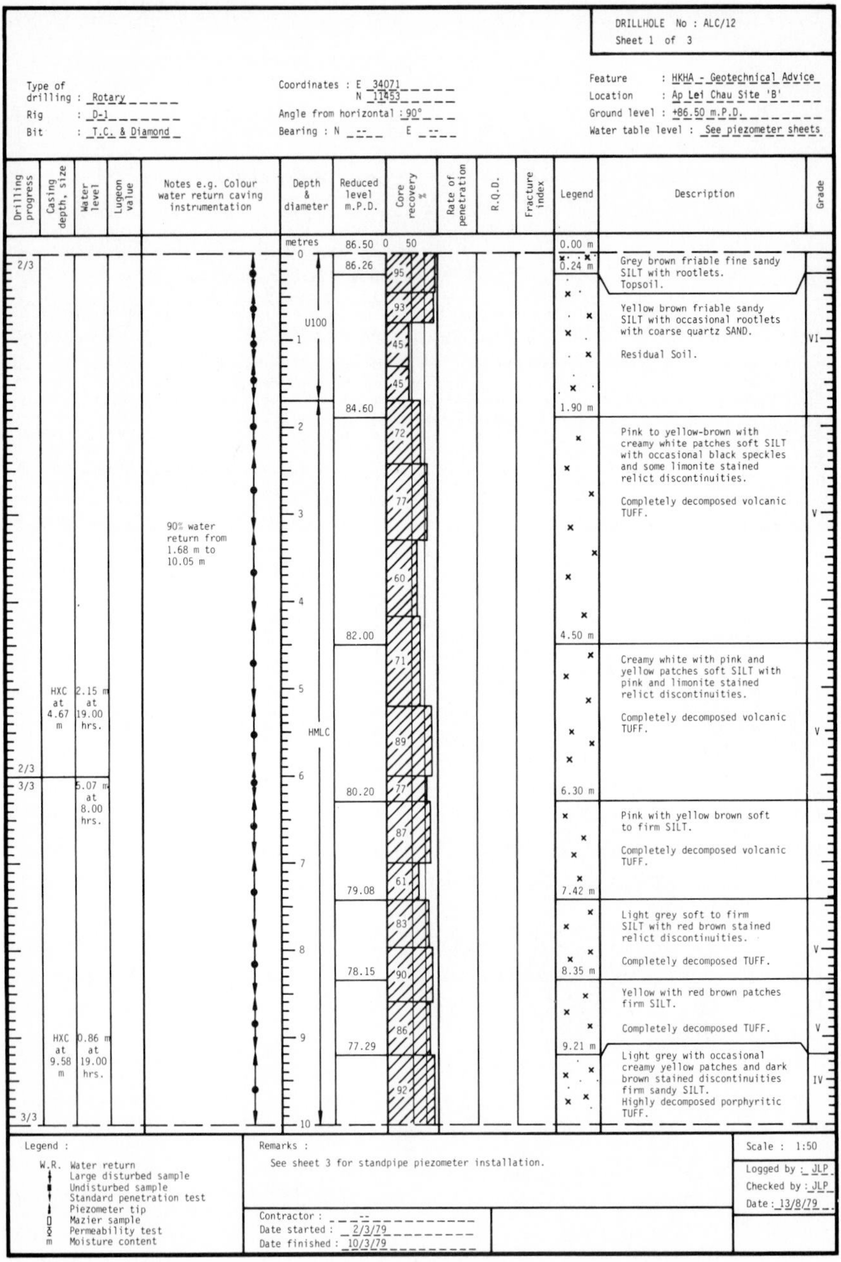

| | | | DRILLHOLE No : ALC/12 |
| | | | Sheet 1 of 3 |

Type of
drilling : _Rotary_ _ _ _ _ _
Rig   : _D-1_ _ _ _ _ _ _ _
Bit   : _T.C. & Diamond_ _

Coordinates : E _34071_ _ _ _ _
              N _11753_ _ _ _ _
Angle from horizontal :_90°_ _ _
Bearing : N _ _ _ _   E _ _ _ _

Feature   : _HKHA - Geotechnical Advice_
Location : _Ap Lei Chau Site 'B'_ _ _ _
Ground level : _+86.50 m.P.D._ _ _ _ _ _
Water table level : _See piezometer sheets_

Drilling progress	Casing depth, size	Water level	Lugeon value	Notes e.g. Colour water return caving instrumentation	Depth & diameter	Reduced level m.P.D.	Core recovery %	Rate of penetration	R.Q.D.	Fracture index	Legend	Description	Grade
					metres	86.50 0 50					0.00 m		
2/3					0	86.26	95				0.24 m	Grey brown friable fine sandy SILT with rootlets. Topsoil.	
							93					Yellow brown friable sandy SILT with occasional rootlets with coarse quartz SAND.	
				U100	1		45					Residual Soil.	VI
							45						
						84.60					1.90 m		
					2		72					Pink to yellow-brown with creamy white patches soft SILT with occasional black speckles and some limonite stained relict discontinuities.	
			90% water return from 1.68 m to 10.05 m		3		77					Completely decomposed volcanic TUFF.	V
							60						
					4	82.00					4.50 m		
	HXC at 4.67 m	2.15 m at 19.00 hrs.			5		71					Creamy white with pink and yellow patches soft SILT with pink and limonite stained relict discontinuities.	
				HMLC			89					Completely decomposed volcanic TUFF.	V
2/3					6	80.20	77				6.30 m		
3/3		5.07 m at 8.00 hrs.					87					Pink with yellow brown soft to firm SILT.	
					7	79.08	61				7.42 m	Completely decomposed volcanic TUFF.	V
					8		83					Light grey soft to firm SILT with red brown stained relict discontinuities.	
						78.15	90				8.35 m	Completely decomposed TUFF.	V
												Yellow with red brown patches firm SILT.	
	HXC at 9.58 m	0.86 m at 19.00 hrs.			9	77.29	86				9.21 m	Completely decomposed TUFF.	V
							92					Light grey with occasional creamy yellow patches and dark brown stained discontinuities firm sandy SILT. Highly decomposed porphyritic TUFF.	IV
3/3					10								

Legend :
W.R. Water return
  ↧   Large disturbed sample
  ●   Undisturbed sample
  ↥   Standard penetration test
  ▮   Piezometer tip
  ▯   Mazier sample
  ⊗   Permeability test
  m   Moisture content

Remarks :
See sheet 3 for standpipe piezometer installation.

Contractor : _ _ _ _ _
Date started : _2/3/79_ _ _ _ _ _ _ _ _ _
Date finished : _10/3/79_ _ _ _ _ _ _ _ _

Scale : 1:50

Logged by :_JLP_
Checked by :_JLP_
Date :_13/8/79_

Figure 6.10   Typical form of core log. (*Geotechnical Control Office, 1984.*)

with numbers and also with the direction of the base or the top of the drillhole. Soft and weathered samples are preserved by wrapping in plastic and then aluminum foil. Further sample numbers are added on the outside of the wrap so that the samples may later be selected according to testing requirements.

These procedures prevent desiccation but do not protect the rock mechanically. Adequate packing materials are needed to separate the samples from each other and from the outer surfaces of the box. They must be insulated from extreme temperatures during shipment and storage. For extra mechanical protection the samples can be packaged in polyurethane foam (Stimpson et al., 1970). The two-component foam spreads around the samples, enveloping and protecting them. Substantial pressures are developed during foaming so that the containing boxes must be sturdy and open ended to allow foam to overflow. The foam is cut away using a hacksaw blade when the specimens are to be removed for testing.

6.6 Downhole Observations

6.6.1 Observations and Impressions

6.6.1.1 Periscopes. Information on rock jointing patterns is also available from the borehole walls, which are disturbed to a lesser extent than the core. Not only are joint orientations left in their original configuration, but also the apertures between joint faces and the natural infillings of soft clay, sand, and broken and crushed rock usually remain in place. Downhole observations are, however, more difficult than those on core and require special and often expensive equipment (Zemanek, 1968).

The simplest device for downhole observations is the borehole periscope, a tube with a telescope at one end and an inclined mirror and light at the other. Borehole periscopes are usually hand-held and quite short, being limited to applications in the upper 2–3 m of hole, although with lens systems they have occasionally been used to depths of 30 m. They are convenient, for example, for observing rock characteristics at shallow depths beneath foundations.

6.6.1.2 Cameras. Observations at greater depth require suspended drillhole cameras or, more often, television cameras. These use either an inclined planar mirror for a sideways view of the drillhole wall, a conical mirror giving a distorted doughnut image of the complete drillhole wall circumference, or a lens to look straight ahead toward the base of the hole (Fig. 6.11). The planar sideways viewing mirror can usually be rotated by a small electric motor, which has an indica-

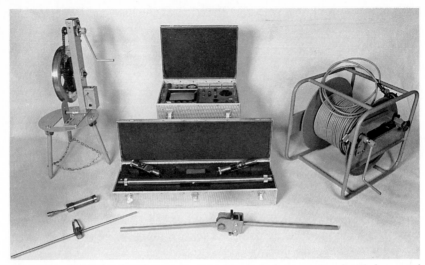

Figure 6.11   Drillhole television camera. The camera is shown equipped with downward and sideways viewing attachments, cable and winding assembly, viewer and video recorder, and alignment tools. (*Courtesy of Solinst Canada Ltd., Burlington, Ont.*)

tor to display the viewing direction. The doughnut-shaped and downward images also require an indication of the viewing direction, usually obtained by a compass pointer and spirit level in the field of view. Television cameras electronically transmit an image of the drillhole wall by closed-circuit television to a viewing monitor at surface. The image may be recorded on videotape, or features of interest may be photographed from the television screen.

The camera is sealed and so can be used below the water table. The hole is flushed after drilling until the water is clear, if necessary adding flocculents to remove turbidity. Locations where water enters or leaves through fissures can be identified and the flow directions and rates observed if a weighted thread is suspended beneath the television camera in the downward-viewing mode. In situations where television cameras cannot operate, a rotating ultrasonic scanner device can be used to map the drillhole wall with remarkable accuracy.

Lau (1980) reviews the development of drillhole television cameras and describes a system developed for use at the underground research laboratory of Atomic Energy of Canada Ltd. in Pinawa, Man. Survey procedures and techniques are presented for converting the apparent orientation of fractures observed in the drillhole to true orientations by means of stereographic projection and spherical trigonometry.

Film cameras are available for taking photographs down drillholes to depths of up to 2000 m. They include downward and sideways viewing models. Lighting is generally supplied by a strobe flash unit (Ash et al., 1974).

6.6.1.3 Impression packers. The impression packer is an alternative technique for obtaining a record of drillhole wall jointing features. A rubber tube is lowered to the required position and then inflated to make contact with the rock. The rubber is sufficiently soft to carry a permanent impression of the fissures against which it is pressed. This replica of apertures and orientations is recovered after the impression packer is deflated and raised to the surface. To preserve the packer for reuse, a removable soft plastic or rubber outer sheath can be employed (Harper and Hinds, 1977).

6.6.2 Downhole geophysical logging

In downhole logging, a string of geophysical testing devices (logging tools) is lowered into a drillhole to take measurements of various properties continuously or at intervals. The method is most commonly used in deep drillholes like those for oil, gas, and coal exploration, where continuous coring is uneconomic, and where the majority of the hole is drilled using a tricone bit that produces cuttings rather than a core sample. However, the techniques are also being applied more and more at the shallower depths of engineering investigations.

Individual types of measurements include sonic velocity, electric resistivity, spontaneous electric potentials in drilling fluid, and nuclear radiation emitted naturally from the rock or backscattered from the rock in response to nuclear radiations from the probe. The drillhole logs can be interpreted in terms of porosity, density, saturation, clay content, and other mechanically relevant information. Caliper logging, which shows variations in hole diameter that occur in rocks of differing competence, is usually included (ISRM, 1981).

Olsson et al. (1985) describe a drillhole radar system to detect the positions and orientations of fracture zones from reflections caused by differences in resistivity at the edges of the zones. Orientations were determined by combining data from several holes in a step-by-step procedure supported by a computer program.

6.6.3 Downhole testing

Various mechanical, geophysical, and hydrogeological downhole testing devices are described elsewhere in this book. If used in conjunction with core logging and drillhole wall observations, they can greatly add to the value of information obtained from the drilling program (Van Schalkwyk, 1976).

Probably the most useful of these is the dilatometer (Chap. 9). Either a rigid or a deformable dilatometer can be employed for hole logging, taking either systematically spaced measurements or measurements within previously identified GMUs. The device is expanded against the walls of the hole, and readings are taken of the expansion pressure against hole dilation to give an estimate of the rock mass deformability, albeit on a relatively small scale.

An index to the shear strength of intact rock can be obtained using a drillhole testing device described in Pitt and Rohde (1984). The tool uses a wedge action to apply simultaneously normal and shear stress to a level sufficient to cause localized shear fracturing in the drillhole walls. Tests can be done in small drillholes such as the 38-mm-diameter holes typically used for rockbolting.

References

Ash, J. L., B. E. Russell, and R. R. Rommel: "Improved Subsurface Investigation for Highway Tunnel Design and Construction," in *Subsurface Investigation System Planning*, vol. 1, Rep. FHWA-RD-74-29, Fed. Highway Admin., Washington, D.C., 398 pp. (1974).

Barr, M. V., and E. T. Brown: "A Site Exploration Trial Using Instrumented Horizontal Drilling," *Proc. 5th Int. Cong. Rock Mech.* (Melbourne, Australia, 1983), vol. A, pp. 51–58.

Belesky, R. M., and H. R. Hardy, Jr.: "Seismic and Microseismic Methods for Cavity Detection and Stability Monitoring of Near-Surface Voids," *Proc. 27th U.S. Symp. Rock Mech.* (Tuscaloosa, Ala., 1986), pp. 248–257.

Canadian Geotechnical Society: *Canadian Foundation Engineering Manual*, 2d ed. (BiTech Publ., Vancouver, B.C., 1985), 456 pp.

Church, R. H., and W. E. Webb: "Evaluation of a Ground Penetrating Radar System for Detecting Subsurface Anomalies," Rept. Invest. RI9004, U.S. Bureau of Mines, 21 pp. (1986).

Deere, D. U.: "Engineering Geologists' Responsibilities in Dam Foundation Studies," *Proc. Conf. Foundations for Dams* (Pacific Grove, Calif., 1974), Am. Soc. Civ. Eng., New York, pp. 417–424.

Franklin, J. A.: "Size-Strength System for Rock Characterization," *Proc. Symp. Application of Rock Characterization Techniques to Mine Design* (1986), Am. Soc. Min. Eng. Ann. Mtg., New Orleans, La., pp. 11–16.

Geotechnical Control Office: *Geotechnical Manual for Slopes*. (Geotech. Control Office, Eng. Develop. Dept., Hong Kong Govt., 1984), 295 pp.

Gordon, R., R. Bender, and G. T. Herman: "Algebraic Reconstruction Techniques (ART) for Three-Dimensional Electron Microscopy and X-Ray Photography," *J. Theor. Biol.*, **29**, 471–481 (1970).

Griffiths, D. H., and R. F. King: *Applied Geophysics for Engineers and Geologists*, 2d ed. (Pergamon, London, 1981), 223 pp.

Harper, T. R., and D. V. Hinds: "The Impression Packer, a Tool for Recovery of Rock Mass Fracture Geometry," in *Storage in Excavated Rock Caverns, Rockstore '77*, vol. 2 (Pergamon, Oxford, 1977), pp. 259–266.

Hatheway, A. W., and F. B. Leighton: "Trenching as an Exploratory Method," in *Geology in the Siting of Nuclear Power Plants. Rev. in Eng. Geol.*, 4 (Geol. Soc. Am., 1979), pp. 169–195.

Hunter, J. A., S. E. Pullan, R. A. Burns, R. M. Gagne, and R. L. Good: "Shallow Seismic Reflection Mapping of the Overburden-Bedrock Interface with the Engineering Seismograph—Some Simple Techniques," *Geophysics*, **49**(8), 1381–1385 (1984).

ISRM: "Suggested Methods for Geophysical Logging of Boreholes," in "Suggested Methods for Rock Characterization, Testing and Monitoring," ISRM Commission on Testing Methods, E. T. Brown, Ed. (Pergamon, Oxford, 1981), pp. 53–70.

Keary, P., and M. Brooks: *An Introduction to Geophysical Exploration* (Blackwell Scientific Publ., 1984).

Krinitzsky, E. L.: "Seismic Stability Evaluation of Alben Barkley Dam and Lake

Project," vol. 2, Tech. Rep. GL-86-7, U.S. Army Corps of Engineers Waterways Experiment Station, 231 pp. (1986).

———and W. F. Marcuson III: "Principles for Selecting Earthquake Motions in Engineering Design," *Bull. Assoc. Eng. Geol.*, **20**(3), 253–265 (1983).

Lau, J. S. O.: "Borehole Television Survey," *Proc. 13th Can. Rock Mech. Symp.* (Toronto, Ont., 1980), pp. 204–210.

Majtenyi, S. I.: "Horizontal Site Investigation Systems," *Proc. Rapid Excavation and Tunneling Conf.* (1976), Am. Inst. Min. Eng., New York, pp. 64–79.

McEldowney, R. C., and R. F. Pascucci: "Application of Remote Sensing Data to Nuclear Power Plant Site Investigations," *Rev. in Eng. Geol.* (Geol. Soc. Am.), vol. 4, pp. 121–139 (1979).

Mollard, J. D.: "Photo Analysis in Interpretation in Engineering Geology Investigations: A Review," *Rev. in Eng. Geol.* (Geol. Soc. Am.), vol. 1 (1962).

Murphy, P. J., J. Briedis, and J. H. Peck: "Dating Techniques in Fault Investigations," *Rev. in Eng. Geol.* (Geol. Soc. Am.), vol. 4, *Geology in the Siting of Nuclear Power Plants* (1979), pp. 153–168.

Nelson, P. P., and T. D. O'Rourke: "Tunnel Boring Machine Performance in Sedimentary Rock," Geotech. Eng. Rep. 83–3, Cornell Univ. School of Civ. and Environm. Eng., Ithaca, N.Y., 438 pp. (1983).

Olsson, O., L. Falk, O. Forslund, S. Lundmark, and E. Sandberg: "Radar Investigations of Fracture Zones in Crystalline Rock," *Proc. Int. Symp. Fundamentals of Rock Joints* (Björkliden, Sweden, 1985), pp. 515–523.

Pitt, J. M., and J. R. Rohde: "Rapid Assessment of Shear Strength and Its Variability," *Proc. 25th U.S. Symp. Rock Mech.* (Evanston, Ill., 1984), pp. 428–436.

Proctor, R. J.: "Mapping Geological Conditions in Tunnels," *Bull. Assoc. Eng. Geol.*, **8**(1), 1–43 (1971).

Rocha, M., and M. Barroso: "Some Applications of the New Integral Sampling Method in Rock Masses," *Proc. Int. Symp. on Rock Fracture* (Nancy, France, 1971), paper I-21, 12 pp.

Rubin, L. A., and J. C. Fowler: "Ground Probing Radar for Delineation of Rock Features," *Eng. Geol.*, **12**, 163–170 (1978).

Stimpson, B., F. G. Metcalfe, and G. Walton: "A New Technique for Sealing and Packing Rock and Soil Samples," *Quart. J. Eng. Geol.*, **3**(2), 127–133 (1970).

Stimpson, W. E.: "Determining Bedrock Elevation by Acoustic Sounding Technique," in *Rock Engineering for Foundations and Slopes, Proc. ASCE Specialty Conf.* (Boulder, Colo., 1976), vol. 1, pp. 1–12.

Telford, W. M., L. P. Geldhart, R. E. Sheriff, and D. A. Keys: *Applied Geophysics.* (Cambridge Univ. Press, New York, 1976), 860 pp.

U.S. Department of the Army: "Recommended Guidelines for Safety Inspection of Dams," National Program of Inspection of Nonfederal Dams, DAEN-CWE Circular 1110-2-188, Office of the Chief of Engineers, Washington, D.C. (1977).

———: *Geophysical Exploration,* Engineers Manual EM 1110-1-1802, U.S. Dept. of the Army, Washington, D.C., 309 pp. (1979).

U.S. Nuclear Regulatory Commission: "Code of Federal Regulations, Title 10 Energy; Part 100 [10 CFR 100], Reactor Site Criteria; Appendix A, Seismic and Geologic Siting Criteria for Nuclear Power Plants," NRC, Washington, D.C. (1978).

Van Schalkwyk, A. M.: "Rock Engineering Testing in Exploratory Boreholes," *Proc. Symp. Exploration for Rock Eng.* (Johannesburg, S. Africa, 1976), Z. T. Bieniawski, Ed., vol. 1 (A.A. Balkema, Rotterdam and Boston, 1976), pp. 37–55.

West, G., P. G. Carter, M. J. Dumbleton, and L. M. Lake: "Site Investigation for Tunnels," *Int. J. Rock Mech. Min. Sci. Geomech. Abstr.*, **18**, 345–367 (1981).

Wong, J., P. Hurley, and G. F. West: "Crosshole Seismology and Seismic Imaging in Crystalline Rocks," *Geophys. Res. Lett.*, **10**(8), 686–689 (1983).

Zemanek, J.: "The Borehole Televiewer—A New Logging Concept for Fracture Location and Other Types of Borehole Inspection," Soc. Petrol. Eng. (Sept. 1968).

Measurement, Prediction, and Monitoring of Rock Behavior

Design methods are the techniques used to predict the behavior of rock excavations and support systems, and *monitoring* methods check how closely the predictions agree with reality. These are the first and last topics to be addressed in Part 2. In between are sandwiched four chapters that examine how rock behaves in the laboratory and in situ.

Part 2 is mainly about the testing needed to define how intact and jointed rock deform under the action of stresses, time, and temperature. The choice of design method determines which tests are needed. Numerical modeling simulates behavior of the rock mass in terms of "constitutive" equations, which mathematically relate applied stresses to strains in the rock mass. Before a constitutive law can be used in design, it must be completed by "filling in the gaps," in other words, by inserting values for the materials parameters that pertain to the rocks on site. *Design* tests have to be conducted in order to measure these properties. On the other hand, *index* tests are required if the design is empirical.

Of the various design approaches available, empirical methods allow the engineer to proceed directly from a rock classification based on index tests to the design of rock excavations and support systems. They are used more often than not, but can give misleading results if applied out of context. The construction of physical models is labor-intensive, time-consuming, and expensive. The alternative of numerical modeling becomes increasingly attractive with developments in the methods and in microcomputer technology.

In spite of remarkable improvements in our ability to model and predict rock behavior, design methods remain quite unreliable, and even more so, the data on which they are based. Hence the need for surveillance and monitoring of engineering works. The three most common reasons for monitoring are first, to ensure safety of construction personnel and the public who later use the facility; second, to safeguard the environment; and third, to check the assumptions and predictions of design. Chapter 12 describes the methods and instruments available for monitoring rock and support behavior, and how monitoring systems are designed and installed.

7

Design Methods

7.1 Design Principles and Phases

Design in rock is the process of making engineering decisions on such
matters as the locations, alignments, sizes, and shapes of excavations
and their stabilization and support systems. Design is also applied
to the techniques themselves: a blast, for example, needs to be
"designed" to achieve fragmentation while attempting to preserve the
stability of the rock walls.

The usual approach is to select provisionally the configuration and
support system that from experience seem the best; to check for stability by predicting stresses and displacements in the rock and the
likelihood of collapse; and then to try alternatives and modify the design until it satisfies criteria of cost, stability, and safety.

The design is often completed in three or more phases of increasing
detail and complexity. In *conceptual* design, project components are
identified and related to each other and to a forecast of ground conditions. An example of conceptual design is the continuing debate over
siting of nuclear waste disposal facilities. Various alternative modes
of geological storage have been conceived since the late 1950s (IAEG,
1986). Once feasibility is established, targets and constraints are defined to which later stages of design must conform. This is followed by
preliminary design, often carried out simultaneously for several alternatives evaluated with respect to their relative merits and costs. *Final*
design is completed for just one selected alternative and produces detailed drawings from which the project can be constructed. Even final
designs may be provisional, embodying flexibility to accommodate the
variations in rock, water, and stress conditions that in rock engineering can seldom be guaranteed in advance.

The design phases are accompanied by increasingly detailed and localized exploration, site investigation, and testing of mounting complexity and cost. A concept based on a broad appreciation of soil, rock, groundwater, and ground stress conditions thereby progresses to a thoroughly evaluated and costed specific plan of construction.

7.1.1 Uncertainties of design

Whereas some sites are quite simple and intensively explored, usually little precise information is available on ground conditions at the time rock works are designed. Mechanical characteristics and even boundaries between strata remain uncertain, and important faults or cavities may have been missed even by a thorough investigation. Analytical methods can therefore be applied rigorously to idealized and simplified situations only. This contrasts sharply with the analysis of structures made of concrete and steel, where material properties are known and can even be controlled, and the precise geometry of the structure is defined.

Uncertainties that remain even after analysis lead to reliance on monitoring of behavior, and to adjustments during construction. Flexibility is written into the specifications, with provisions for on-site monitoring and inspections during construction, and for design-as-you-go changes to the methods of excavating and support. The main purpose of preconstruction studies is to eliminate major surprises, accidents, and cost escalations, and to reduce design changes to a few that can be made easily.

A fully monitored experimental opening in the rock provides a potentially very practical method for the design of underground works such as tunnels (Sakurai, 1983). The method combines empirical and analytical approaches to design, involving measurement and back-analysis of displacements in order to determine bulk rock mass characteristics such as modulus of elasticity, Poisson's ratio, and initial stresses. The materials properties that are determined are not necessarily real and precise characteristics of the rock mass but may be regarded as equivalent values for evaluating the behavior of the full-size prototype tunnel.

Inherent uncertainties in design can be quantified mathematically in terms of risk and the probabilities of failure. Recent years have seen rapidly growing research into applied probability and an increasing number of applications in geotechnical engineering practice (Whitman, 1984).

7.1.2 Modeling

7.1.2.1 Concept of modeling. Design aims at solving a simplified version or *model* of the real problem. Modeling, a process of simplification, is needed because the full complexity of the real geometry and geology, even if fully known, would present a picture so complex as to be unsolvable by currently available methods.

In precomputer days, the complex cross section of an opening was often approximated as a square or circle, jointing was ignored, and elastic behavior assumed. Models have become much more realistic because of computers, and can accommodate geometries and materials properties of almost any degree of complexity. In practice, however, analytic models are still oversimplified because of the impracticality of obtaining a full set of real data. Moreover, each analytic method requires its own type of model and set of simplifying assumptions.

7.1.2.2 Model type. The types of models include conceptual, mathematical (analytical or numerical), analog, physical, and statistical models. At the very least, a conceptual model is required for empirical and all other types of design. Mathematical models are superior when we have the data and the techniques to solve them. Analog models offer an alternative when the governing equations are similar to those of another, more easily solved physical system. Testing of a realistic physical model is usually a last resort because of high cost, but can give excellent results, even when the basic laws governing rock mass behavior are uncertain, provided that the modeling materials and methods of loading are carefully selected. Finally, a quantitative statistical model of behavior can be assembled based on events in the past and probabilities of occurrence.

7.1.2.3 Continuum and discontinuum alternatives. A rock mass can be adequately modeled as a "true" continuum when it falls into one of four categories: it is free from jointing (a very few, very massive rock formations); its joints are practically as strong as the intact rock because they are rough, discontinuous, and widely spaced, or the rock is extremely weak; its joints are strengthened by high general compressive stresses at depth (such as in the South African deep-level gold mines); or it is so closely jointed that the joints can be adequately sampled on the scale of a large laboratory or field test specimen (some highly jointed rocks and materials such as coal).

In other cases, and this includes the majority of mining and construction situations, the mathematical model used to simulate rock mass behavior must explicitly include the major joints and faults as separate entities, and also the effects of the more closely spaced joints. The latter is achieved either by testing on a very large scale, or by making somewhat arbitrary reductions in rock mass strength and rigidity to take into account the effect of joints not present in the test specimens (the *equivalent continuum* approach), or by including the joints individually or as sets in the model (the *discontinuum* approach). Soil materials, for example, contain discrete grains bounded by discontinuities but are nearly always modeled as equivalent continua, and successfully so. Numerical methods are not limited to one or the other type; they can be formulated as hybrid continuum/discontinuum models.

7.1.2.4 Permissible forces and displacements. An important step in modeling is to decide how to express the behavior of the rock in terms of the stresses, forces, or displacements imposed at its free faces (boundaries) or within the mass itself. In continuum mechanics we assume that the rock mass is a continuous deformable medium. In discontinuum mechanics we assume that it is composed of blocks that themselves are relatively rigid, bounded by deformable joints or surfaces along which sliding can occur.

Analyses of stresses and displacements require that conditions of equilibrium and compatibility of displacements be satisfied everywhere. In other words, rock blocks must not be allowed to overlap, nor must adjacent blocks exert different forces at their points of contact. Newton's law that action and reaction are equal and opposite must be satisfied unless, as in some types of analysis, rock blocks are being analyzed as projectiles that are permitted to accelerate.

7.1.2.5 Representation of the materials. A continuum mechanics analysis requires that a constitutive equation (a relationship between stress, strain, and, if necessary, time for the rock mass) be formulated for each geological or fabricated material (Chaps. 9 and 10). This relationship is in theory established by testing, although in practice tests are rarely precise emulators of mass behavior. A complete constitutive equation should describe rock behavior before, during, and after rupture of intact material, but this calls for such an exhaustive test program for even one material that simplifications are essential. Often behavior is "forced" to conform to an idealized type of behavior, such as linear elasticity.

7.1.3 Parametric studies

A *parametric study* plays a particularly important role in the earlier phases of design, when conditions are least well known. Its purpose is to examine a wide range of input data assumptions, to obtain a "feel" for the relative importance of each variable or parameter in the model. After setting up the model, each parameter is assigned a range of possible values on the basis of experience or other available information. All parameters but one are fixed (usually close to mid-range), and the remaining parameter is varied over its expected range to determine its relative effect on cost, safety, or stability.

The parametric study is a powerful design tool. First it draws attention to those variables that have the greatest impact so that additional testing can be done to refine the estimates of these parameters. Second, overall patterns of behavior are revealed, allowing trends to be understood and the underlying principles of design established. For example, a parametric study during the early stages of tunnel design might well show that stability is affected hardly at all by increasing the thickness of a liner, or that the tunnel is more stable with a thin liner than with a thick one, or even that long-term stability can under no circumstances be achieved. Hence, the validity of the model itself can be checked against known solutions or just common sense. Alternative models should be compared, when available.

7.1.4 Methods and their limitations

The following aids to design are discussed in this chapter: empirical methods, analytical methods, analog modeling, and physical modeling.

Analytical methods are powerful and are continuing to improve. Nevertheless, design in rock engineering projects continues to rely heavily on empirical methods based on experience and judgment, on an appreciation of site geology and rock characteristics, and on an understanding of the capabilities and limitations of the available excavating and stabilizing techniques.

Two types of analytical method are available: closed-form solutions and numerical approximation techniques. Closed-form methods give exact results (within the limits of the sometimes grossly oversimplified assumptions) by the solution of one or a set of equations. Their use is limited to simple geometries (e.g., a horizontal foundation or circular tunnel), few materials (usually just one), and simple constitutive laws (often linear elastic or viscoelastic). In contrast, numerical methods can solve problems with quite complex geometries, several layers of rock and support materials, and intricate constitutive relationships (Hornbeck, 1975). The solution is obtained by iterations of increasing exactitude, stopping when the degree of precision is considered sufficient. In practice, the numerical techniques give results that are usually more "exact" (realistic) than those obtained by simplistic closed-form solutions, which they are replacing for all but the simplest of geomechanics problems.

7.1.5 The empirical method

Judgment and experience are major ingredients in rock engineering design, and are the sole ingredients in the absence of other more formal methods. However, our ability to judge and forecast is always limited by the extent and quality of our experience. For example, to estimate the support needed in a tunnel we relate our impression of the ground conditions to our knowledge of the performance of support systems on previous projects in similar ground. We are at a loss if either the particular ground or tunneling itself is unfamiliar.

Empirical design solves the problem of our own limited experience by making available the accumulated experience of others. It requires three steps beyond simple judgment:

1. Description of *ground quality* by a quantitative classification system, to provide a universal language whereby the experience gained globally working in ground of many different qualities can be related to future projects

2. Description of *ground performance* by a formalized quantitative system which defines such parameters as unsupported stand time and support requirements, bearing capacities, excavating methods

3. *Correlation of ground quality to performance* by a compilation and comparison of results from a variety of projects over a full spectrum of ground conditions (case histories)

Empirical design makes use of empirical formulas, "rules of thumb," or design charts. Correlations can be shown graphically and are gradually improved by plotting successes, partial successes, or failures. Contours separating successes from failures form the basis for predicting future performance. A familiar example is the use of allowable bearing pressure tables for structural foundations, often incorporated into building codes. These specify how much pressure can be applied to a rock, depending on its strength and (sometimes) its degree of jointing. Codes vary from region to region, and are usually at first excessively conservative. Accumulated experience, testing, and monitoring lead to a progressive relaxation of requirements. In Toronto, for example, permissible foundation bearing pressures have increased by an order of magnitude over the past 30 or 40 years.

Tunneling is an application in which empirical methods are mostly used. Various rock mass and performance classifications have emerged specifically for tunneling (Barton et al., 1974; Bieniawski, 1984). Other situations where empirical design is the only available method include prediction of drilling and boring rates, assessment of rock aggregate and building-stone quality, and evaluation of the susceptibility of rock to weathering and erosion.

Empirical rules are safe to apply only in the context for which they were formulated, and extrapolation can be dangerous if a method has no theoretical basis. Semiempirical methods are perhaps better, in which rock quality and performance are related through a "theoretically sound" formula where only the constants of the formula are determined from experience. However, even the soundness of a theory must eventually be judged by results.

7.1.6 Computer graphics

Although not strictly a design procedure in its own right, the power of three-dimensional *interactive computer graphics* is increasing daily. These modern methods are a far cry from the glass plate models of just a few years ago in their ability to portray the complex geometries of mining and civil works. Many problems of rock mass stability can be solved, at least intuitively, when the three-dimensional configuration of jointing can be visualized in relation to the three-dimensional geometry of the openings (Fig. 7.1).

Detailed perspective views can readily be obtained with the help of a display package called CAD (computer-assisted design). Versions offer color shading which lends depth and clarity. Images can be rotated for viewing in any direction, can be magnified, contracted, and sectioned. Hard-copy prints can be obtained of the screen display. Twin images can be produced for stereoscopic viewing. Soon, even animation will be available.

Interactive graphics can be a great help to any analytic design, allowing properties, boundary conditions, loads, excavation sequences,

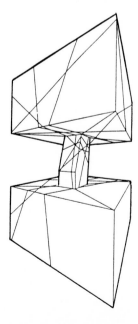

Figure 7.1 Computer-drawn mine pillar with structure produced by five dominant joints, forming key blocks. The graphics program allows removal of key blocks. (*Warburton, 1983.*)

and so on, to be modified easily and quickly in response to the results obtained. In some systems, the user can draw lines on the screen to represent joints, faults, and other boundaries and can edit the picture by erasing or adding lines (Cundall et al. 1975; Cundall, 1976). The computer then converts these lines into a system of blocks or elements.

7.1.7 Closed-form solutions

A *closed-form solution* is one in which the behavior of a model can be expressed by simple equations for which a unique solution has been found. Most assume that the medium is elastic, isotropic, and homogeneous. Some more complex closed-form solutions allow for the development of a plastic zone, for elastic orthotropic or linearly viscoelastic materials, or for the layering and anisotropy of stratified rock.

7.1.7.1 Mine pillar example. The concept of a closed-form solution can be introduced by the simple but practical example of an underground coal mine excavated using the "room and pillar" method, in which the mine roof is supported by equally spaced pillars of square cross section (Fig. 7.2). The target is to predict the size of pillar required for stability, and hence the *extraction ratio,* defined as the ratio of excavated coal to total original coal in the rooms and pillars combined. What

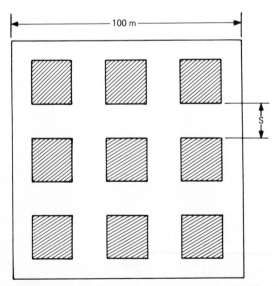

Figure 7.2 Worked example of simple room and pillar design.

cross-sectional area of pillar is needed to support the overburden loading? How much coal must be left in the pillars to provide this support?

The overburden loading can be calculated as the total plan area A of the mine workings multiplied by the depth z of the mine roof and by the unit weight γ of the rock (Sec. 5.3.1.2). Dividing this load by the combined cross-sectional area of pillars a gives the average vertical stress U in these pillars:

$$U = \frac{Az\gamma}{a}$$

We can now compare this stress with the strength S of the pillars to determine the factor of safety F against failure:

$$F = \frac{S}{U}$$

$$U = \frac{S}{F} = \frac{Az\gamma}{a}$$

$$a = \frac{FAz\gamma}{S}$$

The example in Fig. 7.2 with $F = 1.2$, $A = 10,000$ m^2, $z = 100$ m, $\gamma = 25$ kN/m^3, and $S = 10$ MPa, gives $a = 3000$ m^2 and an extraction ratio of $(10,000 - 3000)/10,000$, or 70% (compared with 75% for $F = 1.0$). Each of the nine pillars measures 18 by 18 m and is separated from the next by a roof span s of 15.3 m.

In practice this simple solution is modified to account for the pillar being weaker than the test specimen because of its size, and for stress modifications that depend on the slenderness of the pillar. The roof would also be checked for stability against bending and buckling, making use of another simple closed-form solution. More complex closed-form solutions have been formulated to predict the behavior of pillars in materials such as salt or potash, which can be approximated as ideal elastoviscoplastic materials.

7.1.7.2 Two-dimensional analysis of tunnels. The mine pillar example given above is one of the few for which a uniaxial or one-dimensional representation is close to reality. Extending this type of calculation to two dimensions, a tunnel may be modeled as a circular hole in an "infinite elastic plane." The Kirsch solution for two-dimensional distribution of stress around a circular opening in an elastic medium was developed in 1898. Mindlin in 1940 added the effects of body forces and variable horizontal boundary forces while still assuming a homogeneous elastic material (Savin, 1961; Obert and Duvall, 1967).

Tunnel geometries that can be solved by closed-form elastic methods include circles and ovals (Goodman, 1980), ellipses (Jaeger and

Cook, 1979), and nearly square or rectangular openings (Szechy, 1966). Elastic solutions can be superimposed to simulate multiple openings.

Kestner developed a plastic zone theory in 1949 (Szechy, 1966). Elastoplastic solutions allow the development of a plastic zone near the wall of an opening, controlled by either a Tresca or a Mohr-Coulomb yield criterion. At some distance R away from the wall, the plastic zone ends and a zone of rock exhibiting elastic behavior begins. Bray expanded Kestner's work and included rock strength parameters for both the plastic and the elastic zones (Goodman, 1980). This made the solution much more realistic as broken rock can have strength parameters considerably different from those of intact rock. Vandillen and St. John (1984) have extended these solutions to include the effects of rockbolts, represented as a uniform internal pressure on the tunnel. Solutions have been developed for dilatant and nondilatant materials.

This class of solution is limited largely to circular openings in hydrostatic stress fields, but other conditions can be approximated by the superposition of various solutions and careful adjustments of the results, giving powerful design aids.

Solutions are also available for inelastic materials, although only for simple geometries and simple constitutive laws such as linear viscoelasticity (Baoshen and Dezhang, 1981), or a Hookean spring combined with a Kelvin-Voight cell (Secs. 10.2.4.2 and 10.2.4.3). A particular case of analysis in a viscoplastic medium has been developed by Mraz (1984). Ideal Tresca plasticity is assumed for the zone adjacent to the opening (but beyond the broken zone), and the rest of the mass is assumed to behave viscoelastically.

Sofianos (1986) has derived an elastoplastic closed-form solution for symmetric and asymmetric wedges in a tunnel roof. He employs a nonlinear stress-strain law for the joints forming the wedge. An elastic analysis is carried out to determine the horizontal stress in the crown of the excavation. The joints are then given some finite values for shear and normal stiffnesses, at which point the factor of safety for conditions of limiting equilibrium may be computed. The formulation has been checked by laboratory testing (Crawford and Bray, 1983).

7.1.7.3 Foundations and externally applied loads. Simple elastic closed-form solutions are available to predict displacements and stresses beneath a loaded foundation. Boussinesq in 1885 and Cerruti in 1887 [described in Jaeger (1979) and other texts] developed solutions for normal and shear point loading on a three-dimensional elastic half-space that can be used to build up, by direct superposition, the distribution of stresses and displacements for any system of noninteracting applied loads. The solutions can be solved for particular cases either directly, using the appropriate equations, or by graphic techniques such as Newmark charts for foundation design. Extension to a layered half-space exists for the Boussinesq but not for the Cerruti solution.

These solutions can be integrated to give other useful ones. For example, a line load (strip footing) solution is obtained by mathematical integration of the Boussinesq solution under plane strain boundary conditions. Circular, elliptical, and other footing shapes can be solved by similar methods (Ahlvin and Ulery, 1962; Milovic, 1972; Nyak, 1973). Nonlinear stress-strain relationships can also be formulated (Hobbs, 1973). Poulos and Davis (1974), in an excellent compendium, give equations and charts for many relevant elastic solutions in soil and rock mechanics.

The unit displacement of a line section of unit length on the boundary of a medium is solved. These point and line solutions are extremely important to rock engineering, not just in foundation applications, but because the solutions can be numerically integrated, and in a sense form the basis for all of our displacement discontinuity, boundary element, and boundary integral methods.

7.1.7.4 Internal forces in the rock mass. Solutions are available for a force acting in an arbitrary direction at a point in an infinite elastic space (e.g., Mindlin, 1936). Again, this can be extended to a line load or to a circular load. The latter can be looked upon as a model of a pressurized penny-shaped crack (Sneddon and Lowengrub, 1969). If a unit volume change is applied at a point in a medium, a stress and displacement field is set up, and a solution exists to this problem (Mindlin and Cheng, 1950). It can be integrated to give the stresses, displacements, and surface subsidence above a compacting reservoir (Geertsma and van Opstal, 1973).

7.2 Numerical Methods of Analysis

7.2.1 Overview

Stress analysis problems for which no closed-form solution is available can be solved by methods of numerical approximation. These are far more powerful than closed-form solutions, but take a little more time to set up and run. Most of this time is to define the model and to prepare the input data. The time requirements are being reduced by automated mesh generators and user-friendly input control software, which are opening up numerical approaches to the nonspecialist engineer. The further time required for a computer program to iterate and converge on a solution is being greatly reduced with the advent of true desktop dedicated processors.

The trend to more and more complex and hence realistic models has created difficulties. Programs which can address simple linear elastic or linearly viscoelastic problems are widely understood and widely distributed. However, the complex formulations are often idiosyncratic and inadequately documented, and their behavior is only understood in detail by the person who wrote and uses the program regularly. The casual user runs a serious risk that the results may be misleading. Complex problems should therefore be referred to an analytical specialist.

There are three major categories of numerical techniques: finite-difference, finite-element, and boundary-discretization methods. The concepts were developed many years ago but, because of the need for repetitive calculations and the storage of large quantities of data, everyday use dates from the advent of the computer.

7.2.2   Finite-difference method

The *finite-difference* (FD) method has long been used, particularly for solving water flow problems. The partial differential equations that stipulate physical behavior are replaced by approximations between discrete points, nodes, or block centers throughout the rock mass. Special forms of the difference equations are required to accommodate different boundary conditions. For flow, the two common ones are stipulated flux or no flow. For stress, the conditions are usually constant stress, no strain, or a specified reaction coefficient such as a stiff spring. The resulting simultaneous equations are then solved directly or by successive approximation (relaxation).

Time solutions require discretization along a time axis as well. There are three basic solution techniques, explicit, implicit, and central difference solutions, as well as exotic combinations of these. Each requires time boundary conditions dependent on the order of the differential equation to be solved, that is, a second-order differential equation requires two boundary conditions.

In the *explicit,* or forward averaging, method, three known values at the j level or time step are used to determine one central value at the $j + 1$ time step or level. In the *implicit,* or backward averaging, method, sets of equations are solved along a grid line. Even though only one value is known at most of the time steps, solving the equations simultaneously with the known boundary conditions allows the next grid point to be solved. In the *central difference,* or Crank-Nicholson, method, the forward and backward averaging techniques are combined. New values are determined at times halfway between two nodes. The solution using the forward averaging method can become unstable if the grid spacing ratio becomes too large, whereas the backward and central averaging methods are stable for any grid spacing ratio, but require the solution of a large matrix at each step. Depending on the application, a trade-off exists. The equations for the explicit, implicit, and central finite-difference solutions are all similar.

Finite-difference methods are limited mainly by the requirement for equally sized, square, rectangular, or perfectly axisymmetric grid spacings. To resolve the high stress and strain gradients near an opening, a finely spaced mesh is needed throughout the model, which leads to a waste of computer time in regions of low stress gradient. Also, variations of properties are difficult to model. For stress-strain problems, use of the finite-difference method is mainly limited to linear elastic homogeneous materials with rectilinear or axisymmetric boundaries. The time-stepping method described above is, however, highly general, and is used in most time-dependent analyses. For example, in a typical viscoelastic problem, the solution for stresses and strains is achieved by a finite-element formulation, but the time step to a new set of conditions is achieved by the Crank-Nicholson methods along a time line.

7.2.3 Finite-element method

The finite-element (FE) method (Zienkiewicz and Cheung, 1967) is the most powerful and versatile technique for general stress analysis of surface and underground works in rock. It can accommodate two- and three-dimensional situations, elastic, plastic, and viscous materials, and can incorporate "no-tension" zones, joints, faults, and anisotropic behavior. The method is also used to solve problems of seepage (Wittke et al., 1985), consolidation, and heat flow. If the correct physical coupling differential equations can be written (e.g., coupling water flow, heat flow, effective stresses, volume changes), the finite-element method can generate a solution.

The continuum to be analyzed is separated by imaginary lines or surfaces into a number of finite elements, which are assumed to be interconnected at a discrete number of points on their boundaries, termed *nodes*. Irregularly shaped, triangular, prismatic, or curvilinear elements can be used (Fig. 7.3). Their size can be varied to allow a close distribution of nodes near the source of stress, heat, and so on, and a larger grid spacing further from the source. The elements can be arranged to fit within predefined geological boundaries so that differ-

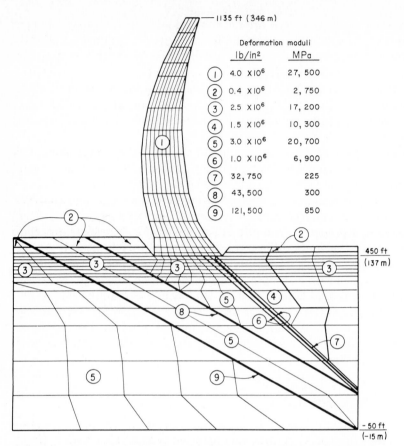

—1135 ft (346 m)

Deformation moduli

	lb/in²	MPa
①	4.0 ×10⁶	27, 500
②	0.4 ×10⁶	2, 750
③	2.5 ×10⁶	17, 200
④	1.5 ×10⁶	10, 300
⑤	3.0 ×10⁶	20, 700
⑥	1.0 ×10⁶	6, 900
⑦	32, 750	225
⑧	43, 500	300
⑨	121, 500	850

450 ft
(137 m)

- 50 ft
(-15 m)

Figure 7.3 Finite-element mesh used for design of the Auburn Dam, California, by the U.S. Bureau of Reclamation. Note modeling of the arch dam and its foundation by a complex pattern of rectangular, trapezoidal, and triangular elements, and the simulation of layers of different elastic moduli. (*Frei, 1976.*)

ent rock materials can be assigned different properties. Application of the method is aided by a skillful selection of element types, sizes, and distributions.

Displacements at the nodal points are the unknown parameters. A function (called a *displacement* or *strain interpolation function*) is chosen to define uniquely the state of displacement and strain within each element in terms of the nodal displacements. A system of forces is applied at the nodes to equilibrate the boundary forces and any internally distributed forces. The stiffness of the whole model is then expressed as the sum of the contributions of individual elements. The response of the structure to the combination of internal and external loads is then computed by the solution of a set of simultaneous equations, to determine the distribution of strains throughout the model.

The solution minimizes a work function in the volume analyzed. Stresses in each element can be determined from the corresponding strains, knowing the preexisting state of stress if any, and using the appropriate constitutive equation for the material of which the element is composed. The matrix to be solved at any step is sparse, banded along the diagonal, and is usually diagonal-dominant. Highly efficient solution techniques have been generated for these types of matrices, avoiding direct solution.

Modeling of joints or any other planar feature requires the development of joint elements (Goodman et al., 1968). Various types have been developed, such as the novel three-dimensional one proposed by Beer et al. (1985), that can be assigned zero thickness and permits complete separation of joint surfaces. Desai et al. (1983) describe some recent refinements, including the use of a thin-layer element for joints and interfaces between strata, and between rock and concrete liners, and the use of various constitutive equations to represent the behavior of the rock mass and of individual discontinuities.

Finite-element models can also cope with yielding behavior. Elastoplastic problems are investigated by stipulating a yield function (yield criterion) for each material. The solution checks for yield and when detected, applies out-of-balance forces at the nodal points to restore the stresses in the element to the yield limit. These forces of course increase the stresses on adjacent elements (stress redistribution), so the mesh must be resolved iteratively, each time allocating new properties to the yielded elements according to some criterion usually based on volumetric shear strain.

Comparing finite-element and finite-difference methods, the former permits the use of a much more realistic model and thus is capable of giving a more realistic result, particularly for complicated geomechanics situations with intricate geological conditions and openings of complex geometry. It is more economical in computer time and storage because larger elements can be used in regions of lesser interest.

The finite-element method, although powerful, requires the entire continuum to be divided into elements, which is slow and expensive for larger models. Complex mine layouts, for example, are seldom analyzed by finite elements. They would require a prohibitively large number of elements and considerable time and computer storage to achieve a realistic solution. This limitation in many cases can be overcome by one of the boundary-discretization methods discussed below.

7.2.4   Boundary-discretization methods

7.2.4.1   Principles, advantages, and disadvantages. Stresses, strains, and displacements at or near the free faces of excavations are usually the greatest and the most detrimental to stability. Boundary discretization methods are unusually efficient because they discretize only these exposed rock surfaces, and not the entire rock mass (Fig. 7.4). Generating the meshes for typical problems is very simple compared with the meshes needed for a finite-element analysis, and many variations of design can be studied rapidly if stresses are required only at a few selected points in the rock mass. Large systems of equations are avoided (Brebbia and Walker, 1980; Banerjee and Butterfield, 1981; Crouch and Starfield, 1983).

The methods, introduced in the 1960s, are now used extensively. Peirce and Ryder (1983) report that the boundary-element program MINSIM and its successors currently account for nearly 1500 problem runs per year in South Africa alone for the design of deep tabular mining layouts and sequences.

The procedure is to subdivide two-dimensional boundaries into linear segments, and three-dimensional ones into surface elements. A closed-form solution is then applied at each element. The sum of the resulting stresses and displacements at a specified point yields the desired stress-displacement information at that point (Fig. 7.4). Usually the matrices to be solved are full, requiring direct solution methods, so

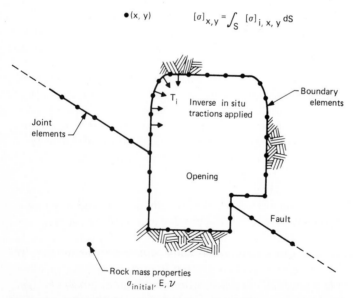

Figure 7.4   Discretization using boundary elements.

the number of boundary elements must be kept to a minimum. This limits the precision obtained. For example, 12 elements will give very reasonable solutions at several diameters from a tunnel wall, but very imprecise results close to the wall.

Rapidly developing boundary-discretization methods are able to address increasingly complex nonlinear problems. There remains the fundamental limitation that since discretization is on the boundary only, multiple materials of realistic geology and complex geometry cannot easily be analyzed.

There are three variations of the method: boundary-element (BE), displacement-discontinuity (DD), and boundary-integral (BI), but the basic approach is similar in all three (Crouch and Starfield, 1983).

7.2.4.2 Boundary-element method. The boundary-element method uses a closed-form solution for the stresses and displacements caused by a unit stress applied at the midpoint of the line or surface segment. Compatibility of deformation fields between elements is enforced by making the slopes continuous at the element boundaries (nodes in the case of two-dimensional problems). Slope continuity is required for each pair of elements, not just adjacent ones, and this leads to a full matrix. Boundary-element methods can be used, for example, to analyze tabular ore bodies (Brady and Bray, 1978).

7.2.4.3 Displacement-discontinuity and boundary-integral methods. The displacement-discontinuity method uses a closed-form solution for the stresses and displacements that result from a unit displacement of a boundary segment. A continuity criterion similar to the above is enforced.

The method has been developed into a three-dimensional formulation for modeling multiple parallel, tabular openings, either flat-lying or inclined (Wardle and Enever, 1983; Wardle, 1984). Each pillar in a level is replaced by a number of elastic elements represented by pairs of points, called *strain nuclei,* at the upper and lower extremities of the pillar element. The sill between openings is modeled as an elastic beam. The solution gives the load in each pillar element for the entire model. A limitation is that the plan area grid divisions, hence the displacement-discontinuity elements, must all be uniform in size. Recent extensions of this procedure allow the modeling of multiple levels with cross anisotropy, layering, and continuous or frictionless interfaces between layers. These complexities require iterative solution. The program MINAP (Crouch, 1976) is restricted to two dimensions but otherwise more general in that excavations need not all be tabular, and inhomogeneities, fault sliding, and other quasi-nonlinearities can be studied (Deering, 1980).

The boundary-integral method uses an integral formulation rather than a direct solution, but works in a similar manner.

7.2.4.4 Comparison of methods.

Although each approach has wide application, the boundary-element method is most suited for openings such as odd-shaped tunnels, displacement discontinuity for models with sharp cracks or reentrant corners such as those for the solution of hydrofracture problems, and boundary integral for heat or fluid flow problems. Hybrid models incorporating two or more methods can be used: for example, a hydrofracture propagating from a drillhole can be analyzed using the boundary-element method for the drillhole, displacement discontinuity for the crack, and boundary elements for the fluid flow problem.

Boundary-element and finite-element methods can be combined to give an economical and elegant hybrid way of solving complex geomechanics problems. The boundary stresses are resolved in detail, and the stresses remote from boundaries are determined in only sufficient detail to evaluate the overall behavior of the rock mass (Lorig and Brady, 1982; Dowding et al., 1983). By combining methods, some of the limitations of each are overcome. Efficiency is enhanced because of the ability to use any material law for each finite element, the ease of boundary segmentation, and the smaller numbers of elements.

7.2.5 Discontinuum models

7.2.5.1 Distinct-element method.

The distinct-element dynamic relaxation method allows a problem to be formulated assuming rigid blocks, with deformation and movement occurring only at the joints. For this type of analysis no information is needed on the deformability and strength of intact rock. When the joint stiffness (ratio of normal force to normal displacement) is much lower than the stiffness of the intact rock, the assumption of block rigidity is a reasonable one.

Cundall (1971) was the first to apply this type of analysis to a stack of rectangular rock blocks simulating a rock slope. Blocks were allowed to consolidate and then released to move under gravity. The ensuing pattern of displacements simulated, with an unprecedented degree of realism, the "toppling" mode of rock failure accompanied by the formation of a hinge zone and "tension" cracks (Fig. 7.5).

Cundall's distinct-element method (Cundall, 1976) considers assemblies of rigid blocks or particles that interact through stiffnesses (springs) acting at the contact points. Normal contact forces are proportional to the small overlap permitted between blocks, and shear forces are incrementally proportional to shear displacements up to the limit imposed by a shear strength criterion. Forces become zero when

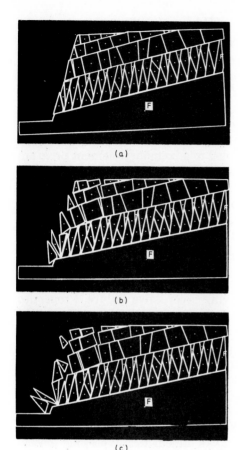

(a)

(b)

(c)

Figure 7.5 Distinct-element method, using Newton's second law to simulate the movement of blocks in a rock slide. (*Cundall, Voegele, and Fairhurst, 1975.*)

contact between blocks is lost. Newtonian laws of motion permit the model to simulate gravitational forces and movements realistically, including large displacements and rotations, by solving at a number of time steps. At each time step, each block is only allowed to interact with its immediate neighbors, but over many time steps, all effects propagate throughout the entire block system. Energy-dissipating mechanisms are introduced to damp oscillations, and to simulate energy losses resulting from plastic deformation on impact.

These methods require considerably less computer storage than a finite-element program, but are demanding in their requirements for computer time, because of the need for extremely large numbers of solutions. The computational effort expands by approximately the square of the number of blocks. A microcomputer version FLAC is now available.

Lemos et al. (1985) discuss further development of the distinct-element method for modeling jointed rock. A recent version, the university distinct-element code (UDEC), provides most capabilities that existed separately in previous programs. It can model variable rock deformability, nonlinear elastic behavior of the joints, plastic behavior and fracture of intact rock, and fluid flow and fluid pressure generation in joints and voids. An automatic joint generator in the program produces joint patterns based on statistically derived joint parameters. UDEC simulates the influence of the far-field rock mass for both static and dynamic conditions. It is coupled to a boundary-element program that represents the effects of a static, elastic, far-field response. Nonreflecting boundary conditions are available for dynamic simulations. A three-dimensional version, 3-DEC, is now available.

This method has several advantages. It automatically identifies mechanisms that are kinematically possible, and identifies the most critical, whereas other methods simply solve a stipulated mechanism. Large movements of fragments and blocks can be modeled, which is not possible with the classical methods of continuum mechanics. The method is ideally suited for studying progressive failure. Properties such as the coefficient of friction on selected joint sets can be varied incrementally to explore limiting conditions. The method has been used to study the stability of slopes and underground excavations of many kinds in jointed rock, even flow of angular material through a hopper to study blockage when the throat becomes too small with respect to the particle size.

7.2.5.2 Key block method. The key block concept (Goodman and Shi, 1981) identifies potentially loose blocks (kinematically possible mechanisms) in exposed rock faces and underground excavations. In any jointed rock mass, certain "key" blocks can be removed or fall from rock faces without breaking intact rock, whereas others are locked in place until key blocks are removed. Goodman and Shi have developed a powerful approach that uses analytical geometry to identify the key blocks from jointing statistics. Rockbolting and other support requirements can thus be planned in the most economical and efficient manner.

The method automatically identifies the removable blocks formed by a planar or a cylindrical cut, and computes the forces needed to maintain these blocks in place. Shaft and tunnel orientations can be optimized with regard to support requirements. A *block reaction curve* relates block displacements and the forces needed to maintain equilibrium. The key block method in its present form assumes that the rock joints are continuous, and no fracturing or deformation of rock is permitted.

Warburton (1983) has developed methods for three-dimensional graphic display that allow an examination of the kinematic stability of key blocks (Fig. 7.1). The program can reconstruct the hidden three-dimensional block structure around a complex excavation, eliminate selected blocks, and analyze the stabilities of individual blocks. It can also be used to examine and analyze the stabilities of blocks that would be exposed if the excavation were modified or extended.

7.2.6 Limiting equilibrium method

7.2.6.1 Overview. The limiting equilibrium method is commonly used for analyzing the stability of earth and rock slopes and dam foundations and, to a limited extent, for the design of underground excavations. It also is pertinent to the prediction of earthquakes, which are caused mainly by slippage along faults.

Unlike the techniques of continuum mechanics, the limiting equilibrium method does not assess stresses, strains, or displacements. Instead, it considers the equilibrium of forces acting on a potentially unstable mass such as a rockslide or dam. "Resisting" forces that tend to prevent sliding are compared with "disturbing" forces that tend to promote sliding, to determine whether or not sliding will occur.

7.2.6.2 Block and slope example. The method can be illustrated by the example of a rock block resting on a flat surface (Fig. 7.6). We wish to calculate the angle of tilt at which sliding starts, and the factor of safety against sliding for any lesser degree of tilt.

The principle of *limiting equilibrium* states that at some critical inclination of limiting equilibrium, when the block is on the point of sliding, the up-slope resistance and down-slope forces are exactly in balance. The ratio of up-slope resistance to down-slope forces is a measure of the *factor of safety* against sliding. If it is greater than 1.0, a resultant up-slope component will keep the block in stable equilib-

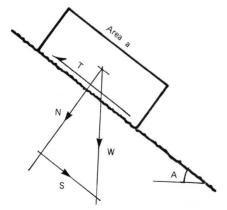

Figure 7.6 Worked example of limiting equilibrium analysis, showing a block sliding on an inclined plane.

rium. If it is less than 1.0, an out-of-balance down-slope force will cause acceleration, as predicted by Newton.

First, the model is sketched to show the geometry of the problem (contact area a, slope angle α) and the directions of forces acting on the block (weight W, frictional restraining force T). For equilibrium, forces along the slide plane direction must balance. To equate these forces, the weight W must first be resolved into two components, S acting parallel to the slide plane and N acting perpendicular. By sketching the triangle of forces, the following relationships are established:

$$\frac{N}{W} = \cos \alpha \qquad N = W \cos \alpha$$

$$\frac{S}{W} = \sin \alpha \qquad S = W \sin \alpha$$

The frictional restraining force T is now calculated as a function of the normal force N using some law of friction. For simplicity we will use the Mohr-Coulomb criterion with zero cohesion (Chap. 11):

$$\frac{T}{a} = \frac{N}{a} \tan \phi$$

$$T = N \tan \phi$$

where ϕ is the angle of shearing resistance.

In this case there is just one disturbing force, S, and one resisting force, T, so for any angle α the safety factor is given by the relationship

$$F = \frac{S}{T}$$

$$= \frac{W \sin \alpha}{W \cos \alpha \tan \phi}$$

$$= \frac{\tan \alpha}{\tan \phi}$$

This is the required value of the safety factor for angles α up to the point of sliding. For this case of zero cohesion, the safety factor does not depend on the size or weight of the block. Now consider the special case of limiting equilibrium:

$$F = 1.0$$

$$\tan \alpha = \tan \phi$$

$$\alpha = \phi$$

The block starts to slide when the angle of tilt reaches the angle of shearing resistance. This result permits the determination of the (peak) angle of shearing resistance by a tilt test (Chap. 11).

7.2.6.3 Advantages and limitations. The limiting equilibrium method has the advantage that the engineer need not measure the strength or elastic properties of the rock material. Requirements include forecasts of the locations of potential surfaces of sliding and the shear strengths along these surfaces; the unit weight of rock; and the water pressures, seismic accelerations, and restraining forces of anchors and other support systems. These data are usually easier to obtain than bulk rock characteristics such as deformability and strength.

The method gives a yes-no answer with regard to stability, which, rightly or wrongly, has considerable appeal when engineering and planning decisions are to be made.

The advantages are to some extent offset by the method's inability to predict the strains and displacements that precede or accompany sliding. Other methods predict displacements and permit useful comparisons with the results of monitoring during construction. The precision and reliability of the limiting equilibrium method are limited first by the difficulty in forecasting the persistence of discontinuities, which has a very great influence on shear strength; second by scale effects in extrapolating shear strength values from the laboratory to the field; and third by the possibility of modifications to strength and water pressures as movements develop. These uncertainties are usually taken into account by increasing the factor of safety, guided by a parametric study and judgment, or by probabilistic methods (Whitman, 1984).

Vector analysis techniques have been developed, initially for simple slabs or wedges, and more recently for a multifaceted block (Warburton, 1981). Three-dimensional *wedge slides* are analyzed using limiting equilibrium methods, generating design charts that permit the rapid analysis of simple slide geometries. More rigorous calculations require computation based on measurements of shear strength or back-analyses of existing slides (Hoek and Bray, 1973).

7.3 Analog and Physical Models

7.3.1 Analog models

7.3.1.1 Concept of the analog. Equations such as those governing electric potential differences and currents, with suitable transformations, are identical in form (analogous) to those governing stress distributions and water or heat flow. This enables stress, water, or heat flow problems to be simulated and solved by electric analog methods. Largely superceded by numerical computations, analogs remain a useful aid to teaching because they are easily controlled and visualized, and are still employed in some practical applications.

7.3.1.2 Electric analogs. The conducting-paper analog method employs a sheet of paper coated or impregnated with conducting graphite or aluminum. Boundary voltages are applied that generate currents and electric potentials, which vary across the surface of the conducting paper. Excavations such as tunnels can be cut into the paper to examine the stress distributions around them. The applied boundary voltages are analogous to boundary stresses, and the interior voltages are analogous to the sum of principal stresses within an elastic material. This principal stress sum can be combined with the principal stress difference, obtained using a photoelastic analog as described below, to "separate" the individual principal stresses at any number of points of interest. Flow lines are in this case analogous to stress trajectories.

The same analog can be used to simulate water flow, in which case, boundary voltages become analogous to hydrostatic heads (Sec. 4.4.2). Three-dimensional problems can be solved by transformation of the problem into two dimensions, or by using an electrolytic tank.

7.3.1.3 Base friction models. A base friction model is an attempt to qualitatively simulate gravity in the laboratory. A two-dimensional model of blocks, representing the blocks in a rock mass, is constructed on a continuous loop of sandpaper. Gravity is simulated by moving the sandpaper belt beneath the model. The base friction on each block is proportional to its contact area (weight) and acts through its centroid. Although block momentum is ignored, the models are useful because they permit the designer to study quickly and economically the effects of rock mass displacements under gravity, and the kinematically possible mechanisms (Ergunvanli and Goodman, 1972; Bray and Goodman, 1981; Goodman, 1980).

As an alternative to strong and rigid blocks, a modeling material can be used such as a weak mixture of sand, oil, and flour (Stimpson, 1970). The blocks are then weak enough to fail through the intact material as well as by sliding along joint planes. A layer of material 1 or 2 cm thick is troweled within a frame to produce a compacted homogeneous slab. Simulated rock joints can be cut into the slab, and the sandpaper belt displaced a short distance to consolidate the newly jointed model. Slopes or underground excavations can be constructed by "excavating" the modeling material before applying the simulated gravitational forces. The sandpaper belt is then started, observing the progressive displacements of blocks and the ways in which rupture develops.

7.3.1.4 Photoelastic models. A photoelastic model is a type of analog that displays and permits the measurement of the pattern of stresses in an elastically behaving rock mass (Hoek, 1967). As a teaching aid it illustrates vividly the concept of stress distribution, and it is still used for solving some practical three-dimensional problems of elasticity.

The model is machined from a stress-birefringent material like glass or plastic; for example, a tunnel is represented as a circular hole in a plate. When polarized light is projected through a stressed, stress-birefringent material, patterns of colored or black fringes are produced (Fig. 7.7). Plane-polarized light gives black fringes (*isoclinics*) along the trajectories of major principal stress. In other words, the fringes give the directions of principal stress. Circularly polarized light produces colored fringes (*isochromatics*) along contours of constant shear stress. The colored fringes can be used to determine the magnitude of the principal stress difference, that is, the shear stress acting at any point in the model.

Models made of rigid materials such as glass, or moderately rigid ones like polyurethane rubber or epoxy resin, have to be loaded at the boundaries to simulate the *far-field* stresses acting at a remote distance. Models constructed from a highly deformable material such as gelatin develop photoelastic patterns under their own weight. Photoelastic models can also be constructed out of blocks or grains of

Figure 7.7 Photoelastic study of stresses around a propagating crack. (*Camponuovo et al., 1980.*)

photoelastic material to simulate a jointed rock or a granular soil. The method is time-consuming and requires considerable skill and experience to select an appropriate modeling material, which is usually annealed to remove residual stresses.

For two-dimensional studies, photoelastic patterns can be observed directly, photographed, and measured at different stress levels. The alternative of *stress freezing* is available to solve problems in three dimensions. The model is first loaded over the faces of the block, then placed in an oven while maintaining the load at the required level. With the appropriate choice of modeling material the photoelastic patterns will be "frozen" if the temperature is gradually reduced to room temperature while maintaining the loads. The model is then sliced in the required directions of observation and viewed in polarized light to determine three-dimensional distributions of stresses such as around a central opening.

As an example, Camponuovo et al. (1980) employed three-dimensional photoelastic models together with holography and finite-element analysis to study crack propagation by hydraulic fracturing (Fig. 7.8). For the photoelastic analysis, the classical technique of stress freezing and slicing was used. Cracks were propagated by air pressure applied through drillholes in an epoxy block.

7.3.2 Physical models

7.3.2.1 Overview. The purpose of a physical or "realistic" model is to simulate, in the laboratory, the behavior of the full-scale prototype. Careful selection of modeling materials and loading methods allows the observation and measurement of not just elastic behavior, but also plastic deformation, viscous flow, and fracture (Fig. 7.8). Physical models have been used, for example, to investigate the behavior of rock slopes, dams, and mining and civil engineering underground excavations at various stages of construction (Fumagalli, 1968; Stillborg et al., 1979; Everling, 1982) and even geological processes (Hubbert, 1937).

When properly designed and constructed, the physical model can be a valuable aid to design, such as in the design of arch dams, which are a complex three-dimensional combination of rock and concrete. To analyze such dams in combination with their foundations by numerical methods is hardly possible even nowadays. The physical model has a further important advantage (again, when properly constructed) in that it determines its own mode of deformation and failure. Sometimes this follows an unexpected and informative pattern. The major disadvantages of the physical model are that it must be large for fi-

(a)

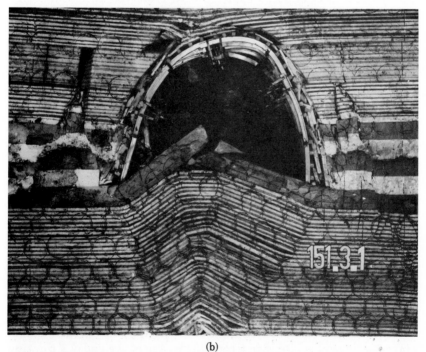

(b)

Figure 7.8 Realistic physical modeling of a mine roadway in stratified rock. (*a*) Large biaxial loading frame containing the model. (*b*) Mine roadway model with simulated steel sets, after loading. Floor heave and buckling of roof strata are apparent. (*Everling, 1982; photos courtesy of Jürgen te Kook, Bergbau-Forschung GmbH, Essen, West Germany.*)

delity, requires considerable time and labor to construct, and must be rebuilt for each new study.

7.3.2.2 Model materials. A physical model is constructed of weak fabricated materials, and loaded with real forces to simulate those that will be acting in the prototype structure. Modeling materials can be blended to give any desired combination of properties to simulate stratification, jointing, and other realistic geological features. The models are built from materials chosen according to principles of dimensional similitude to scale down prototype properties such as density, strength, and deformability.

A scale model behaves similarly to the full-scale prototype only if the modeling materials and applied loads are carefully selected. Otherwise the results are impressive but often grossly misleading. The scaling ratios from prototype to model, not only of dimensions but also of properties such as strength and deformability, must be selected to comply with modeling theory as expressed in Buckingham's pi theorem (Roxborough and Eskikaya, 1974; Bakhtar et al., 1986). Often the model materials are loaded under their self-weight, that is, normal gravitational accelerations apply in both prototype and model. This requires a modeling material that is as heavy as possible but much weaker and more deformable than the prototype material. Most rock structure models use compacted mixtures of plaster and lead oxide, or of sawdust and oil.

A realistic modeling material used by Barton (1972) to simulate open-pit slopes consisted of a mixture of lead oxide and gypsum plaster. Joints guillotined in the model had roughnesses and shear strengths that scaled well with rock joints in the open-pit prototype. Two-dimensional slab models were constructed horizontally and then rotated vertically to apply self-weight loading. Extreme weakness of the jointed model allowed the open-pit mine to be "excavated" with a small vacuum cleaner that enlarged the pit and steepened its side slopes. The progressive development of pit wall failures could be observed through the glass walls confining the model.

7.3.2.3 Model loading. Realistic scaling of stresses in the model and prototype means increasing the self-weight of the model in proportion to its reduction in size. High densities are difficult to achieve, at least while at the same time correctly scaling the strengths and deformabilities.

One alternative is to increase the self-weight by centrifuge loading. The entire model is placed in a centrifuge cage at the end of a rotating arm and subjected to centrifugal forces of several times normal gravitation (Sugawara et al., 1983). Centrifuge model testing is expensive, first because of the cost of the centrifuge, and also because of the difficulty of taking measurements while the model is rotating at high speed. In most countries there is but one geomechanically oriented centrifuge for research, not normally used for routine design work.

A further alternative is to apply loads at the boundaries of the model to simulate those assumed to be acting at similar locations in the prototype, and to ignore the stress gradients that result from actual gravitational loading. This approximation is less realistic but adequate in most cases, particularly for underground excavations in rock at moderate to great depth. Most large-scale models, such as the dam models constructed at LNEC in Portugal and at ISMES in Italy, have body forces simulated by external loading. They are using many small jacks or inflatable rubber balloons for a more uniform pressure application.

References

Ahlvin, R. G., and H. H. Ulery: "Tabulated Values for Determining the Complete Pattern of Stresses, Strains and Deflections beneath a Uniform Circular Load on a Homogeneous Half-Space," *Highway Res. Board Bull.* 342 (1962).

Bakhtar, K., A. H. Jones, and R. E. Cameron: "Use of Rock Simulators for Rock Mechanics Studies," *Proc. 27th U.S. Symp. Rock Mech.* (Tuscaloosa, Ala., 1986), pp. 219–223.

Banerjee, P. K., and R. Butterfield: *Boundary Element Methods in Engineering Science* (McGraw-Hill, London, 1981).

Baoshen, L., and L. Dezhang: "Linear Visco-Elastic Analysis in Rock-Support System of Roadway," in M. Borecki and M. K. Wasniewski, Eds., *Application of Analytic Methods to Mining Geomechanics (World Mining Congress)* (A. A. Balkema, Rotterdam, 1981), pp. 117–124.

Barton, N. R.: "A Model Study of Rock Joint Deformation," *Int. J. Rock Mech. Min. Sci.,* 9, 579–602 (1972).

———, R. Lien, and J. Lunde: "Engineering Classification of Rock Masses for Design of Tunnel Supports," *Rock Mech.,* 6, 189–236 (1974).

Beer, G., R. Cowling, and S. Bywater: "A Novel Three-Dimensional Joint Element with Application to the Modelling of Mining Excavations," *Proc. Int. Symp. Fundamentals of Rock Joints* (Björkliden, Sweden, 1985), pp. 357–365.

Bieniawski, Z. T.: *Rock Mechanics Design in Mining and Tunneling* (A. A. Balkema, Rotterdam and Boston, 1984).

Brady, B. H. G., and J. W. Bray: "The Boundary Element Method for Elastic Analysis of Tabular Ore Body Extraction, Assuming Complete Plane Strain," *Int. J. Rock Mech. Min. Sci. Geomech. Abstr.,* 15, 29–37 (1978).

Bray, J. W., and R. E. Goodman: "The Theory of Base Friction Models," *Int. J. Rock Mech. Min. Sci. Geomech. Abstr.,* 18, 453–468 (1981).

Brebbia, C. A., and S. Walker: *Boundary Element Techniques in Engineering* (Newnes-Butterworths, London, 1980), 210 pp.

Brown, E. T., Ed.: *Analytical and Computational Methods in Engineering Rock Mechanics* (Allen and Unwin, London, 1987), 259 pp.

Camponuovo, G. F., A. Freddi, and M. Borsetto: "Hydraulic Fracturing of Hot Dry Rocks. Tri-dimensional Studies of Crack Propagation and Interaction by Photo-

Elastic Methods," Rep. 136, Inst. Experim. Models and Structures (ISMES), Bergamo, Italy, 12 pp. (1980).

Crawford, A. M., and J. W. Bray: "Influence of the in situ Stress Field and Joint Stiffness on Rock Wedge Stability in Underground Openings," *Can. Geotech. J.,* **20,** 276–287 (1983).

Crouch, S. L.: "Solution of Plane Elasticity Problems by the Displacement Discontinuity Method, Pts. 1 and 2," *Int. J. Numer. Methods Eng.,* **10,** 301–343 (1976).

—— and A. M. Starfield: *Boundary Element Methods in Solid Mechanics* (Allen and Unwin, London, 1983), 322 pp.

Cundall, P. A.: "A Computer Model for Simulating Progressive, Large-Scale Movements in Blocky Rock Systems," *Proc. Int. Symp. Rock Fracture* (Nancy, France, 1971), paper II-8, 12 pp.

——: "Computer Interactive Graphics and the Distinct Element Method," in *Rock Engineering for Foundations and Slopes, Proc. ASCE Specialty Conf.* (Boulder, Colo., 1976), vol. 2, pp. 193–199.

——, M. Voegele, and C. Fairhurst: "Computerized Design of Rock Slopes Using Interactive Graphics for the Input and Output of Geometrical Data," *Proc. 16th U.S. Symp. Rock Mech.* (Minneapolis, Minn., 1975), pp. 5–14.

Deering, J. A. C.: "Simulation of Mining in Non-homogeneous Ground Using the Displacement Discontinuity Method," *J. South Afr. Inst. Min. Metall.,* **80,** 225–228 (1980).

Desai, C. S., I. M. Eitani, and C. Haycocks: "An Application of Finite Element Procedure for Underground Structures with Nonlinear Materials and Joints," *Proc. 5th Int. Cong. Rock Mech.* (Melbourne, Australia, 1983), vol. F, pp. 209–216.

Dowding, C. H., T. B. Belytschko, and H. S. Yen: "Coupled Finite Element–Rigid Block Method for Transient Analysis of Rock Caverns," *Int. J. Numer. Anal. Methods Geomech.,* **7** (1), 117–127 (1983).

Ergunvanli, K. A., and R. E. Goodman: "Applications of Models to Engineering Geology for Rock Excavations," *Bull. Assoc. Eng. Geol.,* **9** (2), 89–104 (1972).

Everling, D.: "Models for Strata Control and Support Dimensioning: Successes and Present Achievements," Glukauf transl. 118, no. 1, pp. 9–12; German version, pp. 16–23 (1982).

Frei, L. R.: "Auburn Dam Foundation Investigation, Design and Construction," *Field Trip Guide Book, Assoc. Eng. Geologists' 18th Ann. Mtg.* (Lake Tahoe, Calif., 1975), 20 pp., rev. Apr. 1976.

Fumagalli, E.: "Model Simulation in Rock Mechanics Problems," in K. G. Stagg and O. C. Zienkiewicz, Eds., *Rock Mechanics in Engineering Practice* (Wiley, New York, 1968), pp. 353–384.

Geertsma, J., and G. van Opstal: "A Numerical Technique for Predicting Subsidence above Compacting Reservoirs, Based on the Nucleus of Strain Concept," *Verhand. Kon. Ned. Geol. Mijnbouwk Gen.,* **28,** 63–78 (1973).

Goodman, R. E.: *Introduction to Rock Mechanics* (Wiley, New York, 1980), 478 pp.

—— and G. H. Shi: "Geology and Rock Slope Stability—Application of a 'Keyblock' Concept for Rock Slopes," *Proc. 3d Int. Conf. Stability in Surface Mining* (Vancouver, B.C., 1981), pp. 347–373.

——, R. L. Taylor, and T. L. Brekke: "A Model for the Mechanics of Jointed Rock," *J. Soil Mech. Found. Eng. Div., ASCE,* **94** (SM3), 637–659 (1968).

Hobbs, N. B.: "Effects of Non-linearity on the Prediction of Settlements of Foundations on Rock," *Quart. J. Eng. Geol.* (London), **6** (2), 153–168 (1973).

Hoek, E.: "Photoelastic Technique for Determination of Potential Fracture Zones in Rock Structures," *Proc. 8th U.S. Symp. Rock Mech.* (Minneapolis, Minn., 1967), pp. 94–112.

—— and J. W. Bray: *Rock Slope Engineering* (Inst. Mining and Metall., London, 1974), 309 pp.

Hornbeck, R. W.: *Numerical Methods* (Quantum Publ., New York, 1975), 310 pp.

Hubbert, M. K.: "Theory of Scale Models as Applied to the Study of Geologic Structures," *Bull. Geol. Soc. Am.,* **49,** 1459–1520 (1937).

IAEG: *Engineering Geology Related to Nuclear Waste Disposal Projects,"* Colloq. vol., *Bull. Int. Assoc. Eng. Geol.,* 34, 131 pp. (1986).

Jaeger, C.: *Rock Mechanics and Engineering,* 2d ed. (Cambridge Univ. Press, London, 1979), 523 pp.

—— and N. G. W. Cook: *Fundamentals of Rock Mechanics* (Chapman and Hall, London, 1979), 593 pp.

Kemeny, J., and N. G. W. Cook: "Frictional Stability of Heterogeneous Surfaces of Contact," *Proc. 27th U.S. Symp. Rock Mech.* (Tuscaloosa, Ala., 1986), pp. 40–46.

Lemos, J. V., R. D. Hart, and P. A. Cundall: "A Generalized Distinct Element Program for Modelling Jointed Rock Mass—A Keynote Lecture," *Proc. Int. Symp. Fundamentals of Rock Joints* (Björkliden, Sweden, 1985), pp. 335–343.

Lorig, L. J., and B. H. G. Brady: "A Hybrid Discrete Element Boundary Element Method," *Proc. 23d U.S. Symp. Rock Mech.* (Berkeley, Calif., 1982), pp. 628–636.

Milovic, D. M.: "Stresses and Displacements in an Anisotropic Layer Due to a Rigid Circular Foundation," *Geotechnique* (London), 22 (1), 169–173 (1972).

Mindlin, R. D.: "Forces at a Point in the Interior of a Semi-infinite Solid," *Physics,* 7, 195–202 (1936).

—— and D. H. Cheng: "Nuclei of Strain in the Semi-infinite Solid," *J. Appl. Phys.,* 21, 926–930 (1950).

Mraz, D. Z.: "Solutions to Pillar Design in Plastically Behaving Rocks," *Bull. Can. Inst. Min. Metall.,* 77 (868), 55–62 (1984).

Nyak, M.: "Elastic Settlement of a Cross-anisotropic Medium under Axi-symmetric Loading," *Soils and Foundations* (Tokyo), 13 (2), 83–90 (1973).

Obert, L., and W. I. Duvall: *Rock Mechanics and the Design of Structures in Rock* (Wiley, New York, 1967).

Peirce, A. P., and J. A. Ryder: "Extended Boundary Element Methods in the Modeling of Brittle Rock Behavior," *Proc. 5th Int. Cong. Rock Mech.* (Melbourne, Australia, 1983), vol. F, pp. 159–167.

Poulos, H. G., and E. H. Davis: *Elastic Solutions for Soil and Rock Mechanics* (Wiley, New York, 1974).

Roxborough, F. F., and S. Eskikaya: "Dimensional Considerations in the Design of a Scale Model for Coal Face Production System Research," *Int. J. Rock Mech. Min. Sci. Geomech. Abst.,* 11, 129–137 (1974).

Sakurai, S.: "Field Measurements for the Design of the Washuzan Tunnel in Japan," *Proc. 5th Int. Cong. Rock Mech.* (Melbourne, Australia, 1983), vol. A, pp. 215–218.

Savin, G. N.: *Stress Concentration around Holes* (Pergamon, Oxford, 1961).

Sneddon, I. N., and M. Lowengrub: *Crack Problems in the Classical Theory of Elasticity* (Wiley, New York, 1969).

Sofianos, A. I.: "Stability of Rock Wedges in Tunnel Roofs," *Int. J. Rock Mech. Min. Sci. Geomech. Abstr.,* 23 (2), 119–130 (1986).

Stillborg, B., O. Stephansson, and G. Swan: "Three-dimensional Physical Model Technology Applicable to the Scaling of Underground Structures," *Proc. 4th Int. Cong. Rock Mech.* (Montreux, Switzerland, 1979), vol. 2, pp. 655–662.

Stimpson, B.: "Modelling Materials for Engineering Rock Mechanics," *Int. J. Rock Mech. Min. Sci.,* 7 (1), 77–121 (1970).

Sugawara, K., M. Akimoto, K. Kaneko, and H. Okamura: "Experimental Study on Rock Slope Stability by the Use of a Centrifuge," *Proc. 5th Int. Cong. Rock Mech.* (Melbourne, Australia, 1983), vol. C, pp. 1–4.

Szechy, K.: *The Art of Tunneling* (Akademiai Kiado, Budapest, 1966), 891 pp.

Vandillen, D. E., and C. M. St. John: "An Analysis of Ground Support through Fully Grouted Rockbolts," *Proc. 25th U.S. Symp. Rock Mech.* (Evanston, Ill., 1984), pp. 231–238.

Warburton, P. M.: "Vector Stability Analysis of an Arbitrary Polyhedral Rock Block with Any Number of Free Faces," *Int. J. Rock Mech. Min. Sci. Geomech. Abstr.,* 18 (5), 415–427 (1981).

——: "Applications of a New Computer Model for Reconstructing Blocky Rock Geometry—Analysing Single Block Stability and Identifying Keystones," *Proc. 5th Int. Cong. Rock Mech.* (Melbourne, Australia, 1983), vol. F, pp. 225–230.

———: "Displacement Discontinuity Method for Three-dimensional Stress Analysis of Tabular Excavations in Nonhomogeneous Rock," *Proc. 25th U.S. Symp. Rock Mech.* (Evanston, Ill., 1984), pp. 702–709.

Wardle, L. J., and J. R. Enever: "Application of the Displacement Discontinuity Method to the Planning of Coal Mine Layouts," *Proc. 5th Int. Cong. Rock Mech.* (Melbourne, Australia, 1983), vol. E, pp. 61–69.

Whitman, R. V.: "Evaluating Calculated Risk in Geotechnical Engineering," *J. Geotech. Eng. Div., ASCE*, **110** (2), 145–188 (1984).

Wittke, W., C. Erichsen, and M. Kleinschnittger: "Influence of Seepage and Uplift Forces on Stresses in the Rock Foundations of Arch Dams," *Proc. 5th Int. Conf. Numerical Methods in Geomech.* (Nagoya, Japan, 1985), pp. 1787–1793.

Zienkiewicz, O. C., and Y. K. Cheung: *The Finite Element Method in Structural and Continuum Mechanics*, 3d ed. (McGraw-Hill, London, 1977).

Chapter

8

Strength

8.1 Introduction

8.1.1 Strength concepts

8.1.1.1 Failure, rupture, strength, and yield. Terminology is best explained at the outset, because many terms are used loosely and can be confusing. Adoption of a consistent terminology is strongly recommended.

Failure is the loss of capacity to perform the stipulated function for which a structure was designed. The term does not refer to load-carrying capacity exclusively; it can refer to other functions, such as the ability to retain water, as in the case of a dam or a pressure tunnel.

Rupture is the breakage or fracturing that occurs when rock is stressed beyond its limit of strength. The term failure must not be used to mean rupture, because many excavations remain stable and serviceable in spite of the development of ruptured zones in the roof or wall rock. These zones can be retained by surrounding rock and support systems, and continue to carry a portion of the load.

Strength refers to the stress state at which a rock specimen or rock mass element ruptures. Further definitions of strength are introduced in Chap. 11, in the context of rock joint behavior.

Yield refers to the stress level at which rock ceases to behave elastically, so that when unloaded, it no longer returns to its original shape. Further discussion of yield is deferred until Chaps. 9 and 10, where the topic of rock deformation is introduced.

In order to measure strength, the point of rupture must be identified. When it occurs abruptly, rupture is accompanied by a sudden drop in the applied load, usually by a noise, and sometimes by a loud bang. However, it may be gradual and difficult to detect when testing weak rocks at high confining pressures, and when the specimen is hidden inside a testing device. In such cases, the concept of rupture no longer has physical significance and is replaced by one of yield, or of strain as a continuous function of stress.

8.1.1.2 Relevance of strength to engineering design. Much rock strength testing is done for the design of deep underground mining projects, where even strong rocks can burst or yield. In weaker rocks such as shales, strengths can be exceeded even at quite shallow depths, so strength testing is relevant in civil projects also.

Design in brittle rocks, where rupture predominates over yield, relies mainly on predicting stress distributions around openings and comparing these with strengths. Wherever stresses exceed strengths, the rock will rupture. The extent of probable ruptured zones can readily be mapped, and appropriate precautions can be taken.

Design methods in more ductile types of rock, in which yield predominates, explore the extent of yielding zones around an excavation. However, the computations become more complex, because not only stresses but also strains and displacements must be predicted, often as a function of time.

8.1.2 Strength criteria

A *strength criterion* is an equation used to check whether rupture will occur under the combination of the three principal stresses predicted at a particular location.

The strength criterion would be very simple if rock strength were a single, unique value, but this is not the case. For example, uniaxial compressive strength is usually at least eight times the tensile strength, and compressive strength itself is greatly increased by confinement. A realistic strength criterion takes the form of a graph or equation that gives all combinations of stress under which the rock will rupture. This leads to a need for various strength tests to obtain the parameters for full definition of all possible rupture conditions.

8.1.2.1 Stress space. The full strength criterion for a material can be plotted as a surface in three-dimensional *stress space* with the principal stress components as cartesian axes (Fig. 8.1). This *strength surface* is the locus of all combinations of principal stresses just sufficient to cause rupture. The surface lies dominantly in the octant where all three stresses are compressive, because rock is weak in tension.

Suppose that a cubic rock specimen is subjected to increases in the

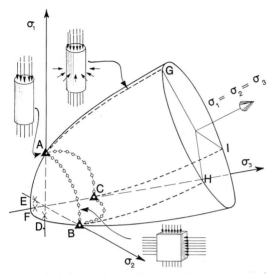

Figure 8.1 Strength surface in three-dimensional stress space. Different parts of the surface are explored by different tests. Points *A, B,* and *C* = uniaxial compressive tests; points *D, E,* and *F* = uniaxial tensile tests; curves *AG, BH,* and *CI* = axisymmetric triaxial tests; curves *AB, BC,* and *CA* = biaxial compression tests. (*Franklin, 1971.*)

principal stress components applied normal to each face. The stress history of the cube can be plotted as a *stress path* that wanders through stress space. While the path lies within the strength envelope, the cube remains intact, but as soon as the envelope is touched, rupture occurs. The element cannot be stressed to a condition outside the strength envelope. It may continue to bear the rupture stresses, but usually experiences a drop in load-carrying capacity.

As shown in Fig. 8.1, different types of test are used to explore and define different parts of the strength envelope in stress space.

The uniaxial compressive and tensile strength tests apply stress only to two opposing faces of a cube or ends of a cylindrical specimen. The stress path must therefore move along a principal stress axis, because the other two principal stresses are zero. The points of rupture are *A, B,* and *C* of uniaxial compressive strength, and *D, E,* and *F* of uniaxial tensile strength.

Other tests are needed to map the more extensive and important part of the envelope. The easiest to perform is the axisymmetric triaxial compression test, described later, in which two principal stresses are maintained constant and equal by a confining fluid pressure, and the third is increased until rupture occurs. The stress path moves in the planes where two principal stresses are equal, to define three curves on the strength envelope, *AG, BH,* and *CI.*

The biaxial test on a plate or cube of rock gives *plane stress* conditions where one principal stress is zero, defining the rupture curves *AB, BC,* and *CA.*

Simple tests are the main focus of this chapter. More complicated tests, mentioned only briefly, are needed for research, to delineate other regions of the strength envelope, and to check various assumptions and theories of rupture. Because it is nearly symmetric, the complete strength surface for an isotropic material can be approximated, usually with sufficient accuracy, from the curves *AG, BH,* and *CI.*

We can define, again by testing, a *yield surface* or *yield envelope* that lies entirely within the strength envelope, and is often subparallel to it. The yield envelope marks the boundary between elastic and plastic modes of deformation, whereas the strength envelope delineates the boundary between possible and impossible states of stress on rock specimens. Identification of yield requires strain measurements, particularly residual strain after unloading. Yield and the stress-strain behavior of rock are the topics of Chaps. 9 and 10.

8.2   Strength Testing Methods

8.2.1   Tests and testing precautions

Tests can be conducted on small or very large cores or blocks, either in the laboratory or in the field. The choice is also available between slow, moderate, and rapid rates of loading, and between specimens that are wet or dry. The factors affecting the choice of one or the other alternative remain much the same for each of the many types of test available.

8.2.1.1   Joints in the specimen, and the effects of scale.   For designs based on continuum mechanics, specimens must adequately represent an element of that continuum. Either they must contain many joints, or intact rock can be tested and the measured properties adjusted to include the anticipated effects of jointing. Tests on rock containing just two or three joints are difficult to interpret, and should be avoided. Tests on individual joints are required for discontinuum types of analysis, including limiting equilibrium calculations, discrete block numerical modeling, and hybrid methods that combine the behavior of joints with a continuum approach. The behavior and testing of individual joints is the subject of Chap. 11.

The choice of specimen size and type of test is determined by design requirements, rock conditions, and cost. Evidently, small specimens of intact rock are the least expensive to obtain and test.

Larger specimens are found to be weaker than small ones, perhaps because they are more likely to contain flaws (joints and cracks) at unfavorable angles (the "weak link" theory). The drop in strength is more pronounced at smaller sizes. The test results of Fig. 8.2 for diorite and coal indicate an approximately threefold decrease in strength for specimen sizes of 50–200 mm, and only a twofold decrease between 200 mm and 1 m. The strength of in situ rock must be estimated either by large-scale testing on specimens that include joints, or by extrapolating the results of small-scale testing in some manner such as suggested by Herget and Unrug (1976).

8.2.1.2 Influence of stress rate. Rates of applying stress to a specimen should more or less match those expected in the prototype to be analyzed, although it is difficult or impossible to simulate the very rapid stress changes associated with blasting, or the slow changes of geological events or even of rock excavation. Test methods are therefore usually standardized, and the results adjusted if necessary. Routine tests take no more than 10 min and usually less. Completing the test in less than a second (dynamic loading) may double the strength. Slowing the test down sufficiently for rupture to take several days can reduce the strength by up to 30%. If a constant uniaxial stress is maintained at a level greater than about two-thirds of short-term (10-min) strength, most rocks will eventually rupture.

The time-dependent variation of strength is probably caused by stress concentrations at the tips of flaws: prolonged loading permits local time-dependent crack growth. Rapid loading precludes time-dependent crack growth and gives higher strengths.

8.2.1.3 Effects of moisture. Specimens are weaker when moist. Colback and Wiid (1965) found that specimens tested after storage in an environment of 50% relative humidity were between 19 and 33% stronger than those tested in the water-saturated condition. The weakening effect of water is caused by a reduction of surface energy at grain boundaries and at the tips of internal flaws. In silicates, the water molecule (H–O–H) tends to hydrolize silicon-oxygen bonds (–Si–O–Si–) to give rise to hydroxyl groups (–Si–OH), and thereby weaken the bonds at microcrack tips. The energy required to generate new crack areas is therefore lower because of the presence of the water. An even more substantial strength increase is observed if specimens are artificially dried in an oven before testing. Similar effects hold true for all kinds of strength tests.

A further cause of strength decrease is the weakening effect of pore-water pressures in saturated rocks according to the principle of effective stress (Sec. 4.2.1.3). The strength criteria used in rock engineering are understood to be expressed with strength as a function of the *effective* stress between mineral grains, and not of the *total* stress applied to the specimen or rock element.

Pore pressures are hardly ever measured when testing rock specimens. The measurements are very difficult and imprecise in rocks with a porosity in the range of 0–5% and correspondingly low permeability. Rocks are usually tested at a moisture content as close to the field condition as possible, and slowly to prevent generation of excess pore pressures that would cause premature rupture and unrealistically low strength values. When applying the resulting strength criteria to rock stability calculations, pore-water or joint-water pressures must be subtracted from the total stress because they reduce the stress acting normal to a potential plane of sliding.

8.2.2 Uniaxial compressive strength

8.2.2.1 Outline of the method. The uniaxial or "unconfined" compressive strength test (ISRM, 1981) is used either in conjunction with triaxial testing to map the strength envelope or, more commonly, as an index test. In the latter application (Chap. 2) its value is limited by time-consuming specimen preparation requirements, giving an economic limit to the number of tests that can be performed.

To carry out a uniaxial compressive strength test, a right circular cylinder of rock is compressed between the platens of a testing machine. A load cell or calibrated pressure gauge is used to determine the applied peak load *P,* and the strength value σ_c is obtained by dividing *P* by the cross-sectional area *A:*

$$\sigma_c = \frac{P}{A} = \frac{4P}{\pi D^2}$$

Corrections to account for the increase in cross-sectional area are negligible if rupture is reached before 2–3% strain.

8.2.2.2 Platen conditions and specimen dimensions. Frictional restraint results in triaxial confinement close to the solid steel platens, which for relatively short specimens (length-to-diameter ratio $\approx$ 1.0) gives an unrealistically high strength value. Researchers have tried various measures, such as polishing the platens, lubricating them with graphite, or using special "brush" platens. These consist of a brush of closely spaced small steel bars, allowing the specimen to expand freely near its ends.

A more common method of avoiding the end constraint problem is to use specimens with a length-to-diameter ratio of between 2.5 and 3.0, in which case the central one-third of the specimen is in a "true" uniaxial state of stress, and it is in this region that rupture initiates. The platens should be of smoothly finished hardened steel. Specimen end-capping materials such as the sulfur used in concrete testing are not permitted when testing rock.

Specimen ends should be flat to 0.02 mm, and within 0.001 rad of perpendicular to the cylinder axis. The specimen diameter should be at least 50 mm (requiring NX or larger size core), and at least 10 times the size of the largest grain in the rock. Load should be applied through a spherical seat to accommodate any lack of parallelism of the end faces, and to prevent bending and nonuniformity of stress. The stress application rate is controlled to within 0.5–1.0 MPa/s, such that rupture occurs within 5–10 min (Fig. 8.2).

These strict specifications can be relaxed slightly for index tests and when testing soft rock types. The degree of flatness of the end faces, for example, is most critical when testing strong, stiff rocks, because then even small irregularities can give high local tensile stresses and premature rupture (Pells and Ferry, 1983).

Testing the uniaxial compressive strength of small cubes used to be standard practice, particularly in the coal mining, aggregate, and building-stone industries. Because of the platen constraint problems, which lead to unrealistically high strengths for small, squat specimens, the standardized alternative of cylinder testing has become almost universal. Cubic specimens continue to be preferred for field testing, where the much larger sizes of cube make these tests somewhat less affected by end constraint problems.

8.2.2.3 Anisotropy in the uniaxial compressive strength test. To evaluate the strength anisotropy of a rock, several uniaxial compressive strength tests are performed on core oriented at various angles to any foliation or plane of weakness. Strength is usually least when the foliations (or joints in the field) make an angle of about 30° to the direction of loading along the cylinder axis, and greatest when foliation is either parallel or perpendicular to the axis (Fig. 8.3). Only for joints or foliations within 20° of perpendicular to the direction of major principal stress is the strength approximately constant.

8.2.3 Triaxial strength

The familiar *triaxial* test is a convenient one in which equal intermediate and minor principal stresses are applied by fluid pressure to the curved surface of a rock cylinder. This therefore is a special case of general triaxial compression (Fig. 8.4c), and is more rigorously known as the *axisymmetric triaxial* test. It will be discussed after introducing other less familiar members of the triaxial testing family.

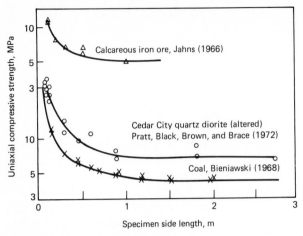

Figure 8.2   Effect of specimen size on its uniaxial compressive strength. (*After Bieniawski*, 1974.)

8.2.3.1 Hydrostatic compression test.   Rock stress is said to increase *hydrostatically* when all three principal effective stress components are equal (Fig. 8.4a). Referring back to Fig. 8.1, the stress path in reaching all hydrostatic stress states follows the "stress space diagonal," the line $\sigma_1 = \sigma_2 = \sigma_3$.

Dense rocks can sustain virtually unlimited increases in hydrostatic stress without rupture. However, some porous rocks compress under sufficient hydrostatic loads, with permanent disruption of the fabric. Collapse-prone rocks include chalk, porous coal, pumice lavas, and weathered igneous rocks with decomposed and dissolved mineral grains.

Hydrostatic tests are performed on jacketed rock specimens surrounded by a pressurizing fluid in a pressure vessel. Specimens are usually instrumented with strain gauges to measure compressibility as well as the yield or collapse point (Chap. 9).

8.2.3.2 Biaxial tests.   In a *biaxial* test (Fig. 8.4b), a plate (or cube) of rock is stressed along its edges. Brush platens or fluid-filled cushions give uniform edge loading. The stress path is limited to the "axial" planes of stress space that contain pairs of principal stress axes, because the stress perpendicular to the plate is zero. The strength data so obtained plot along curved lines on the three-dimensional rupture surface where it is cut by the axial planes (Fig. 8.1). However, biaxial loading has more commonly been used to apply smaller loads in the elastic range, to test models for photoelastic stress analysis (Sec. 7.3.1.4).

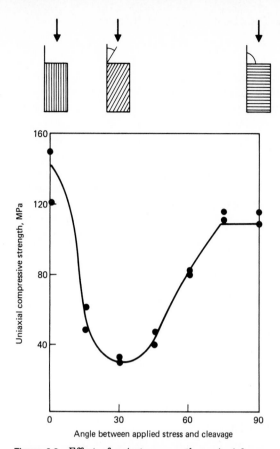

Figure 8.3 Effect of anisotropy on the uniaxial compressive strength of slate. Similar trends are found when testing anisotropic rocks in triaxial compression. (*Test data from Hoek, 1965.*)

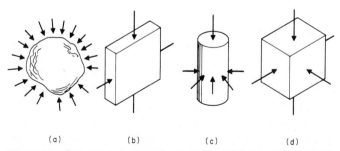

Figure 8.4 Compressive strength tests. (*a*) Hydrostatic (equal all round) compression. (*b*) Biaxial compression (minor principal stress is zero). (*c*) Axisymmetric triaxial test (intermediate and minor principal stresses are equal). (*d*) Polyaxial test (all three principal stresses can be varied independently).

Biaxial conditions are relevant to many rock engineering situations because they simulate stress conditions in a free unloaded surface of rock such as in the wall of an underground excavation or the face of a rock slope. However, they are much more difficult to perform and are seldom used except for research purposes.

A *plane stress* test can be conducted using a biaxial loading frame, either with the faces of the plate unstressed or with a constant stress applied using flat jacks. An alternative *plane strain* test is one in which the faces of the plate are confined to prevent changes in thickness, either by an extremely rigid test frame or by an electronic servomechanism. The state of stress is then triaxial rather than biaxial, because stresses in the third direction are developed by confinement even though not applied by the loading machine itself.

8.2.3.3 Polyaxial tests. In a polyaxial test, otherwise called a *multiaxial* or *true triaxial* test (Fig. 8.4d), three independently variable principal stresses are applied to pairs of opposing faces of a cube of rock. This is the only test capable of exploring the entire strength surface while maintaining a uniform and homogeneous state of stress in the specimen. Unfortunately it is difficult to perform, or it would be used more often. Adjacent platens tend to touch, resulting in "platen interference" as the specimen deforms. Even if platen contact is avoided, great care is needed to maintain the applied stresses uniform across the faces of the rock cube.

The polyaxial test can be used to check the influence of intermediate principal stress on the strength of the rock. Normally the strength criterion is expressed in terms of major and minor principal stresses only, on the assumption that the intermediate stress level has a negligible influence on strength. Gau et al. (1983) describe such an investigation based on polyaxial test data for 10-cm cube specimens of red sandstone loaded in polyaxial compression by both rigid and flexible platens.

A review of polyaxial testing is given in Amadei et al. (1984), and extensive results are reported in Amadei and Robison (1986). They describe testing under conditions of combined compression and tension, using flexible fluid-filled cushions for compressive loading and brush platens for tensile loading. Their results demonstrate that biaxial strength can be significantly higher than uniaxial strength, so the magnitude of the intermediate principal stress has an important effect on rock strength. Strength criteria such as the Mohr-Coulomb criterion (Sec. 8.4.2.1), which assumes independence of intermediate principal stress, appear inadequate. This also means that rock in the walls of an underground excavation is likely to rupture at stresses greater than its uniaxial compressive strength. Fortunately these factors lead to conservative rather than nonconservative errors in design.

8.2.3.4 Testing blocks in the field. Uniaxial, biaxial, and even polyaxial testing are most convenient in the field, where cubic specimens are easier than cylinders to prepare, and where large specimen sizes make platen interference less of a problem. For most really large-scale testing it is better to take the equipment to the rock than vice versa. Large cubic specimens of jointed rock are prepared in situ by line drilling or with a circular, chain, or wire saw. If the specimens are inclined, they must be securely clamped before they lose support from the surrounding rock mass, otherwise they may disintegrate before they can be tested.

Flat jacks are a convenient method of loading. They can be manufactured to any required size or shape, are lightweight, and can be handled and installed easily. Because of their large area and flexibility, they can apply high, uniform loads to the specimen. They have a limited capability to expand, but this problem can be overcome by stacking and connecting several flat jacks, one on top of the other. They can be installed in slots to apply biaxial load to a specimen cube cut into a rock wall or floor. Loading in the third direction requires a reaction frame to transfer loads to the walls or roof of the test adit.

When both deformability and strength are of interest, displacements are measured by dial gauges or electric transducers fixed to a reference beam securely anchored well outside the areas where load is applied.

8.2.4 Axisymmetric triaxial test

8.2.4.1 Conventional test method. The conventional *axisymmetric triaxial* compression test is similar to the uniaxial compressive strength test. However, the rock cylinder is loaded not only axially through spherical seats and platens, but at the same time radially by a *confining* pressure, or *cell* pressure, usually oil pressure acting on a synthetic rubber membrane (Fig. 8.5). The confining pressure $\sigma_2 = \sigma_3$ is usually held constant while the axial stress σ_1 is increased until the specimen ruptures. The strength data so obtained allow the mapping of curves where the planes $\sigma_2 = \sigma_3$ meet the strength surface (Fig. 8.1).

In the alternative and less common axisymmetric triaxial extension test, rupture is achieved by increasing the lateral pressure or decreasing the axial stress. This gives a condition $\sigma_1 = \sigma_2 > s_3$.

Axial loading is provided by a compression testing machine, preferably a stiff hydraulic one that allows nonexplosive testing past the

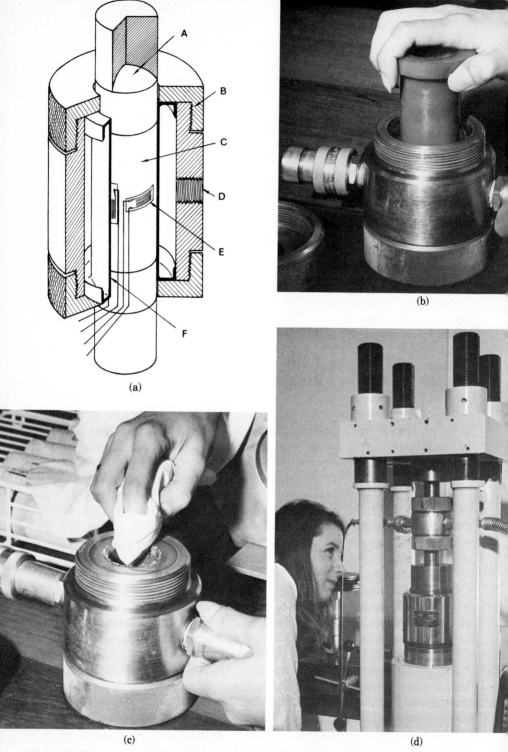

Figure 8.5 Triaxial test cell. (a) Cell components including spherical seats A, end caps B, specimen C, fluid port D, strain gauges E (optional), and polyurethane rubber sleeve F. (b) Insertion of sleeve. (c) End cap removed to clean accumulated rock fragments. (d) Triaxial test with cell in position, showing four-column load frame, 200-t hydraulic jack, load cell, and pressure transducer connected to cell for automatic plotting of stress path. (*Franklin and Hoek, 1970.*)

point of rupture. Confining pressure is provided by one of several designs of the triaxial cell. The cell designed by Hoek and Franklin (Fig. 8.5) uses a one-piece polyurethane rubber membrane that simultaneously prevents the cell from leaking and oil from reaching the rock. A large number of specimens can, one at a time, be inserted, tested, then extruded without delays to replace membranes or oil (Franklin and Hoek, 1970).

Cell pressure is maintained to within 2% by a mechanical or electrically operated hydraulic pump, or by air pressure acting on the oil through an intensifier and accumulator. Cell pressure and axial loads are monitored visually using pressure gauges, or electrically using pressure transducers. Pore pressures can, in research applications usually, be controlled or measured. Electronic measurements of applied loads and pressures allow data to be plotted as the test proceeds, but can be less reliable than direct-reading gauges.

Axial strain, or displacement between the platens, can also be monitored during the triaxial strength test. Such measurements help identify rupture points when testing the more ductile types of rock, and to prevent overstraining and damage to the membrane or the cell. They are also necessary when carrying out the continuous failure state type of test discussed below.

Usually tests on at least five specimens, each at a different confining pressure, are needed to define the *peak strength envelope* for the rock sample, a graph of σ_1 versus σ_3 at rupture. A residual strength cannot be determined by triaxial testing, because this requires at least tens of millimeters of shear displacement, more than enough to cause membrane rupture and specimen instability.

8.2.4.2 Multiple and continuous failure state testing.

An approximation to the peak strength envelope can be obtained using just a single specimen by a *multiple failure state* or *continuous failure state* type of test (Kovari, Tisa, and Attinger, 1983). Axial strain is monitored to ensure that the specimen is stressed up to but not beyond rupture. Axial and confining pressures are increased together so that the state of stress in the specimen repeatedly touches the strength envelope (multiple failure state alternative) or actually follows the envelope (continuous failure state alternative).

These methods have the advantage of requiring only one specimen, and the disadvantage of requiring strain instrumentation, automatic plotting of stress versus strain, and very careful control. There is also a degree of risk in basing a design on just one specimen, as opposed to several that may better represent the variability of the rock unit being tested. Also, if the specimen is overstressed, the recorded strength values at subsequent stages of the test will be lower than the true peak strength values. This conservative underestimate of strength may be acceptable, although often the preparation of additional test specimens may be preferred.

8.2.4.3 Large-scale axisymmetric triaxial tests in the laboratory.

The size of laboratory triaxial tests is limited only by the available space and the problems of obtaining large specimens. Techniques developed at the University of Karlsruhe, West Germany (Natau et al., 1983), allow sampling of jointed, undisturbed rock cylinders, 0.6 m in diameter and 1.6 m high, and testing under triaxial compression in the laboratory. The procedure is an excellent one for closely jointed rock masses that can be sampled to include a representative number of joints in all sets that are present.

The specimen is prestressed by a central rockbolt to prevent loosening during drilling. It is then isolated from the rock mass either by perimeter line drilling or, preferably, by coring with a large-diameter calyx or a thin-walled coring barrel. The surface of the core is sealed in situ with thick gypsum plaster. Steel casing is secured to it using a thin slurry of plaster, and then the cased rock cylinder is lifted by crane from the hole.

Because of the large size of the specimen, testing has some special features. Water is more convenient than oil as the pressurizing fluid. Preparation of the end faces of the specimen is usually by careful chiseling, and cavities need to be filled with a cement-plaster mix with a stiffness similar to that of the rock. Displacement measurements are often made by optical "surveying" rather than by more sensitive electronic means.

8.2.5 Tensile strength

8.2.5.1 Cases where tensile strength is relevant.

If a rock mass is closely jointed, its overall tensile strength is zero. Conditions of tensile stress, and therefore a need to know tensile strength, occur only seldom in rock engineering situations. Examples include the tensile stresses that develop along the underside of horizontally bedded roof strata, those produced during hydraulic fracturing of a borehole, and those developed during solar heating and cooling of exposed architectural facing stone. Very localized tensile stresses govern breakage of rock during all drilling and blasting operations.

8.2.5.2 Direct tensile test.

In the direct tensile strength test, a cylinder of rock, diameter D, is stressed along its axis by a uniaxial tensile load (Fig. 8.6a). Tensile strength is calculated as the rupture load T divided by the cross-sectional area A:

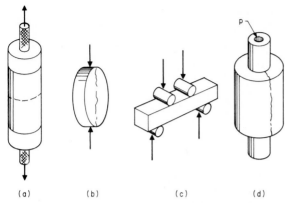

(a)   (b)   (c)   (d)

Figure 8.6 Tensile strength tests. (*a*) Direct tensile test. Specimen is glued to steel platens by epoxy resin adhesive. (*b*) Brazilian test. (*c*) Four-point beam-bending test. (*d*) Sleeve-fracture test, with pressure applied within a hollow rock cylinder through a flexible rubber tube.

$$\sigma_t = \frac{T}{A} = \frac{4T}{\pi D^2}$$

Before the development of adhesives stronger than rock, premature rupture was common because of stress concentrations where the rock specimen was gripped. Two alternatives were available to overcome this problem. The most reliable but time-consuming was to machine the specimen into a "dumbbell" shape similar to that used when testing metals. This ensured that rupture would occur only at the reduced cross section, and it provided strong end sections for gripping. The easier but less reliable alternative was to use an "indirect" tensile test, as outlined below, in which tension was no longer "pure," but combined with compressive stress components and stress gradients throughout the specimen. Interpretation of these tests required certain assumptions that were not always valid.

Nowadays one can perform a direct tensile test with little difficulty using a cylindrical specimen similar to that for a uniaxial compressive strength test. Steel platens are glued with epoxy adhesive to the end faces of the rock cylinder, and tension is applied axially through flexible steel cables to eliminate bending stresses.

8.2.5.3 Indirect tensile strength tests. When subjected to a biaxial stress field ($\sigma_3 = 0$), rocks are found to rupture at their uniaxial tensile strength even when only one of the principal stresses is tensile and the other is compressive, provided that the magnitude of the compressive component remains less than three times the tensile stress. This permits the measurement of tensile strength using simple specimen and loading configurations such as a disk loaded across its diameter (the Brazilian test described below).

Two characteristics (and limitations) of such tests are that the maximum tensile stress varies through the specimen rather than being uniform as in a direct tensile test, and that the maximum value must be calculated assuming elastic behavior of the specimen. This assumption is sometimes unjustified. For example, softer rocks yield when subjected to stresses close to rupture. Inaccuracies may render the test of limited value for design, but since it is simple, it may still be used to provide a strength index.

8.2.5.4 Brazilian and disk-with-hole tests.   In the Brazilian test (Mellor and Hawkes, 1971) a rock disk is loaded diametrically between platens (Fig. 8.6b). The disk diameter should be at least 50 mm, and the ratio of diameter D to thickness t about 2:1. The disk circumference is wrapped in tape to reduce contact irregularities. A loading rate of 0.2 kN/s is recommended, such that the specimen ruptures within 15–30 s, usually along a single tensile-type fracture aligned with the axis of loading. The rupture load P is measured, usually with the help of automatic recording equipment. The tensile strength of the rock is calculated from the following formula:

$$\sigma_t = \frac{2P}{\pi Dt} = \frac{0.636P}{Dt}$$

The direction of any visible planes of weakness with respect to the applied load must be noted. If the fracture plane deviates significantly from a straight line between platen contacts, the test is considered invalid. The Brazilian test is particularly convenient for investigating strength anisotropy, by comparing the strengths of disks rotated to give a range of angles between the rock fabric and the direction of loading. The same test can be used to measure anisotropic elastic behavior, as discussed in Chap. 9.

Alternative disk-with-hole, or *annulus,* tests were developed in an attempt to overcome a tendency for mixed modes of rupture in the Brazilian test, caused by confinement at the platen contacts. A hole drilled in the center of the disk acts as a stress raiser and induces crack propagation from the central area of the specimen which is uninfluenced by platen friction. Tensile strength can be calculated from an elastic analysis of stress distribution in the disk. Such tests are now seldom conducted.

8.2.5.5 Point-load strength test.   The point-load strength test, described in Chap. 2 in its role as an index test, is one member of the indirect tensile test family. It can also be used for purposes of design, to give the tensile strength of rock materials. The point-load strength I_{s50} is approximately 0.8 times the uniaxial tensile strength.

8.2.5.6 Beam-bending tests. Tensile strength can be determined indirectly by a beam-bending test, using either a three-point or a four-point loading system (Fig. 8.6c). In the three-point method, tensile rupture (crack propagation) starts at the location of maximum tensile stress immediately beneath the point of application of load P. The initiation point and the direction of crack propagation are affected by local inhomogeneities, such as crystal contacts, microflaws, or surface defects, such that the tensile stress in the lowermost fiber can be estimated only as an average value:

$$\sigma_t = \frac{1.5PL}{bt^2}$$

where t is the beam height, b the width, and L the span between supports.

The four-point beam test is better because lower-fiber stress is constant between the two top load application points. The beam usually ruptures between these points: a "critical flaw" is very likely to be found in this uniformly stressed area, so the measured strength should be more representative than in the case of three-point bending. The maximum tensile stress in the lowermost fiber at the instant of rupture is given by

$$\sigma_t = \frac{2PL}{bt^2}$$

Both equations are derived assuming elastic stress distribution. The central part of the beam may be notched or reduced in area to ensure fracture initiation at that location (see Sec. 8.2.6).

8.2.5.7 Torsion and hollow cylinder tests. In a variation of axisymmetric triaxial testing, a rock cylinder with an axially bored concentric hole can be loaded internally, externally, and axially, and also with torque if required. Many stress states and stress paths can be explored (Jaeger and Cook, 1969; Dusseault, 1981), including axial plane strain, cylinder burst or collapse, decreasing or increasing mean stress to rupture, and combinations of tension, shear, and various compressive states.

The test is useful in some research applications. However, the stress distributions in the cylinder must be assessed from the theory of elasticity and are nonuniform unless the cylinder walls are very thin and the rocks fine-grained. Often it is difficult to define exactly when and where rupture begins during stressing, leading to difficulties of interpretation.

8.2.5.8 Sleeve-fracturing tests. A dilatometer, mainly a field-testing instrument for measuring the deformability of rock (Chap. 9), can also be used in the laboratory to measure both tensile strength and modulus of elasticity in tension by internally pressurizing intact rock cylinders (Singh, 1974; Stephansson, 1983). The sleeve-fracturing device (dilatometer) consists of a flexible tube, a pump to inflate the tube, and instruments to measure pressure and volumetric expansion. A small hole is drilled along the axis of a piece of core, and the dilatometer is inserted and expanded (Fig. 8.6d). The pressure P required to rupture the core is related to the tensile strength σ_t and the internal and external radii of the core r and R by the following formula for a thick-walled elastic cylinder:

$$\sigma_t = P \left(\frac{R^2 + r^2}{R^2 - r^2} \right)$$

8.2.5.9 Field measurements of tensile strength. When inserted in a drillhole, a dilatometer can be used to measure not only the deformability of the rock, which is its main application (Chap. 9), but also its tensile strength. There are two types of dilatometer, a flexible inflatable type and a rigid type also known as a *borehole jack,* consisting of two steel half-cylinders that are expanded against the drillhole wall. Using either type, measurements are made of drillhole expansion (dilation) versus applied pressure. Yield characteristics can also be determined from the nonlinear pressure-dilation curve (Hughes and Ervin, 1980).

With sufficient pressure, a crack is generated and propagates outward in a stable manner. The crack initiation pressure P_1 is measured, and then released and reapplied, to measure the pressure P_2 needed to reopen the crack. P_1 is a function of the tensile strength and the compressive stress state around the hole. P_2 is a function only of the ground stress component. Tensile strength can be obtained by subtraction, with back-calculation to account for the stress-concentrating effect of the drillhole.

The principle is similar to that of stress measurement by the hydrofracturing method, discussed in Chap. 5. A borehole jack fracture test measures tensile strength in a particular direction affected by the alignment of the platens, whereas a flexible dilatometer or hydrofracture test measures the lowest tensile strength of planes intersecting the drillhole, corrected for the stress distributions.

Flat jacks can be used in a similar manner to propagate a fracture and measure not only rock deformability (Chap. 9) and stress (Chap. 5), but also tensile strength.

8.2.6 Fracture toughness tests

8.2.6.1 Objectives and test types. Whereas most other types of test are designed to investigate the large-scale, quasi-continuum behavior of a specimen, the *fracture toughness* tests measure the energy needed to propagate a single crack. They are designed to initiate a crack at a precise, predetermined location, and to propagate it in a fully stable manner, for which purposes the stress intensity must be made to decrease as the fracture length increases. Fracture toughness tests are needed for studies of rock cutting, hydrofracturing, and explosive fracturing (Nemat-Nasser and Horii, 1982; Ouchterlony and Zongqi, 1983; Rossmanith, 1983; Senseny and Pfeifle, 1984; Ouchterlony, 1986).

If fracture is accompanied by much plastic behavior, linear elastic fracture mechanics approaches are inapplicable. Bazant (1984) points out that most rocks fracture with a zone of distributed cracking ahead of the fracture front. Instead of the classical approach, Bazant represents fracture propagation in materials such as rocks, concrete, and sea ice by a strain-weakening finite-element model.

For cores, several test configurations are typically used: a core disk with a radial notch; a short rod (i.e., core) with an axial V-shaped or "chevron" notch, and a longer core beam in bending, with a central notch sawn radially into its underside (Fig. 8.7). The longer beam test can provide specimens for subsequent tests of the other two types. In the disk and short-rod tests, the notch is forced to open by applying opposing tensile forces to the expanded notch end or through aluminum pieces cemented to the rock. In the notched beam test, three-

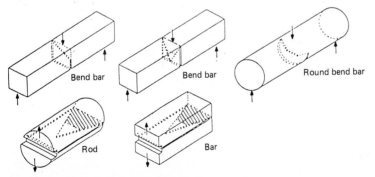

Figure 8.7 Various configurations of fracture toughness test with chevron-notched specimens. (*Karfakis et al., 1986.*)

point bending is used, similar to the beam bending described. The chevron slots must be thin and sharp for plane strain conditions to prevail at the crack tip (Karfakis et al., 1986). Fracture toughness can also be determined as a function of axial and confining pressure using a variation of the hollow cylinder test. An internally prenotched specimen is subjected to internal pressure until it bursts (Abou-Sayed, 1977).

8.2.6.2 Parameters of fracture toughness. *Fracture toughness* is a measure of the critical stress intensity $K_{\mathrm{Ic}'}$ at the crack tip required to initiate and propagate the fracture (Ingraffea et al., 1982). It is a function of applied load and stress concentration, which is related to Poisson's ratio and geometry.

Fracture toughness can also be expressed as a *critical energy release rate* $G_{\mathrm{Ic}'}$, which is the energy required to create new surface area. It is a function of $K_{\mathrm{Ic}'}$, the elastic modulus, and Poisson's ratio:

$$G_{\mathrm{Ic}'} = \frac{K_{\mathrm{Ic}'}^2(1 - \nu^2)}{E}$$

and is also known as the *crack driving force,* which has been shown to be correlatable with the cutting performance of disks on a tunnel-boring machine (Nelson and Fong, 1986).

8.3   Brittle Behavior of Rocks

8.3.1   Griffith crack theory

The idea of a *critical flaw* was first conceived by Griffith (1924) to explain why glasses rupture at much lower stress levels than predicted from their theoretical molecular bond strength. This theory has been extended to polycrystalline rocks. An overview of the mechanisms of crack growth will help explain how a test specimen ruptures and, more importantly, the processes of hydrofracturing and rockbursting.

Natural flaws, called *Griffith cracks,* are present as grain boundary cracks in igneous rocks, as pores in sedimentary rocks, and as flaws within crystals.

Griffith cracks act as stress raisers. At their tips, the stress can be much higher than the average for the specimen as a whole, and is often tensile even though the average stress on a larger scale is compressive. Cracks perpendicular to an applied uniaxial tensile stress are the most "critical," whereas in compressive conditions, oblique cracks have the greatest tensile stress concentrations at their tips. Thus when the level of uniaxial stress on the specimen is increased, an obliquely oriented critical crack is the first to "initiate" (start extending).

The cracks that grow in a uniaxial compressive strength test speci-

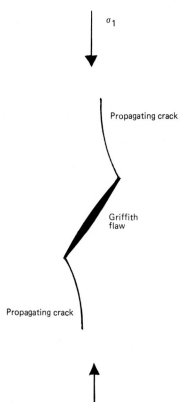

Figure 8.8 Cracks that propagate from a "Griffith flaw" tend to curve toward the direction of major principal stress, and into a plane perpendicular to the minor principal stress.

men curve toward the direction of applied loading, which leads to a rapid increase in the diameter of the specimen. Cracks in a polyaxially compressed specimen or element of rock curve into a plane parallel to the major and perpendicular to the minor principal stress (Fig. 8.8).

Individual cracks usually stabilize when their growth has reached a critical level, allowing stress to be redistributed to adjacent flaws, which then propagate and coalesce in a cascading process until rupture occurs. Crack growth distributed throughout the specimen or rock element results in a progressive weakening and increase in volume. Propagation of cracks can be heard as *microseisms* or *acoustic emissions* by a geophone or accelerometer attached to the rock surface.

8.3.2 Modes of rupture and yield in triaxial compression

An increase in confining stress greatly increases the strength of a rock by inhibiting the growth of cracks. Even small amounts of confinement can have a significant effect; for example, a mine pillar is greatly strengthened by rockbolting or by wrapping with wire cable. The mechanisms of rupture change according to the condition of stress in the specimen or element. Stress space can be divided, approximately, into five regions in which different modes of rupture and yield predominate (Hoek, 1968; Franklin, 1971):

Region I (Fig. 8.9) is that of pure tensile rupture in which flaws perpendicular to the applied tensile stress are the first to initiate cracks. The cracks, once initiated, are unstable and propagate in their own plane.

Region II is that of axial cleavage rupture, predominant at low confining stresses. Flaws oblique to σ_1 are the first to initiate cracks. Crack propagation is unstable. Cracks curve toward σ_1 so the final plane of rupture is parallel to the σ_1 direction. In region IIa the stress normal to the critical flaw is tensile, whereas in region IIb it is compressive.

Region III is that of brittle compressive shear rupture. Friction between the crack walls must be overcome. The microscopic rupture process is usually abrupt, audible, and easily detected. Individual cracks grow incrementally until there are enough of them to weaken the specimen and permit generation of an oblique surface of shearing.

Region IV is that of ductile rupture. Shear planes, once generated, are stronger than adjacent material, so multiple shears result. Ductile behavior is manifested by "barreling" of cylindrical core. Multiple slip planes inclined at about 45° to the axis of the cylinder are

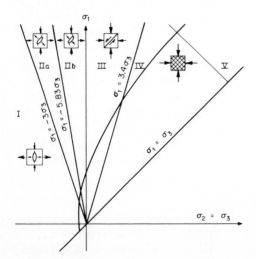

Figure 8.9 Rupture mechanisms in different regions of stress space. (*Franklin, 1971.*)

visible on a specimen's surface. Rupture is usually gradual and inaudible. Strength is often defined as σ_1 at a strain of 1%.

Region V is that of pure plastic yield. Crack propagation and associated dilatency are entirely suppressed by internal friction. Yield occurs by generalized slippage, gliding, and twinning along crystal planes. At these very high stresses, the strength surface probably becomes parallel to the space diagonal, a condition for "perfect plasticity" observed in materials such as clays at much lower levels of stress.

Regions I, II, and III are all characterized by brittle rupture behavior. The *brittle-ductile transition* boundary for the axisymmetric condition $\sigma_2 = \sigma_3$ has been proposed by Mogi (1966) as corresponding to the equation

$$\sigma_1 = 3.4\sigma_3$$

The transition from brittle to ductile and then plastic behavior occurs gradually with increasing stress, and depends on more than just stress. Brittle-ductile transition occurs at lower stress levels if the rock is moist or hot, or if it is compressed very slowly. It also depends on the minerals that are present. Some rock-forming minerals, such as clays, halite, and even calcite, become ductile at relatively low temperatures and pressures, whereas the quartz and feldspars characteristic of harder rocks become ductile only at depths of tens of kilometers (Rutter, 1974).

8.4   Strength and Yield Criteria

8.4.1   Overview

8.4.1.1   The need for, and types of, strength criteria.   Each type of rock has its own strength surface which, if a suitable form of equation were found, could be represented algebraically as a strength criterion. Much research effort has been expended in the search for a suitable form of equation which, with the insertion of "strength parameters" appropriate for each specific material, could be used to represent all types of rock, and might even accommodate the behavior of soils. A useful summary of the many alternatives is provided in Johnston (1985).

The earlier attempts were mainly theoretical and evolved from yield criteria for metals and Griffith crack theory for the more brittle rock materials. They had the common goal to express bulk behavior of rock in terms of its constituent minerals and flaws. They led to an improved understanding of micromechanics, but none of them fit the experimental data particularly well (Jaeger and Cook, 1969).

Therefore the strength criteria most used in practice are empirical ones. The approach is to fit a curve (or a surface in three-dimensional stress space) to a real set of experimental data. The axes may be those of principal stress or of principal stress sums and differences. The form of equation may have a theoretical justification, but is chosen mainly to give a good fit and for its simplicity. The need for a simple equation is less pressing than it used to be, when only closed-form solutions were available, since modern numerical design techniques can accommodate any form of criterion, no matter how complex, and can even make automatic changes to the criterion to suit different stress levels while the computations are in progress.

Several of the better known criteria are reviewed below. The reader is referred to the original publications for further details of their derivation and merits. The real behavior of rock is probably much simpler than the multiplicity of artificial strength criteria might suggest.

8.4.1.2 Attributes of the strength surface. What basic conditions must a strength criterion fulfill? Referring back to Fig. 8.1, the following can be inferred, on theoretical and experimental grounds, as being common attributes of the strength surfaces for rock materials of all types. Any strength criterion must therefore conform to these requirements:

Isotropic rock has a strength surface that has 120° symmetry about the space diagonal $\sigma_1 = \sigma_2 = \sigma_3$ with the three compressive planes $\sigma_1 = \sigma_2$, $\sigma_2 = \sigma_3$, and $\sigma_3 = \sigma_1$ defining the rounded apices of a triangular cross section.

The surface is shaped like a paraboloid (half an egg), which is closed at a "nose" or "apex" in the $(-, -, -)$ octant of stress space, indicating that rock cannot withstand great tension.

The surface opens out in the $(+, +, +)$ octant and continues to open. It probably becomes asymptotic to a line parallel to the space diagonal at large values of hydrostatic stress (pure plastic behavior). A hydrostatic compressive stress is unlikely to be sufficient to cause rupture, although it can cause yield by collapse.

The positive intercepts on the stress axes correspond with the uniaxial compressive strength of the material and often are 8 to 10 times the negative intercepts of uniaxial tensile strength.

The intermediate principal stress has only a moderate influence on strength, such that the strength criterion can, with some approximation, be expressed in two dimensions, in terms of major and minor principal stresses only. Recent results of polyaxial testing (Sec. 8.2.3.3) indicate that the degree of approximation may be greater than expected, and perhaps not always acceptable.

8.4.2   Theoretical criteria

8.4.2.1   Mohr-Coulomb criterion.   The Mohr-Coulomb criterion is one of the earliest and simplest, although the least accurate in its representation of rock behavior. It can be expressed in principal stress coordinates by the equation

$$\frac{\sigma_1 - \sigma_3}{2} = \left(S_0 \cot \phi + \frac{\sigma_1 + \sigma_3}{2} \right) \sin \phi$$

where the *inherent shear strength* S_0 and the *angle of shearing resistance* ϕ are shear strength parameters that characterize the particular material being tested (Jaeger and Cook, 1979). The criterion can be reduced to the simple linear form

$$\sigma_1 = A + B\sigma_3$$

where A and B are combinations of S_0 and ϕ. This equation can be normalized with respect to uniaxial compressive strength σ_c:

$$\frac{\sigma_1}{\sigma_c} = 1 + C \frac{\sigma_3}{\sigma_c}$$

where

$$C = \frac{1 + \sin \phi}{1 - \sin \phi}$$

When plotted in principal stress space, this forms a cone rather than the required paraboloid. The Mohr-Coulomb equation, when fitted to data in the compressive region, grossly overestimates the tensile strength of the rock, but this is unimportant because tensile strength can be arbitrarily stipulated as zero (the *tension cutoff* approach). The Mohr-Coulomb criterion is often used in soil mechanics, because many soil types have nearly conic strength envelopes with zero tensile strength. The curvature of the true envelope is more pronounced for rocks.

The most common representation of the Mohr-Coulomb criterion, employing the properties of a Mohr plot, is a graph whose axes are shear stress τ and normal stress σ_n (Fig. 8.10),

$$\tau = c + \sigma_n \tan \phi$$

where τ is the shear strength, $\tau = (\sigma_1 - \sigma_3)/2$; σ_n is the normal stress on the shear plane, $\sigma_n = (\sigma_1 + \sigma_3)/2$; and c is the cohesion intercept on the τ axis.

The classical Mohr method of graphic construction is to draw semicircles defined by each pair of major and minor principal stresses at rupture. The (τ, σ_n) axes must be to equal scales. The shear and normal stress components acting on planes at various orientations at a point can be rapidly determined graphically. The semicircles corre-

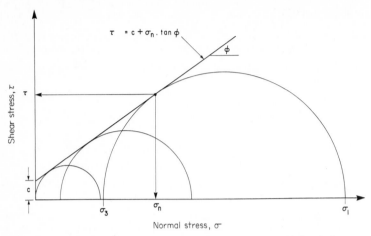

Figure 8.10   Mohr-Coulomb linear strength criterion with Mohr circles.

sponding to rupture are joined by a tangent curve or *Mohr envelope,* which defines the upper limit of all possible nonrupture states of stress. The straight-line version of this envelope is the Mohr-Coulomb strength criterion.

Some typical values of the Mohr-Coulomb parameters c and ϕ for intact rock are given in Table 8.1.

8.4.2.2   Griffith criterion and extensions.   The Griffith criterion and extensions of it relate to stresses when microcracks begin to propagate in a purely brittle manner, and do not necessarily relate to rupture on a large scale. Griffith's theory leads to the following criterion:

$$(\sigma_1 - \sigma_3)^2 = 8T(\sigma_1 + \sigma_3) \qquad \text{if } \sigma_1 + 3\sigma_3 \geq 0$$

$$T = \sigma_3 \qquad \text{if } \sigma_1 + 3\sigma_3 < 0$$

This is a parabolic envelope, a moderately good fit to real rock strength data, although strength is again assumed independent of intermediate principal stress.

Murrell (1966) showed that if the uniaxial compressive strength is 8 times the uniaxial tensile strength, the Griffith criterion is modified as follows:

$$(\sigma_1 - \sigma_3)^2 - 8T(\sigma_1 + \sigma_3) = 16T^2$$

McClintock and Walsh (1962) considered a flawed system containing many noninteracting cracks which close under compressive stresses and exhibit frictional slip across the closed crack surfaces.

TABLE 8.1 Triaxial Strength Data Compilation, with Typical Values for
Mohr-Coulomb Peak Shear Strength Parameters from Laboratory Tests on
Intact Rock Specimens*

	c, MPa		ϕ, deg	
Crystalline limestones	35	(15–50)	26	(17–35)
Porous limestones	30	(7–60)	26	(11–45)
Chalk	1	(0–2)	39	(11–45)
Sandstones	30	(1–70)	31	(6–54)
Quartzite	42	(30–50)	50	(42–54)
Slates and high-durability shales	30	(1–6)	39	(17–42)
Low-durability shales	0.2	(0–1)	26	(17–35)
Coarse-grained igneous rocks	56	(10–170)	45	(26–56)
Fine-grained igneous rocks	35	(10–80)	26	(6–63)
Schists	46	(15–70)	26	(17–48)

*Typical values (mode of between 10 and 30 reported test results) are given together
with ranges of reported values in parentheses. Because of the considerable variations in
test methods and in the rocks themselves, and the error of fitting a straight line to
curved data, the data should be taken only as a very rough guide to rock properties.

Making assumptions as to the nature and distribution of the crack
network, the following expression was generated:

$$4T = [(\sigma_1 - \sigma_3)(1 + \nu^2)^{1/2}] - \nu(\sigma_1 + \sigma_3)$$

The original Griffith theory and subsequent modifications are all
similar, and contain just one measureable strength parameter such as
tensile strength T. Hoek (1968) showed that analysis based on a single
flaw in an all-around compressive stress field will always lead to this
parabolic relationship if the rupture mode is tensile.

8.4.3 Empirical criteria

An empirical criterion, by definition, is one where an arbitrary equation is fitted, usually statistically, to a real set of experimental data.
This gives a sufficiently accurate prediction of strength for most practical purposes, in spite of the equation having no theoretical justification. One must be careful not to extrapolate far beyond the range for
which data are available, where the equation is likely to give poor predictions. Franklin (1971) compared various alternative formulations:

$$\sigma_1 = A + B\sigma_3 \tag{1}$$

$$\sigma_1 = A + B\sigma_3{}^C \tag{2}$$

$$\sigma_1 = A \log (B + \sigma_3) \tag{3}$$

$$\sigma_1 - \sigma_3 = A + BC^{\sigma_3} \tag{4}$$

$$\sigma_1 - \sigma_3 = \frac{A(\sigma_1 + \sigma_3) + B}{\sigma_1 + \sigma_3 + C} \tag{5}$$

$$\sigma_1 - \sigma_3 = A + B(\sigma_1 + \sigma_3)^C \tag{6}$$

$$\sigma_1 - \sigma_3 = A(\sigma_1 + \sigma_3)^B \tag{7}$$

where A, B, and C are empirical curve-fitting parameters.

Equation (7) was found to simulate best the basic attributes of the strength surface, described in Sec. 8.4.1.2, and also to give the best fit to the experimental data for intact rock (Fig. 8.11). The parameters are determined from a log-log plot of experimental data. B is the slope of the straight line and A the antilogarithm of the intercept. The parameter A (megapascals) was found to vary in the range of 1–8, whereas the parameter B (dimensionless) typically varied in the range of 0.6–0.9. The two parameters are related by the equation

$$A = \sigma_c^{1-B}$$

8.4.3.1 Bieniawski criterion. Bieniawski (1974) developed the following empirical power law strength criterion:

$$\frac{\sigma_1 - \sigma_3}{2\sigma_c} = 0.1 + B\left(\frac{\sigma_1 + \sigma_3}{2\sigma_c}\right)^a$$

This is similar to Eq. (6), but is normalized with respect to uniaxial compressive strength σ_c. The exponent a represents the curvature of the strength surface and assumes values in the range of 0.85–0.93.

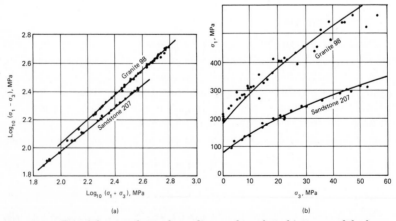

Figure 8.11 Triaxial strength trends are linear when plotted in terms of the logarithms of principal stress differences and sums. Data are for peak strengths. In principal stress coordinates, data plot as curves. (*Franklin, 1971.*)

The constant B controls the position of the envelope, and varies from 0.7 to 0.8 for most rock types.

8.4.3.2 Hoek and Brown criterion. Hoek and Brown (1980) underlined the need to extend strength criteria to the prediction of rock mass behavior. They proposed the following criterion for rock mass strength:

$$\frac{\sigma_1}{\sigma_c} = \frac{\sigma_3}{\sigma_c} + \sqrt{m\frac{\sigma_3}{\sigma_c} + s}$$

where m and s are dimensionless parameters that depend on the degree of interlocking between blocks in the jointed rock mass. Here m controls the curvature of the σ_1 versus σ_3 curve, and s is a material constant which controls the location of this curve in stress space. Both m and s have been correlated with, and can be predicted from the rock mass quality index Q and the rock mass rating (RMR) (Secs. 3.3.3.4 and 3.3.3.5). For intact rock $s = 1$, and it drops very rapidly to small values for broken rock masses; m has been found to range from 5.4 for limestone to 27.9 for granite.

8.4.3.3 Yudhbir criterion. Yudhbir et al. (1983) tested in triaxial compression 122 specimens of limestone, sandstone, granite, and model material made from a mixture of gypsum and polyester resin, both in solid form and containing fractures. Attempting to fit various empirical strength criteria to the data, they found that although the Hoek and Brown criterion was satisfactory in the brittle range, it gave a poor fit to the data in the ductile range. They proposed an alternative criterion:

$$\frac{\sigma_1}{\sigma_c} = A + B\left(\frac{\sigma_3}{\sigma_c}\right)^a$$

where a ranges from 0.65 to 0.75 and A and B are functions of the rock type. Having three parameters instead of two, as in the equations quoted earlier, makes a reasonable fit easier to obtain. The physical implication of the parameters and their relation to other rock properties must, however, be assessed.

8.4.3.4 Kim and Lade criterion. The three-parameter criterion proposed by Kim and Lade (1984) is expressed in terms of the first and third stress invariants:

$$\left(\frac{I_1^3}{I_3} - 27\right)\left(\frac{I_1}{P_a}\right)^m = n_1$$

where $I_1 = \sigma_x + \sigma_y + \sigma_z$, $I_3 = \sigma_x \sigma_y \sigma_z$, and P_a is the atmospheric pressure expressed in the same units as the applied stresses; n_1 and m are two parameters that are obtained by regression analysis. To include the effects of tension and cohesion sustained by rocks, an axis translation parameter a is introduced and a constant P_a is applied to stresses σ_x, σ_y, and σ_z. In Amadei and Robison's research (1986), Mohr-Coulomb, Hoek and Brown, and Lade criteria were compared with the test data. All three showed various deficiencies as predictors, although each of them performed well in various parts of stress space.

8.4.3.5 Johnston criterion. Johnston (1985) has proposed the following strength criterion to describe the behavior of materials ranging from clays to extremely strong rocks:

$$\frac{\sigma_1}{\sigma_c} = \left[\left(\frac{m}{B} \right) \left(\frac{\sigma_3}{\sigma_c} \right) + s \right]^B$$

For intact materials, $s = 1$, as in the Hoek and Brown criterion. The parameter B describes the nonlinearity of the strength envelope and decreases from about 1.0 for normally consolidated clays to about 0.5 for rocks with strength $\sigma_c = 250$ MPa. The parameter m describes the slope of the strength envelope at $\sigma_3 = 0$ and increases from about 2.0 ($\phi = 20^0$) for normally consolidated clays to between 7 and 21 for strong rocks.

Johnston concluded that despite the apparently radical differences between slightly overconsolidated clays and extremely hard rocks and all the materials in between, their intact strength variations may be a matter of degree rather than of fundamental nature.

8.4.3.6 Selection of a strength criterion. Empirical criteria can hardly be criticized if they fit the data, and why ignore data to accommodate a theory? The choice of criterion should therefore depend on its ability to predict strengths. Over a broad range of stresses and rock types, the Johnston criterion appears suitable. The Hoek and Brown criterion has the merit of being the first to attempt a simulation of large-scale rock behavior, but like all others, it requires large-scale test data to verify its predictions.

Some criteria are "normalized" by dividing by a strength such as σ_c, so that the parameters become dimensionless. A significant disadvantage is that great weight is then attached to a single point on the strength surface, that of uniaxial strength, for which data are known to be variable and unreliable. Nonnormalized criteria, on the other hand, allow all tests to be given equal weight, and the dimensional parameters they contain can be determined by statistical curve fitting.

8.4.3.7 Future directions. The above criteria express strength in terms of stress level. This has proven quite satisfactory in most applications. However, the concept of a critical level of strain has been proposed, particularly for predicting tensile rupture. The criterion states that local tensile rupture will occur when extensional strain reaches a critical amount, usually on the order of 0.10 for rocks.

Another possibility would be an energy criterion, which combines both stress and strain. It implies that fracturing will occur when the energy density in a specimen or element surpasses some critical value. The concept is currently under research, but may prove to be a practical tool. The strain energy required to fracture rock can be estimated from laboratory tests. In the uniaxial case, the threshold (critical) energy density is

$$W = 0.5\epsilon\sigma_c = 0.5\frac{\sigma_c^{\,2}}{E}$$

which is the area under the stress-strain curve. For triaxial loading there are added strain energy components contributed by the intermediate and minor principal stresses (Finley and Farmer, 1986). This approach is closely related to the concept of fracture toughness as outlined above.

References

Abou-Sayed, A. S.: "Fracture Toughness K_{Ic} of Triaxially Loaded Indiana Limestone," *Proc. 18th U.S. Symp. Rock Mech.* (Keystone, Colo., 1977), pp. 2A3-1–2A3-8.

Amadei, B., V. Janoo, M. Robison, and R. Kuberin: "Strength of Indiana Limestone in True Biaxial Loading Conditions," *Proc. 25th U.S. Symp. Rock Mech.* (Evanston, Ill., 1984), pp. 338–348.

———— and J. Robison: "Strength of Rock in Multi-axial Loading Conditions," *Proc. 27th U.S. Symp. Rock Mech.* (Tuscaloosa, Ala., 1986), pp. 47–55.

Bazant, Z. P.: "Mathematical Modelling of Progressive Cracking and Fracture of Rock," *Proc. 25th U.S. Symp. Rock Mech.* (Evanston, Ill., 1984), pp. 29–37.

Bieniawski, Z. T.: "Estimating the Strength of Rock Materials," *J. South Afr. Inst. Min. Metal,* **74,** 312–320 (1974).

Colback, P. S. B., and B. L. Wiid: "The Influence of Moisture Content on the Compressive Strength of Rocks," *Proc. 3d Can. Symp. Rock Mech.* (Toronto, Ont., 1965), pp. 65–83.

Dusseault, M. B.: "A Versatile Hollow Cylinder Triaxial Device," *Can. Geotech. J.,* **18** (1), 1–7 (1981).

Finley, R. E., and I. W. Farmer: "Tunnel Design Using Strain Energy Concepts and Serviceability Limit States," *Proc. 27th U.S. Symp. Rock Mech.* (Tuscaloosa, Ala., 1986), pp. 834–839.

Franklin, J. A.: "Triaxial Strength of Rock Materials," *Rock Mech.* (Springer), **3,** 86–98 (1971).

———— and E. Hoek: "Developments in Triaxial Testing Technique," *Rock Mech.* (Springer), **2,** 223–228 (1970).

Gau, Q. Q., H. T. Cheng, and D. P. Zhuo: "The Strength, Deformation and Rupture Characteristics of Red Sandstone under Polyaxial Compression," *Proc. 5th Int. Cong. Rock Mech.* (Melbourne, Australia, 1983), vol. A, pp. 157–160.

Griffith, A. A.: "Theory of Rupture," *Proc. 1st Int. Cong. Appl. Mech.* (1924), vol. 1, p. 55.

Herget, G., and K. Unrug: "In situ Rock Strength from Triaxial Testing," *Int. J. Rock Mech. Min. Sci. Geomech. Abstr.,* **13,** 299–302 (1976).

Hoek, E.: "Rock Fracture under Static Stress Conditions," Rep. MEG 303, CSIR, Pretoria, S. Africa, 228 pp. (1965).

———: "Brittle Failure of Rock," in K. G. Stagg and O. C. Zienkiewicz, Eds., *Rock Mechanics in Engineering Practice* (Wiley, New York, 1968).

——— and E. T. Brown: "Empirical Strength Criterion for Rock Masses," *J. Geotech. Eng. Div., ASCE,* **106** (GT9), 1013–1035 (1980).

Hughes, J. M. O., and M. C. Ervin: "Development of a High Pressure Pressuremeter for Determining the Engineering Properties of Soft to Medium Strength Rocks," *Proc. 3d Australian–New Zealand Conf. Geomech.* (Wellington, New Zealand, 1980), vol. 1, pp. 243–247.

Ingraffea, A. R., K. L. Gunsallus, J. F. Beech, and P. Nelson: "A Fracture Toughness Testing System for Prediction of Tunnel Boring Machine Performance," *Proc. 23d U.S. Symp. Rock Mech.* (Berkeley, Calif., 1982), pp. 463–470.

ISRM: "Suggested Methods for Rock Characterization, Testing, and Monitoring," ISRM Commission on Testing Methods, E. T. Brown, Ed. (Pergamon, Oxford, 1981), 211 pp.

Jaeger, J. C., and N. G. W. Cook: *Fundamentals of Rock Mechanics* (Methuen, London, 1969), 513 pp.

Johnston, I. W.: "The Strength of Intact Geomechanical Materials," *J. Geotech. Eng. Div., ASCE,* **3** (6), 730–749 (1985).

Karfakis, M. G., K. P. Chong, and M. D. Kuruppu: "A Critical Review of Fracture Toughness Testing of Rocks," *Proc. 27th U.S. Symp. Rock Mech.* (Tuscaloosa, Ala., 1986), pp. 3–10.

Kim, M. K., and P. V. Lade: "Modelling Rock Strength in Three Dimensions," *Int. J. Rock Mech. Min. Sci.,* **21,** 21–33 (1984).

Kovari, K., A. Tisa, and R. O. Attinger: "The Concept of Continuous Failure State: Triaxial Tests," *Rock Mech. Rock Eng.,* **16** (2), 117–131 (1983).

McClintock, F. A., and J. B. Walsh: "Friction on Griffith Cracks in Rocks under Pressure," *Proc. 4th U.S. Nat. Cong. Appl. Mech.,* vol. 2 (Am. Soc. Mech. Eng., New York, 1962), pp. 1015–1021.

Mellor, M., and I. Hawkes: "Measurement of Tensile Strength by Diametral Compression of Discs and Annuli," *Eng. Geol.,* **5,** 173–225 (1971).

Mogi, K.: "Pressure Dependence of Rock Strength and Transition from Brittle Fracture to Ductile Flow," *Bull. Earthquake Res. Inst.* (Tokyo), **44,** 215–232 (1966).

Murrell, S. A. F.: "The Effects of Triaxial Stress Systems on the Strength of Rocks at Atmospheric Temperatures," *Geophys. J., R. Astron. Soc.,* **10** (3), 231–281 (1966).

Natau, O. P., B. O. Frohlich, and T. O. Mutschler: "Recent Developments of the Large-Scale Triaxial Test," *Proc. 5th Int. Cong. Rock Mech.* (Melbourne, Australia, 1983), vol. A, pp. 65–74.

Nelson, P. P., and F. L. C. Fong: "Characterization of Rock for Boreability Evaluation Using Fracture Material Properties," *Proc. 27th U.S. Symp. Rock Mech.* (Tuscaloosa, Ala., 1986), pp. 846–852.

Nemat-Nasser, S., and H. Horii: "Compression-Induced Nonplanar Crack Extension with Application to Splitting, Exfoliation and Rockburst," *J. Geophys. Res.,* **87** (B8), 6805–6821 (1982).

Ouchterlony, F.: "A Core Bend Specimen with Chevron Edge Notch for Fracture Toughness Measurements," *Proc. 27th U.S. Symp. Rock Mech.* (Tuscaloosa, Ala., 1986), pp. 177–184.

——— and S. Zongqi: "New Methods of Measuring Fracture Toughness on Rock Cores," *Proc. 1st Int. Symp. Rock Fragmentation by Blasting* (Lulea, Sweden, 1983), vol. 1, pp. 199–223.

Pells, P. J. N., and M. J. Ferry: "Needless Stringency in Sample Preparation Standards for Laboratory Testing of Weak Rocks," *Proc. 5th Int. Cong. Rock Mech.* (Melbourne, Australia, 1983), vol. A, pp. 203–208.

Rossmanith, H. P., Ed.: *Rock Fracture Mechanics* (Springer, New York, 1983), 484 pp.

Rutter, E. H.: "The Influence of Temperature, Strain Rate and Interstitial Water in the Experimental Deformation of Calcite Rocks," *Tectonophysics,* **22,** 311–334 (1974).

Senseny, P. E., and T. W. Pfeifle: "Fracture Toughness of Sandstones and Shales," *Proc. 25th U.S. Symp. Rock Mech.* (Evanston, Ill., 1984), pp. 390–397.

Singh, B.: "Reliability of Dilatometer Tests in the Determination of the Modulus of Deformation of a Jointed Rock Mass," in *Field Testing and Instrumentation of Rock,* STP 554 (Am. Soc. Test Materials, Philadelphia, Pa., 1974), pp. 52–72.

Stephansson, O.: "Rock Stress Measurement by Sleeve Fracturing," *Proc. 5th Int. Cong. Rock Mech.* (Melbourne, Australia, 1983), vol. F, pp. 129–137.

Yudhbir, Lemanza, W., and F. Prinzl: "An Empirical Failure Criterion for Rock Masses," *Proc. 5th Int. Cong. Rock Mech.* (Melbourne, Australia, 1983), vol. B, pp. 1–8.

Deformability

9.1 Concepts of Strain

9.1.1 Rock deformation

9.1.1.1 Deformability, stiffness, and strain. *Deformation* is defined as change in shape (expansion, contraction, or other forms of distortion). It occurs usually in response to an applied load or stress, but it also may result from a change in temperature (thermal expansion or contraction) or water content (swelling or shrinkage). *Deformability* describes the ease with which rock can be deformed, and its inverse, *stiffness,* the resistance to deformation. Deformability, like strength (Chap. 8), depends mostly on the porosity and the degree of jointing of the rock under test. Pores and joints are the weakest and most deformable elements in the rock.

Whereas deformation is measured in units of length, for convenience we usually express it in dimensionless form as *strain,* the ratio of the change in length of an element to its original length. Changes of length are usually small, so strain is expressed in units of microstrain (the strain value $\times 10^{-6}$).

Strain, like stress, is a directional phenomenon. A cubic element of rock in general experiences different levels of strain perpendicular and parallel to each face. The total state of strain in the cube can be stated in the form of a *strain tensor* (Fig. 9.1) analogous to the *stress tensor* described in Sec. 5.2.1.2.

9.1.1.2 Why measure and predict deformations? Most design studies in rock have the prime objective of predicting strains and displacements. The thin concrete shell of an arch dam will crack if the abutments are

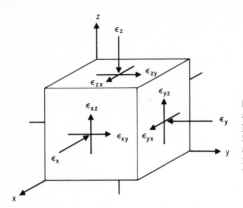

Figure 9.1 Strain components at a point. The shear strains disappear when the elemental cube is rotated into a unique direction, and at the same instant the normal strain components reach their maxima.

allowed to displace more than a few decimeters. The rock walls of tunnels, mine rooms, and stopes start to converge the moment they are excavated, and distress to pillars and supports, rather than any sudden catastrophic rockfall or burst, often governs the design. Large buildings on shale or closely jointed limestone require a foundation design based on settlements more often than one that guards against complete rupture of the foundation rock. Dangerous wastes can be stored in underground repositories only with a knowledge of the time-dependent deformation properties of the host rock.

9.1.1.3 Stress-strain behavior and constitutive models. In Chap. 8 rock strength alone was discussed, and deformation was ignored. These two aspects are, however, completely intertwined, because any rock must be stressed, hence strained, before its strength can be mobilized. A *strength envelope* was defined as the limit, in three-dimensional stress space, to the stresses that a rock element can tolerate before rupture. Now we discuss how to measure and predict the deformations and strains experienced by the element as it responds to some arbitrary stress path that travels within the limits imposed by this envelope.

The *stress-strain behavior* of rock is measured by deformability tests on rock materials in the laboratory, or on rock masses in the field. The results can be portrayed either graphically as a *stress-strain curve,* or algebraically as an equation that matches the curve. This is termed a *constitutive equation,* or a constitutive model, law, or relationship. Essential steps in the design of rock structures (Chap. 7) are first to assume a form of constitutive equation that will predict stress-strain behavior with sufficient accuracy, and then to obtain by testing the materials parameters that, when entered in the equation, make it match and predict the particular behavior of the rocks on site. Various alternative choices of constitutive models are discussed in Sec. 9.3.2

and also in Chap. 10, after a review of the testing methods and the behavior of rock.

The most general form of constitutive equation predicts strains as a function not only of stresses, but also of other controlling variables such as stress rate, time, and temperature. Fortunately the strains experienced by most rocks in most engineering situations are unaffected by time and temperature. Thus the constitutive relationships used in most designs assume behavior to be elastic, such that the rock returns to its original shape after applying and removing a system of stresses or loads. There are, however, situations when this is not the case, and time and temperature are important, notably when predicting the behavior of weaker rocks and those at great depth. Time and temperature effects are described in Chap. 10.

9.1.2 Measurement of strain

Tests that measure stress-strain behavior require the measurement of strain either directly or from monitored displacements. The following are the more commonly used techniques.

9.1.2.1 Strain from displacements.

Field tests, in particular, require that strain be measured over a long gauge length, often of 1 m or more, to allow averaging or "smoothing" of the strains that occur across individual rock blocks and the much larger strains across joints. Strain is obtained by dividing the displacement measured between widely spaced reference points by the distance separating these points.

The methods used for measuring strains in large-scale tests are identical to those of displacement monitoring, described in Chap. 12. The diameter of a pressurized tunnel is measured mechanically by a tape extensometer, or electrically by an inductive displacement transducer. Drillhole extensometers permit the measurement of strains at depth, and can be installed through the center of a test block or beneath a loaded plate (see Fig. 9.6b and c and Sec. 12.2.5.2).

On the scale of a laboratory test specimen, the simplest method is to clamp a mechanical dial gauge or an electrical displacement transducer to gauge points attached to the specimen or to the platens of the test machine. Measurements between platens can be unreliable because of nonuniform strains near the steel-to-rock contacts and the risk of movements of the clamps or of the test machine itself. Precise mechanical strain indicators (± 0.002 mm) make use of mechanical levers to amplify the measured displacements, whereas the optical levers of rotating mirror and telescope systems give an even greater accuracy (0.0002 mm), at least in theory.

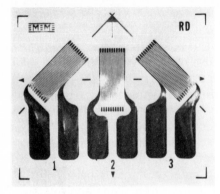

Figure 9.2 Three-component electric resistance strain gauge rosette.

9.1.2.2 Electric resistance strain gauges. Being small, moderately inexpensive, and reliable, the electric resistance strain gauge (Fig. 9.2) is now practically the only device used to measure strains in the laboratory (Window and Holester, 1982). It consists of an etched metal foil in a zigzag pattern with a plastic backing strip. Epoxy resin or another form of adhesive bonds the gauge to the rock surface. The foil resistance strain gauge is subjected to the same strains as those in the rock surface to which it is bonded. Compressive strain has the effect of shortening the length of foil and decreasing its resistance to the flow of electric current. The decrease is measured using a resistance bridge called a *strain meter* and is proportional to strain. Manufacturers supply the gauges together with a calibration certificate giving the *gauge factor,* which is the constant of proportionality between change in resistance and strain (change in length).

The lengths of available strain gauges range from a few millimeters to several centimeters. Gauge lengths of at least 10 times the maximum grain diameter in the rock are selected so as to average out the grain boundary displacements, yet they must be sufficiently short to monitor strains at a "point" if the strain field is nonuniform. Strain gauge "rosettes" contain two or three gauges mounted in a rectangular or triangular configuration on a single backing, and are used in preference to several individual gauges for measuring several strain components at a point.

9.1.2.3 Grids and coatings. Whereas the electric resistance strain gauge measures strain at a point, various grid and coating methods are available to measure, with a more limited precision, the strain field over a surface.

In the Moiré method (Kostak and Popp, 1966), a square, circular, or spiral grid (grating) of lines attached to the surface about to be strained is viewed through an identical unstrained master grid. Superimposing the strained and unstrained grids produces an interference pattern of *Moiré fringes* that can be evaluated in terms of fringe pattern, position, and orientation to give measurements of displacements or strains over the extent of the surface. Grids or *diffraction gratings* can be ruled at high density, on the order of 100 lines per millimeter, to improve the precision of measurement.

A brittle coating method (Durelli, Phillips, and Tsao, 1958) was developed in Germany and in 1941 became available in the United States under the name Stresscoat. A lacquer consisting of a natural or synthetic resin, solvent, and plasticizer is applied to the surface. When the surface is strained, the coating cracks along lines perpendicular to the maximum tensile strain, and at a spacing proportional to the magnitude of strain. The method has a sensitivity on the order of 500 microstrain, compared with about 1–10 microstrain for a resistance strain gauge system.

Photoelastic coating methods make use of the birefringent properties of certain types of transparent material (Sec. 7.3.1.4). A thin coating of plastic backed by reflective paint is attached to the surface of interest. When strained, a pattern of black and white fringes (isoclinics) becomes visible along the principal strain directions if the coating is viewed in plane-polarized light. A different pattern of colored fringes (isochromatics) can be observed in circularly polarized light, which permits measurement of the surface shear strain magnitudes.

9.1.2.4 Optical interferometry and holography. Holography was invented in 1948 but did not become practical until the early 1960s, when it was used in combination with laser light. Interferometric holography is a promising technique for measuring small amounts of movement and strain in the surfaces of rock specimens or the walls of drillholes (Khair, 1983; Park, 1986; Schmitt et al., 1986). It does not require any gauge emplacement on the specimen, and the material to be tested need not be ground or polished. Very small displacements can be measured, equal to half the wavelength of the laser radiation used, typically on the order of one-third of a micrometer (3×10^{-7} m).

There are two methods of achieving holographic images. The double exposure method requires an exposure first of the undeformed, then of the deformed surface, before developing the film. The second, real-time, holographic method requires just one exposure of the undeformed configuration to be taken and developed. The object is then viewed through this hologram so that any subsequent three-dimensional movement can be detected as interference fringes, which are contours of equal displacement.

9.2   Deformability Testing

9.2.1   Overview

9.2.1.1   Quasi-static elastic parameters.   The test methods described in the initial and largest part of this chapter are called *quasi-static* ones. This means that the full test load is applied in a period of minutes to a day, in contrast to the *dynamic* tests described at the end of the chapter, in which load is applied in fractions of a second by impact or vibration, and the *creep* tests (Chap. 10), in which load may be maintained for weeks or months. The term *elastic* implies that the aim is to measure an equivalent *static elastic modulus* as defined later in this chapter, even though elasticity may be an approximation to the true behavior of the rock.

9.2.1.2   Scale effects and selection of a test method.   The stress-strain curve obtained in a static elastic test has to represent as closely as possible the behavior of an element of the rock mass, represented as an *equivalent continuum* (Sec. 7.1.2.3). The curve obtained by testing a specimen of intact material usually is quite different from that obtained for a larger specimen or an in situ test, because of the presence and influence of joints. The rock mass is more deformable (less stiff) because the closure of joints contributes greatly to the overall compressive strain of the rock mass element. A progressive decrease in stiffness, a *scale effect*, is observed as the test "specimen" increases in size from centimeters for a laboratory test to decimeters or meters for small- or large-scale field tests.

In the following outline of testing techniques, the methods are presented in a sequence of increasing size, from the small-scale laboratory ones to progressively larger ones conducted in situ. The volume of stressed rock depends on the magnitude of the load and the extent of the surface of load application (Fig. 9.3). Laboratory tests are at the lower end of the scale, followed by drillhole tests, plate and flat-jack tests, tests on large in situ blocks, and tests conducted by pressurizing a tunnel. Evidently the cost increases, and the number of tests that can be done within a limited budget decreases as the tests become larger. Choice depends on several factors, notably:

The project budget, which reflects the importance of precise modeling and the cost of an imprecise prediction resulting in structural damage or failure.

The nature and influence of jointing. The scale of testing is unimportant when the joints are very widely spaced and tight, or the rock is relatively soft and deformable such that the influence of jointing is minimal. Small specimens can then be tested for economy. In all

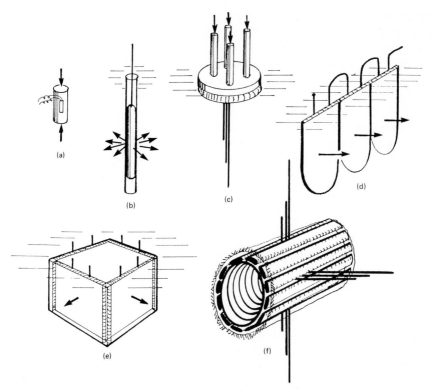

Figure 9.3   Deformability test configurations in order of increasing size. (*a*) Uniaxial and triaxial tests in the laboratory. (*b*) Dilatometer test in a drillhole. (*c*) Plate-loading test. (*d*) Large flat-jack test. (*e*) Block test. (*f*) Radial jacking or tunnel pressurization test.

other situations, either the test specimens must include many joints, or the stiffnesses measured on unjointed rock must be scaled down to account for the missing jointing.

Laboratory deformability tests are nearly always done, if only to check on measurements in the field. Usually the laboratory test is a uniaxial one, although modulus values obtained by loading in triaxial compression may be different, particularly if the rock is porous. Field modulus values can be estimated approximately from laboratory values by applying a reduction factor that is a function of joint spacing and aperture (Sec. 9.3.1.3).

9.2.2   Uniaxial stress-strain test

9.2.2.1   Outline.   Laboratory measurements of deformability employ the same methods as those for strength testing, described in Chap. 8, except that the specimens must now be instrumented to measure strains. A deformability test is often followed immediately by a strength test on the same specimen.

Tests on intact rock use cylindrical core with flat ends perpendicular to the axis of the cylinder. Specimens with a length-to-diameter ratio of between 2 and 3 are loaded between spherical seats in a compression testing machine (Fig. 9.3a). Larger cores of jointed rock can be tested using the method described in Sec. 8.2.4.3.

Uniaxial stress is increased at a rate of 0.5–1.0 MPa/s, while taking at least 10 readings of stress and corresponding axial and circumferential strain at evenly spaced stress intervals. The results are plotted on a graph (e.g., Fig. 9.10). Young's modulus E is obtained as the slope of the axial stress-strain curve, and Poisson's ratio v is obtained by dividing E by the slope of the circumferential stress-strain curve.

9.2.2.2 Strain measurement.

The specimen is instrumented with electric resistance strain gauges, typically of 20–30 mm gauge length (at least 10 times the maximum grain diameter or microjoint spacing in the rock), or with dial gauges or transducers that measure strains within the central one-third of the specimen. The alternative of measuring between the loading platens is easier, but can introduce unacceptable errors, as discussed earlier. Resistance strain gauges are expensive and can be used once only, whereas clamp-on transducers (e.g., Schuler, 1978) are removed and used many times. Clamp-on gauges and even optical methods of "surveying" become increasingly reliable as the specimen size increases.

Determination of Poisson's ratio in addition to Young's modulus requires the measurement not only of longitudinal strain, but also of either diametral strain by an electric transducer and clamping system, or of (equal) circumferential strain using an electric resistance strain gauge cemented with its measuring axis along the curved specimen circumference at mid-height. To simplify and speed up the mounting of strain gauges, a strain gauge rosette can be used as an alternative, which comprises two or three differently oriented gauges on a single backing strip or patch (Sec. 9.1.2.2).

9.2.2.3 "Stiff" test machines to measure postpeak behavior.

The commonly experienced explosive-type rupture of a uniaxial or even a triaxial test specimen under small confining pressures is often caused as much by the nature of the testing apparatus as by the brittleness of the rock being tested. As the stress level increases, strain energy is stored in the frame and hydraulics of the machine and is released violently when peak strength is surpassed. The microfractured rock material disintegrates very rapidly, and the stress-strain curve shows an abrupt downward trend as the load falls to zero with little if any additional straining. The same rock, when tested in a stiff testing machine, will show a gradual, not an abrupt, decrease in load at rupture, until complete disintegration occurs after considerable straining and at quite low levels of applied stress.

Test machines that permit explosive rupture are called *soft,* and can be used only to explore the part of the stress-strain curve that precedes rupture. This is sufficient when one wishes to measure just the static elastic parameters at relatively low stress levels. However, any attempt to plot the stress-strain curve up to and past the point of rupture, the *postpeak* behavior of the specimen, requires a loading system that precludes sudden transfer of energy from the machine to the rock.

There are three alternatives: (1) a stiff and sturdily constructed machine that stores little energy and contains a minimum of compressible hydraulic fluid in the ram; (2) a servo-controlled machine that removes load very rapidly in response to sudden increases in strain, and one that controls strain rates rather than stress rates; or (3) a test configuration in which the specimen is loaded in parallel with instrumented steel cylinders that share a measured proportion of the applied load, a proportion that increases sharply as the specimen starts to rupture. Although expensive, the most effective system is a machine that is both stiff and servo-controlled.

9.2.3 Triaxial stress-strain tests

9.2.3.1 Axisymmetric triaxial test. Triaxial stress-strain behavior is explored using a method similar to that described for uniaxial testing, and an extension of that used for triaxial strength measurement (Chap. 8). An axisymmetric triaxial cell is most often used, which can even be sufficiently large to test specimens containing joints (Natau et al., 1983, discussed in Sec. 8.2.4.3). The strains in these large specimens can be measured directly by optical means. Strains in smaller specimens are usually measured by electric resistance strain gauge rosettes mounted at mid-height, either at opposite ends of a diameter or at third points around the circumference, or with lateral and longitudinal gauges mounted alternately at quarter points. A small axial stress holds the specimen in place; then cell pressure is applied and held constant at the required level. Axial stress is increased steadily, taking stress-strain readings either continuously or at increments.

These tests are much more difficult and expensive than simple strength tests, because of the need for strain gauging. The gauge wires pass along the surface of the specimen or through the fluid and special self-sealing ports, depending on the design of the cell. Wires become pinched or broken by movements. A further problem results from sensitivity of the strain gauge to confining pressures. Gauges can be calibrated to allow pressure compensation, by mounting and testing similar gauges on a metal specimen of known elastic parameters. Errors remain if a gauge is underlain by a void, such as in the surface of a porous sandstone, because confining pressure forces the gauge, or part of it, into the hole. The problem is prevented by applying a sealing coat of adhesive to the rock surface, which is sanded smooth before applying the gauge.

9.2.3.2 Hydrostatic compression test. Testing under conditions of equal all-around compression tends to be much simpler than under full triaxial stress with a deviatoric component. The test cell is a pressure vessel in which the only openings are those needed to inject fluid and transmit the gauge wires, so applied pressures can approach those at great depths in the earth's crust where the ambient stresses are close to hydrostatic. These tests are most often used to investigate the collapse of porous rock types in oil and gas reservoir investigations, where a loss of formation permeability can evidently have serious consequences.

9.2.4 Laboratory measurements of elastic anisotropy

Deformability is affected by mineral types and their alignments and shapes (Brace, 1965). For example, metamorphic rocks with a pronounced grain elongation tend to deform more readily if loaded perpendicular to the schistosity.

When testing a block of rock, cores are usually taken in three orthogonal directions, including those of greatest and least deformability, parallel and perpendicular to any planes of foliation or bedding. Alternatively, cubes can be cut and tested by loading across different pairs of faces. When just a single core direction is available, the following special techniques are needed to measure the 5 or 21 independent elastic parameters (Sec. 9.3.2.3).

Not only the strength but also the elastic parameters of intact rock can be measured by a Brazilian-type test on a diametrally loaded rock disk. Experimental and interpretation methods have been developed for both isotropic and anisotropic elastic cases (Amadei et al., 1983).

In the laboratory, biaxial tests can be run using plates of rock loaded uniformly along their edges in a condition of *plane stress*. Faces of the plate are easily accessible for strain gauging. Polyaxial (multiaxial) types of "true" triaxial test can be run on rock cubes (Amadei et al., 1983). The apparatus tends to be complex and expensive, not only because of platen interference and nonuniform boundary condition problems, but also because of further problems of strain measurement when the specimen cube is completely surrounded by the loading device.

9.2.5 Dilatometer tests

9.2.5.1 Overview. The exploratory drillhole allows testing on a scale slightly larger than in the laboratory. Not only is the undisturbed, jointed rock made accessible at all horizons of interest, but also the walls of the hole provide reaction for very high levels of loading. Similar tests but on a larger scale can be conducted in tunnels where reaction is provided by the walls, roof, and floor. Reaction becomes more of a problem when testing in open excavations.

A dilatometer, sometimes known as a pressuremeter, is a testing instrument that is expanded in a drillhole to apply pressure to the drillhole walls (ISRM, 1987). Dilation (i.e., expansion) of the hole is measured as a function of pressure, from which the deformability parameters of the ground are calculated (Figs. 9.3*b* and 9.4).

The dilatometer is a versatile instrument. In addition to its main application to deformability measurements in the elastic range, the high-pressure types can be used to measure yield and creep properties of weak or plastic rock, such as oil shale, potash, or rock salt (see Fig. 10.5). By in situ fracturing of harder rocks, the instrument can be used to measure both tensile strength and in situ stress (Chaps. 8 and 5). It can be used to test hollow cylinders of rock in the laboratory for measurements of elastic modulus and tensile strength.

The dilatometer for soil deformability measurement was first developed in Germany in the 1930s and came to be widely used following development of the Ménard pressuremeter in France in the late 1950s. In rock mechanics, the first high-capacity flexible dilatometers were

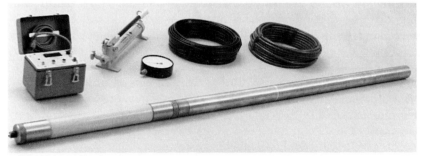

Figure 9.4 Flexible dilatometer (Probex-1 by Roctest, Canada), showing the dilatometer itself, and electric and hydraulic lines, hand pump, and pressure and volume gauges. In this model the probe volume is monitored at probe level by measuring the displacement of a piston. (*Courtesy of Roctest, Montreal, Que.*)

designed in 1964 by Panek at the U.S. Bureau of Mines (Panek et al., 1964), and in 1966 by Rocha at the National Civil Engineering Laboratory (LNEC) in Portugal.

9.2.5.2 Types and advantages. There are two types of dilatometer, rigid and flexible. The rigid type (*borehole jack* or *Goodman jack*) consists of a split rigid cylinder whose halves are thrust outward against the walls of the hole by the action of multiple hydraulic rams within the probe. Separation of the halves is measured by electric transducers. Flexible types, such as the Ménard pressuremeter, the Colorado School of Mines (CSM) cell, and the Roctest version shown in Fig. 9.4, use an inflatable synthetic rubber probe (Baguelin et al., 1978; Hustrulid and Hustrulid, 1975). Dilation pressure is measured hydraulically, and dilation itself either from the volume of injected fluid, or directly by displacement transducer. Only the direct-measuring (Lisbon) type can be used to determine the anisotropy of deformability as a function of radial direction within the drillhole (Fig. 9.5): Volume change types give an average value for the deformability modulus.

In comparison with other methods for measuring deformability, the dilatometer is the quickest and least expensive of the field techniques. Therefore it is used not just for isolated measurements, but for complete logging of deformability variations from stratum to stratum along a drillhole, thus combining the functions of an index test and a design test. It is particularly useful for exploring variations caused by weathering, and for the rapid index logging of drillholes in fragile, clayey, or closely jointed rocks that yield poor core recovery and inadequate specimens for laboratory testing. It can be a valuable aid to selecting foundation bearing depths.

A disadvantage is that the volume of rock stressed by a dilatometer is usually less than one-third of a cubic meter and often too small for direct application of the results to design problems. Correlation of the dilatometer modulus with that obtained by plate loading, radial jacking, or flat-jack methods, or still better, by back-calculation from observations on real rock structures, allows an extrapolation of the dilatometer test results to the large scale (Panek, 1970). It has the further limitation of directionality. Measurements can only be made radial to the drillhole, usually horizontally in a vertical hole. This can be a problem when, as is often the case, the rocks are anisotropic and more deformable in the vertical than in the horizontal direction. Vertical settlements of a foundation on stratified rock, for example, cannot be predicted from horizontal measurements.

Rigid types of dilatometer can apply a somewhat higher range of pressure, are less vulnerable to damage caused by drillhole irregularities, and can also measure radial variations in modulus as can the

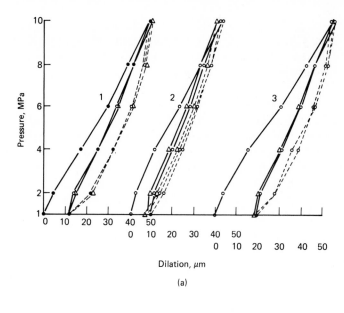

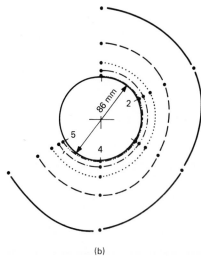

Figure 9.5 Dilatometer test results. The example is for a Lisbon-type dilatometer that can measure not only average deformability, but also anisotropy by recording dilations in different directions. (*ISRM, 1987.*)

directional type of flexible dilatometer. Flexible types, however, have the advantage of applying a much more uniform and well-defined contact pressure, whereas the rigid probes often suffer from poorly defined boundary conditions.

9.2.5.3 Procedure. The procedure, with both types, is to lower the probe to the required depth of measurement and then expand it against the drillhole walls. Readings of dilation are taken at increments of pressure. The pressure is maintained at selected increments to assess creep rates. After making corrections for calibration errors, the results are presented as a graph of applied pressure versus dilation, and of dilation versus direction when measurements are in more than one direction (Fig. 9.5).

9.2.5.4 Calculations. For flexible probes, the dilatometric shear modulus G is calculated from

$$G = M_r \frac{\pi L D^2}{x}$$

where, in metric units, L and D are the length and the diameter of the drillhole test section, x is the pressure generator constant (= 3.6×10^{-7} m^3/turn for the CSM system), and M_r is the pressure-volume relationship (stiffness) of the rock. M_r for the rock alone, is expressed as follows, in terms of megapascals, per turn of the screw pump:

$$M_r = \frac{M_s M_t}{M_s - M_t}$$

where M_s, the pressure-volume relationship for the system, is given by

$$M_s = \frac{M_c M_m}{M_c - M_m}$$

with M_c being the pressure-volume relationship for the calibration cylinder, given by the expression based on the theory of elasticity for a thick-walled cylinder, M_m that for the pressurizing system plus calibration cylinder measured during calibration, and M_t that for the pressurizing system plus rock, obtained as the slope of the test result graph.

If Poisson's ratio v is known or assumed, the dilatometric modulus of deformability E may then be calculated from

$$E = 2(1 + v)G$$

For rigid probes, such as the Goodman jack, the modulus is calculated as

$$E = 0.8DT \frac{dP}{dD}$$

where dP/dD is the slope of the linear part of the graph of hydraulic pressure versus dilation, D is the hole diameter, and T is a coefficient with values between 1.1 and 1.5, depending on Poisson's ratio. A further graphic correction is needed to account for bending of the "rigid" probe at higher pressures.

9.2.6 Plate-loading tests

9.2.6.1 Overview. *Plate testing* (Fig. 9.3c) is the most commonly used and often the least expensive of the larger-scale methods for determining rock mass deformability, particularly in foundation engineering applications (ISRM, 1981).

In its simplest form, a rigid plate, often built up from several concentric disks of heavy-gauge steel, is pushed against a prepared rock surface, and settlements are measured at the surface of the plate. The load is increased in increments, at each stage measuring the resulting plate *settlement* or *penetration* and checking for creep. The applied stress levels are chosen to be similar to those expected in the full-scale structure, and load is often cycled to check for nonlinearity and differences in loading and unloading behavior. More reliable types of plate-load test, described below, employ flat-jack cushions to distribute the load evenly, and drillhole extensometers to measure strains at depths beneath the area of load application.

9.2.6.2 Tests at surface and underground. Tests at the surface often use a dead-load system or *kentledge*, such as steel ingots or concrete blocks stacked on a trestle or rail car. A water tank is an alternative for loading a large area to a relatively small stress level. More commonly, the reaction is provided by a system of rock anchors.

Plate tests at the base of drillholes or caissons can be used to investigate modulus variations with depth, and the properties of various potential foundation strata. The drillhole is advanced to a new horizon after each test has been completed. Similar but larger-diameter tests can be done at the base of a caisson, using a rock anchoring system at the ground surface with a braced trestle to transmit loads from the plate to the anchors. Alternatively, flat jacks can be installed at the base of the caisson shaft, which is then backfilled with a concrete plug to provide the reaction. The shear strength of the bond between plug and surrounding rock must be checked as sufficient to resist flat-jack thrust, and several flat jacks may need to be stacked on top of each other to give sufficient expansion capability to allow for ground displacements during loading.

The advantages of testing in a tunnel include easier and stronger reactions and better access to the undisturbed rocks of interest. The tests usually are in pairs, loading two identical plates at opposite ends of a reaction column. Displacements are measured either with multiple extensometers beneath the plates or, less precisely, by monitoring tunnel convergence from plate to plate. Measurements in several directions radial to the tunnel allow determination of the rock anisotropy caused by bedding or jointing. Test equipment can be mounted on a rotating axis to facilitate rapid tests in various directions at the same location.

9.2.6.3 Rigid and flexible plates. A derivation of the required static elastic parameters from the results requires the use of a closed-form solution or numerical analysis assuming either a perfectly rigid plate (uniform boundary displacement) or a perfectly flexible one (uniform applied stress). A rigid plate is best for testing soft foundations such as on soil, but when testing harder rocks, particularly with loaded diameters of 1 m or more, the assumption of plate rigidity is not often justified. Even a reinforced plate can bend.

The alternative *flat-jack cushion* method of applying a uniform loading, as opposed to a uniform boundary displacement, is more reliable (Figs. 9.6 and 9.7). Flat-jack cushions are made from thin steel plates that are circumferentially welded to form a doughnut with a small central hole (Fig. 9.6b). The hole gives access to a drillhole extensometer to measure strains beneath the loaded area (Fig. 9.6c). The flat jack can be used simply as a cushion to equalize pressure beneath conventional hydraulic rams, or it can be inflated by hydraulic pressure to apply the load directly.

In this improved version of the plate-loading test, a multiple-position extensometer is installed in a small hole drilled in the center of the zone to be tested. Displacements are monitored at various depths beneath the area of loading. This ensures that the computed deformability modulus values are free from the influence of superficial irregularities of loading, plate tilting, and near-surface rock loosening. Variations, usually increases, in modulus as a function of increasing depth can also be determined.

Large-scale plate-load tests of this type were performed for the Rocky Mountain pumped hydro project (Bakhtar and Barton, 1986), applying a load of 250 t over 1 m^2 to siltstones, sandstones, and organic shales. Tests were conducted in a 10.5-m-diameter tunnel, with reaction provided by deeply anchored diverging rockbolts and fabricated steel frames. Pressure buildup to 2.5 MPa was provided by circular flat jacks, and rock displacements were measured beneath the loaded area at various depths.

(a)

(b)

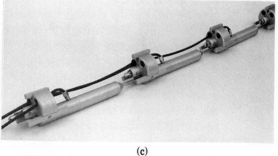

(c)

Figure 9.6 A 500-t plate-load test for the La Grande 2 underground power house, James Bay, Que., Canada. (*a*) Testing in progress. (*b*) Details of the flat jacks for a horizontal test. Ends of the multirod drillhole extensometer appear through the central hole in the flat jacks and cover plate. (*c*) The BOF-EX multielectric extensometer eliminates the need for a hole through the flat jacks. (*Courtesy of Roctest, Montreal, Que.*)

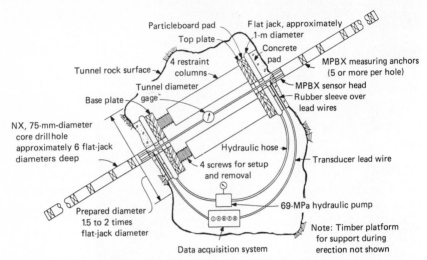

Figure 9.7   Test arrangement for the plate test with load applied by flat jack, conducted underground. (*ISRM, 1981.*)

9.2.6.4 Procedure and interpretation. Pressure is applied in increments using a hydraulic pump, and the corresponding rock displacements are recorded. Load is usually cycled several times from near-zero to the required full test load to investigate hysteresis and creep. Graphs are plotted either of plate settlement, or of strain as a function of depth beneath the loaded area, for each increment of load.

When measuring displacements at the surface only, the initial portion of the force-displacement curve is concave upward as a result of near-surface cracking of the rock foundation and the incomplete contact between the loaded plate and the rock surface. The contact area increases as load is applied, and the stress-strain curves steepen for successive load-unload cycles. Near-surface effects are usually irrelevant to foundation performance and must be ignored or avoided. They may be reduced by careful rock surface preparation, avoiding damage and loosening, and using a thin layer of grout between the plate and the rock to improve the contact.

The engineer has to decide which part of the stress-strain curve to use in design, not only which level of stress, but also which load or unload cycle. The design of different types of structure may call for the use of loading or unloading modulus values, and for values from the first or subsequent cycles.

For the design of small footings, displacements can be scaled directly in terms of foundation settlements by applying a *scaling factor* to account for the size difference between the test plate and the diameter of the prototype footing. Larger foundations and a more rigorous analysis require computation of the deformation modulus E from the following rigid-plate formula:

$$E = \frac{dQ}{dS} \frac{\pi}{4} D(1 - v^2) I_c$$

where dQ/dS is the gradient of the applied pressure versus settlement graph, D is the plate diameter, v is Poisson's ratio, and I_c is a depth correction factor, obtained graphically, which varies between 0.5 and 1.0 depending on the ratio of test depth to plate diameter and on Poisson's ratio.

For tests using a flat-jack cushion and multiple extensometers beneath the area of loading, the modulus of deformability E can be calculated for each layer of rock beneath the plate, using the following expression:

$$E = q \frac{K_{z1} - K_{z2}}{W_{z1} - W_{z2}}$$

where q is the average applied pressure, $W_{z1} - W_{z2}$ is the measured closure between a pair of extensometer measuring points, and K, at each point, is given by the following formula in terms of Poisson's ratio v, the radius of loaded area a, and the distance z from the hole collar to the extensometer reference point:

$$K = 2(1 - v^2)(a^2 + z^2)^{1/2}(a_1^2 + z^2)^{1/2}$$
$$+ z^2(1 + v)[(a_1^2 + z^2)^{-1/2} - (a_2^2 + z^2)^{-1/2}$$

where a_2 is the outer radius of the flat jack and a_1 is the radius of the hole (ISRM, 1979).

9.2.7 Flat-jack tests

The *large flat-jack test* (ISRM, 1986) was developed by the National Civil Engineering Laboratory (LNEC) in Lisbon, Portugal. In contrast to the *cushioned plate test* described above, load reaction is provided by inserting the flat jacks in sawn or line-drilled slots (Fig. 9.3d). Heavy columns and reaction systems are thus avoided. Pressures as high as 20 MPa can be applied over areas of several square meters by installing several flat jacks in line.

A conventional circular saw can cut only to a depth somewhat less than the radius of the saw blade, only a fraction of a meter even using a large blade. This would be a severe limitation, because the near-surface rock is usually unrepresentative, and because the rock can spall and fracture if high pressure is applied parallel and near to the surface. However, a sawing technique invented at the LNEC allows cutting of a slot to any required depth (Fig. 9.8). This is achieved by

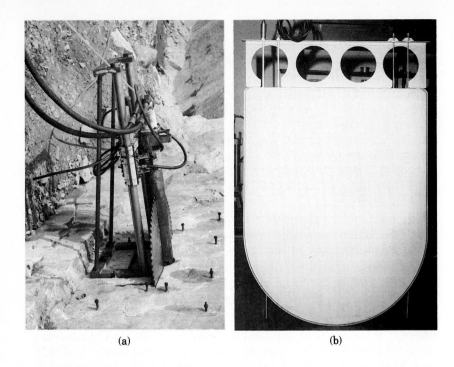

(a)   (b)

(c)

Figure 9.8 Large flat-jack test for measuring rock mass deformability. (*a*) Special saw for slot cutting. (*b*) Typical flat jack with spacer flange. (*c*) Flat jacks installed side by side in a single slot, showing hydraulic pump, pressure gauge, and readout for measuring the amount of slot expansion. (*Courtesy of LNEC, Lisbon, Portugal.*)

mounting the blade on a column and powering it with a chain drive. A large hole is predrilled along the centerline of the slot to accommodate the drive column.

Alternative slot cutting methods include wire sawing and line drilling with overlapping percussive-drilled holes. Sawn slots are better because they are smooth and flat and allow insertion of the flat jack without grouting. The grout quality can affect the results. Flat jacks have been known to burst if there are air or water pockets in the grout.

When testing over an extended area, the slots are sawn in line, close to each other, but not overlapping. Slot imperfections are limited to ± 0.5 mm. A 250-mm-deep spacer flange welded to the edge of the jack ensures adequate embedment. The flat jacks are inserted. Slot centerline holes are backfilled with grout. The flat jacks are filled with hydraulic fluid and bled to remove all air. Flat jacks containing high-pressure air will explode if ruptured, and are dangerous to use.

Slot expansion displacements, in the LNEC technique, are measured by an electric transducer located centrally within the flat jack. The alternative of measuring between drillholes to either side of the slot is less convenient and less precise. Pressure is cycled, taking readings of slot expansion at selected pressure increments. It is maintained at the highest point of each cycle to check for creep. The results are presented graphically, and the modulus of deformability E is calculated for any given point on the pressure-displacement curve from

$$E = k(1 - v^2) \frac{dP}{dD}$$

where k is a coefficient, determined graphically, that depends on the shape and number of jacks, the test chamber configuration, and so on; v is Poisson's ratio, usually approximated by laboratory testing; and dP/dD is the slope of the pressure versus slot opening curve at the selected tangent point.

9.2.8 Block tests

Polyaxial tests are common in the field, where large cubic or rectangular blocks are the easiest to prepare, and where measuring need not be so precise because of the larger scale. These *block tests* (Fig. 9.3e) have become popular for characterizing large-scale thermomechanical behavior of rock to obtain data for modeling radioactive waste repository chambers. The large faces of the blocks can easily be uniformly stressed, to a high level of stress, by flat jacks in slots. Zimmerman et al. (1984) describe biaxial testing of a heated block of volcanic tuff at the Nevada test site, with a volume of 8 m^3 at biaxial stresses of up to

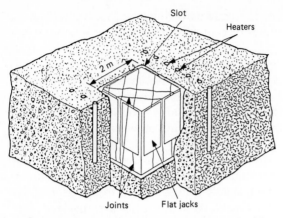

Figure 9.9 Deformability test on a 2- by 2- by 2-m block equipped with heaters, to study the thermomechanical behavior of jointed rock. (*Zimmerman et al., 1984.*)

10.6 MPa (Fig. 9.9). The tests were uneventful except for one in which a flat jack ruptured by expansion into a pumice-filled cavity. The cracked region was chipped loose and filled with a grout before replacing the flat jacks.

9.2.9 Radial jacking and pressure tests in tunnels

9.2.9.1 Overview. In a tunnel pressurization test (Fig. 9.3*f*), of which there are several variations (ISRM, 1981), the deformability of the rock mass is measured by applying an internal pressure to a test chamber of circular cross section, and measuring the resulting radial expansions. This large-scale method, although relatively expensive, usually gives better and more reliable values for jointed rock conditions than can be obtained with the smaller-scale alternatives. It is most often used in hydroelectric power applications for the design of high-pressure penstock tunnels and to obtain modulus values for the design of dam foundations.

The test chamber, usually about 2.5 m in diameter and of the same loaded length, is often an extension of an exploration gallery driven beneath the foundation or into an abutment. The rock must be carefully excavated to avoid loosening, and is either machine-bored or lined with concrete to give a smooth finish. The liner is segmented to limit its resistance to radial expansion if the modulus of the rock alone is of interest, but not when one wishes to evaluate the combined liner-rock behavior.

9.2.9.2 Loading systems. Internal pressure is applied by fluid in one of three ways. In the more traditional form of test, the tunnel is filled with water and pressurized. This alternative suffers from the limitation that one needs to thoroughly seal the rock and to provide load-bearing mass-concrete bulkheads. Also, the displacement-monitoring instruments have to be constructed to operate in a high-pressure water environment, by remote control.

The second alternative is to inject high-pressure water between the tunnel and a waterproof and pressure-resistant loose-fit liner. Compared with the completely filled alternative, the bulkheads are replaced by smaller water seals, and access is available to instruments which nevertheless must be sealed where they pass through the steel liner plate.

The third and usually the simplest method is to use flat jacks, installed side by side and connected together to form a continuous pressurized cushion around the tunnel perimeter. In this *radial jacking test*, reaction is provided by a ring beam of steel, usually with an intermediate timber liner to provide a smooth surface and prevent rupture of the flat jacks. Not only are water seals and expensive liners avoided, but also access is unobstructed within the tunnel. The small gaps between flat jacks give room for the insertion of displacement measuring instrumentation.

9.2.9.3 Procedure and interpretation. In the flat-jack version, tunnel diameter (convergence) measurements are made between the flat-jack locations and often also deep within the rock using drillhole extensometers (Chap. 12). Pressure is cycled, taking readings and plotting displacements against pressures. On reaching the maximum pressure for each cycle, the pressure is maintained while creep movements, if any, are recorded as a function of time.

Elastic modulus E and deformation modulus V are calculated from the following expressions:

$$E = \frac{2p_r}{D_e}\left(\frac{v+1}{v} + F\right)$$

$$V = \frac{2p_r}{D_t}\left(\frac{v+1}{v} + F\right)$$

where p_r is the pressure applied at radius r. When this pressure is applied by nontouching flat jacks, it must be calculated by multiplying the measured flat-jack pressure by a reduction factor S/C, where S is the sum of the flat-jack widths and C is the excavated circumference. D_e and D_t are the recoverable (elastic) and total (elastic plus plastic) tunnel convergences obtained from the experimental graphs after final unloading, and v is an estimated value for Poisson's ratio. F is taken as zero when calculating the modulus of the as-excavated tunnel walls. Alternatively, the modulus of the rock beneath a superficially damaged zone limited by radius R may be calculated by taking $F = \ln (R/r)$, where ln is the Naperian (natural) logarithm.

The dimensions of pressure tunnel liners can be determined graphically directly from the results of this type of radial jacking test. Oberti et al. (1983) give procedures for carrying out and interpreting the test under conditions of rock mass anisotropy.

9.3   Static Elastic Deformation Behavior of Rocks

9.3.1   Stress-strain curves for rocks

9.3.1.1   Longitudinal strain.
Figure 9.10 shows a uniaxial stress-strain curve for a typical specimen of porous or jointed rock. Load-displacement curves for field tests are similar in shape to stress-strain curves measured in the laboratory, so both can be reviewed together.

During the initial application of load, rock progressively becomes denser as pores, cracks, and joints close, and as a result, becomes stiffer and less deformable. The load-displacement curve thus shows an upward concavity (region A) reflecting this increasing stiffness at higher loads.

At intermediate stress levels, usually between one-third and two-

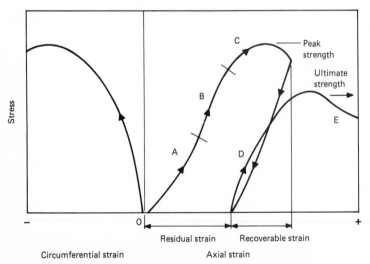

Figure 9.10   Stress-strain curve for typical porous or jointed rock.

thirds of the uniaxial compressive strength, the behavior of most rocks becomes linear elastic (region *B*). All or the majority of pores and joints have closed, so strain increments become proportional to those of applied stress.

At stresses higher than this, but before reaching peak strength, the joints start to shear, and cracks start to propagate and coalesce (Sec. 8.3.1). The stress-strain curve becomes concave downward (region *C*). If the applied load is removed during this phase of testing, there is a pronounced residual or plastic strain. If the load is maintained, the specimen is likely to creep and eventually to rupture.

Any unloading followed by reloading is accompanied by hysteresis (curve *D*) caused by internal friction along the surfaces of closed pores, cracks, or joints. The walls of cracks slip frictionally as the stress increases, and in the reverse direction as it decreases. The work done, and the energy lost during the load-unload cycle, are measurable as the area within the hysteresis loop.

Postpeak stress-strain behavior (region *E*) can only be investigated using a stiff system of loading (Sec. 9.2.2.2). Joints in the mass undergo large shear displacements, and shearing develops through intact rock specimens according to a brittle or ductile mode of rupture, as discussed in Sec. 8.3.2. As a whole, the rock specimen or element exhibits a "strain softening" behavior in which displacements and strains continue under conditions of decreasing load and stress, reaching an *ultimate* strength at some lower stress level than the *peak* strength value. Note that the ultimate strength depends on the amount of displacement that can be achieved, and in turn on the type of test and equipment. It differs from the true *residual* strength, which applies to larger displacements and tests on individual joints (Sec. 11.2.5.2).

9.3.1.2 Lateral strain.   At the same time as the uniaxially loaded specimen is shortening, it is expanding. When of interest, lateral expansion is plotted on the negative strain axis, as shown in Fig. 9.10. Note that the circumferential strain is identical to the lateral (diametral) strain, because circumference is proportional to (π times) diameter.

At low stress levels the lateral strain is usually less than one-quarter of the measured longitudinal strain. The ratio of lateral to longitudinal strain is called *Poisson's ratio*. At stresses up to about 50–60% of uniaxial compressive strength, a Poisson's ratio in the range of 0.1–0.25 is typical. At higher stress levels approaching rupture, longitudinal cracks appear that result in lateral strains much greater than those occurring longitudinally. Poisson's ratio therefore increases dramatically. Poisson's ratio was conceived as an "elastic" measure. Therefore when rock behaves inelastically, as it does when it dilates, the measured value of v can be much higher than 0.5, which is the theoretical maximum for purely elastic deformation.

9.3.1.3 Behavior of various rock types.

Although the above pattern of behavior is quite typical, the different rock types behave differently (Fig. 9.11). Dense, unjointed igneous and sedimentary rocks have the most linear stress-strain curves and retain linearity at higher stress levels. Their postpeak behavior, however, typically shows pronounced strain weakening, with a sharp drop from peak to ultimate strength.

Closely jointed rocks such as coal, and also porous materials such as sandstones and fragmental limestones, exhibit the greatest degree of stiffening during initial loading. Usually they have lower strengths but also a lesser degree of strain weakening, with only a subdued drop from peak to ultimate strength.

Weak yet dense materials such as clay shale approach the closest to "elastic perfectly plastic" behavior (Sec. 9.3.2.4). Their initial stress-strain curve is close to linear, and on yielding they exhibit little if any peak. The linear elastic phase tends to be followed by a plastic one in which further strain occurs at constant stress. The two segments are joined, however, by a rounded nose rather than a sharp corner as in the ideal plastic model.

Moderate increases in confining pressure result in stiffer and more linearly elastic behavior, with less tendency to expand volumetrically as rupture approaches. The specimens also become much stronger and undergo a greater strain before rupture.

Substantial increases in confining pressure and/or temperature (the two are complementary) result in a transition from brittle to ductile behavior (Sec. 8.3.2). Hard rocks start to behave like soft ones do at more normal pressures and temperatures. The peak of the stress-strain curve is suppressed, leaving a smooth elastic-plastic transition.

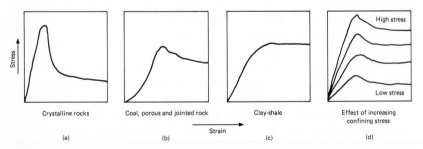

Figure 9.11 Stress-strain behavior of typical rock types. (*a*) Igneous and dense sedimentary rocks. (*b*) Coal, porous and jointed rocks. (*c*) Clay shale. (*d*) Effect of increasing the level of confining stress.

9.3.1.4 Comparison of laboratory and field results. Because of the influence of jointing, the stiffness (deformability modulus) of a rock specimen is often much greater than that of an element of jointed rock in situ. The ratio of laboratory modulus to field modulus of deformability assumes values close to 1 when the rock is massive, but may be as high as 10 when the rock is closely jointed. This ratio is affected not only by joint spacing but also by the filling and aperture of individual joints.

9.3.2 Idealized behavior and constitutive models

9.3.2.1 Overview. Constitutive equations are obtained by fitting linear or curved "models" to the real data for the rock. Equations can be selected with some theoretical justification or just empirically; the data themselves are the judge of whether the equations are good predictors. The process of curve fitting is exactly the same as that for strength criteria (Chap. 8), where equations predict rupture and the shape of the strength surface in stress space. A strength or yield criterion forms a part of the constitutive relationship for a material. One set of equations describes strain as a function of stress, time, and temperature at points within the space enclosed by the strength envelope, and another set defines strain behavior at the limit, in the envelope's surface.

Many of the common models are idealized ones derived from early studies of the behavior of diverse materials such as metals and clays. Their main advantage is simplicity. They give satisfactory predictions for some types of rock under certain limited conditions of stress, strain, or temperature.

9.3.2.2 Elastic behavior. The strain in a specimen of solid material such as steel or rock increases with stress during loading, and then decreases during unloading. The material is said to be *elastic* (or to behave elastically) if the specimen returns exactly to its original shape on unloading, that is, there is no residual strain. Materials that deform permanently are called *inelastic*.

There are three commonly defined types of ideal elastic behavior (Fig. 9.12): *perfectly elastic* if the stress-strain relationship follows a single curve regardless of shape, and for every stress there is a unique level of strain; *elastic with hysteresis* if the unloading path differs from the loading path; and *linearly elastic* if the load-unload path is not only reversible but also a straight line. Most metals and many hard and dense rocks are linearly elastic at least until cracks start to form

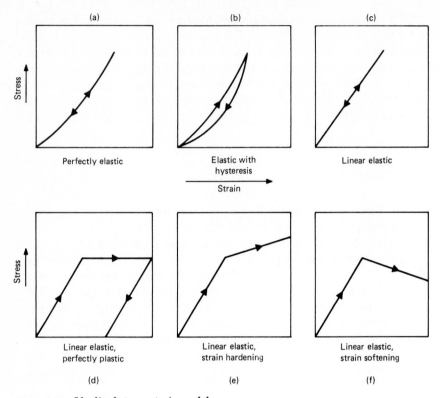

Figure 9.12   Idealized stress-strain models.

at stress levels of perhaps 60–70% of rupture. Other hard yet porous rocks tend to behave elastically with hysteresis at stress levels of up to two-thirds of their compressive strength, after which they suffer permanent strain.

9.3.2.3   Linear elastic parameters.   Hooke in the seventeenth century was probably the first to note the linear proportionality between stress and strain in metals, and this linear elastic relationship is termed *Hooke's law:*

$$\epsilon = \frac{\sigma}{E}$$

where ϵ and σ are, respectively, the axial strain and stress in a uniaxially loaded specimen, and E, the constant of proportionality or gradient of the curve, is termed *Young's modulus.* It is independent of both stress and strain, and is a constant property for a given linear elastic material.

Young's modulus is not constant if the material is nonlinearly elastic or inelastic, but nevertheless can be defined approximately as a function of stress for any material that approximates elastic behavior. The equivalent deformability modulus (defined below in the context of plastic behavior) must be specified in relation to the stress-strain curve, either as a *tangent modulus* at a given stress level, or as a *secant modulus* between the origin and a specified stress, or as a *chord modulus* between two specified stresses (Fig. 9.13).

The ratio of lateral to longitudinal strain in a uniaxial compressive test is called *Poisson's ratio*. It has values close to 0.25 for many metals, but is substantially lower for most rocks and higher for some others.

The two elastic parameters, Young's modulus and Poisson's ratio, are sufficient to completely describe the behavior of linear elastic and isotropic materials. An isotropic material, by definition, is one that has the same elastic properties in all directions. Various other elastic parameters can be defined, but are interrelated such that knowing

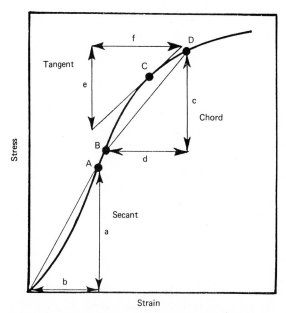

Figure 9.13 Definitions of equivalent linear elastic parameters. Secant modulus at A is a/b; chord modulus between B and D is c/d; tangent modulus at C is e/f.

any two, one can calculate any or all of the others. The interrelationships, in terms of E and ν, are as follows:

$$G = \frac{E}{2(1 + \nu)}$$

$$K = \frac{E}{3(1 - 2\nu)}$$

$$\lambda = \frac{E\nu}{(1 + \nu)(1 - 2\nu)}$$

The elastic parameter G is termed the *shear modulus of rigidity* (or simply the modulus of rigidity), and is the ratio of applied shear stress to measured shear strain in a pure-shear test. The parameter K is called the *bulk modulus of elasticity* (or simply bulk modulus), and is the ratio of applied hydrostatic pressure to volumetric strain (contraction per unit volume) in a hydrostatic compression test. The parameter λ, termed *Lamé's constant,* has no direct physical significance.

Typical values for the static, quasi-elastic parameters for various rock types are summarized in Table 9.1.

Materials that are linear elastic but completely *anisotropic* require not 2 but 21 independent elastic parameters to describe their behavior, although this number reduces to 5 for the special case of a *transversely isotropic* material such as the typical sedimentary rock, which has identical elastic properties in all directions in the plane of the bedding.

9.3.2.4 Plastic behavior. Plastic deformation is defined as permanent deformation that remains after the stress is returned to the initial conditions. All rocks show permanent deformation if subjected to a high enough stress level. In general, some component of strain, termed the *elastic* strain, is recoverable, whereas the remaining *residual* strain is not (Fig. 9.10). Total strain is the sum of elastic and plastic components. When the plastic component becomes significant, the material is termed *ductile,* and when not, *brittle.*

The total slope of loading curves that contain some element of plastic deformation, as is often the case particularly in field testing, is termed the *deformation modulus* (stress divided by total strain), in contrast to the *elastic modulus* (stress divided by elastic-recoverable component only).

Three special idealized cases of plastic deformation deserve mention (Fig. 9.12). The *perfectly plastic solid* deforms linear elastically up to a yield point, beyond which no further stress can be sustained, and further strains occur at a constant level of applied stress. A *strain-hardening material* has a similar yield point, but the stress-strain

TABLE 9.1 Static and Dynamic Elastic Parameters for Various Rock Types from Laboratory Tests on Intact Rock Specimens*

	Static elastic parameters		Dynamic elastic parameters	
	E, GPa	ν	E, GPa	ν
Crystalline limestones	60 (17–100)	0.25 (0.06–0.50)	50 (17–103)	0.27 (0.01–0.45)
Porous limestones	45 (10–100)	0.24 (0.15–0.29)	48 (12–71)	0.20 (0.09–0.24)
Chalk	2 (0.1–12)	0.10 (0.05–0.15)		
Salt	26 (5–44)	0.26 (0.06–0.73)		
Sandstones	18 (1–100)	0.15 (0.02–0.51)	20 (6–55)	0.15 (0.05–0.47)
Quartzite	62 (11–119)	0.18 (0.10–0.40)	65 (21–98)	0.15 (0.10–0.20)
Slates and high-durability shales	40 (12–96)	0.22 (0.02–0.38)	46 (13–94)	0.15 (0.10–0.20)
Low-durability shales	5 (2–30)		15 (10–48)	0.35 (0.33–0.36)
Coal	3 (1–30)	0.42 (0.17–0.49)		
Coarse-grained igneous rocks	56 (8–125)	0.20 (0.05–0.39)	47 (3–119)	0.19 (0.01–0.37)
Fine-grained igneous rocks	62 (7–117)	0.22 (0.07–0.38)	76 (9–109)	0.23 (0.09–0.41)
Schists	40 (5–98)	0.15 (0.01–0.40)	73 (18–116)	0.21 (0.12–0.44)

*Typical values (mode of between 10 and 30 reported test results) are given together with ranges of reported values in parentheses. Because of the considerable variations in test methods and in the rocks themselves, the data should be taken only as a very rough guide to rock properties.

curve continues to rise, whereas a *strain-softening material* has a reducing capacity to sustain load after yield has occurred.

The stress-strain curves for different rock types under different conditions of confining stress and temperature show elements of both strain-hardening and strain-softening behavior. Attempts have been made to model rocks as perfectly plastic and frictional materials with a generalized form of Mohr-Coulomb law (Drucker and Prager, 1952), and as perfectly plastic and nonfrictional materials (Talobre, 1957). More recently, the yielding behavior of porous reservoir sandstones has been modeled by Chin et al. (1986).

9.3.2.5 Viscous behavior (creep). Elastic and plastic constitutive models ignore the effects of time and temperature, which is acceptable only if the rocks are brittle and under moderate levels of stress. To predict the behavior of weak rocks, and even of stronger ones under high stresses and temperatures, such as found deep below the earth's surface, both time and temperature need to be included as variables. *Viscous* behavior occurs when the rock continues to deform under conditions of constant stress. This time-dependent behavior, and the *creep* tests for measuring it, are the subject of Chap. 10.

9.4 Dynamic Elastic Behavior

9.4.1 Concept of the stress wave

9.4.1.1 Overview. In the theory of quasi-static elasticity, the material is considered as in equilibrium under the action of applied forces, and the elastic deformations are assumed to have reached their static values. The forces of earthquakes or blasting, however, are applied only for very brief periods and are changing rapidly. The effects must be considered in terms of the propagation of stress waves similar to those associated with sound traveling through air (Kolsky, 1963).

9.4.1.2 Stress waves and their velocities. Different types of stress wave travel at different rates through different types of media. The velocities of propagation can be predicted assuming that the materials behave elastically at the very low levels of strain involved. Fluids have no shear strength, so just one type of wave travels at velocity v_f through a fluid of density ρ and dynamic bulk modulus of elasticity K_d:

$$v_f = \sqrt{\frac{K_d}{\rho}}$$

In contrast, two types of wave travel through a rigid, infinite elastic solid: waves of dilation (compressive waves, or P waves) and the slower waves of distortion (shear waves, or S waves). The respective velocities v_P and v_S through a solid of dynamic shear modulus of rigidity G_d are given by

$$v_P = \sqrt{\frac{K_d + \frac{4}{3} G_d}{\rho}}$$

$$v_S = \sqrt{\frac{G_d}{\rho}}$$

In addition, *Rayleigh* waves (or *surface* waves) propagate along a free surface when available. They travel at a fraction F of the velocity of the shear waves, where F is given by the Rayleigh equation

$$F^6 + 8F^4 + (24 - 16A^2)F^2 + 16A^2 = 16$$

and A is the following function of the dynamic Poisson's ratio v_d:

$$A = \sqrt{\frac{1 - 2v_d}{2(1 - v_d)}}$$

Furthermore, *shock* waves and *plastic* waves propagate through a medium if the elastic limit is exceeded. Shock waves are generated by very large stress impulses and are characterized by wave fronts that become steeper as they propagate, limited only by the strength of the material. Plastic waves are generated in a material that is elastic up to a certain yield point and then exhibits plastic deformation. Under these conditions, elastic waves are followed by plastic waves at slower velocity.

A rock joint slows the passage of a stress wave, and seismic velocities are slower when the joints have large apertures and are closely spaced (Tanimoto and Ikeda, 1983), hence the value of seismic velocity measurements as an index to rock mass quality, as discussed in Chap. 3.

9.4.2 Measurement of dynamic elastic parameters

9.4.2.1 When not to use dynamic elastic parameters. The above relationships suggest the possibility of determining elastic parameters from measurements of sonic velocity, which are both easy and nondestructive. However, the dynamically measured elastic parameter values turn out to be up to 10 times higher than static ones; they therefore cannot be used to replace the quasi-static elastic properties determined, at greater cost and with more difficulty, by the methods described earlier in this chapter. If dynamic moduli are used in static design, such as of a dam foundation, the strains that occur are likely to be much greater than predicted.

The large discrepancy between quasi-static and dynamic elastic parameters results mainly from the amplitude of the stress wave being insufficient to cause even partial closure of pores, cracks, or joints. The rate of strain application by an elastic wave is also much more rapid than that beneath the loaded plate in a plate test. Rocks subjected to stress waves, except shock waves, appear much stiffer than those loaded by jacks and buildings.

To avoid any possibility of confusion, it is much better to use the sound velocities themselves as an index to rock character, and not even to compute dynamic elastic parameters when the application is a static one. Also to avoid confusion, the dynamic elastic parameters are distinguished by giving their symbols a suffix d.

Dynamic elastic parameters retain an important role in relation to wave propagation calculations, notably those of blasting, mechanical vibration, and earthquake studies, for which purposes these properties must be determined experimentally, using one of the following laboratory or field techniques.

9.4.2.2 Test methods. Laboratory methods, reviewed in detail in Kolsky (1963), include those based on observing the free vibration or resonance of prismatic or core specimens (e.g., Obert et al., 1946); measuring dynamic strains using electric resistance strain gauges, optical interferometry or other devices with a sufficiently rapid response time to record the undistorted stress wave; or measuring the velocity of wave propagation (ISRM, 1981). Only the latter will be outlined here. Other methods are common in the field of earthquake engineering. For example, Mishi et al. (1983) describe a cyclic triaxial testing machine for determining shear modulus and damping ratio under dynamic conditions of load application.

Sonic velocities determined in the field by surface methods or drillhole logging (Chap. 6) can be used to give the equivalent dynamic elastic properties of the rock mass. More often, the compressional and shear wave velocities v_P and v_S of a core specimen are measured ultrasonically in the laboratory, such as by the method given in Sec. 3.3.2.3. Also required for the calculations is the air-dry density ρ, obtained from the weight and the measured volume of the core.

The dynamic shear modulus G_d may then be determined from

$$G_d = \rho v_S^2$$

G_d is obtained directly in gigapascals (10^9 Pa) if ρ is expressed in units of grams per cubic centimeter and v_S in units of millimeters per second.

The dynamic Poisson's ratio ν_d is determined from

$$Q = \frac{\nu_S}{\nu_P}$$

$$\nu_d = \frac{1 - 2Q^2}{2(1 - Q^2)}$$

The dynamic Young's modulus E_d is determined from

$$E_d = 2G_d(1 + \nu_d)$$

The dynamic bulk modulus (modulus of rigidity) K_d is determined from

$$K_d = \frac{E_d}{3(1 - 2\nu_d)}$$

9.4.2.3 Typical dynamic elastic properties. Cregger and Lamb (1984) give typical values for the dynamic Poisson's ratio, measured by a borehole logging technique, which range from 0.29 for sandstones and siltstones through 0.34 for mudstones. Representative values of dynamic elastic parameters for various rock types are given in Table 9.1.

References

Amadei, B., J. D. Rogers, and R. E. Goodman: "Elastic Constants and Tensile Strength of Anisotropic Rocks," *Proc. 5th Int. Cong. Rock Mech.* (Melbourne, Australia, 1983), vol. A, pp. 189–196.

Baguelin, F., J. F. Jezequel, and D. H. Shields: *The Pressuremeter and Foundation Engineering* (Trans Tech Publ., Clausthal, 1978), 617 pp.

Bakhtar, K., and N. Barton: "In situ Rock Deformability Tests at the Rocky Mountain Pumped Hydro Project," *Proc. 27th U.S. Symp. Rock Mech.* (Tuscaloosa, Ala., 1986), pp. 949–953.

Brace, W. F.: "The Relation of Elastic Properties of Rocks to Fabric," *J. Geophys. Res.,* **70,** 5657–5667 (1965).

Chin, H. P., Z. P. Duan, and M. M. Carroll: "Plastic Deformation of Boise Sandstone," *Proc. 27th U.S. Symp. Rock Mech.* (Tuscaloosa, Ala., 1986), pp. 591–598.

Cregger, D. M., and T. J. Lamb: "Poisson's Ratio as a Parameter for Determining Dynamic Elastic Modulus," *Proc. 25th U.S. Symp. Rock Mech.* (Evanston, Ill., 1984), pp. 417–427.

Drucker, D. C., and W. Prager: "Soil Mechanics and Plastic Analysis or Limit Design," *Quant. Appl. Math.,* **10** (2), 157–165 (1952).

Durelli, A. J., E. A. Phillips, and C. H. Tsao: *Introduction to the Theoretical and Experimental Analysis of Stress and Strain* (McGraw Hill, New York, 1958), 498 pp.

Hustrulid, W., and A. Hustrulid: "The CSM Cell—A Borehole Device for Determining the Modulus of Rigidity of Rock," E. R. Hoskins, Ed., *Applications of Rock Mechanics, Proc. 15th U.S. Symp. Rock Mech.* (Rapid City, S. Dak., 1975), pp. 181–225.

ISRM: "Suggested Methods for Determining in situ Deformability of Rock," ISRM Commission on Standardization of Laboratory and Field Tests, *Int. J. Rock Mech. Min. Sci. Geomech. Abstr.,* 16(3), 195–214 (1979).

————: "Suggested Methods for Rock Characterization, Testing, and Monitoring," ISRM Commission on Testing Methods, E. T. Brown, Ed. (Pergamon, Oxford, 1981), 211 pp.

————: "Suggested Method for Deformability Determination Using a Large Flat Jack Technique," *Int. J. Rock Mech. Min. Sci. Geomech. Abstr.*, **23** (2), 131–140 (1986).

————: "Suggested Methods for Deformability Determination Using a Flexible Dilatometer," *Int. J. Rock Mech. Min. Sci. Geomech. Abstr.*, **24** (2), 125–134 (1987).

Khair, A. W.: "Analysis of Interaction between Models of Mine Roof Pillar-Floor Using Holographic Interferometry and Analytical Techniques," *Proc. 24th U.S. Symp. Rock Mech.* (Texas A. & M. Univ., College Station, 1983), pp. 107–117.

Kolsky, H.: *Stress Waves in Solids* (Dover, New York, 1963), 213 pp.

Kostak, B., and K. Popp: "Moiré Strain Gauges," *Strain,* vol. 2 (1966).

Mishi, K., T. Kokusho, and Y. Esashi: "Dynamic Shear Modulus and Damping Ratio of Rocks for a Wide Confining Pressure Range," *Proc. 5th Int. Cong. Rock Mech.* (Melbourne, Australia, 1983), vol. E, pp. 223–226.

Natau, O. P., B. O. Frohlich, and T. O. Mutschler: "Recent Developments of the Large-Scale Triaxial Test," *Proc. 5th Int. Cong. Rock Mech.* (Melbourne, Australia, 1983), vol. A, pp. 65–74.

Obert, L., S. L. Windes, and W. I. Duvall: "Standardized Tests for Determining the Physical Properties of Mine Rock," Rep. Invest. 3891, U.S. Bureau of Mines (1986).

Oberti, G., L. Goffi, and P. P. Rossi: "Study of Stratified Rock by Means of Large-Scale Tests with an Hydraulic Pressure Chamber," *Proc. 5th Int. Cong. Rock Mech.* (Melbourne, Australia, 1983), vol. A, pp. 133–141.

Panek, L. A.: "Effect of Rock Fracturing on the Modulus Determined by Borehole Dilation Tests," *Proc. 2d Int. Cong. Rock Mech.* (Belgrade, Yugoslavia, 1970), paper 2–16, 5 pp.

————, E. E. Hornsey, and R. L. Lappi: "Determination of the Modulus of Rigidity of Rock by Expanding a Cylindrical Pressure Cell in a Drillhole," *Proc. 6th U.S. Symp. Rock Mech.* (Rolla, Mo., 1964), pp. 427–449.

Park, D. W.: "Holographic Testing Method of Rock," *Proc. 27th U.S. Symp. Rock Mech.* (Tuscaloosa, Ala., 1986), pp. 192–199.

Schmitt, D. R., C. L. Smither, T. J. Ahrens, and B. L. Jensen: "Holographic Measurement of Elastic Moduli," *Proc. 27th U.S. Symp. Rock Mech.* (Tuscaloosa, Ala., 1986), pp. 185–191.

Schuler, K. W.: "Lateral-Deformation Gage for Rock-Mechanics Testing," *Experim. Mech.*, 477–480 (Dec. 1978).

Talobre, J. A.: *Rock Mechanics and Its Application,* in French (Dunod, Paris, 1957).

Tanimoto, C., and K. Ikeda: "Acoustic and Mechanical Properties of Jointed Rock," *Proc. 5th Int. Cong. Rock Mech.* (Melbourne, Australia, 1983), vol. A, pp. 15–18.

Window, A. L., and G. S. Holester: *Strain Gauge Technology* (Applied Science Publ., 1982).

Zimmerman, R. M., M. P. Board, E. L. Hardin, and M. D. Voegele: "Ambient Temperature Testing of the G-Tunnel Heated Block," *Proc. 25th U.S. Symp. Rock Mech.* (Evanston, Ill., 1984), pp. 281–295.

10

Viscous, Thermal, and Swelling Behavior

10.1 Introduction

10.1.1 Effects of time, temperature, and moisture

Time, temperature, and moisture, acting on their own or in combination, are often sufficient to deform or even rupture a rock. Prolonged application of loads leads to creep (deformation under conditions of constant stress, also termed *viscous,* or *rheid,* behavior), particularly when the ambient stresses or temperatures are already high. Increases of temperature on their own cause thermal expansion, sometimes spalling, and changes of moisture content lead to swelling or shrinkage.

10.1.2 "Ambient" conditions

Most engineering projects are constructed close to surface where the rocks are typically at stresses in the range of 0–15 MPa, temperatures in the range of −5 to +30°C, and water saturations in the range of 20–100%. Changes of temperature and degree of saturation usually amount to no more than a few percent. Changes of stress are greater, sometimes several hundred percent close to the walls of an excavation or beneath a footing. The durations of change of stress, from pre- to postconstruction at any given location, such as at a tunnel heading or in a foundation, are on the order of days to months, and the new stresses remain acting for the expected operating life of the structure, often 20–100 years. These are considered *normal,* or *ambient,* engineering conditions.

For most design purposes, stress-strain behavior and strength are measured using tests at these ambient conditions. Fortunately, variations in measured rock behavior within these limited ranges are minimal in most cases, and are usually ignored or taken into consideration by including a factor of safety in the design.

10.1.3 Where time, temperature, and moisture are important

10.1.3.1 Engineering projects. At normal ambient stress levels most rock types behave quasi-elastically. Special consideration must be given, however, to viscous and swelling effects when designing in soft rocks, including shales and salt rocks, whose behavior, even at shallow depths, can be sensitive to small variations in stress, moisture, time, or temperature. Stresses and temperatures increase with depth below the earth's surface. A vertical *stress gradient* of 23 kPa/m (1 lb/in^2 per foot) is typically accompanied, in nonthermal regions, by a *geothermal gradient* of 25–30°C/km. In geothermal regions, the gradient is greater than this.

Stress, moisture, time, and temperature interact with each other. For example, in an oil reservoir subjected to fireflood or steamflood for enhanced oil recovery, thermal expansion can generate stresses sufficient to shear the weaker strata. In contrast, when a shale is heated, a moderate initial expansion is often followed by shrinkage as moisture is driven off. When heat is extracted from a geothermal well by injecting cool water and extracting hot water or steam, the thermal contraction is accompanied by a substantial reduction in the compressive stress field. Other examples of stress-thermal interaction are to be found in the storage of cryogenic liquids such as liquified natural gas in underground rock caverns, prolonged heating by high-level radioactive wastes in storage repositories, and mine development in permafrost areas (Berest and Weber, 1988).

Thermal and time-dependent effects can even change the geometry of the structures being analyzed. Caverns can change in shape, shrink, or expand as the result of prolonged creep or spalling in soft shales, salt rocks, or ice-rich rocks.

Changes in stress, temperature, or moisture content often alter the properties and behavior of the rock, particularly if weak in the first place. When shales dry out, they become stiffer and stronger, as do porous sandstones when the stress level increases. The viscous behavior of salt rock is sensitive to temperature changes, as is the behavior of an ice-rich fractured rock mass in a permafrost area.

10.1.3.2 Geological processes. Time and temperature play a dominant role in geological processes at substantial depths, where elevated stresses and temperatures combine to make competent rocks behave like weaker and softer ones do closer to the surface. High pressure and temperature together permit crystalline and intercrystalline deformation through viscoplastic mechanisms. Geological deformation is then by folding rather than by faulting. For example, below depths of 150 km, large earthquakes simply do not occur because the rocks are almost fluid and no longer display strain-weakening or stick-slip behavior. Shear stresses are released rapidly by creep at these depths.

10.2 Creep Behavior

10.2.1 Time and temperature dependence of strength and strain

10.2.1.1 Overview. The strengths and deformations of intact and jointed rock depend to some extent on the rate of application of stress. Another way of saying the same thing is that the stress and available strength at any time depend to some extent on the rate of strain. If stress is applied rapidly, the rock appears stronger, mainly because cracks have little time to develop and grow, and asperities have little time to shear through. If stress is applied gradually, the rock appears weaker, mainly because cracks have time to coalesce and to propagate through asperities. If stress is maintained constant at a high enough level, as in a creep test, strains may continue indefinitely, and rupture can occur at stresses as low as one-half the short-term uniaxial compressive strength of the rock material.

However, creep can be beneficial. Rock bursts happen because massive and brittle rocks cannot creep, and instead carry more and more stress until they rupture explosively.

The effects of time are difficult to divorce from those of temperature. A constitutive equation expressed just in terms of time (or strain rate) is valid only for a single temperature, or approximately valid within a narrow temperature range. However, to define a generalized constitutive law coupling stress, strain, time, and temperature would require a prohibitively expensive amount of testing, even for a single material. Rock salt, for example, has been studied exhaustively because of its potential value as a host for radioactive and toxic wastes. Even in this case, researchers have yet to agree on its behavior over the full range of stresses and temperatures of interest.

10.2.1.2 Creep micromechanisms. Creep can result from various internal mechanisms. At high shear stresses, crystal planes slip past one another, and dislocations in the crystal lattice also generate movement. Mechanisms that are important at the high temperatures and pressures of the mantle, but not in engineering, include vacancy diffusion, recrystallization, and ionic transfer in grain boundary fluid films.

In metallic and ionic (e.g., halide) bonds, movements can occur without crack generation if the stresses and rates are appropriate. In contrast, damage to the strongly covalent oxygen-silicon bonds of silicate rocks is permanent and cumulative. Cracks are generated, most of which originate at the interfaces between grains, where stresses are concentrated because of different material moduli and flaws. Strain cracking either reaches a level sufficient to cause rupture (unstable crack growth), or terminates (stable crack growth), perhaps after a very long time (Dusseault et al., 1987).

In porous rock and jointed rock, particularly if coated with clay, chlorite, molybdenum disulfide, or graphite, creep can occur at a local or microscopic level rather than throughout the mass.

Elevated stresses at contacts or asperities can result in local viscoplastic deformation. The rock mass displays time-dependent behavior even though the great majority of the material is below the stress level associated with creep. Creep may be the external manifestation of many small stick-slip events on many surfaces.

In the field, materials that are confined, such as the rock beneath the base of a pile or around a tunnel, creep and redistribute their high stresses to materials around them. This usually leads to a condition of stability, and to a strain rate that decreases steadily with time. Continuing steady deformation should be taken as a warning of worsening conditions because of accumulating damage and block loosening.

10.2.2   Laboratory tests

10.2.2.1   Uniaxial creep tests.   Most tests that provide rheological properties for insertion in constitutive equations are uniaxial ones, and the creep test is no exception. Uniaxial stress is applied to a specimen and maintained constant while monitoring strain as a function of elapsed time, usually at constant temperature.

Because of the duration of each test, several must be run simultaneously, and each specimen has its own load-maintaining device consisting of a screw-loaded spring (Fig. 10.1), deadweight, and lever, or hydraulic actuator. Rock core specimens are often no larger than 1 or 2 cm in diameter, permitting the application of high stresses using quite moderate loads and small testing machines. The conventional loading machines designed for short-term testing are uneconomic for

Figure 10.1   Uniaxial creep tests.

creep testing, because several would be needed, tied up for at least weeks, often months.

Uniaxial strains, and lateral strains if desired, are measured periodically or recorded continuously, usually by transducers or mechanical dial gauges between platens. Because of debonding problems, electric resistance strain gauges are seldom used for long-term measurements. Strain measurements between platens are permissible in creep tests (not in short-term tests, see Secs. 9.1.2.1 and 9.1.2.2) because platen bedding-in movements occur mainly during initial loading, after which they become negligible.

Because of the sensitivity of the creep rate to temperature and humidity, test conditions must be controlled and monitored within close tolerances. When testing silicate rocks, $\pm$ 2°C is acceptable. For ductile rocks such as shales or halite, $\pm$ 0.25°C is desirable. Temperature and humidity control are achieved in a climate-controlled room, or by constructing environmental chambers around individual or multiple specimens.

If temperature dependence of creep is of interest, tests are carried out at different constant temperatures. A single specimen can be "stage-tested" through a range of temperatures, but only if the properties of the material are known to be independent of time and accumulated strain.

10.2.2.2 Triaxial creep testing. Creep rates vary according to the level of confining stress, so creep data often must be obtained under triaxial confinement. The apparatus is similar to that for an axisymmetric short-term triaxial test (Chaps. 8 and 9), except that to permit several concurrent tests, the axial stress usually is applied by hydraulic jacks through pressure intensifiers and accumulators that maintain the applied load constant within specified limits.

When testing porous rocks, the magnitude of pore pressure can affect the creep rate. Pore pressure control is achieved using a specially equipped triaxial cell. Careful monitoring of the volume of pore water expelled gives another means of determining creep behavior.

The effects of a series of stresses or temperatures can be explored by stage loading a single specimen, but only if the specimen has not been seriously damaged by the accumulation of creep strain. The method usually is acceptable for salt rocks but perhaps not for more brittle materials. Strain rates predicted by stage tests on silicate rocks are equal to or larger than those predicted from using a new specimen each time. Therefore the results are likely to be conservative and appropriate for engineering design. One of the more reliable methods is to achieve steady state at one temperature and a given shear stress, return the stress to hydrostatic, change the temperature and allow it to equilibrate, and then reimpose the same shear stress as before.

Figure 10.2a gives time-deformation data for a multistage triaxial creep test series on intact potash ore. At each stage, the stresses were held constant for several weeks before new conditions were imposed.

10.2.2.3 Stress relaxation tests. Stress relaxation tests require a very stiff testing frame and a load cell. Stress is applied and the system "locked" in place to impose a zero-strain condition on the specimen. With time, the material slowly creeps and the level of stress drops off. Usually it becomes asymptotic to some limit related to the ability of the material to sustain load at a given confining stress level. Stress relaxation tests can be carried out at several stress levels on a single specimen, but again, only if the specimen is known or assumed to have been unaffected by previous testing.

Joints, filled or unfilled, can be tested for time-dependent properties as described in Chap. 11. Usually these tests are conducted with a modified version of the direct shear test machine. However, a triaxial relaxation test is more convenient if pore-water pressures need to be measured.

10.2.2.4 Strain-rate controlled tests. An alternative to constant stress or relaxation tests is to conduct conventional uniaxial or triaxial tests, but at strain rates much slower than those of routine testing (e.g., Griggs et al., 1960). The strain rates, and often temperatures, are selected to give a family of stress-strain curves and strengths (Fig. 10.3).

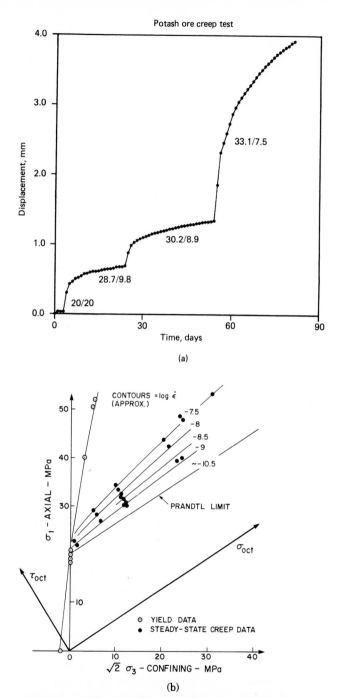

Figure 10.2 (*a*) Stage loading creep tests. First number is σ_1; second number is σ_3. (*b*) Strain rate contours and yield criteria for a domal rock salt. The Prandtl limit defines a characteristic shear stress where the dominant creep mechanism changes.

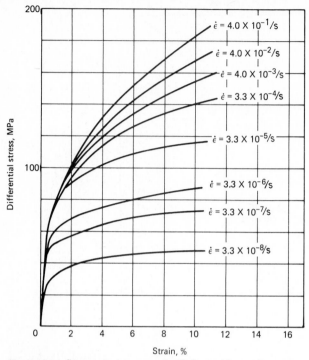

Figure 10.3 Stress-strain curves for Yule marble specimens under triaxial confinement at 500°C and at various slow strain rates. (*Heard, 1963.*)

The required constant rates of strain are obtained using mechanical jacks with reduction gears, in place of the hydraulic jacks more normally employed in rock testing. Axial stress levels are measured independently using a load cell, and if required, lateral strains can be monitored using electric resistance strain gauges.

A serious problem with this type of testing is that the slowest reliable laboratory strain rates are on the order of 10^{-8}/s or faster, whereas typical field rates may be from 2 to 4 orders of magnitude slower. Also, conditions of constant strain are seldom found in the field. Constant stress or relaxation are more common, so tests simulating constant or relaxing stress tend to be easier to interpret.

10.2.2.5 Presentation of creep test results. Creep curves are plotted as a family to demonstrate that creep is faster at higher stresses and temperatures. A typical curve (Fig. 10.4) displays three characteristic sections known as the *primary, secondary,* and *tertiary* phases. Primary creep occurs for a short time after initial loading and is marked by a decelerating rate of strain. Secondary or steady-state creep fol-

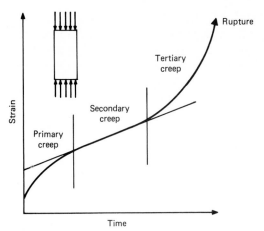

Figure 10.4 Primary, secondary, and tertiary phases of creep.

lows, and is characterized by a steady rate of strain with no acceleration or deceleration. After this comes tertiary creep, characterized by accelerating strain rates leading to rupture, usually associated with "necking" in tensile tests on metal specimens, or crack coalescence in compressive tests on rock. Rock specimens loaded to stress levels less than about 0.6 times the uniaxial compressive strength may never get beyond the primary creep phase.

Perhaps the most contentious issue in creep testing is to decide if steady-state creep has been reached, and whether unstable tertiary creep will ensue. The issue is largely academic, because the concept of primary, secondary, and tertiary creep is an idealized one in the first place, and because laboratory testing gives only an approximation of the behavior of an element of rock in the field. Any persistent acceleration in strain rate must be interpreted as a sign of eventual rupture. If strains are carried on beyond 10%, bulging, ram punching, end cracking, and other problems arise, making the test results questionable at best.

Test information should not be extrapolated far beyond the conditions under which it was obtained, in the absence of evidence that the mechanisms are the same. For instance, in salt rocks, the dominating creep mechanism changes at about 10^{-11}/s to 10^{-10}/s. The same may be said of geological processes, which are much slower even than this (slower than 10^{-13}/s). Geological processes cannot be reliably tested in the laboratory, so indirect evidence is used to determine the constitutive behavior.

Results that explore several variables such as stress, strain, time, and temperature are difficult to show graphically. Figure 10.2*b* shows contours for the characteristic steady-state strain rates of a domal salt rock from Louisiana. This is an isothermal plot, performed under conditions of triaxial compression ($\sigma_{axial} = \sigma_1$, $\sigma_{radial} = \sigma_2$, σ_3), and contain no information as to the change of strain rate with temperature. To use this method to present temperature data as well would require a series of similar plots at different constant temperatures.

10.2.3 Field tests

The dilatometer method described in Sec. 9.2.5 for short-term measurements of deformability can also be used for the measurement of time-dependent deformation (Fig. 10.5). Either stage-loaded creep tests or step-strained relaxation tests can be carried out (Ladanyi and Gill, 1983).

Wafer cells (Sec. 5.4.5) are similar to dilatometers but being flat, can apply and measure pressure in only one direction. Their main purpose is to measure changes of stress. In spite of a very limited expansion capability, they can also be used for creep testing. After grouting the device in the drillhole, the pressure is increased by a small amount, and the time-pressure relationship is monitored and inter-

Figure 10.5 Creep testing with a high-pressure rock dilatometer in an underground potash mine, western Canada. Pressure was held constant by a hydraulic maintainer system, and deformations were recorded automatically. (*Courtesy of Roctest, Canada.*)

preted in terms of creep behavior. A rigorous analysis is available for the case of ideally viscoplastic rocks.

Tests in drillholes using a dilatometer or a wafer cell are simple and convenient, but suffer from all the disadvantages of smallness of scale. In particular, the influence of jointing on the behavior of the rock mass can only be evaluated by a test on a scale of several meters. Reliable creep data are therefore obtained only by monitoring and back-analyzing a full-scale excavation, such as a tunnel or an underground chamber. To help in designing a system of mine openings in a viscoplastic material such as salt, a trial adit can be excavated, then, with the help of monitoring instrumentation, back-analyzed to measure the parameters of assumed behavioral laws. Multiple-point extensometers and other forms of instrumentation suitable for this application are described in Chap. 12.

10.2.4 Rheological models

10.2.4.1 Concept of a rheological model. Time- and temperature-dependent materials require constitutive equations to represent these aspects of their behavior, which replace the elastic and plastic equations introduced in Chap. 9. Rheological models are used extensively to assist in visualizing the mechanisms that govern the constitutive equations, and to help in formulating the equations themselves. They are built, conceptually, from a combination of spring, dashpot, slider, and rupture elements (Fig. 10.6), which can be assembled into a combined "rheological body." Each basic element or combination of elements has a unique and readily defined stress-strain-time law.

Real physical processes of deformation, although analogous in some respects, may be quite different from those of the models, and also the models are one- rather than three-dimensional. In spite of these disadvantages, rheological models help in gaining some "feel" for stress-strain-time behavior, and the more complex and better designed ones lead to constitutive equations that simulate quite accurately the behavior of real rocks under many conditions (Christian and Desai, 1977).

10.2.4.2 Rheological elements. Linear elastic *spring,* or *Hookean,* elements deform instantaneously in response to loading, and give the linear elastic load-displacement or stress-strain relationship predicted by Hooke's law.

Newtonian liquid *dashpot,* or *viscous,* elements are characterized by a viscosity μ such that the stress level and the strain rate $d\epsilon/dt$ are linked by an equation of the form

$$\sigma = \mu \frac{d\epsilon}{dt}$$

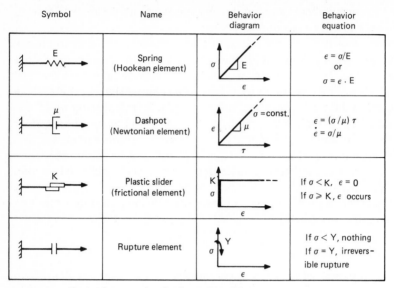

Figure 10.6 Basic elements for rheological modeling.

Dashpots behave as rigid bodies when loaded instantaneously, after which they introduce time-dependent behavior into the models. (In everyday life, dashpots are leaky cylinders such as used for shock absorbers.)

Frictional elements (*plastic slider,* or *yield,* elements) allow no motion below a yield stress K, above which the stress is maintained at K. The slider introduces plastic yield and hysteretic behavior, assuming that the yield strength of the element is independent of direction. In equation form

$\sigma < K$ no motion possible

$\sigma \geq K$ motion permitted, flow rule must be specified

The *rupture* element is rarely discussed in classical rheological modeling, but is essential if the model is to simulate the strain-weakening behavior of rock. This element can withstand any stress up to a limit Y, beyond which it ruptures irreversibly and can sustain no further stress.

10.2.4.3 Simple rheological "bodies."

Figure 10.7 shows several of the better known combinations of the fundamental elements, with the corresponding stress-strain-time graphs and constitutive equations of these compound bodies when subjected to a constant stress or an initial strain.

When the *Maxwell* body (Fig. 10.7a) is subjected to an instantaneous stress, it displays an instantaneous strain followed by a continued delayed strain with time:

$$\epsilon = \frac{\sigma}{E} + \frac{\sigma}{\mu}(t - t_i)$$

where σ/E is the instantaneous strain upon application of the stress σ, and the second term is the viscous strain after some elapsed time t from an initial time t_i. If the body is subjected to an instantaneous strain, both components carry precisely the stress ϵ/E. Thereafter the viscous element imparts an overall strain rate that decays exponentially, the stress relaxes, and the rate of strain is continually adjusting to respond to the new stress. The Maxwell model is acceptable only for steady-state conditions in materials such as ice, salt, and deep crustal rocks at elevated temperatures. Creep testing permits evaluation of the steady-state viscosity parameter. No transient effects are modeled, and all time-dependent strain is plastic.

A *Kelvin-Voight* body (Fig. 10.7b) contains the same two elements, but in parallel rather than in series. Instantaneously applied external stress is at first carried entirely by the dashpot, after which, stress is progressively transferred to the spring. The strain rate decreases along with the stress carried by the dashpot until, at infinite time, all stress is taken by the spring, which stabilizes at a strain σ/E. If external stress is removed, the body will relax with time in an identical manner because of the strain energy stored in the spring. Consequently the Kelvin-Voight body is a purely viscoelastic one, with full internal strain recovery. One cannot impose an instantaneous strain on a Kelvin-Voight body, because the dashpot reacts with infinite rigidity. This body only models recoverable transient creep, and cannot simulate steady-state creep or instantaneous elastic deformation.

The body in Fig. 10.7c is an *elastoplastic* one with a spring and a frictional element in series, which displays perfectly elastic behavior up to the level K, then perfectly plastic behavior. Reversal of the stress direction results in hysteretic behavior. Note that the change in stress must be $2K$ if the direction is reversed, although the absolute magnitude of the frictional stress remains the same. This model reasonably represents the gross behavior of smooth rock joints subjected to shear stress reversals.

A rupture element can be added to "suddenly" permit a mechanism to occur. When added to an elastoplastic body, the rupture element simulates shearing of asperities along a rough joint, and introduces the concept of peak and ultimate shear strength, where the peak strength Y exceeds the ultimate strength R.

The *Bingham* body (Fig. 10.7d) is a viscoplastic one with viscous and plastic elements in parallel. It has the very important property

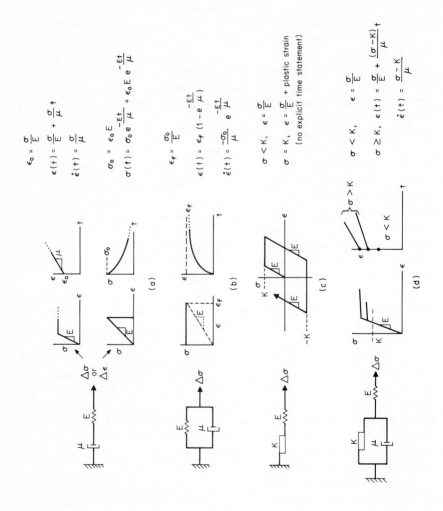

$$\epsilon_0 = \frac{\sigma}{E}$$

$$\epsilon(t) = \frac{\sigma}{E} + \frac{\sigma}{\mu} t$$

$$\dot{\epsilon}(t) = \frac{\sigma}{\mu}$$

(a)

$$\sigma_0 = \epsilon_0 E$$

$$\sigma(t) = \sigma_0 e^{-\frac{Et}{\mu}} = \epsilon_0 E e^{-\frac{Et}{\mu}}$$

$$\epsilon_f = \frac{\sigma_0}{E}$$

$$\epsilon(t) = \epsilon_f \left(1 - e^{-\frac{Et}{\mu}}\right)$$

$$\dot{\epsilon}(t) = \frac{-\sigma_0}{\mu} e^{-\frac{Et}{\mu}}$$

(b)

$$\sigma < K, \quad \epsilon = \frac{\sigma}{E}$$

$$\sigma = K, \quad \epsilon = \frac{\sigma}{E} + \text{plastic strain}$$
(no explicit time statement)

(c)

$$\sigma < K, \quad \epsilon = \frac{\sigma}{E}$$

$$\sigma \ge K, \quad \epsilon(t) = \frac{\sigma}{E} + \frac{(\sigma - K)}{\mu} t$$

$$\dot{\epsilon}(t) = \frac{\sigma - K}{\mu}$$

(d)

320

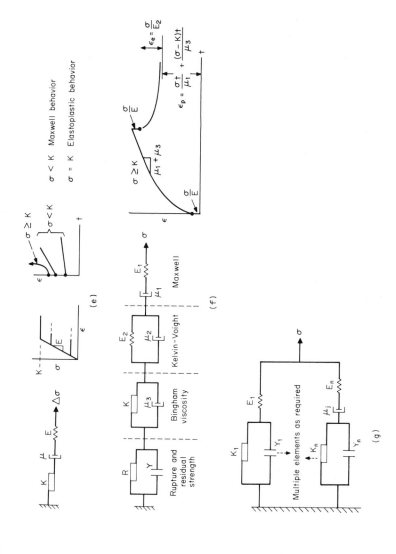

Figure 10.7 Rheological bodies, seven models. (*a*) Maxwell body. (*b*) Kelvin-Voight body (viscoelastic). (*c*) Elastoplastic body (hysteretic behavior). (*d*) Bingham body (with elastic component). (*e*) Elastoviscoplastic body with elements in series (simplest form). (*f*) General rock model (suitable for salt rocks). (*g*) Distributed-element model for harder rocks. (*Kaiser and Morgenstern, 1981.*) Gives quite realistic simulations of rock behavior depending on element and parameter distributions.

that no strain occurs until a yield stress Y is exceeded, after which the strain rate depends on the difference between the imposed and the yield stresses. The Bingham body is a realistic model for the flow of some thixotropic liquids and for gliding mechanisms in crystals.

The body in Fig. 10.7e is also a viscoplastic one, but with the elements in series. It simulates purely viscous behavior at low stresses, but both creep and plasticity above a certain threshold or "trigger" stress. To specify its deformation rate requires that two flow rules be given: one for plastic and one for viscous flow. The total strain rate is the sum of the two component rates at any given stress level.

The bodies in Fig. 10.7f and g are complex and more realistic rheological bodies, and are discussed later.

10.2.4.4 Variable elastic, plastic, and viscous parameters. To this point we have assumed that the elastic moduli, yield strength, and viscosity parameters are univalued and constant. With care in deriving the constitutive equations, however, any or all of these can be stipulated to be functions of the stress, strain, or strain rate. For example, in an attempt to simulate the curvature of the stress-strain relationship for a fissured rock, the stress-stiffening behavior of a spring element can be stipulated in the form

$$E(\sigma) = E_1 + A\sigma$$

where A is a constant. To give shear-thickening or shear-thinning behavior, the viscosity of a dashpot can be considered as a function of the strain rate:

$$\mu(\epsilon) = \mu_i + B\dot{\epsilon}$$

To simulate a material in which a number of different mechanisms are acting, two or more of the bodies are placed in parallel or in series. Behavior rules for these compound bodies can be extracted by careful application of the rules for the individual elements. Few facets of rock behavior cannot be simulated in this way. The mechanics of deformation (the constitutive equations) are easily derived from the models.

10.2.5 Creep behavior of rocks

Mineralogy, texture, and jointing of the rock being tested affect its deformation characteristics.

10.2.5.1 Soft rocks. Porous rocks and those containing minerals such as calcite or halite, which deform more readily, for example, limestones and salt rocks, show time- and temperature-dependent behavior to a much greater extent, and at lower stress levels, than do dense and quartzitic rock types. Behavior of rock salt, for example, is dominated by creep at all shear stress levels. There seems to be no lower "Bingham" limit to viscous behavior.

The body in Fig. 10.7f is a model that, with reasonable choices for the element parameters, simulates actual behavior of salt (Dusseault et al., 1987). The model gives instantaneous recoverable deformation, primary recoverable creep, slow steady-state creep rates below K, more rapid creep rates above K, and strain-weakening behavior above certain stress levels. The major observed aspects of intact salt rock behavior are quite well reproduced, although certain features have been omitted such as an ultimate strength limit for the fractured rock salt.

10.2.5.2 Hard rocks. Dense rocks composed only of high-strength minerals such as quartz and feldspar, for example, granites and quartzitic sandstones, tend to behave in a linear elastic manner up to the point of rupture. Normally they are nonviscous and show negligible creep even when subjected to sustained loads at stress levels approaching their strength. However, Ito (1983) found that even granites can creep when subjected to beam bending over periods as long as 24 years.

Simulation of the behavior of hard rock, including strength in a fractured state after peak strength is surpassed, requires that the following aspects of behavior be modeled (Kaiser and Morgenstern, 1981): elastic deformation over a limited stress range, some terminating time dependence below peak strength, a strain-rate-dependent peak strength, a strain-dependent peak strength, frictional and cohesive strength components, and drop-off to ultimate strength which is history dependent.

The body in Fig. 10.7g shows a model that replicates the major features listed above (Kaiser and Morgenstern, 1981). Peak and ultimate strengths depend on the applied rates of stress and strain, as observed in testing. Test results are similar to those predicted by the rheological model. An even closer matching of model to prototype behavior could be achieved by increasing the number of rheological cells of various types in the composite model, and by making the elastic modulus strain-rate dependent.

10.2.5.3 Future directions. Information on the high-temperature deformation and strength behavior of rock is only starting to be collected. Most of this is to be found in the geophysical literature, and usually relates to temperatures and pressures beyond the engineering range of interest. Data are even scarcer for low temperatures. Typical values for the viscothermal characteristics of rocks are given in Table 10.1.

TABLE 10.1 Viscothermal Stress-Strain Behavior of Geologic Materials*

Material	C, MPa$^{-n} \cdot$ s^{-1}	n	E_a, kJ $\cdot$ mol^{-1}	Test range temperature T, °C	Shear stress, MPa
Ice[†]	8.8×10^5	3.0	60.7	$-40 - -3$	Low stress
Halite[†]	9.5×10^{-1}	5.5	98.3	$40-200$	5–30
Dry quartzite[†]	6.7×10^{-12}	6.5	268	>800	High stresses
Wet quartzite[†]	4.4×10^{-2}	2.6	230	>800	High stresses
Limestone[†]	4.0×10^3	2.1	210	>800	High stresses
Diabase[†]	5.2×10^2	3	356	>800	High stresses
Olivine[†]	4.2×10^5	3	523	1400	10–100
Lithographic limestone[‡]	2.35×10^3	4.7	297	$600-900$	70–200
Solenhofen limestone[‡]	2.45×10^4	1.7	213	$600-900$	1–7
Yule marble (calcite)[‡]	1.14×10^{-4}	8.3	259	$400-800$	16–112
Dolomite marble[‡]	1.1×10^{-13}	9.1	348	$700-900$	>300
Anhydrite[‡]	2.9×10^1	2	152	$350-450$	20–60

*To express thermal and stress effects on creep rates, it is necessary to use a widely accepted empirical equation:

$$\dot{\epsilon} = C(\sigma_1 - \sigma_3)^n \, e^{-\frac{E_a}{RT}}$$

where $\dot{\epsilon}$ is strain rate, s^{-1}; C is a material specific constant, MPa$^{-n} \cdot$ s^{-1}; $\sigma_1 - \sigma_3$ is principal stress difference, MPa; n is an empirical exponent; E_a is activation energy, kJ $\cdot$ mol^{-1}; R is the universal gas constant, 8.314 J $\cdot$ mol$^{-1} \cdot$ K^{-1}; T is temperature, K. This equation is valid only over the T and $\sigma_1 - \sigma_3$ range for which it was empirically derived.

[†]From Turcotte and Schubert (1982).
[‡]From Carmichael (1984).

Rheological models can be powerful tools. A recent review is provided in Hardy and Sun (1986), who derive a linear viscoelastic Burger's model as a function of the level of stress. The new model is claimed to be capable of predicting the creep of rocks over a wide range of stress levels and strain rates. It is a nonlinear viscoelastic one because the stress-strain relationship is nonlinear at any specified time point. The analysis is also different from that of classical viscoplasticity because no postloading information is considered, and no yield point or yield surface is introduced.

The engineer must identify the important mechanisms that act within the pertinent ranges of pressure and temperature, and select an appropriate assemblage of elements, bodies, and generalized bodies to give the desired behavior. If the simple element rules are followed carefully, then the behavior of the compound model can be expressed quite simply in the form of equations and logical statements (if-then-else logic). This logic and the behavioral laws can be coded and studied by computer, such as by using a general finite-element formulation (Fritz, 1982).

The shortcoming of rheological models is that they are capable only of matching known behavior, and have little predictive power into regions of stress and temperature beyond those for which they were derived. The process of curve fitting remains empirical in spite of various analogies that can be drawn between behavior of the rheological elements and of rock. Whether based on rheological models or not appears largely irrelevant. Empirical curve fitting is the only approach available that comes close to simulating real rock behavior. Soundly based theoretical modeling suitable for application lags far behind.

10.3   Heat Flow and Thermal Properties

10.3.1   Heat flow calculations

Heat flow calculations are required in certain specialized applications, such as for the design of geothermal heat extraction systems, radioactive waste isolation vaults, borehole stability, and mine ventilation systems. In extreme cases, heat applied to rock can be sufficient to spall the exposed surfaces (Fig. 10.8). In most cases, changes in temperature are sufficient only to generate thermally induced stresses, and to modify the properties of the rock to some extent (Berest and Weber, 1988).

To investigate and forecast these effects, first the temperature distribution (*temperature field*) within the rock mass must be predicted. It depends on the imposed boundary temperatures, heating, and cooling, and on the thermal characteristics of the rock.

The calculations are very similar to those used to predict groundwa-

Figure 10.8 Thermal spalling in a tunnel exposed to hot gases.

ter equipotential fields as a function of drainage or injection of water (Chap. 4), and stress fields as a function of boundary stresses (Chap. 7). Solutions are derived from the governing differential equation for flow problems, such as the following simple equation for conduction with no heat sources or sinks, which assumes isotropic thermal properties:

$$K\nabla^2 T = \frac{dT}{dt}$$

$$K = \frac{k}{C_p m}$$

where K is the thermal diffusivity, C_p is the specific heat, m is the mass, k is the thermal conductivity of the rock, and $\nabla^2 T$ is the Laplacian of the temperature field. Thermal conductivity and specific heat are analogous to hydraulic conductivity and storativity for porous media flow. The heat flow properties are assumed independent of the stress state, which is a reliable assumption for intact, but not for jointed rock. For limited temperature ranges, such as those encountered in studies of mine ventilation, the parameters are assumed constant with temperature, usually a valid assumption.

10.3.2 Thermal conductivity

Thermal conductivity is a measure of the rate at which heat will travel by conduction through a rock. It is measured by the quantity of heat transmitted across a unit cross section in a unit time for a unit temperature gradient.

10.3.2.1 Measurements on intact rock. Thermal conductivity can be measured in the laboratory using a conductivity comparator (Birch and Clark, 1940; Sibbitt et al., 1979). Rocks of low porosity are easier to test than porous saturated ones, because porous samples must be kept confined, and because pore fluids migrate under a thermal gradient. Specimens that are partially saturated are difficult to test because thermal gradients give rise to internal variations in the degree of saturation. Even more difficult are tests under confining stress with pore-fluid back-pressure. Comparative tests can be run dry and saturated, making sure that the test configuration precludes convection.

Scott and Seto (1985) addressed these problems for oil sand specimens by using a radially symmetric test configuration, measuring heat flux between an inner central point and the exterior. Because of the length of the specimen, end effects were negligible, and a cylindrical heat conduction solution (Carslaw and Jaeger, 1959) permitted back-calculation of the bulk conductivity. Because pore fluid in an oil sand is essentially immobile, there were no problems with convection.

The alternative of larger-scale in situ testing is much more expensive, although more reliable, particularly when the effects of porosity, jointing, and saturation are to be taken into account.

10.3.2.2 Measurements on jointed rock. Sandford et al. (1984) describe laboratory measurements of the thermal conductivity of intact rock into which joints had been progressively introduced by fracturing. Joints reduced the thermal conductivity in all cases, by approximately 3.5–2.5 W/m · K from the intact to the broken condition. Conductivity decreased systematically with increasing joint roughness, and increased with increasing stress normal to the joint plane. Tests such as this are aimed at allowing the thermal behavior of individual joints to be combined with that of the intact rock for a simulation of bulk rock behavior, thus avoiding the expense and difficulty of large-scale testing.

In the field, thermal conductivities are obtained by careful one-dimensional or cylindrical heating tests that can be interpreted using closed-form solutions or by numerical analysis. Kuriyagawa et al. (1983), for example, used an electric "line heater," 600 mm long and 60 mm in diameter, placed at a depth of 3 m in a 100-mm drillhole. The heater was held at 440°C for about three months, during which time five thermocouples in a line at distances of up to 5 m from the heat source were used to monitor the resulting thermal gradient. Thermal conductivity of the granite gneiss rock, determined by finite-element analysis, was found to decrease with increasing temperature.

Also, laboratory measurements gave values more than twice as high as those determined in the field. Similar *heater tests* have been conducted in other rock types and other countries as part of the worldwide study of sites for radioactive waste disposal (Fig. 9.9).

10.3.3 Specific heat

The specific heat of a rock is the amount of thermal energy needed to raise the temperature of a unit volume of the rock by 1 degree. Specific heats of individual minerals are obtained using accurate calorimeters, and are usually about one-quarter the specific heat of water. Because the specific heat of a mass of minerals is the volume fraction sum of the specific heats of the individual minerals, a quite precise estimate can be made of the bulk specific heat of any rock whose mineralogy is known. If a porous rock contains different phases such as water, minerals, and air, the fractional volume calculation again applies, provided that the phase concentrations do not change because of flow or phase changes.

10.3.4 Thermal coefficients of expansion

Thermomechanical analysis requires not only a study of the effects of the predicted temperature field on the elastic, plastic, and viscous, yielding and rupture behavior of the rock mass under applied or excavation loadings, but also a study of the strains and stresses produced by the temperature changes on their own. For these thermal stresses and strains to be analyzed, the coefficient of thermal expansion of the rock must also be measured.

The thermal coefficient of linear expansion α describes the one-dimensional strain caused by a given change in temperature:

$$\alpha = -\frac{1}{L}\frac{dL}{dT}$$

where dL/L is the thermally induced strain and dT the change in temperature. Units in the SI system are those of strain per degree Celsius. The thermal coefficient of volumetric expansion is the volumetric strain per degree Celsius, and is 3 times the linear coefficient.

10.3.4.1 Measurement. Thermal expansion can be measured conveniently using electric resistance strain gauges cemented to the faces of cubic or prismatic specimens (Finke and Heberling, 1978; Poore and Kesterson, 1978; Senior and Franklin, 1987). Strain gauges are quicker and more convenient than dilatometers, interferometers, or electric transducers, which are traditionally used in physics for mea-

suring changes in dimensions. Because they are small and inexpensive, they can be mounted on more than one cube face to allow the measurement of anisotropic expansion. The strains in a large number of specimens can be monitored simultaneously by a data logger, which scans at frequent intervals while temperature is cycled through the range of interest.

To study thermal expansion of 11 rock types, Senior (1985) cut the rocks into 45 cubic specimens of 25-mm side length, ground the faces smooth, cleaned and etched them, and applied the strain gauges (Fig. 10.9). Curing at 65°C ensured dimensional stability of the adhesive. The specimens were placed in an insulated plywood box along with standards of quartz, titanium silicate, and various metals for calibration. Air in the box was heated by thermofoil elements and circulated by a fan. The entire box was placed in a freezer. Thermocouple sensors turned the heaters on at $-20°C$ and turned them off at $+70°C$. The thickness of insulation was adjusted to give approximately daily cycles. Different initial moisture conditions were tested by allowing the rock specimens to equilibrate in an environment of controlled relative humidity before testing.

10.3.4.2 Thermal expansion characteristics of rocks. Experiments such as these show most rocks and minerals to have nonlinear thermal expansion characteristics, such that α is constant only within a small range of temperatures. Values of α for rocks are lower than those for metals, less than 10 microstrain/°C compared with steel (12.2), copper (17.5), and aluminum (23.5) (Table 10.2). The coefficient for granite increases from about 3 to 6 microstrain/°C over the temperature range of -20 to $+50°C$. Nonlinearity is even more marked for calcite marbles that can display negative coefficients, from -5 at lower temperatures to $+10$ at higher temperatures.

Other peculiarities of the thermal expansion behavior of rock include anisotropy, hysteresis, and "growth." Thermal expansion anisotropy is a well-known characteristic of minerals such as quartz, calcite, dolomite, and feldspars. However, it is also exhibited by polymineralic rocks. Some rocks in Senior's test program had anisotropy coefficients of up to 2.0. Hysteresis was noted in the behavior of calcite marbles, in the form of higher α values during cooling than during heating in the upper range of temperatures. Growth was also most evident for the marbles, which retained a permanent expansion of as much as 500 microstrain after 56 cycles of heating and cooling.

Combinations of nonlinear, anisotropic, and irreversible behavior result in grain boundary stresses in polymineralic rocks and monomineralic ones containing highly anisotropic minerals, and also

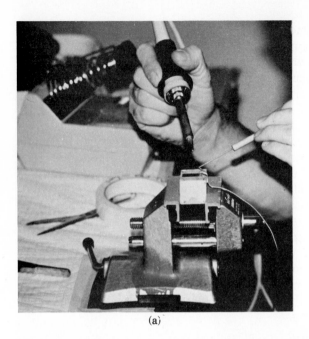

(a)

(b)

Figure 10.9   Thermal expansion measurements. (*a*) Close-up of strain gauging. (*b*) Cube specimens and reference standards ready for thermal cycling. (*Senior, 1985.*)

give rise to intergranular stresses and sometimes to cracking in concretes. Ehara et al. (1983) noted that anisotropic expansion during the heating and cooling of granite was accompanied by acoustic emission, caused by intergranular stress and the development of grain boundary cracks. The coefficient of expansion was high during the first heating cycle, but further cycles were accompanied by hysteresis and residual strain. Simmons and Cooper (1978) reported cracking of granite and

TABLE 10.2   Typical Thermal Properties of Various Metals, Rocks, and Minerals

	Coefficient of linear thermal expansion, microstrain/°C	Specific heat, kJ/kg · °C	Thermal conductivity, W/m · °C
Aluminum	21–25	0.88	17–22
Brass	19–21	0.38	10
Steel	11–13	0.46	1–5
Salt (halite)	38–40	0.88	5–7
Ice	44	2.01	2
Granite	5–11	—	2–3
Diabase	6–9	—	2–3
Quartz	11–15	0.75	4–11
Quartzite	11	—	6–7
Sandstone	9–11	—	3–4
Calcite	6	0.80	2–5
Limestone	0–10	—	2–3
Marble	−3–+8	—	2–3
Dolomite	7–10	0.84	4–6
Coal	—	—	0–5

SOURCE: Data mainly from Buntebarth (1984), Clark (1966), and Senior and Franklin (1987).

diabase specimens when heated to temperatures in the range of 500–800°C.

10.3.4.3 Frost action on rocks. Weathering in arctic and high alpine climates is dominated by physical degradation as the result of frost action. An example is the "felsenmeer" (literally sea of rock), large areas of fractured rock generated by freezing stresses and movements caused by ice formation.

On a smaller scale, volumetric expansion of water in cracks causes high tensile stresses at the tips of the cracks. Freezing usually progresses from the open end of the crack, sealing the tip, and allowing the full force of the freezing water to act. Because rocks are weak in tension, they are quite susceptible to frost splitting (Tsytovich, 1975). Cyclic freeze-thaw tests have been developed for quality control of mineral aggregates subject to cold climates, as noted in Sec. 2.3.3.4.

10.4 Swelling of Rocks

10.4.1 Swelling mechanisms

Swelling is a time-dependent volumetric expansion caused by physicochemical reaction with water. This definition excludes, for example, "squeeze" which is sometimes accompanied by a volumetric expansion, but is caused entirely by stress. Several types of swelling result from the reactions of different minerals with water (Linder,

1976). Clay swelling is the most common, and occurs through moisture migration alone, without chemical changes. Other mechanisms create by-products that occupy a larger volume than the original materials. Included are the oxidation of sulfides to sulfates, hydration such as of anhydrite to gypsum, creating a new crystal structure, and alkali-aggregate reactions between certain rock types and cement paste chemicals.

The entire rock may consist of swelling minerals, or the minerals may be present as joint infillings, individual beds or veins, or even as a minor disseminated constituent. In all cases, severe consequences can ensue for the entire rock mass. Swell potential must be recognized in advance, as failure to design for it or avoid it can result in serious damage to structures. Particular problems are the uplift of foundations (Katzir and David, 1968) and excessive pressures and damage to tunnel liners.

Access to air and water for drying and wetting is needed for swelling to occur. Often swelling can be avoided by sealing the rock from air, or by spraying with water to maintain an excavated face in a saturated condition. However, the physical and chemical processes associated with swelling are usually quite complex, and a treatment that may prevent swelling in one set of conditions can induce it in another. Swelling is most pronounced when the confining stress on a rock is diminished, allowing joints to open and giving easier access to air and water.

10.4.2 Swelling of clays

10.4.2.1 Swelling clay minerals.
Older shale formations with less than 2% swelling clay minerals are mainly composed of chlorite, illite, and kaolinite clays. Although not classified as swelling types, these shales usually shrink and expand with changes in water content. Expansion is aided by capillary suction that draws water into fissures which, being weakly bonded, tend to open. If the individual clay platelets are well cemented with a mineral such as calcite, or even organic matter, swelling is much reduced.

The most expansion-prone shales contain substantial proportions of smectite, a family of clays of high surface area that can adsorb considerable amounts of water. The best known member of this family is montmorillonite, usually formed by the degradation of siliceous volcanic ash. Smectite is most abundant in shales and clays of the geologically younger formations, and is rare in sediments older than perhaps 200 million years (Sec. 2.2.1.3).

A dry smectite can absorb water equal to several times its volume before it behaves as a plastic clay. For instance, smectitic shales of

Cretaceous age from the western part of North America (Yukon-Alberta-Montana-Colorado-Mexico, east of the Rockies) can have uniaxial compressive strengths greater than 20 MPa, yet they can have very high water contents. The Bearpaw Shale at Gardiner Dam, for example, contains up to 35% water. The unusual water adsorbtion is caused by the very fine-grained nature of the mineral, usually smaller than 0.2 μm, and a lack of chemical bonding so that water can easily enter between platelets, expanding the mineral grains into individual thin sheets. The cation on the smectite surface also greatly affects the properties. Perhaps the worst of all smectites is sodium montmorillonite. The variety calcium montmorillonite is much less troublesome, because the divalent calcium cations tend to bond the platelets together.

10.4.2.2 Clay swelling mechanisms. Swelling of any shale is minimal when the rock is maintained in a confined saturated condition. It is most pronounced when the shale is allowed to dry and is then rewetted.

Depending on clay mineral types, proportions, and microstructure, various physicochemical mechanisms contribute to the swelling that occurs to some extent in nearly every clay-bearing rock. These include cation hydration, negative electric force fields, and surface wetting of particles; the action of capillary forces, the van der Waals force field, and air pressure; and the effects of osmotic pressures and double-layer expansion (Ladd, 1959). These effects are generally lumped together in a macroscopic measure of swelling potential. There is no satisfactory way of predicting the swelling behavior of shales from mineralogical or fabric analysis: tests must be carried out.

Osmotic swelling occurs as a result of a difference in salinity between water in the pores of the rock, which is often somewhat saline, and water surrounding the rock, which is often fresh. Swelling is usually greatest with respect to distilled water, because the pore fluid tends to draw the purer water into the rock by the process of osmotic suction.

Matrix swelling results from geological overconsolidation. The strain potential is locked into the rock fabric because of preexisting high stresses, and is released only when excess water is available after the stress on the rock has been relieved.

10.4.3 Anhydrite-gypsum reaction

Hydration of anhydrite to gypsum can yield an expansion of up to 62.6%, whereas the reverse dehydration can cause a shrinkage of up to 38.5% (Zanbak and Arthur, 1984). Laboratory studies show that completion of the swelling process requires from three months to a year. Phase transitions in calcium sulfate minerals are controlled by pressure, temperature, and composition of coexisting aqueous solutions. Gypsum tends to be the stable phase at low temperatures.

The relationship between expansion and pressure depends on physicochemical equilibrium criteria, such that the pressure generated by confined expansion can limit the extent of the phase transitions. Expansion pressures produced by the hydration of anhydrite could theoretically reach gigapascal levels, but the rock ruptures long before these stresses are attained. Confined expansion can result in the partial or complete loss of porosity in a previously porous rock, of particular consequence in the oil industry.

10.4.4 Pyrite and marcasite reactions

Certain black shales such as found in Oslo, Norway; Ottawa, Ont., Canada; and Cleveland, Ohio, are well known for foundation swelling problems. Expansion is caused by the growth of crystals of gypsum and a related mineral, jarosite, resulting from the oxidation of pyrite to iron oxide in the shale. The volume of the minerals produced is appreciably greater than that of the original constituents.

The reaction appears at least partly biochemical, in that iron- and sulfur-oxidizing bacteria, ferrobacilli and thiobacilli, play an active role. Sulfuric acid is also produced, which attacks calcite in the shale to form gypsum, and can attack concrete in the foundation. The reactions can be inhibited or stopped by preventing drying, for example, by spraying exposed surfaces immediately with asphalt (Grattan-Bellew and Eden, 1975).

10.4.5 Swelling tests

10.4.5.1 Index tests for durability and swelling potential. Swelling behavior is closely related to the slake durability of the rock, its capacity to resist drying and wetting cycles. The slake-durability index test and simple free-swell tests on either intact or powdered rock (Sec. 2.3.3.3) can help identify materials likely to swell. Parameters describing their swelling behavior can then be measured for use in design, using the tests described below, or can be predicted empirically from the measured indexes. X-ray diffraction and differential thermal analysis (DTA) methods for identifying clay mineral types and percentages are described in Sec. 2.2.1.8.

A simple and quick way to detect rocks prone to severe swelling, suitable mainly for poorly cemented smectite shales, is immersion in water and visual observation of expansion, slaking, and dispersion of clays in the water (Dusseault et al., 1983). To closely predict real behavior, simulated groundwater rather than pure tap water should be employed. The water chemistry should reproduce faithfully the expected salinity and also the valence level of the saturation cations, as rocks that contain swelling minerals can behave differently when exposed to different cations (Singh and Cummings, 1983).

10.4.5.2 Unconfined swelling tests.

In the simple uniaxial or triaxial-unconfined swelling test (Fig. 10.10a), a cylinder or cube of rock is immersed in water in a container and is monitored in one or several directions by dial gauges or electric transducers that measure changes in the dimensions of the specimen. These are plotted as a function of time, usually on a log-time graph. Results plotted versus log-time usually approach a straight line after an initially curved trend, and can be expressed in terms of strain per log-time cycle (Fig. 10.10b).

Unconfined swelling tests allow the classification of rocks into *swelling* and *nonswelling* types, but give only an indication of the swelling pressures that will develop under conditions of confinement.

10.4.5.3 Oedometer tests.

High swelling pressures can cause considerable damage to structural members such as tunnel liners and retaining walls. Swelling strains are reduced if the specimen is confined, but at the expense of the development of sometimes considerable pressures on the confining system.

Measurement requires a confined swelling test (ISRM, 1981). This form of test commonly makes use of an oedometer similar to that used for the consolidation testing of soils. A rock disk specimen is inserted into a confining ring and subjected to an axial stress, then flooded to permit swelling. In the *constant axial surcharge* test, rates of swelling are measured using several specimens of the same rock, each at a different stress. Graphs are plotted to show the rates and magnitudes of swelling, which decrease as the level of axial surcharge increases.

In an alternative *zero axial swelling* test using similar equipment, a single specimen is tested under conditions of zero axial strain by progressively increasing the applied stress to prevent any axial expansion. A graph is plotted to show the swelling pressures that develop under conditions of complete restraint.

10.4.5.4 Ring swell tests.

Oedometer swelling tests are one-dimensional and incapable of measuring strains or pressures in more than one direction. An alternative, the *ring swell* test, can be employed when three-dimensional strains are important to predict (Franklin, 1984).

A disk-shaped rock core specimen is grouted into an aluminum ring of thickness usually between 2 and 8 mm, selected to control the rate

(a)

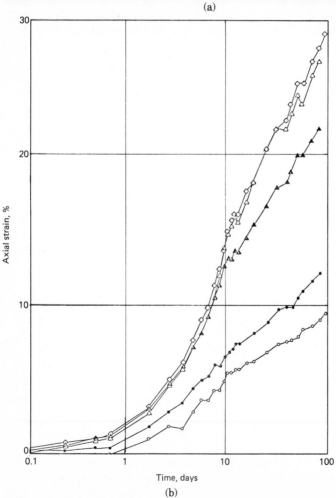

Time, days

(b)

Figure 10.10 Three-dimensional free swell tests. (a) Test cells containing rock speci-mens immersed in water. (b) Example of unconfined swelling strains developed by siltstone. (*Data from Harper et al., 1979.*)

of diametral confinement as a function of swelling strain. The outer surface of the aluminum ring is instrumented with electric resistance strain gauges so that circumferential strains may be measured. Confining pressure that develops as a result of swelling cannot be controlled, but the amount of confinement can readily be calculated from the strains measured in the aluminum. The specimen with its confining ring is placed in a loading frame that applies an axial surcharge. The rock can be dried or wetted by introducing dry nitrogen or water through perforations in the platens. Axial pressures are either maintained constant or increased to inhibit swelling, in the same way as in the conventional oedometer swelling test.

References

Berest, P., and Ph. Weber, Eds.: "La Theromécanique des Roches," *Manuels et Méthodes,* vol. 16, Bureau de Recherches Géologiques et Miniéres, Orléans, France, 1988, 327 pp.

Birch, F., and H. Clark: "The Thermal Conductivity of Rocks and Its Dependence upon Temperature and Composition, Parts 1 and 2," *Am. J. Sci.,* **238,** 529–558, 613–635 (1940).

Buntebarth, G.: *Geothermics* (Springer, Berlin, 1984), 144 pp.

Carmichael, R. S., Ed.: *Handbook of Physical Properties of Rocks,* vol. 3 (CRC Press, Boca Raton, Fla., 1984), 340 pp.

Carslaw, H. S., and J. C. Jaeger: *Conduction of Heat in Solids* (Oxford Univ. Press, New York, 1959).

Christian, J. T., and C. S. Desai, Eds.: "Constitutive Laws for Geologic Media," in *Numerical Methods in Geological Engineering* (McGraw-Hill, New York, 1977), pp. 65–115.

Clark, S. P., Jr., Ed.: *Handbook of Physical Constants,* rev. ed. (Geol. Soc. Am., New York, 1966), Memo. 97, 587 pp.

Dusseault, M. B., P. Cimolini, H. Soderberg, and D. W. Scafe: "Rapid Index Tests for Transitional Materials," *ASTM Geotech. Test. J.,* **6** (2), 64–72 (1983).

———, D. Z. Mraz, and L. Rothenburg: "The Design of Openings in Saltrock Using a Multiple Mechanism Viscoplastic Law," *Proc. 28th U.S. Symp. Rock Mech.* (Tucson, Ariz., 1987), pp. 633–642.

Ehara, S., M. Terada, and T. Yanagidani: "Thermal Properties of Stressed Rocks," *Proc. 5th Int. Cong. Rock Mech.* (Melbourne, Australia, 1983), vol. E, pp. 137–140.

Finke, T. E., and T. G. Heberling: "Determination of Thermal-Expansion Characteristics of Metals Using Strain Gages," *Experim. Mech.,* 155–159 (Apr. 1978).

Franklin, J. A.: "A Ring Swell Test for Measuring Swelling and Shrinkage Characteristics of Rock," *Int. J. Rock Mech. Min. Sci. Geomech. Abstr.,* **21** (3), 113–121 (1984).

Fritz, P.: "Numerical Solution of Rheological Problems in Rock," *Proc. Int. Symp. Numerical Models in Geomechanics* (A. A. Balkema, Rotterdam, 1982), pp. 793–803.

Grattan-Bellew, P. E., and W. J. Eden: "Concrete Deterioration and Floor Heave Due to Biogeochemical Weathering of the Underlying Shale," *Can. Geotech. J.,* **12** (3), 372–378 (1975).

Griggs, D. T., F. J. Turner, and H. C. Heard: "Deformation of Rocks at 500° to 800°C," Geol. Soc. Am., Memo. 79, pp. 39–104 (1960).

Hardy, H. R., and X. Sun: "A Non-linear Rheological Model for Time-Dependent Behavior of Geologic Materials," *Proc. 27th U.S. Symp. Rock Mech.* (Tuscaloosa, Ala., 1986), pp. 205–212.

Harper, T. R., G. Appel, M. W. Pendleton, J. S. Szymanski, and R. K. Taylor: "Swelling

Strain Developed in Sedimentary Rock in Northern New York," *Int. J. Rock Mech. Min. Sci.*, **16** (5), 271–292 (1979).

Heard, H. C.: "The Effect of Large Changes in Strain Rate in the Experimental Deformation of Yule Marble," *J. Geol.*, **71** (2), 162–195 (1963).

ISRM: "Suggested Methods for Rock Characterization, Testing, and Monitoring," ISRM Commission on Testing Methods, E. T. Brown, Ed. (Pergamon, Oxford, 1981), 211 pp.

Ito, H.: "Creep of Rock Based on Long-Term Experiments," *Proc. 5th Int. Cong. Rock Mech.* (Melbourne, Australia, 1983) vol. A, pp. 117–120.

Kaiser, P. K., and N. R. Morgenstern: "Phenomenological Model for Rock with Time-Dependent Strength," *Int. J. Rock Mech. Min. Sci. Geomech. Abstr.*, **18**, 153–165 (1981).

Katzir, M., and P. David: "Foundations in Expansive Marls," *Proc. 2d Int. Research in Eng. Conf. Expansive Clay Soils* (Texas, 1968).

Kuriyagawa, M., I. Matsunaga, and T. Yamaguchi: "An in situ Determination of the Thermal Conductivity of Granitic Rock," *Proc. 5th Int. Cong. Rock Mech.*, (Melbourne, Australia, 1983), vol. E, pp. 147–150.

Ladanyi, B., and D. E. Gill: "In situ Determination of Creep Properties of Rock Salt," *Proc. 5th Int. Cong. Rock Mech.* (Melbourne, Australia, 1983), vol. A, pp. 219–225.

Ladd, C. C.: "Mechanisms of Swelling by Compacted Clay," *Highway Res. Board Bull.* 245 (1959).

Linder, E.: "Swelling Rock: A Review," in *Rock Engineering for Foundations and Slopes, ASCE Specialty Conf.* (Boulder, Colo., 1976), vol. 1, pp. 141–181.

Poore, M. W., and K. F. Kesterson: "Measuring the Thermal Expansion of Solids with Strain Gages," *J. Testing and Evaluation*, **38** (2), 98–102 (1978).

Sandford, T. C., E. R. Decker, and K. H. Maxwell: "The Effect of Discontinuities, Stress Level and Discontinuity Roughness on the Thermal Conductivity of a Maine Granite," *Proc. 25th U.S. Symp. Rock Mech.* (Evanston, Ill., 1984), pp. 304–311.

Scott, J. D., and A. C. Seto: "Thermal Property Measurements on Oil Sand," *Preprints, 36th Ann. Tech. Mtg., Petrol. Soc. of Can. Inst. Min. Metall.* (Edmonton, Alta., 1985), pp. 177–188.

Senior, S. A.: "Thermal Expansion of Concrete Aggregates," M.Sc. thesis, Univ. of Waterloo, 105 pp. (1985).

——— and J. A. Franklin: "Thermal Characteristics of Rock Aggregate Materials," Res. Rept. RR241, Ont. Ministry of Transport. and Commun., Toronto, 68 pp. (1987).

Sibbitt, W. L., J. G. Dodson, and J. W. Tesler: "Thermal Conductivity of Crystalline Rocks Associated with Energy Extraction from Hot Dry Rock Geothermal Systems," *J. Geophys. Res.*, **84** (B3), 1117–1124 (1979).

Simmons, G., and H. W. Cooper: "Thermal Cycling Cracks in Three Igneous Rocks," *Int. J. Rock Mech. Min. Sci. Geomech. Abstr.*, **15**, 145–148 (1978).

Singh, M. M., and R. A. Cummings: "Predicting Moisture-Induced Deterioration of Shales," *Proc. 5th Int. Cong. Rock Mech.* (Melbourne, Australia, 1983), vol. E, pp. 87–95.

Tsytovich, N. A.: *Mechanics of Frozen Ground* (McGraw-Hill, New York, 1975), 426 pp.

Turcotte, D. L., and G. Schubert: *Geodynamics* (Wiley, New York, 1982), 450 pp.

Zanbak, C., and R. C. Arthur: "Rock Mechanics Aspects of Volume Changes in Calcium Sulfate Bearing Rocks Due to Geochemical Phase Transitions," *Proc. 25th U.S. Symp. Rock Mech.* (Evanston, Ill., 1984), pp. 328–337.

11

Behavior of
Discontinuities

11.1 Introduction

This chapter discusses the strength and deformability of joints, and tests for measuring these properties. As elsewhere in the book, the simpler term *joints* replaces *discontinuities,* whose use is reserved for situations where we wish to stress that the arguments apply not only to joints in the geological sense, but also to faults, planes of schistosity, and cleavage.

Sliding along individual joints often governs stability, and this is recognized by the models used in design. The method of limiting equilibrium, for example, calls for measurement of the shear strength of surfaces of potential sliding in terms of shear strength parameters, using direct shear tests. Strength values may be needed for joints, faults, bedding planes, the interfaces between soil and rock, or concrete and rock in a dam foundation. One reason for choosing the limiting equilibrium method is that it requires only information on shear strength, and none at all on deformability, whether of the joint or of the intact rock (Sec. 7.2.6). In contrast, the options of continuum mechanics are to perform a sufficiently large-scale field test on jointed rock using a block specimen containing many joints, or to test smaller pieces of intact rock and make adjustments to account for the reductions in rock mass strength and stiffness caused by jointing.

More recently, methods of discontinuum mechanics have been developed to include the modeling of individual joint "elements" that are allowed to slide, compress, and in some cases to separate; in other words, to behave like real joints. These techniques can model much more realistically the behavior of rock containing a few extensive and weak features such as clay-filled faults, which were always difficult to include as a part of a continuum. Discontinuum methods of analysis require data not only on bulk behavior, but also on the characteristics of individual joints or faults. Constitutive equations express the strength (shearing resistance) of the joint in terms of normal stress, and normal and shear displacement, and give the components of displacement across and along the joint in terms of its normal and shear stiffnesses or compliances.

11.2   Determination of Shear Strength

The shear test on rock joints is probably the most useful and commonly performed test in rock mechanics. It provides essential data for the analysis of a broad range of engineering projects, including the stability of slopes, dam foundations, and sometimes also underground works. Laboratory and field aspects are therefore covered in greater detail below than they were in Chap. 8 for tests on intact rock, which quite often are of secondary interest in rock engineering.

11.2.1   Tilt tests

11.2.1.1   Tilt test on blocks in the field.   The simplest form of shear test is to take a block of rock sitting on a natural, rough joint surface, and to tilt it until it starts to slide under its own weight (Fig. 11.1a). The smallest angle at which sliding occurs is measured, from which can be calculated the *peak shear strength* of the base of the block, defined as

(a)   (b)

Figure 11.1   Tilt tests for shear strength measurement. (a) Block tilt test. (b) Core tilt test.

the maximum shear stress that the surface can generate to resist sliding (Barton and Choubey, 1977).

11.2.1.2 Tilt test on core in the laboratory. A similar test can be conducted on core in the laboratory (Stimpson, 1981). Two pieces of core in contact with each other are fixed to the surface of a tilting table, and the third, free to slide, is placed on top (Fig. 11.1b). The table is slowly inclined until sliding starts along the lines of contact, at which point the angle of tilt α is measured.

Stimpson showed that the base friction angle ϕ_b is given by

$$\phi_b = \tan^{-1}(1.155 \tan \alpha)$$

He used the test to demonstrate the effects of moisture on the surfaces of contact, obtaining friction angles of 30° for a dry limestone, and of between 41 and 48° for the same cores when wet.

The *base friction angle* is defined as the frictional resistance of rock surfaces that are neither so smooth as to exhibit stick-slip oscillations, nor so rough as to interlock and dilate during shearing (Sec. 11.3.4). Textures such as this are provided by either sawn sandblasted surfaces or, as in this case, by the curved surfaces of drill core.

11.2.1.3 Limitations of tilt tests. Tilt tests suffer from the limitation that the normal stress is fixed at a value somewhat less than the weight of the block. The simple method therefore can be used only with caution when investigating the sliding of much larger and heavier masses of rock, such as landslides. Extrapolation from very low to quite high levels of normal stress can incur substantial error, because shear strength is known to increase substantially, and often in a nonlinear manner, with increasing levels of normal stress. The equations permitting this extrapolation are known as *shear strength criteria,* and are discussed in Sec. 11.3.

Because of the limitations of the tilt test, a *direct* shear test is needed in which the shear and normal components of stress can be varied independently of each other and of the weight of the block. Instead of tilting and relying on gravity, hydraulic jacks are used to apply loads in the normal and shearing directions.

11.2.2 Sampling for shear testing in the laboratory

11.2.2.1 Taking and preserving samples of jointed rock. Special precautions are needed when taking samples of joints for testing in the laboratory, to avoid relative displacement (preshearing or separation) of the opposing joint faces and softening, drying out, or loss of joint filling materials (Goodman, 1970). Samples can easily be damaged while in transit to the laboratory, or later during storage, preparation, or testing. For this reason, and also for reasons of scale, in situ testing is often preferred, particularly for the more fragile types of rock and for joints containing fillings, or faults containing gouge.

Block samples containing mating joint surfaces can be pried loose from an outcrop, or cut by line drilling or with a circular saw, a wire saw, or a chain saw (see Chap. 6).

Core drilling is the only available method when samples are required at depth, and sometimes it is more convenient for shallow sampling also. When the rock is durable, relatively undisturbed test specimens can usually be found in exploratory drill core. When it is more fragile, a suitable joint can be identified in outcrop, and overcored with a large-diameter thin-walled coring bit and a portable drill rig secured by bolts to the rock face. A drill of the sort used for taking concrete cores from walls and pavements is suitable. Drilling can be done with the core barrel normal, parallel, or oblique to the plane of the joint, depending on the size and shape of the sample required, and on the attitudes of joints at the available sampling locations.

Other useful techniques include prebolting of the core using a small-diameter centrally installed rockbolt, a technique similar to that of integral core sampling (Sec. 6.5.3). Alternatively, the outside surface of the core can be sealed into a tubular steel casing, using a grout material such as gypsum plaster, as when taking large-diameter samples of jointed rock for triaxial testing (Sec. 8.2.4.3). Samples containing joints can be held together with binding wire or metal bands, or by encapsulating in polyurethane foam (Sec. 6.3.2.2). Most rock types require wrapping and waxing as a protection against drying out.

A special split-cylinder specimen holder has been developed to give increased protection for shear test specimens during transportation and in storage (Franklin, 1985). Molten sulfur is poured to fill the space between rock and steel holder, and the two faces of the joint are clamped together. Each specimen has its own holder, which is uncoupled only after the specimen has been inserted in the test machine and the initial normal load applied. Encapsulation in the field has the further advantage of allowing uninterrupted testing, by eliminating downtime while waiting for specimens to be cut and encapsulated in the laboratory.

11.2.2.2 Sampling and testing filling materials. Thick joint fillings can in theory be sampled and tested without their surrounding rock. However, in practice, because of sampling disturbance, peak shear strength can only be measured reliably by testing the filling with its surrounding rock in place. McMahon (1985) reports that the residual strengths of detritus removed from sheared joints and tested in a soil shear box are often substantially lower than "ultimate" friction angles for the filled joint. To remove hard rock containing a joint filled with a soft clay requires very careful sampling and preparation.

11.2.2.3 Taking and testing replicas. A replica can be made by applying latex rubber, dental plaster, or epoxy resin to the rock joint surface in the field. In the laboratory, the replica can be used as a mold to obtain "positive" and "negative" impressions in some other substance such as sulfur or cement mortar.

The replicas are useful for examining and measuring surface textures and roughness (Chap. 3). Goodman (1974) suggests that the cast surfaces, if made of a material of similar strength to that of the rock, can be tested, even to the degree that joint infill of the proper thickness can be placed in the surface before testing.

11.2.3 Triaxial test

The laboratory triaxial test (Chaps. 8 and 9) is not at all ideal for testing individual rock joints, and its use is almost exclusively restricted to measurements on intact rock. This is because even small amounts of shear displacement bring the moving pieces of rock into contact with the cell walls and platens, and can rupture the confining membrane.

The triaxial test retains a role for measuring the strengths of coherent planes of weakness, such as the cleavage planes in a slate. Only the initial "peak strengths" can be measured, and then only when they occur after very small shear displacements. The specimen is drilled with the intended shear plane oriented at between 30 and 45° from the core axis.

11.2.4 Direct shear test

The purpose of this test is to measure the *direct shear strength* of a joint or plane of weakness by applying a constant stress normal to the plane, and then a steadily increasing stress tangential to it until sliding occurs. Intact rock is usually too strong to be tested in this manner, except for materials such as clay shales that are weak enough to be sheared along their bedding.

11.2.4.1 Laboratory apparatus. By far the majority of tests are performed in the laboratory, on blocks or cores containing joints. A specimen containing a plane of weakness, the *shear plane* to be tested, is mounted in a *shear box* using epoxy resin, plaster, or cement to hold the two halves securely in the box. Care is needed to keep the adhesive clear of the plane to be tested.

The machine (Fig. 11.2a and b) includes a *normal* loading system to apply and maintain a constant level of stress perpendicular to the shear plane, and a second, *shear*, loading system to cause one-half of the specimen to slide over the other. When testing rock, in contrast to soil, the normal and shear loading systems usually are hydraulic, although a pneumatic or a deadweight system sometimes provides the normal loading, and a mechanical gear drive provides the shear loading.

The moving half (usually the top) of the specimen must be allowed to slide freely. Therefore normal load is usually applied through roller bearings or a freely pivoting rod or wire rope. The resultant normal load must act through the centroid of the specimen, and the shear load must act in and along the plane of shearing. To ensure this, the upper and sometimes lower halves of the shear box have cantilever attachments for pulling or pushing along the shear horizon, and spherical seats or pivots to ensure freedom from transmitted moments. The applied shear and normal forces are measured either hydraulically, using simple Bourdon tube pressure gauges or pressure transducers connected into the hydraulic systems, or by electrical or mechanical load cells (Chap. 12).

Measurements of shear and normal displacement are incidental to the strength test, in that shear strengths can be determined from force readings alone. However, they are useful and allow graphs to be plotted for a better definition of peak and residual strengths, and also permit the measurement of normal and shear stiffnesses. Dial gauge or electric transducer systems can be used. Often a data acquisition device is employed to scan, measure, and record the forces and displacements. Displacement measurements are essential if one is to determine compliances in the shear test (Sec. 11.4).

11.2.4.2 Field-portable apparatus. Portable versions of the laboratory shear box can be employed in a field laboratory, as a compromise between laboratory testing and true in situ (field) testing (Fig. 11.2c). The much more expensive alternative of true in situ testing (outlined below) is used only if necessary to avoid scale effects or problems of sample disturbance.

A field-portable tester eliminates sample shipping costs and delays. Results can immediately be checked and further testing done if needed. "Instant" data are obtained for on-site stability calculations. Disadvantages include problems of testing in inclement weather. Tests under adverse conditions are usually less reliable than those done in the comfort of a well-prepared laboratory.

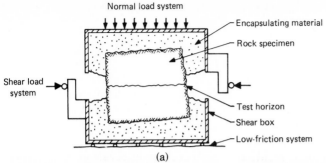

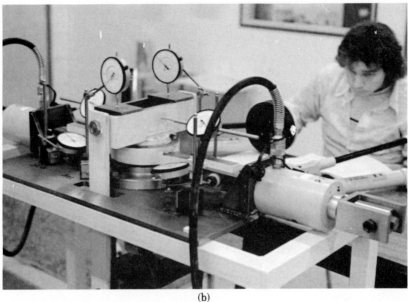

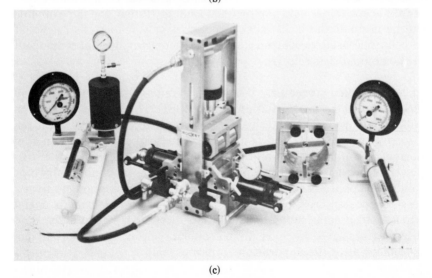

Figure 11.2 Laboratory direct shear test. (*a*) Laboratory shear box configuration. (*b*) Laboratory test machine. (*Franklin, 1985.*) (*c*) Portable shear box. (*Roctest, Canada.*)

11.2.4.3 In situ test blocks and equipment. Large-scale direct shear tests (Fig. 11.3) are performed on in-place rock blocks whose base is a natural joint, fault, or bedding plane. Typical blocks have dimensions of 0.7 by 0.7 by 0.35 m, but specimens with dimensions of several meters have also been tested. The procedures are in most respects similar to those for laboratory testing, although with important practical differences.

Access to the test location often requires excavation of a trench, adit, or test chamber, that must be stabilized to make it safe for testing. The block is isolated from the rock mass by sawing or line drilling. Weak or closely jointed rock is encapsulated in reinforced concrete that substitutes for the laboratory shear box. An inclined block must be fully supported by jacks while it is being cut, to prevent premature sliding along the horizon to be tested (Franklin et al., 1974).

The walls and roof of the test chamber provide reactions for the applied loads, often of several hundred tonnes. Flat jacks are convenient for load application because they are lightweight and easily handled, considering the very substantial pressures they can apply. Similar loading and reaction systems were discussed in Chap. 9 in the context of plate-load testing.

The configuration of the test chamber makes it difficult to apply the shear force along the plane to be tested. A satisfactory compromise is to incline the shear jacks so that they act through the centroid of the shear plane, which avoids undesirable moments that would tend to tilt the block. (This is the arrangement shown in Fig. 11.3.) The applied thrust must then be resolved into its shear and normal components to give true magnitudes of shear and normal stresses. To maintain a constant normal stress using an inclined loading system, the force applied by the normal load jacks must be progressively reduced as the shear force is increased, to compensate for the increasing normal component of the applied shear force.

11.2.4.4 Testing procedure. Whichever version of the shear test is selected, the methods are similar (ISRM, 1981). The specimen is encapsulated leaving a zone of at least 5 mm above and below the shear horizon free from encapsulating material. The first selected increment of normal stress is applied and held constant, monitoring normal displacement until the change is less than 0.05 mm in 10 min. This "consolidation" phase of testing (Sec. 4.2.1.4) allows dissipation of any excess pore-water pressures in the surface to be tested. It is essential when testing soils, but less so when testing most rocks. Only clay-filled surfaces give any measureable amount of continuing normal displacement after the application of normal load.

The shear stress level is then increased until shearing starts. Small

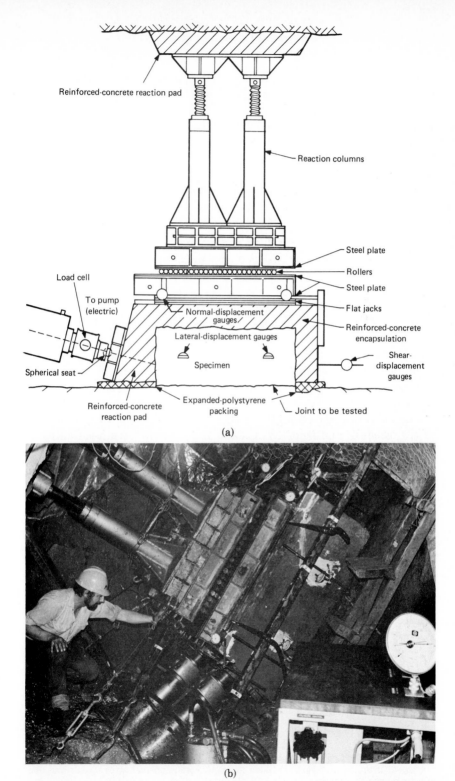

Reinforced-concrete reaction pad

Reaction columns

Steel plate

Rollers

Steel plate

Flat jacks

Reinforced-concrete encapsulation

Load cell

To pump (electric)

Normal-displacement gauges

Lateral-displacement gauges

Shear-displacement gauges

Specimen

Spherical seat

Reinforced-concrete reaction pad

Expanded-polystyrene packing

Joint to be tested

(a)

(b)

Figure 11.3 Field shear test. (*a*) Typical arrangement of testing equipment. (*b*) The same equipment in Greece, testing bedding planes for a reservoir perimeter slope stability study. (*Franklin et al., 1974; ISRM, 1981.*)

displacements occur in proportion to the applied shear force until *peak* strength is reached at a displacement of a few millimeters (Fig. 11.4*a*). Displacement readings are taken at about 10 increments of shear stress up to the peak shear strength value (the maximum shear stress reached in the test), and thereafter at increments of shear displacement. Shearing is continued past the peak to measure the steadily reducing shear stress needed to maintain sliding. After several centimeters of displacement, a more or less stable *ultimate* strength value is reached.

11.2.5   Plotting and interpretation of shear test data

11.2.5.1   Determination of peak and ultimate strengths.
A graph of shear stress versus shear displacement is plotted to identify and measure the peak and ultimate strength values (Fig. 11.4*a*). Peak strength is the maximum shear stress that can be sustained by the specimen. Ultimate strength is defined very loosely as the lowest strength that can be obtained with the maximum available shear box displacement, which should be at least 10% of the length of the specimen. Larger specimens permit greater amounts of shear before the block becomes unstable or the stresses too nonuniform for the test to be usefully continued.

The ratio of peak strength to ultimate strength for rock joints seldom exceeds 4, and is always greater than unity ("strain-softening or strain-weakening" behavior) unless the joint has a thin infilling or is highly polished at the start of the test (Barton, 1973). The ratio decreases with increasing normal stress.

11.2.5.2   Residual strength.
In nature, shear displacements of several meters are common beneath landslides, and sometimes of several hun-

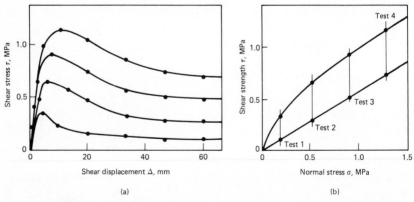

Figure 11.4   Method of presenting shear test results. (*a*) Shear stress versus shear displacement graph (four tests). (*b*) Shear strength versus normal stress (peak and ultimate strength envelopes).

dred meters along faults, further reducing the *ultimate* strength to a true *residual* strength by prolonged grinding and polishing.

Residual strength can be approximately determined in the laboratory by reversing the direction of shearing after reaching the maximum permissible displacement in any one direction, and continuing with reversals until the shear strength no longer is decreasing. Two shearing jacks are needed, one to each side of the specimen (Fig. 11.2b). There is some risk of loss of gouge at each reversal, and of progressive deterioration of the specimen. Patton (1966) suggested that laboratory measurements of the ultimate strength of wet, rough sawn surfaces closely reproduce the field values of the residual strength of joints.

The torsional shear testing alternative, somewhat more difficult and expensive, requires a doughnut-shaped specimen with a central hole. Shearing is rotational instead of linear and can be continued for any required number of revolutions, therefore for any amount of displacement.

11.2.5.3 Which strength value to use in design? An important and often difficult question to resolve is whether to use peak, ultimate, or residual strength values for design. The decision depends on the geological history of the joints and on whether peak strength could have been exceeded in the past, producing an ultimate or residual strength condition along even a single, critical surface. To help answer this question in the case of a slope design, a survey of regional slope stability can be very instructive. Many slopes would not be standing today if even one of the adversely oriented joints were at residual strength. In such cases, design based on ultimate or residual strength values could be too conservative. Often a compromise solution is needed.

Some researchers are of the opinion that "peaked" behavior is essentially a scale effect, which reduces or even disappears when very large surfaces are sheared (Bandis et al., 1981).

11.2.5.4 How many tests for a strength envelope? The *strength envelope* for a joint is a graph relating the shear strength of the joint to the normal stress across it (Fig. 11.4b). The greater the normal stress, the greater the strength. Envelopes can be plotted, one beneath the other, for peak, ultimate, and residual strengths, and these in turn lie beneath the envelope obtained for intact rock (Chap. 8).

Because of irreversible damage to the shear plane, only one peak strength can be measured per specimen, and several specimens are needed to define the envelope for a single joint surface. In contrast, an ultimate or residual strength envelope can be plotted by repeatedly testing a single specimen. The normal stress is increased in stages, if necessary with reversals of shearing.

The ISRM procedure (1981) recommends at least five tests per surface whose strength is to be measured, with each specimen tested at a different but constant normal stress to obtain one peak and one residual strength.

11.2.5.5 How many tests to measure variations of strength? Quite often, joints of different sets have different strengths, and the strengths can also vary greatly within a single set. Bedding planes in a sedimentary rock, for example, often occur in all parts of the spectrum from shaly to sandy. To define these variations, beds need to be sampled and tested according to lithology, and in sufficient numbers for a statistical treatment of the test data. For each lithology, a separate strength envelope is obtained based on at least five tests at different normal stress levels. The substantial amount of testing can be reduced if the strengths of only the weaker beds are of interest.

11.2.6 Estimating shear strength by back-analysis

The method of back-analysis can be a useful supplement to testing. Tests are expensive, particularly in the large numbers often needed, and often are unreliable because of limited scale, problems of extrapolation, and difficulties in estimating the degree of persistence (Sec. 3.2.4). In contrast, estimates of shear strength obtained by back-analysis automatically include the effects of scale and impersistence.

Nearby rockslides similar to the one to be analyzed, when they exist, provide ready-made and very large-scale field tests for the estimation of shear strength. They can be back-analyzed using the method of limiting equilibrium (Sec. 7.2.6). After estimating the various geometric and other characteristics, the only unknowns that remain are the shear strength parameters of the surfaces of sliding, and these may be determined for the factor of safety $F = 1.0$ that evidently pertained when the slide occurred.

However, even if nearby slides exist, the analysis may not be easy. Reliable back-analysis requires reliable data on preslide characteristics, such as slope topography, depth of sliding, and depth of the water table. Many slope failures, particularly the older ones, are not at all well documented, and the parameters that governed a historic slide can be difficult to determine. The procedure first requires careful measurement of the geometry of the surfaces along which sliding occurred. Next, the weight of the slide is evaluated, usually with some approximation because the material is no longer in place. Finally, and even more difficult, it is necessary to estimate the groundwater pressures that may have been acting in the slide surfaces at the instant of failure.

Dinis da Gama (1983) gives worked examples of back-analysis. He used interactive graphics to optimize the slope angles in Brazilian open-pit mines. Even 1 or 2° of variation in the slope angle of an open-pit wall can have a major impact on the economics of mining, hence the importance of obtaining the best possible estimates.

11.3   Shear Strength Trends and Criteria

11.3.1   Trends in shear strength

11.3.1.1   Effects of increasing normal stress level. Clearly, shear strengths, particularly peak strengths, increase substantially with increasing normal stress, so joints at greater depth appear stronger. Shear tests must be conducted at normal stress levels appropriate for the problem at hand to avoid tenuous extrapolation to stresses for which no results are available. For example, most rock slopes are no higher than about 50 m, so the stresses beneath them are in the range of 0–1 MPa. Many laboratory tests on small specimens are conducted at much higher levels of stress, and are relevant only to deep underground mining operations.

11.3.1.2   Effect of moisture conditions in the shear plane. The rate of shear displacement in any type of shear test must be sufficiently slow to allow water pressures developed within the plane of shearing to be dissipated in the time needed to achieve peak strength. Direct shear tests on soils, particularly clay soils, must be carried out slowly at specified strain rates that can be calculated using the theory of consolidation. This precaution is seldom needed in rock, except when the joint is filled with a low-permeability material such as clay or gypsum gouge. Usually it is sufficient to control the rate of shear stress application manually, and to slow the rate as the rock starts to shear, so as to ensure that the measured peak strength is the lowest that can be determined. Postpeak results are little affected by pore-water pressures because the already sheared surface, held open by granular gouge and noninterlocking asperities, is free-draining.

The frictional coefficients of minerals such as quartz and calcite can actually increase in the presence of water, whereas those of layer-lattice structures such as mica and chlorite can decrease. Coulson (1972) reported that very smooth surfaces in gneiss, granite, and sandstone that were polished rather than damaged during shearing showed a significant increase in strength in the presence of water.

Most natural or rough joint surfaces become weaker when wet. Peak strength reduction can range from 5 to 30% or even more, whereas residual strength tends to fall by between 5 and 10%.

Hassani and Scoble (1985) tested rocks wet and dry to assess the effects of moisture in the shear plane. No pore pressures were allowed to develop, so any effects on shear strength resulted from the presence, not the pressure of water. Using a linear approximation to the data set, the friction angles obtained varied from 27 to 34° in the dry state, and from 24 to 32° in the wet state, indicating a reduction in friction angle of 2–3°. This weakening effect of water, amounting to about 10%, parallels that observed when wet and dry specimens are tested in uniaxial compression (Sec. 8.2.1.3).

11.3.2 Selection of a shear strength criterion

Strength criteria in Chap. 8 describe the behavior of an element of either intact or jointed rock. For individual joints the requirements are similar: one or more simple equations to allow prediction and extrapolation from a few tests to any combination of stresses found in the rock mass. Although criteria with some theoretical basis would be preferred, simple ones if possible, the acid test is that they must match the experimental results and give accurate predictions. In this respect, empirical criteria, obtained by curve fitting, are usually the most reliable.

The various shear strength criteria proposed from time to time can be evaluated only by comparison with the trends of real shear strength data, and given an understanding of the variables that influence these trends.

11.3.3 Linear Mohr-Coulomb criterion

11.3.3.1 Definition of criterion. Shear strength envelopes, when nearly linear, are most often represented by a straight line called the *Mohr-Coulomb strength criterion,* which is given by the equation

$$\tau = c + \sigma_n \tan \phi$$

The two shear strength parameters c and ϕ are, respectively, the intercept of the line with the τ axis (at $\sigma_n = 0$) and the slope of the line. c is called the *apparent cohesion,* and ϕ the *angle of shearing resistance,* or *friction angle.*

The Mohr-Coulomb criterion usually provides an excellent fit to residual strength data, for which the apparent cohesion intercept is often close to zero. The residual friction angle varies from about 10° for a weak clay gouge to a little over 30° for most hard rock surfaces. To design on the basis of residual strength, one shear test is in theory sufficient to define the residual friction angle, from which the designer can extrapolate to obtain shear strengths at higher or lower levels of normal stress. In practice, more tests are needed to account for variability of the joints (Sec. 11.2.5.5).

In contrast, peak strength data curve sharply to meet the origin at low normal stresses (Fig. 11.5). The value of c in the Mohr-Coulomb equation of a tangent to this trend can be much larger than the actual shear strength at zero normal stress, which for rock joints is usually zero. Linear extrapolation from high to low normal stresses can therefore be dangerous, hence use of the term *apparent cohesion* for the straight-line intercept. Barton, whose nonlinear criterion is given below, comments that "if more investigators had been interested in low levels of normal stress or alternatively in very large ranges of normal stress (relative to the strength of the rock), there would be universal acceptance of a fundamentally nonlinear peak shear strength envelope for nonplanar rock joints. The use of a cohesion intercept at low or zero normal stress would then be inadmissible."

For simplicity and because of familiarity with the method, however, a linear envelope continues to be used by fitting a straight line to the curved peak strength data. This gives acceptable predictions over a very limited range of normal stress only, and only if this range is selected to match the stress range relevant to the problem to be analyzed. Use of the linear criterion can be extended by fitting not one but

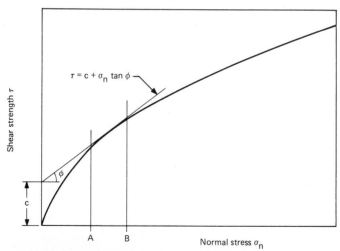

Figure 11.5 Mohr-Coulomb strength criterion applied to a curved trend of peak strength values. A straight-line criterion with apparent cohesion intercept c is an acceptable approximation to the curvilinear strength data only within the range AB of normal stress.

a series of straight lines to the data, switching from one line segment to the next according to the stress level that pertains at any place in the rock mass model being analyzed. A multilinear criterion such as this can easily be handled in a computer program, although a single curvilinear equation is simpler and more rational if one can be found to fit. Several curvilinear alternatives have been proposed, as noted below.

11.3.3.2 Mohr-Coulomb strengths of joints and fillings. Typical shear strengths for joints in various rock types and with various filling materials are shown in Table 11.1. The shear strength of a filled or

TABLE 11.1 Mohr-Coulomb Shear Strength Parameters for Typical Rock Joints and Fillings.*

	Typical ultimate strengths ϕ, deg $(c = 0)$	
Thick joint fillings		
Smectite clays	5–10	
Kaolinite	12–15	
Illite	16–22	
Chlorite	20–30	
Portland cement grout	16–22	
Quartz and feldspar sand	28–40	
Rock joints†		
Crystalline limestones	42–49	
Porous limestones	32–48	
Chalk	30–41	
Sandstones	24–35	
Quartzite	23–44	
Shales	22–37	
Coarse igneous rocks	31–48	
Fine igneous rocks	33–52	
Schists	32–40	

	Typical peak strengths‡	
	c, MPa	ϕ, deg
Sandstone	0.12–0.66	32–37
Siltstone	0.10–0.79	20–33
Pyroclastic rock	0.14–0.36	36–39
Seatearth (mudstone)	0.06–0.18	15–24
Mudstone	0.00–0.46	22–39

*Because of the considerable variations in test methods and in the rocks themselves, and due to the curvature of peak strength envelopes, the data give only a very rough guide to rock properties. The peak strength results are for carboniferous (Coal Measures) rock strata. Compare these data for joints with those in Table 8.1 for the peak strength of intact rock.
†Summary by Lama and Vutukuri (1978).
‡After Hassani and Scoble (1985).

coated joint depends first on the nature of the filling and second on its thickness. Whether the joint as a whole will behave like the strong wall rock or the soft clay filling depends on the thickness of the clay and the height of roughness asperity (Fig. 11.6).

Cording (1976) tested various clay gouge materials from faults intersected by the Washington, D.C., subway system, including plastic clays, and sandy and micaceous clays with liquid limits in the range of 41–65% and plasticity indexes in the range of 10–27%. For clay fillings up to 3 mm thick sheared between two planar rock surfaces, shear strength fell rapidly after peak during about 5 mm of shear displacement, and ultimate strength was reached in less than 50 mm. Ultimate friction angles ranged from 8.0 to 10.5°.

Frictional characteristics of clays and other platy types of filling vary as one would expect from the wide range of mineral types and physical properties (Chap. 2). Even a few micrometers of a smectite filling such as montmorillonite can reduce the angle of shearing resistance to values well below those for clean, hard rock surfaces.

Underwood (1964) describes U.S. Army Corps of Engineers' experi-

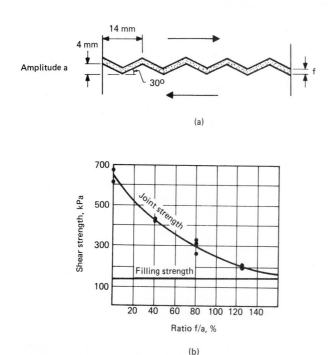

(a)

(b)

Figure 11.6 Effect of filling thickness f and asperity amplitude a on the shear strength of a filled joint or fault. Experimental results reported by Goodman (1970) for sawtooth joint filled with crushed mica.

ence in testing 75-mm-thick seams of bentonite clay in the foundations of the Harlan County dam. Twelve 760- by 760-mm blocks of chalk with the bentonite seam at their bases were isolated and encased in concrete. Vertical loads similar to those imposed by the dam were applied to the top of each block. The bentonite was consolidated for up to 32 h and then sheared at between 0.5 and 2.5 mm/min. The tests gave an average friction angle of 7.4°, and an average cohesion of 15 kPa. This unexpectedly low shear strength required deepening of the heel of the concrete dam by up to 3 m.

Clay mylonites, a term for sheared clay gouge layers in a shale rock, can be particularly troublesome and often give rise to serious problems of slope instability. Stimpson and Walton (1970) measured the strengths of 10–25-mm-thick clay mylonite seams in English Coal Measures shales, and reported residual angles of friction as low as 11°, in spite of a relatively low plasticity index of 21%. The residual friction angle of a highly plastic clay can be as low as 5° and is rarely greater than 12°.

Cement grout is a special case of an artificial joint filling material. Coulson (1972) investigated the effects of filling artificial extension fractures in fine-grained granite with a cement grout ranging in thickness from 0.8 to 6.4 mm. The shear strengths of the grouted joints turned out to be substantially lower than those of the clean, hard rock joints before grouting, probably because of the fine powder generated by shearing of the cement. Residual friction angles ranged from 16 to 22°. Only at very low normal stresses, below about 0.4 MPa, was any increase in strength observed.

11.3.4 Dilatant behavior and i angle

Attempts to explain and predict the shear strength of rough rock joints were inspired in the first instance by earlier studies of the observed dilatant behavior of sand, for which Newland and Allely (1957) proposed the following linear strength criterion:

$$\tau = \sigma_n \tan (\phi_b + i)$$

where i is the average angle of deviation of particle displacements from the direction of the applied shear stress and ϕ_b is the angle of frictional sliding resistance between sand grains.

Withers (1964), Patton (1966), and Goldstein et al. (1966), by testing plaster "joints" with a sawtooth cross section, showed that the inclination of the peak shear strength envelope at very low normal stresses is linear, and is given by $(\phi_r + i)$, where ϕ_r is the residual friction angle and i the *asperity inclination angle* or *peak dilation angle*:

$$\phi_{\text{rough}} = \phi_{\text{smooth}} + i$$

The dilation angle i can be calculated by measuring the ratio of normal to shear displacement at the start of a test:

$$i = \sin^{-1}\left(\frac{\text{normal displacement}}{\text{shear displacement}}\right)$$

The sawtooth model gives a useful illustration of the different mechanisms of shearing at low and high normal stress levels. At low stresses, shearing occurs entirely by the "riding over" of asperities, which remain unbroken. At higher stresses, the asperities start to be sheared and the dilation angle becomes smaller until, at sufficiently high normal stress levels, dilation is entirely replaced by shearing.

A bilinear shear strength criterion can be obtained by combining the dilatant model in terms of i angle for low normal stresses, and the Mohr-Coulomb equation for high normal stresses. However, real rock surfaces produce shear strength envelopes that are curved rather than bilinear, because they contain a broad variety of asperity heights and rise angles. The value of normal stress sufficient to impose complete shearing of asperities, such that $i = 0$, depends on the surface roughness and strength of the rock. The final inclination of the peak strength envelope at high normal stresses approaches ϕ_r as i approaches zero. This concept of the progressive shearing of asperities can be expressed in terms of a *damage coefficient,* as discussed in Sec. 11.3.5.2.

11.3.5 Curvilinear criteria

11.3.5.1 Ladanyi and Archambault's model.
Ladanyi and Archambault (1970) derived a curvilinear, semiempirical shear strength criterion as a function of the strength of the intact rock, the degree of interlocking, dilation, and the residual friction angle. They used principles of thermodynamics and assumed that shear strength is derived from three sources: resistance to sliding along the contact surfaces of asperities, resistance to shearing of asperities, and work performed by the normal load during contraction and dilation of the system:

$$\tau = \frac{\sigma_n(1 - a_s)(\dot{v} + \mu) + a_s\eta C_0[(m - 1)/n](1 + n\sigma_n/\eta C_0)^{1/2}}{1 - (1 - a_s)\dot{v}\mu}$$

where τ = peak shear strength.

σ_n = normal stress

$\dot{v}$ = dilation rate due to shear, = dy/dx

μ = average coefficient of friction for contact surfaces, = $\tan \phi_b$

C_0 = uniaxial compressive strength of unit rock blocks

η = degree of interlocking

n = ratio between uniaxial compressive strength and tensile strength of rock, $= C_0/(-T_0)$

$m = \sqrt{n+1}$

a_s = ratio of area of asperities sheared off to total sheared area

Of these parameters, normal stress and the degree of interlocking represent the basic data of the problem. Any roughness profile can be assumed provided that the degree of interlocking can be estimated. C_0 and n are obtained from laboratory testing, and the base friction angle ϕ_b, by shearing saw-cut specimens.

The dilation rate v and the sheared area ratio a_s are empirically derived. Experimental evidence shows that they depend on the geometry of irregularities and the ratio between the applied normal stress and the transition stress, defined as the stress at which the behavior of rock changes from brittle to ductile. Normal stresses for most engineering applications fall well below the ones required for ductility, in which case the empirical parameters v and a_s can be estimated using the following equations:

$$a_s \simeq 1 - \left(1 - \frac{\sigma_n}{\eta \sigma_T}\right)^K$$

$$\dot{v} \simeq \left(1 - \frac{\sigma_n}{\eta \sigma_T}\right) L \tan i$$

where i is the average dilation angle, σ_T is the transition stress, and K and L are approximately equal to 4.0 and 1.5, respectively. Methods for estimating the transition stress are described in Mogi (1966) and Byerlee (1968). In the absence of sufficient data, the transition stress can be approximated by the uniaxial compressive strength of the rock.

Dight and Chiu (1981) have used the variable dilation equation proposed by Ladanyi and Archambault (1970) in an iterative stochastic procedure to predict shear strength. To accommodate the stochastic analysis, the joint profile was replaced by a simple triangular model. Plesh (1985) considers further the constitutive modeling of rock joints with dilation.

11.3.5.2 Barton's model. Barton (1971, 1973) proposed an empirical nonlinear criterion, suggesting that the peak shear strength of rough joints can be predicted with acceptable precision from a knowledge of only the joint wall compressive strength (JCS). This was extended in Barton and Choubey (1977) to include joint surfaces with varying degrees of roughness (JRC) for both low and high stresses:

$$\tau = \sigma_n \tan \left[\text{JRC} \log_{10} \left(\frac{\text{JCS}}{\sigma_n} \right) + \phi_b \right]$$

The joint roughness coefficient JRC varies from 20 (rough) to 0 (smooth). Values are suggested by Barton as follows: 20 for rough undulating joints, 10 for smooth undulating joints, and 5 for smooth nearly planar joints, such as many foliation and bedding joints (see Sec. 3.2.5.2). As JRC increases, the curvature of the strength envelope decreases, the peak shear strength increases, and the joint wall compressive strength JCS has a greater effect because asperity shearing plays a more important role.

JRC can be obtained (as opposed to estimated) from the measured inclination of a joint at the instant of sliding in a tilt test:

$$\text{JRC} = \frac{\alpha - \phi_r}{\log_{10} (\text{JCS}/\sigma_n)}$$

where α is the angle of tilt at which sliding occurs; ϕ_r is the residual friction angle, $= (\phi_b - 20) + 20(r/R)$; r/R is the ratio of Schmidt rebound numbers (see Sec. 2.3.2.3) r for the saturated joint wall and R for the dry unweathered rock surface; and σ_n is the normal stress on the joint at the instant of sliding.

Alternatively, JRC can be obtained by a simple in situ direct shear "push" test, using a flat jack in a slot to push an in situ rock block under self-weight conditions:

$$\text{JRC} = \frac{\tan^{-1}[(T_1 + T_2)/N] - \phi_r}{\log_{10} (\text{JCS} \times A/N)}$$

where T_1 and T_2 are the self weight and the applied components of the shear force along the slide plane; N is the normal component of the block weight; and A is the contact area.

JCS is equal to the unconfined compressive strength if the joint surfaces are unweathered or the normal stress level is very high, but may reduce to approximately one-quarter of this value if the surface is weathered and the normal stress is moderate or low. Methods for estimating JCS have been reviewed in Chapter 3 (Barton, 1976).

The base friction angle ϕ_b is often assumed to be approximately 30° and varies in the range of 21–38°. More site-specific and reliable values for unweathered joints can be obtained by carrying out residual shear tests on flat surfaces prepared using a diamond saw and sandblasting, or by Stimpson's line-contact tilt test described in Sec. 11.2.1.2. For weathered joints, the base friction angle is replaced by the residual friction angle, which can be estimated using the following equation presented in Barton and Choubey (1977):

$$\phi_r = 10 + \frac{r}{R}(\phi_b - 10)$$

where R and r are the Schmidt rebound values on the unweathered and weathered rock surfaces, respectively.

Barton and Choubey (1977) also investigated the asperity component of the total friction angle. For joints suffering little damage during shear, the following equation was proposed as a first approximation to shear strength:

$$\tau = \sigma_n \tan (\phi_r + d_n)$$

where the peak dilation angle d_n is defined as

$$d_n = \text{JRC} \log_{10}\left(\frac{\text{JCS}}{\sigma_n}\right)$$

By comparing this shear strength equation, which assumes little shear damage, to Barton's empirical shear strength equation, a *damage coefficient M* is obtained:

$$M = \left(\frac{\text{JRC}}{d_n}\right) \log_{10}\left(\frac{\text{JCS}}{\sigma_n}\right)$$

The damage coefficient increases as JRC increases and JCS decreases.

11.3.5.3 Other curvilinear models. Reeves (1985), in his review of Barton's shear strength model, suggests an alternative empirical correlation of the total friction angle and the rms contact gradient of the roughness profile Z_2:

$$\tan \phi_t = C[Z_2(r)]^n$$

where ϕ_t is the total friction angle, C and n are empirical constants, and the root-mean-square roughness Z_2 is defined in Chap. 3. He disagrees with splitting the friction angle into a base friction component and a dilation component, and contends that a simple separation of the components is not always possible.

Denby and Scoble (1984) propose that the curvilinear nature of shear strength envelopes for rock discontinuities can be well represented by a power curve, and that curvature is generally most pronounced over the low normal stress range of 0–1 MPa. Their strength criterion takes the form

$$\tau = A\sigma_n^B$$

where A and B are constants for any given material. A nomogram is given to assist in the prediction of curvilinear shear strength enve-

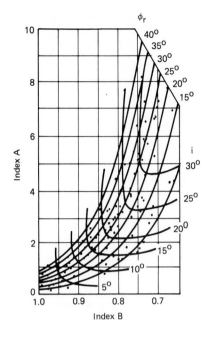

Figure 11.7 Parameters A and B for curvilinear shear strength criterion, as a function of peak dilation angle and residual angle of shearing resistance. (*Denby and Scoble, 1984.*)

lopes, based on empirical studies linking A and B to the residual friction angle and the peak dilation angle (Fig. 11.7). A varies typically in the range of 0–10, and B in the range of 0.65–1. High B values and low A values correspond to the more linear type of shear strength envelope, and are typical for small dilation angles and high normal stress levels.

Kulatilake and Ouyiang (1986) have represented shear test data both by the straight-line Mohr-Coulomb criterion and by a power law as given earlier. They stress the importance of representing strength at any given normal stress level not just as a single average value, but as a probabilistic statistical distribution that can be calculated from the test results using linear or logarithmic regression methods.

11.3.6   Scale and roughness effects

Laboratory direct shear machines accept specimens from about 50 to 300 mm on a side. Larger sizes require field tests that can easily cost $10,000 each, and even then, the largest test is hardly likely to simulate the behavior of field discontinuities with areas of hundreds of square meters. Often the most practical method for design is to carry out small-scale tests, and to extrapolate from these, taking into account the surface roughness measured on a larger scale. When possible, the results should be compared with estimates based on back-analysis (Sec. 11.2.6).

The question of how to extrapolate from a small to a large scale therefore continues to occupy the attention of many researchers. The main focus has been to investigate the influence of roughness on shear strength, and the manner in which roughness changes (if at all) from a small to a large scale.

The dilatancy model implies that the large-scale strengths of joints with rough or wavy surfaces can be predicted by extrapolating from small-scale tests on the same joints, provided that a correction is applied to offset the difference in rise angle i between small and large scales.

McMahon (1985) has pointed out that in applying this theory to practical problems of slope design, the engineer is given too much latitude for interpreting the meanings of "rough," "smooth," and "rise angle." By back-analysis of rockslides he has demonstrated that satisfactory values for the large-scale friction angle of potential rockslide surfaces are obtained by defining ϕ_{smooth} as the *ultimate* friction angle obtained in a laboratory direct shear test, and i as the difference between the flattest dip angle and the mean dip angle measured over a downslope distance of at least 5–10 m, using a geological clinometer. The ultimate friction angle is then corrected for use in a full-scale analysis by adding the dip deviation angle i as described.

Pratt et al. (1972) and Barton and Choubey (1977) predict that the peak strength should reduce with increasing area of joint. They suggest that on a larger scale, wall contact is transferred to the longer, less steeply inclined asperities as the peak strength is approached. These have lower JCS values than the small, steep asperities, and are also less steeply inclined and, hence, give reduced JRC values.

In contrast, Reeves (1985) proposes increasing roughness with scale. He claims that test results from Locher and Rieder (1970) and Brown et al. (1977) support this hypothesis, showing shear strengths significantly higher in the field than in the laboratory.

The studies of Swan (1985*a*, *b*) and Swan and Zongqi (1985) indicate that scale has negligible influence on predicted joint behavior for surfaces of engineering interest.

11.3.7 Time-dependent behavior of joints

11.3.7.1 Joint creep and the effect of strain rate.
When testing smooth joints in hard silicate rocks, Hassani and Scoble (1985) found no discernible difference in strength over 1.5 orders of magnitude of displacement rate. However, the shear strengths of joints or faults con-

taining clay or other types of platy mineral depend on the rate of shearing, even without the contribution of pore-water pressure, and even if the sheared surface is dry or well drained. Clayey discontinuities are stronger when sheared slowly, and long-term creep strengths are greater, as might be expected, when the applied shear stresses are low.

Curran and Leong (1983) carried out direct shear tests on shales at displacement rates of between 0.5 and 256 mm/s. Shear strengths were found to decrease in proportion to the logarithmic increase in the rate of shearing, irrespective of the apparent area of contact. Bowden and Curran (1984) performed creep tests on shales, and found that creep rates increased dramatically according to the ratio of applied shear stress to peak strength. At ratios below approximately 0.8, the creep decreased rapidly with time (primary creep), becoming small after 4 or 5 days of testing. At higher ratios, between 0.8 and 0.9, the creep rate was orders of magnitude higher and led to displacements of centimeters within just a few hours.

Howing and Kutter (1985) investigated the creep behavior of filled rock discontinuities by varying the type of filler material, the filler thickness, the shear stress level, the normal stress, and the size of the specimen. The results indicated a power law for the creep velocity as a function of time. The power m was equal to -1 at lower shear stresses but approached zero at higher stress levels. It was also affected by the clay content, the thickness of the filler, and the stress level. The initial creep velocity was directly proportional to the width of the filler and related to its clay content.

11.3.7.2 A rheological creep model. The same rheological model approach as described in Chap. 10 for simulating the time-dependent behavior of a rock element can be applied to modeling the behavior of a rock joint. A smooth or presheared joint shows small and nearly reversible shear displacements at low levels of shear stress, and starts to slip only after the shear stress has exceeded a certain elastic limit, after which it displays viscoplastic flow such that the displacement rate increases with increasing stress. If shear stress is reversed, the stress change has to be $-2K$ before sliding starts in the opposite direction. Figure 11.8 shows a compound rheological body that simulates this behavior. Expressed in terms of equations,

$$\epsilon = \frac{\sigma}{E} \quad \text{if } \sigma < K$$

$$\epsilon = \frac{\sigma}{E} + \frac{t(\sigma - K)}{\mu} \quad \text{if } \sigma > K$$

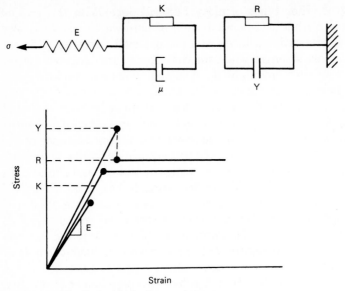

Figure 11.8 Rheological body for a rock joint.

If the stress is reversed at time t but $\sigma \hbar -K$ and flow occurred during the initial load, the sum of the strain (deformation) must include the original plastic flow component:

$$\epsilon = \frac{t(\sigma_i - K)}{\mu} - \frac{\sigma_i - \sigma_j}{E} \quad (= \epsilon_j)$$

and if $\sigma_j < -K$,

$$\epsilon = \epsilon_j - \frac{(t - t_0)(K - \sigma_j)}{\mu}$$

This set of equations is the simplest to simulate the behavior of a joint with no "peak" in the shear stress-displacement curve. With the stipulation of time and stress steps, it can be easily coded into a computer logic program that predicts displacements.

11.4 Joint Deformation, Stiffness, and Compliance

11.4.1 Theory

11.4.1.1 Development of predictive models.
Much of the early emphasis in rock mechanics was focused on measuring the strengths of rock joints, mainly to satisfy the requirements of limiting equilibrium calculations. More recently, with the development of realistic computa-

tional procedures for modeling jointed rock, research has been redirected to measuring joint deformations and to predicting the interrelationships between normal and shear stress components and the corresponding shear strains and changes in joint aperture (e.g., Sun et al., 1985; Yoshinaka and Yamabe, 1986).

A further incentive has been that even quite small changes in aperture have a marked effect on the hydraulic conductivity of the joint, which increases in proportion to the third power of the aperture, as discussed in Chap. 4. Fluid flow and contaminant migration through rock fissures are of increasing concern in relation to underground repositories for radioactive and other wastes. Predictive modeling has therefore included not only stress deformation behavior, but also coupled stress deformation-conductivity problems.

11.4.1.2 Definition of terms. Compressive stress normal to the plane of a joint reduces its aperture. The ratio of applied normal stress to closure is termed the *joint normal stiffness* (gigapascals per meter), and its inverse is called the *joint normal compliance* (meters per gigapascal). Similarly, the shear and normal displacements experienced on the application of shear stress can be expressed in terms of shear stiffness or compliance.

In more general terms, there are three stresses (components of traction) that can be applied to a rock joint, one normal and two shear. Each of these can produce one or more of three possible relative displacements, again one normal and two shear. Sun et al. (1985) suggest the use of compliance rather than stiffness for interpreting and predicting the behavior of joints, and define a 3 × 3 compliance matrix for the joint. The most important components, and the only ones to have been investigated, appear to be the normal compliance component C_{11}, the shear compliance component C_{22}, and the normal/shear compliance component C_{12}.

11.4.2 Testing

11.4.2.1 Joint compliance tests. Sun and coworkers (1985) describe a test machine for measuring normal and shear deformations and hence compliances of a joint. It is a direct shear box 0.5 m long and 0.35 m wide, with a special miniextensometer device for measuring small changes in aperture and shear displacement of the joint, independently of any straining of the intact rock. Errors are minimized by measuring these displacements close to the joint surface, inside a drillhole in the specimen block. The miniature extensometer and shear gauge incorporates two LVDT transducers (Sec. 12.1.2.5) and measures with an accuracy on the order of ± 0.004 mm.

11.4.2.2 Normal compliance. Normal compliance is the simplest to measure, requiring just a compression test with the load perpendicular to the joint. It may be measured during the first, "consolidation," phase of a shear test, although cyclic loading and unloading are usually needed to investigate hysteresis effects. Typically, only about 70% of the initial aperture is recovered after the first unload cycle, increasing to more than 90% recovery during subsequent cycles.

Goodman (1974) suggested a normal pressure versus joint closure relationship in hyperbolic form, in which large amounts of closure are experienced at first, with smaller amounts at higher pressures, when it becomes increasingly difficult to reduce further the joint aperture. The nonlinear behavior results from increasing contact areas and numbers of contacts as the joint is compressed, and probably also from some local crushing, depending on the rock strength and stress levels. Hungr and Coates (1978) found a quite linear relationship, with normal stiffness coefficients on the order of 18 GPa/m for joints in sandstone and limestone loaded in the stress range of 0–2 MPa.

Sun et al. (1985) obtained results for compression tests on granite, as shown in Fig. 11.9. Both an exponential and a power law closely fit the experimental data:

$$d_1 = a_0 + a_1 \ln \sigma_1$$

$$d_1 = b_0 \sigma_n{}^{b_1}$$

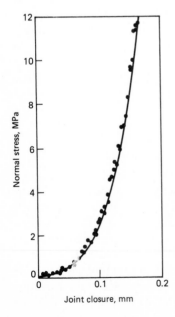

Figure 11.9 Relationship between normal stress and displacement (joint closure) for a joint in granite. Stiffness of the joint increases (compliance decreases) as the level of normal stress is raised. (*Sun et al., 1985.*)

where a_0, a_1, b_0, and b_1 are constants determined experimentally. The normal compliance component C_{11} was found to be proportional to the initial joint aperture.

11.4.2.3 Shear compliance. Shear loads applied to a joint produce both shear displacement and normal displacement (dilatancy), from which the shear compliance component C_{22} and the normal/shear compliance component C_{12} may be calculated.

Sun et al. (1985) distinguish three phases of shearing, all within the first 2 mm of shearing. An initial elastic zone is typified by small displacements, linearly proportional to stress. This is followed at 85–95% of peak strength by a transition zone, and at peak, by a third zone in which sliding is fully underway.

11.4.2.4 Ongoing research. The behavior of rock joints is, at the time of writing, a very popular topic of research, but as yet few data are available and the various models proposed have yet to be explored in detail. Related studies include those of Swan (1985a,b), Zongqi (1985), and Swan and Zongqi (1985). These further describe the model based on elastic Hertzian theory for predicting shear and normal stresses, dilation, closure, joint stiffness, and hydraulic conductivity. Bandis et al. (1983, 1985) and Barton et al. (1985) have proposed a comprehensive numerical model of rock joint behavior, expressed in terms of joint roughness coefficient, joint wall compressive strength, and the initial conducting aperture. It simulates effective normal stress and scale-dependent coupling of shear stress, shear displacement, dilation, and closure. Equations are given relating changes in fluid conductivity to both normal and shear deformation.

References

Bandis, S. C., N. R. Barton, and M. Christianson: "Application of a New Numerical Model of Joint Behavior to Rock Mechanics Problems," *Proc. Int. Symp. Fundamentals of Rock Joints* (Björkliden, Sweden, 1985), pp. 345–355.

———, A. C. Lumsden, and N. R. Barton: "Experimental Studies of Scale Effects on the Shear Behaviour of Rock Joints," *Int. J. Rock Mech. Min. Sci. Geomech. Abstr.*, **18** (1), 1–21 (1981).

———, ———, and ———: "Fundamentals of Rock Joint Deformation," *Int. J. Rock Mech. Min. Sci. Geomech. Abstr.*, **20**, 249–268 (1983).

Barton, N. R.: "A Relationship between Joint Roughness and Joint Shear Strength," *Proc. Symp. Rock Fracture* (Nancy, France, 1971), paper 1-8.

———: "Review of a New Shear Strength Criterion for Rock Joints," *Eng. Geol.*, **7** (4), 287–332 (1973).

———: "The Shear Strength of Rocks and Rock Joints," *Int. J. Rock Mech. Min. Sci. Geomech. Abstr.*, **13**, 255–279 (1976).

———, S. Bandis, and K. Bakhtar: "Strength, Deformability, and Conductivity Coupling of Rock Joints," *Int. J. Rock Mech. Min. Sci. Geomech. Abstr.*, **22**, 121–140 (1985).

——— and V. Choubey: "The Shear Strength of Rock Joints in Theory and Practice," *Rock Mech.,* **10,** 1–54 (1977).

Bowden, R. K., and J. H. Curran: "Time-Dependent Behavior of Joints in Shale," *Proc. 25th U.S. Symp. Rock Mech.* (Evanston, Ill., 1984), pp. 320–327.

Brown, E. T., L. R. Richards, and M. V. Barr: "Shear Strength Characteristics of the Delabole Slates," *Proc. Conf. Rock Eng.* (Newcastle, England, 1977), pp. 33–51.

Byerlee, J. D.: "Brittle-Ductile Transition in Rocks," *J. Geophys. Res.,* **73,** 4741–4750 (1968).

Cording, E. J.: "Shear Strength on Bedding and Foliation Surfaces," in *Rock Engineering for Foundations and Slopes, Proc. ASCE Speciality Conf.* (Boulder, Colo., 1976), vol. 2, pp. 172–192.

Coulson, J. H.: "Shear Strength of Flat Surfaces in Rock," in *Stability of Rock Slopes, Proc. 13th U.S. Symp. Rock Mech.* (Urbana, Ill., 1972), pp. 77–105.

Curran, J. H., and P. K. Leong: "Influence of Shear Velocity on Rock Joint Strength," *Proc. 5th Int. Cong. Rock Mech.* (Melbourne, Australia, 1983), vol. A, pp. 235–240.

Denby, B., and M. J. Scoble: "Quantification of Power Law Indices for Discontinuity Shear Strength Prediction," *Proc. 25th U.S. Symp. Rock Mech.* (Evanston, Ill., 1984), pp. 475–482.

Dight, P. M., and H. K. Chiu: "Prediction of Shear Strength of Joints Using Profiles," *Int. J. Rock Mech. Min. Sci. Geomech. Abstr.,* **18,** 369–386 (1981).

Dinis da Gama, C.: "Interactive Graphics in the Back-Analysis of Slope Failures," *Proc. 5th Int. Cong. Rock Mech.* (Melbourne, Australia, 1983), vol. C, pp. 53–60.

Franklin, J. A.: "A Direct Shear Machine for Testing Rock Joints," *ASTM Geotech. Testing J.,* **8** (1), 25–29 (1985).

———, J. Manailoglou, and D. Sherwood: "Field Determination of Direct Shear Strength," *Proc. 3d Int. Cong. Rock Mech.* (Denver, Colo., 1974), vol. IIA, pp. 233–240.

Goldstein, M. A., B. Goosev, N. Pyrogovsky, R. Tulinov, and A. Turovskaya: "Investigation of Mechanical Properties of Cracked Rock," *Proc. 1st Int. Cong. Rock Mech.* (Lisbon, Portugal, 1966), vol. 1, pp. 521–524.

Goodman, R. E.: "The Deformability of Joints," in *Determination of the in situ Modulus of Deformation of Rock,* STP 477 (Am. Soc. Test. Materials, Philadelphia, Pa., 1970), pp. 174–196.

———: "The Mechanical Properties of Joints," *Proc. 3d Int. Cong. Rock Mech.* (Denver, Colo., 1974), vol. IA, pp. 127–140.

Hassani, F. P., and M. J. Scoble: "Frictional Mechanism and Properties of Rock Discontinuities," *Proc. Int. Symp. Fundamentals of Rock Joints* (Björkliden, Sweden, 1985), pp. 185–196.

Howing, K. D., and H. K. Kutter: "Time-Dependent Shear Deformation of Filled Rock Joints—A Keynote Lecture," *Proc. Int. Symp. Fundamentals of Rock Joints* (Björkliden, Sweden, 1985), pp. 113–122.

Hungr, O., and D. F. Coates: "Deformability of Joints and Its Relation to Rock Foundation Settlements," *Can. Geotech. J.,* **15** (2), 239–249 (1978).

ISRM: "Suggested Methods for Rock Characterization, Testing, and Monitoring;" ISRM Commission on Testing Methods, E. T. Brown, Ed. (Pergamon, Oxford, 1981), 211 pp.

Kulatilake, P. H. S. W., and S. Ouyiang: "Probabilistic Modelling of Shear Strength of Rock Discontinuities Using Direct Shear Test Data," *Proc. 27th U.S. Symp. Rock Mech.* (Tuscaloosa, Ala., 1986), pp. 112–120.

Ladanyi, B., and G. A. Archambault: "Simulation of Shear Behavior of a Jointed Rock Mass," *Proc. 11th U.S. Symp. Rock Mech.* (Am. Inst. Min. Eng., New York, 1970), pp. 105–125.

Lama, R. D., and V. S. Vutukuri: "Handbook on Mechanical Properties of Rocks," *Series on Rock and Soil Mechanics,* vol. 4 (Trans. Tech. Pub., Rockport, Mass., 1978), 515 pp.

Locher, H. G., and U. G. Rieder: "Shear Tests on Layered Jurassic Limestone," *Proc. 2d Int. Cong. Rock Mech.* (Belgrade, Yugoslavia, 1970), vol. 2, paper 3-I, 5 pp.

McMahon, B. K.: "Some Practical Considerations for the Estimation of Shear Strength of Joints and Other Discontinuities," *Proc. Int. Symp. Fundamentals of Rock Joints* (Björkliden, Sweden, 1985), pp. 475–485.

Mogi, K.: "Pressure Dependence of Rock Strength and Transition from Brittle Fracture to Ductile Flow," *Bull. Earthquake Res. Inst.,* **44,** 215–232 (1966).

Newland, P. L., and B. H. Allely: "Volume Changes in Drained Triaxial Tests on Granular Materials," *Geotechnique* (London), **7,** 17–34 (1957).

Patton, F. D.: "Multiple Modes of Shear Failure in Rock," *Proc. 1st Int. Cong. Rock Mech.* (Lisbon, Portugal, 1966), vol. 1, pp. 509–513.

Plesh, M. E.: "Constitutive Modelling of Rock Joints with Dilation," *Proc. 26th U.S. Symp. Rock Mech.* (Rapid City, S.D., 1985), vol. 1, pp. 387–394.

Pratt, H. R., A. D. Black, W. S. Brown, and W. F. Brace: "The Effect of Specimen Size on the Mechanical Properties of Unjointed Diorite," *Int. J. Rock Mech. Min. Sci.,* **9,** 513–529 (1972).

Reeves, M. J.: "Rock Surface Roughness and Frictional Strength," *Int. J. Rock Mech. Min. Sci. Geomech. Abstr.,* **22,** 429–442 (1985).

Stimpson, B.: "A Suggested Technique for Determining the Basic Friction Angle of Rock Surfaces Using Core," *Int. J. Rock Mech. Min. Sci. Geomech. Abstr.,* **18** (1), 63–65 (1981).

—— and G. Walton: "Clay Mylonites in English Coal Measures. Their Significance in Open Cast Slope Stability," *Proc. 1st Int. Cong., Int. Assoc. Eng. Geol.* (Paris, France, 1970), vol. 2, pp. 1388–1393.

Sun, Z., C. Gerrard, and O. Stephansson: "Rock Joint Compliance Tests for Compression and Shear Loads," *Int. J. Rock Mech. Min. Sci. Geomech. Abstr.,* **22,** 197–213 (1985).

Swan, G.: "Determination of Stiffness and Other Joint Properties from Roughness Measurements," *Rock Mech.,* **16,** 16–38 (1985*a*).

——: "Methods of Roughness Analysis for Predicting Rock Joint Behavior," *Proc. Int. Symp. Fundamentals of Rock Joints* (Björkliden, Sweden, 1985*b*), pp. 153–161.

—— and S. Zongqi: "Prediction of Shear Behaviour of Joints Using Profiles," *Rock Mech.,* **18,** 182–212 (1985).

Underwood, L. B.: "Chalk Foundations at Four Major Dams in the Missouri River Basin," *Trans. 8th Int. Cong. Large Dams* (Edinburgh, U.K., 1964), R.2, Q.28, pp. 23–47.

Withers, J. H.: "Sliding Resistance along Discontinuities in Rock Masses," Ph.D. thesis, Univ. of Illinois, Urbana (1964).

Yoshinaka, R., and T. Yamabe: "Joint Stiffness and the Deformation Behaviour of Discontinuous Rock," *Int. J. Rock Mech. Min. Sci. Geomech. Abstr.,* **23** (1), 19–28 (1986).

Zongqi, S.: "Asperity Models for Closure and Shear," *Proc. Int. Symp. Fundamentals of Rock Joints* (Björkliden, Sweden, 1985), pp. 173–183.

Monitoring

12.1 Introduction

12.1.1 Role of instrumentation

12.1.1.1 Definitions. For purposes of this chapter, *monitoring* is defined as the surveillance of ground behavior, either visually or with the help of instruments. Monitoring and testing are similar in that both use instruments. However, monitoring measures the natural behavior of the ground, whereas testing measures induced behavior, such as strain when a specimen is compressed. Contrast, for example, the measurement of tunnel convergence during construction (which is monitoring), with the identical measurements made while pressurizing a tunnel to determine rock mass deformability (testing).

Simple visual inspection is enough on many projects: the human eye is a portable, versatile, and perceptive instrument. However, it is not a measuring device, nor can it remain vigilant for long. Instrumentation assists greatly in the more demanding forms of long-term monitoring, and in the quantitative recording of information, and also at inaccessible locations such as down drillholes. Even when instruments are used, they should not be used alone. Visual inspection figures prominently in any well-designed program.

Monitoring is concerned with changes rather than absolute values. For example, absolute levels and locations are measured when surveying a network of monuments, but only the movements of the monuments are of interest when using the same or similar instruments later in a monitoring role. Stress and groundwater conditions are determined for the first time by testing, and changes thereafter may be monitored by leaving the same instruments in place and taking repeated readings.

The topic of instrumentation is too broad to be covered here in anything but outline. Details such as tabulated comparisons between instruments are given in the recent and very thorough book by Dunnicliff (1988), who assisted with a comprehensive review of the present chapter. Also suggested for background reading are Dunnicliff (1982), Hanna (1985), and Franklin (1977*a*, *b*; 1989).

12.1.1.2 Applications. Monitoring has at least six distinct and important uses in engineering and mining projects, from an initial role in site investigation to a long-term application in the surveillance of completed works:

1. To investigate failures and ongoing instability. Monitoring of landslides and areas of subsidence, for example, helps identify the mechanisms and rates of movement, gives essential data for back-analysis, and assists in planning remedial work.

2. To obtain data for design. Many projects at the design stage require data obtained by monitoring water table fluctuations, ground movements, and seismic events. Data on rock mass deformability and in situ stresses, needed for the design of underground works, can be obtained by monitoring and back-analysis of trial excavations.

3. To verify the data and assumptions used in design. Movements and pressures on support systems are checked during construction and compared with predictions, to give warning of adverse or unpredicted behavior, and to allow timely remedial action. Based on the measurements, the support and excavation methods can be adjusted for economy or to enhance stability.

4. To protect workers during construction. Safety precautions include instrumentation to give warning of rockbursts, roof falls, and landslides, hazardous buildup of groundwater pressures, and overstressing of cable anchors and rockbolts.

5. To control and evaluate ground treatment. Temperatures and movements are measured as a check on ground freezing operations; groundwater pressures and inflow are monitored in works where the ground is being grouted or drained.

6. To check on long-term stability and impact of works on the environment. Checks are often needed to safeguard the public, to monitor groundwater quality near waste storage caverns, patterns of subsidence above mining areas, and potential deterioration of tunnel support systems.

Sometimes a program can be designed to fulfill two or more of these objectives. When instruments are installed well ahead of construction, they can, at least in principle, be used to provide data for design, to check safety during construction, to check the validity of the design itself, and subsequently to determine whether there is a need for further ground support. In practice, long-term instrumentation is usually of more robust design than short-term, and may differ in other respects. By trying to make one set of instruments serve all purposes, there is a danger that it will serve none.

Rock slopes and tunnels cannot be designed with the same degree of precision and reliability as can engineering structures in concrete or steel. Rock properties are variable and difficult to measure, and the design models are not always realistic. One way to overcome these uncertainties is to adopt a conservative approach with correspondingly high factors of safety. Often a better one is to recognize that some design work must be done during construction, with the benefit of observations of the actual rock conditions and the records obtained by monitoring. This "observational method" is no less conservative when properly applied, and is usually less expensive (Peck, 1969). Monitoring costs tend to be more than offset by savings in other areas.

12.1.2 Basic methods of measurement

12.1.2.1 Instrument components.
A monitoring instrument usually consists of three components: a *transducer* or *sensor* to measure the property of interest; a *transmitting system,* such as rods, electric cables, or telemetry devices to transmit the information to the readout location; and a *readout unit,* such as a dial gauge, to give a digital or graphic display of the measured quantity. There are many variations on this theme; for example, the sensor may be fixed in the ground or incorporated in a portable probe to allow measurements at more than one location. The readout may consist of a display only, or may include facilities for recording data on paper, magnetic tape, or disk. The recording may be continuous, or in response to a command signal, or at preset intervals.

The readout may be located close to each instrument station, or it may be remote, transmitting the readings to a central instrumentation house by a pneumatic, hydraulic, or electric transmitting system. For example, the Radiofor by Telemac (a single-position extensometer with an induction-type sensor) is used to measure tunnel convergence and sometimes cliff face movements. Its signal is transmitted by radio. There are no leads to the instrument, a very useful feature in these applications.

Instruments in common use are not nearly as complex as they might at first appear from the many makes and models, and rely on just a few simple measuring sensors. For example, the electric resistance strain gauge (Fig. 9.2) can be used to measure not only strains directly, but also water and ground pressures and the tensions in rockbolts, by including it as part of an appropriately designed measuring transducer.

12.1.2.2 Optical systems. Apart from the human eye itself, the best known optical instrument is the telescope, which although not a measuring device is an essential component of traditional survey theodolites and levels (Fig. 12.1). Magnification of a graduated scale is a direct and simple way of measuring movements to an accuracy often better than 1 mm over substantial distances. The accuracy can be improved using optical levers (rotating mirrors) attached to the moving surface, or by using ruled grids, diffraction gratings, and holographic methods that develop patterns of fringes in response to very small movements (Sec. 9.1.2.4). These devices are more often used in laboratory testing. A similar optical interferometry principle is employed over much greater distances by the Mekometer electro-optic distance measuring device, which compares transmitted and reflected laser beam signals and measures distances in terms of the wavelength of the light source.

Figure 12.1 Leveling, the most common method of the surveyor, provides a simple means of measuring vertical movements. Tunnel in Riyadh, Saudi Arabia.

The principle of photoelasticity is still employed in a few instrument applications. Certain kinds of transparent materials are stress-birefringent: they exhibit a pattern of colored fringes when stressed and viewed in polarized light. The number of fringes and their color are proportional to the magnitude of the applied load (Sec. 7.3.1.4). In the photoelastic rockbolt load cell, the load in the rockbolt is carried across the diameter of a glass disk mounted between steel loading platens. To take readings, the disk is illuminated with polarized light and viewed through a polaroid analyzer, counting the number of fringes.

12.1.2.3 Mechanical systems. Mechanical systems for measuring movement, termed *displacement* or *strain gauges,* include a simple steel measuring tape, wire, or rod fixed to the rock at one end and in contact with a dial gauge micrometer at the other. Mechanical load cells are available based on the principle of the proving ring, where the distortion of a ring or cylinder of steel is measured with a dial gauge and is proportional to the applied load.

Mechanical systems are usually the most reliable, being simple and free from drift and other sources of inaccuracy that often impair the performance of more complicated devices. Their main disadvantage is that they do not lend themselves to remote reading or to continuous recording, which are sometimes essential. Mechanical extensometers cannot, for example, be used readily in the crown of a large underground excavation because of inaccessibility. They can, however, be used during construction while access remains available, and can be converted later to an electric remote-reading system.

12.1.2.4 Pneumatic/hydraulic systems. The Bourdon tube pressure gauge, an early form of hydraulic transducer, was used in the steam engines of the last century, and is still the key component in most types of gauge used to measure oil, water, or gas pressure both in the laboratory and in the field. It operates on the principle that pressure acting in a curved tube, usually of brass or steel, tends to straighten the tube. The amount of straightening is magnified and displayed by a lever and pointer system, and is proportional to the pressure being measured. Accuracy is usually 1% of full-scale, or 0.25% of full-scale in the case of *standard test gauges.*

The pneumatic/hydraulic diaphragm transducer (Fig. 12.2) appears in a number of systems, including those designed for measuring water pressure, rock and soil pressure, settlement of rock structures, and loads on cable anchors. In each of its various applications the transducer is identical and, in its most common form, operates as follows.

The actual measured quantity is fluid pressure, which acts on one

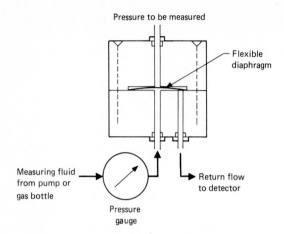

Figure 12.2 Pneumatic/hydraulic diaphragm transducer.

side of a flexible diaphragm made of steel, rubber, or plastic. Twin tubes, which may be up to 500 m long, connect the remote readout instrument to the other side of the diaphragm. To take readings, air, nitrogen, carbon dioxide, or hydraulic oil pressure is supplied from the readout unit through one of the tubes. When the supply pressure is sufficient to balance the pressure to be measured, the diaphragm acts as a valve and deflects to allow flow along the return line to a detector in the readout unit. The balance pressure, then equal to the pressure to be measured, is recorded by a Bourdon or similar pressure gauge in the readout.

The pneumatic/hydraulic diaphragm transducer may be used directly as a piezometer, simply by protecting the measuring side of the diaphragm by a porous plastic or ceramic element, which permits access of fluid pressure but not of earth or rock pressure. It may be used as a pressure cell by connecting the measuring side of the diaphragm to a flat jack, or as a settlement gauge by connecting the measuring side of the diaphragm to a liquid-filled U tube and reservoir, the reservoir being embedded in the structure to be monitored so that settlements are proportional to the head of liquid at the transducer location.

12.1.2.5 Electric systems. Most electric instruments employ electric resistance strain gauges, described in a laboratory application in Sec. 9.1.2.2. They consist of a zigzag wire or, more often, its equivalent in etched metal foil on a plastic backing. The strain gauge is fixed to a rock, steel, or concrete surface by an adhesive and measures strain in the surface to which it is attached. Extension is accompanied by a proportionate increase in electric resistance, measured by a resistance bridge readout calibrated directly in terms of resistance, or in terms of strain, load, or pressure. There are a variety of strain gauging techniques; for example, gauges may be connected to the bridge singly or in pairs so as to compensate for bending or temperature effects.

Many designs of load cell are based on the electric resistance strain gauge principle whereby four strain gauges are cemented to the surface of a steel column or cylinder. The load in the cylinder, applied by a rockbolt or by a steel rib, for example, results in a proportional strain in the steel cell which is measured as a change in the resistance of the gauges. In pressure transducer applications the strain gauges are cemented to the surface of a flexible diaphragm or tube. Fluid pressure acting on the opposite face of the diaphragm or within the tube results in a deflection or strain. Most types of drillhole gauge for measuring ground stresses by overcoring make use of resistance strain gauges, either applied to the rock or cemented to a flexible metallic cantilever or diaphragm to measure its deflection.

A great many other types of electric transducer are in use, the most common of which are probably the *direct current differential transformer* (DCDT) and the *linear variable differential transformer* (LVDT), which operate on an inductive principle. Wire-wound variable resistors (rheostats, linear and rotary potentiometers) monitor strains in some applications, and variable capacitors (proximity transducers) have the singular advantage of being able to monitor displacements without actual contact with the surface being measured.

12.1.2.6 Vibrating wire systems. The tension in a vibrating wire is proportional to the square of its natural frequency of vibration, so if this frequency is measured, the tension in the wire can be found (Fig. 12.3). The vibrating wire load cell is a typical example of the use of such a device. A stainless-steel wire running down the central bore of a steel tube is stretched and held taut by nonslip clamps attached to the tube ends. The load to be measured, applied to the axis of the tube, causes a reduction of tension in the wire proportional to the applied load. An electromagnet plucks the wire, detects its frequency of vibration, and transmits this to a display.

Vibrating wire piezometers work similarly, except that the wire is now attached at one end to a flexible diaphragm in contact with the fluid whose pressure is to be measured. Vibrating wire transducers can also be used to measure settlement, and more directly to monitor strains in steel or concrete supports. They have the advantage of measuring a frequency count that can be transmitted for long distances without distortion of the signal, even through heavy background

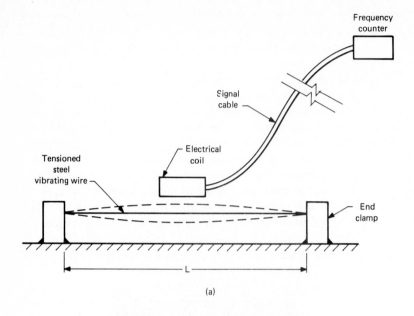

(a)

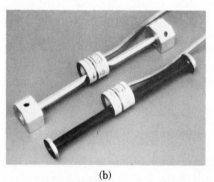

(b)

Figure 12.3 Vibrating wire strain gauge. (a) Schematic of measuring principle. (*Dunnicliff, 1988.*) (b) Vibrating wire strain gauges. (*Courtesy of Geokon Inc., Lebanon, N.H.*)

noise. In contrast, the output from most other electric types of transducers is a current or voltage, which can be attenuated and distorted much more easily when transmitted.

12.2 Monitoring Instruments

12.2.1 Overview

Monitoring instruments can be conveniently subdivided into five main classes: for quasi-static movements; for dynamic movements (vibrations) from rockbursts, blasts, and earthquakes; for groundwater pressures and fluctuations in the groundwater table; for changes in rock stress; and for rock pressures and loads on anchors and other support systems.

The International Society for Rock Mechanics is drafting suggested procedures for carrying out each of these various monitoring activities. Several of these *Suggested Methods* have already been published (ISRM, 1981).

12.2.2 Monitoring of movements

The large movements associated with the instability and failure of engineering structures are nearly always preceded by smaller ones that can be detected by sensitive instruments. Therefore, movement monitoring gives the most direct and useful measurement of impending instability, and is the most commonly employed type of monitoring.

Delay intervals between occurrence and detection of movement, and between detection and collapse of the structure depend on the characteristics of the ground and on the sensitivity of the instruments. In most circumstances, however, a warning period of from several days to several weeks can be achieved.

12.2.3 Geodetic surveying methods

12.2.3.1 Leveling, triangulation, and offset measurements. Conventional surveying techniques provide the simplest and often the most reliable means of measuring the settlement of structural foundations, earth and rock slopes, and embankments, and the pattern of subsidence above an advancing tunnel (Hedley, 1972; Davis et al., 1981). Large areas of ground can be covered quickly, although the measuring accuracy is usually limited to 1 or 2 mm. Lines of sight must be available. Accuracy is adversely affected by atmospheric temperature and pressure variations, particularly over longer measured distances. Lines of sight located in zones of atmospheric turbulence, such as along the crown of a tunnel, give rise to severe problems. Measurements across an open-pit mine are usually made early in the morning when turbulence is least.

Vertical movements can be conveniently monitored by precise leveling (Fig. 12.1). Taking precautions, the standard deviation of measurements over a 1-km line of levels can be held to within 1 mm. Measurements with this high a precision require a level with a parallel-plate micrometer and a telescopic magnification of at least × 40. Leveling rods of Invar are equipped with supporting legs, a circular level, and two scales to permit double readings. Readings are referred

to at least two fixed benchmarks outside the zone of expected movement. The benchmarks and intermediate leveling stations are sturdily
constructed and anchored below the zone of superficial frost action.
Lengths of sightings should be close to equidistant and should not exceed 25 m.

Horizontal movement control is usually less precise, more time-
consuming, and more expensive than vertical control by leveling. Co-
ordination of a network of reference points in the past entailed repeated triangulation using the theodolite, but nowadays this is more
often achieved by trilateration using electro-optic distance measuring
devices as discussed below. Horizontal displacements can also be measured by tape and transit lines, using a theodolite or laser system to
observe departures from colinearity of a line of targets. This procedure
is satisfactory over short distances and where there is access for measurements using an Invar tape.

12.2.3.2 Electro-optic distance measuring instruments. Electro-optic
distance measuring (EDM) instruments employ a modulated light or
laser beam projected onto an array of reflecting targets fixed in the
ground. Distances between the instrument and each target are calculated from the time taken for the light beam to travel from the instrument to the target and back. Target coordinates may then be determined by trilateration calculations. The principal advantage of the
electro-optic method over conventional surveying alternatives is that
many observations can be made much more quickly.

Precise measurements over distances greater than a few tens of
meters (measurements of up to 60 km are possible) require targets in
the form of retrodirective optical reflectors (glass "corner cubes" usually 60 mm in diameter), an expensive item in the survey. These are
best fixed permanently to the ground, but for economy, one or two corner cubes can be moved from place to place and fixed to a reproducible
mounting system, if walking over the area of potential movement is
considered safe. For shorter distances, which are more common in engineering works, it is often sufficient to use the cheaper alternatives
of truck reflectors or adhesive reflecting tape. Problems can develop if
these cheaper targets rotate even slightly from the line of sight, or if
they become coated by dust.

The precision of electro-optic distance measuring instruments is
usually between 1 and 10 mm, depending on the make and model of
the instrument. The Mekometer achieves a measuring precision on
the order of 0.1 mm using a modulated light beam so that distance is
computed in comparison with the wavelength of the light as opposed
to its travel time. This instrument, however, is several times more expensive than its competitors.

12.2.3.3 Photogrammetry. Photogrammetric methods of surface contouring are usually much less precise than either conventional surveying or electro-optic distance measurements, but have the great advantage of covering a complete field of view rather than a set of prelocated targets. Hence one does not need to forecast the locations or directions of potential movements. A sequence of photographs taken at suitable time intervals can be compared to detect movements wherever these might develop. Inaccessible faces can be surveyed without interfering with construction. Ground topography can be quantified, as a basis for engineering geological mapping, and later will permit a back-analysis of slides and rockfalls.

Contour precision is inversely proportional to the distance between the camera and the ground surface, so in monitoring applications, ground and not air photographs are used. Photographs at an object distance of 100 m, for example, when measured in a stereocomparator, permit a measuring precision better than 20 mm in the plane of the photograph, and better than 30 mm in the perpendicular direction. This may be sufficient when attempting to record the progress of major landslide movements over extensive areas, for which purposes a photographic method may be ideal. A much greater precision can be obtained using close-up photographs, but the field of view is smaller and more photographs are needed. The stones of Stonehenge, for example, have been contoured photogrammetrically with a contour interval of just a few millimeters.

Surveying and contouring of absolute positions requires a control survey using conventional methods to locate selected points in the photograph. This can be avoided if movement measurements alone are sufficient, provided that the camera stations are stable and permanent.

12.2.4 Other surface and near-surface measurements

12.2.4.1 Tape extensometers. Tape extensometers measure changes in the distance between a pair of rock anchors. Often called *convergence gauges*, their most common application is underground to measure the inward "convergence" of the walls of a cavern or tunnel (Fig. 12.4), but movements can also be monitored between surface-mounted anchors, such as along the face of a rock slope.

The most common type of convergence meter consists of a stainless-steel tape or wire with a tensioning spring and a mechanical measuring device, usually portable to measure between any number of pairs of fixed targets. Tapes or wires can also be installed permanently, anchored at one end, and tensioned by a freely suspended weight or

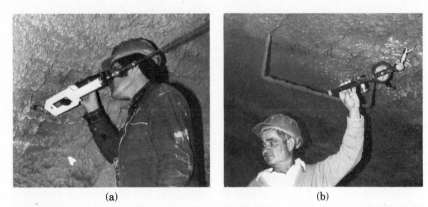

(a)   (b)

Figure 12.4 Convergence measurements in a tunnel; Barcelona, Spain. Instrument by Interfels, Austria. (*a*) Tensioning and reading device. (*b*) Tape reel and pivoting ball attachment.

spring at the other. When selecting a convergence instrument, particular attention should be given to tension reproducibility, which should usually be sufficient to give an overall measuring accuracy of about 0.1 mm. In the Interfels convergence meter, a typical instrument, the tape is punched with holes every 10 cm to give a coarse distance reading. A 1-cm adjustment is also provided, and parts of a centimeter are measured by a dial gauge.

Convergence measurements in excavations up to about 5 m in diameter can be made using a convergence meter with a rod or bar in place of the steel tape, avoiding the need for a tensioning device. The rod may be of fixed length or telescopic, and is fitted with a micrometer or a dial indicator. Rod gauges are more precise for measurements of smaller distances, but are inconvenient in greater lengths when their flexibility leads to a loss of accuracy.

Targets for both rod and tape convergence meters consist of a steel or aluminum anchor bolt secured in the ground by an expanding shell anchor or by grouting. Depending on the design of the instrument, they have a spherical or flat reference surface or a hook to connect with the convergence gauge. They need to be protected against blast damage, by recessing in the walls of a tunnel, and sometimes also against vandalism.

12.2.4.2 Measurements at cracks, joints, and faults. A record of the pattern of cracking in a concrete structure, a shotcrete tunnel lining, or in rock around the crest of an open-pit mine gives very useful information on the mechanisms and directions of movements. Cracks should be marked with spray paint, so that at each subsequent date of observation it will be possible to recognize new cracks and to measure the elongation of old ones. Crack patterns and the history of cracking should be recorded on plans and cross sections just as carefully as when recording more complex types of measurement.

A further step is to fix "telltales" across selected cracks, such as glass microscope slides attached to the rock or concrete surface with an epoxy adhesive. Cracked telltales confirm the continuation of movement and give a useful indication of movement direction and amount. Cement or plaster telltales can be more convenient than glass plates when the surface is irregular, but care is then needed to distinguish between cracks caused by ground movements and those that may be the result of shrinkage of the plaster.

When greater precision is needed, simple portable instruments are available to measure crack expansion and contraction (Fig. 12.5). The Demec gauge (the name is short for "demountable mechanical") was invented by Morice and Base (1953). It consists of an Invar steel bar with two conical gauge points, one fixed and the other pivoting on a knife edge. The dial gauge and lever system measures distances, typically of about 150 mm, between pairs of steel stud targets. Measuring precision, at least under laboratory conditions, is on the order of 3 microstrain (about 1 μm over a 150-mm gauge length).

When the surface is firm, stainless-steel disks with a central indentation are cemented to either side of the crack. Deeply embedded an-

Figure   12.5 Demec   portable gauge, monitoring cracks in primary   crusher;   Milton   quarry, Ontario, Canada.

chor targets are better in soils and weathered or weak rocks where the surface is friable. Target separation and embedment must be sufficient to ensure that the bolts remain securely fixed in the rock or soil. The same bolts can be used with various types of fixed-in-place electric displacement or strain gauges, as discussed below. A portable clinometer can be used to measure differential vertical movements to supplement measurements of changes in distance between targets. The clinometer consists of a spirit level bubble mounted on a bar with micrometer adjustment at one end.

12.2.4.3 Strain gauges. Flanged vibrating wire strain gauges (Fig. 12.3) can be embedded in concrete to measure strains within a tunnel liner, retaining wall, or concrete dam, and nonflanged types can be spot-welded onto steel ribs or reinforcing bars. Before pouring concrete or spraying shotcrete, the gauges are fixed to either the reinforcement or the formwork, and the signal cables are carried to a terminal panel nearby. After hardening of the concrete, any relative displacement between the ends of the tensioned vibrating wire gives a change in the frequency of vibration that can be detected by the readout instrument. Interpretation is not without problems, because of thermal and moisture effects and possible creep in the gauge and in the concrete.

Strains beneath and within natural slopes and embankment dams are often monitored using a strain gauge of much greater travel. One version consists of a wire-wound linear variable potentiometer in a stainless-steel housing, filled with oil to prevent ingress of moisture. The housing and its sliding plunger are anchored so that any relative movement of the anchors can be recorded as a change in potentiometer resistance. Several potentiometers can be installed in line in a trench to give a complete record of movements along the trench axis.

12.2.4.4 Settlement gauges. By definition, a settlement gauge measures vertical movements only, usually on the principle of the liquid-filled U tube. The level of liquid in one end of a plastic tube is compared with the level in the other, one end being mounted alongside a measuring scale in a stable instrument house, and the other in the structure to be monitored. Water, when used, has to be de-aired to prevent the formation of bubbles and in cold climates should contain an antifreeze additive. The simple U-tube method is convenient when the instrument house can be at very nearly the same elevation as the sensor. The permissible difference in elevation can be increased by employing two liquids in the one tube, such as water and mercury, or by providing a measured and constant back-pressure to the end of the tube in the instrument house.

Another type of "overflow" settlement cell is embedded at the monitoring location and connected to the instrument house by three plastic tubes. Water is introduced down one of the tubes and overflows through an orifice in the cell. The second tube is used to drain surplus water from the cell, and the third to equalize the air pressure in the cell and the instrument house. The level of water in the measuring tube at the instrument house is then equal to the level of the overflow orifice.

The *pneumatic settlement cell* is a U tube filled with an aqueous solution, with one end at the monitoring point and the other at the gauge house. The head (elevation) difference between ends of the tube is measured by injecting a gas into a pneumatic transducer (Fig. 12.2 and Sec. 12.1.2.4).

Electric settlement devices use the same U-tube principle, replacing the pneumatic transducer with an electric one, usually of the vibrating wire type. Pneumatic or electric transducers enable measurements to be made in spite of a considerable difference in elevation between the instrument and the gauge house.

The settlement gauges described are installed permanently at selected points in the zone of potential movement. Sometimes more informative and economical are the alternative *traveling probe* types of instruments, which allow the engineer to obtain not just isolated readings, but a continuous profile of settlements beneath a highway embankment or a dam foundation. One or several flexible plastic guide tubes are embedded in the foundation, open at each end. The probe is inserted and pulled or pushed along the tubes, taking readings. The operating principle may be any one of those described; commonly it is overflow or a sensitive pressure transducer connected to a fluid column.

12.2.4.5 Tiltmeters. Tiltmeters are very sensitive instruments for determining inclination or the change in elevation between two points. Resolutions down to 1 second of arc are possible. The U.S. Geological Survey has continuously recorded solid-earth tidal tilting using long-base liquid-level tiltmeters capable of resolving tilt to 10^{-8} rad (Allen et al., 1973). In geotechnical applications, portable tiltmeters usually make use of servo-accelerometer sensors identical to those in the better known borehole inclinometer instrument described in Sec. 12.2.5.5.

12.2.5 Measurements of movement in drillholes

12.2.5.1 Need for subsurface measurements. Surficial measurements are easier and less expensive than deep ones, but may not be enough. In-depth readings are required, for example, when one needs to locate the sliding surface of a landslide, or to examine ground movements as they spread outward and upward from a tunnel or mine stope. Surficial movements can be misleading in that they may reflect the local and insignificant movement of loose blocks, shallow soil creep, frost action, or disturbance caused by construction traffic. Drillhole instrumentation gives extra, more detailed, and more reliable information. Often drillholes are used to gain access to stable strata, to provide a reference for measurements when stable surface ground is a long way off, if available at all.

12.2.5.2 Fixed extensometers. Extensometers measure movements along the axis of the drillhole (for example, settlement when the hole is vertical) in contrast to inclinometers, which measure movements in directions perpendicular to the hole.

Simple *single-position* extensometers (Fig. 12.6a) usually consist of an end-anchored rockbolt, untensioned, and equipped with a measuring facility at the hole collar. Measurements are by portable dial gauge, or by fixed electric transducer to permit continuous monitoring or automatic warning.

Several imaginatively named commercial warning devices that monitor roof sag in underground mines are variations on the single-position extensometer theme. The Glowlarm has a flexible transparent plastic tube containing two liquids which, when mixed, produce a bright yellow warning light that persists for about a day. The tube is attached by a wire to an anchor in a drillhole. Movement fractures an inner glass tube and triggers the alarm. The Guardian Angel also monitors roof sag. A reflector flag drops when a preset amount of movement has been reached. Visual roof sag bolts consist of a standard mechanically anchored bolt with three bands of reflective tape or paint (green, yellow, and red) at the bolt head, and a polystyrene foam plug in the drillhole collar. A quick glance tells mine personnel whether the roof has moved (Bauer, 1985).

The *multiple-position* extensometer, one of the more common and useful types of instrument, records differential movements of anchor points installed at various depths, from which rods or tensioned wires extend to a measuring instrument at the hole collar (Fig. 12.6b). Usually up to eight, but sometimes more anchors can be monitored in any one hole. Usually the holes are drilled sufficiently deep for the deepest anchor to act as a stable reference. As a further check, movement of the drillhole collar is monitored with respect to an external datum.

In multiple-rod installations, the rods are individually protected by low-friction plastic sleeves, sometimes filled with grease or oil, and are installed side by side in the drillhole. The anchor ends protrude and are fixed to the surrounding rock. The remaining lengths of rod

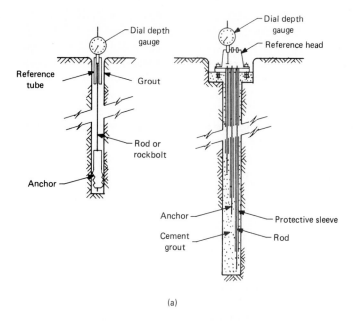

(a)

(b)

Figure 12.6 Drillhole rod extensometer. (*a*) Schematic of single- and multiple-rod installations. (*ISRM, 1981*.) (*b*) Taking readings on three-rod installation above a tunnel, Barcelona, Spain.

are free to move within their sleeves, so measurement at the hole collar gives the anchor movements. Anchoring is achieved either by filling the drillhole with grout, or by a mechanical system such as individual C clips that, by pulling a pin, spring into contact with the rock. Although more difficult, grouting appears a better alternative for reliable measurements in softer rocks and to resist blast vibrations.

Multiple-rod installations require a diameter of hole that increases according to the number of rods, whereas multiple-wire extensometers need only a small drillhole irrespective of the number of wires. The wires are fixed to the rock or soil by mechanical expanding anchors, and must be tensioned by a spring or suspended weight. Measuring precision is limited by loss of tension and the potential for creep and kinking. Multiple-wire instruments are used mainly for economy where considerable movement is expected and some loss of precision can be tolerated, and give good service in such applications as the monitoring of caving above a mine stope (Panek and Tesch, 1981).

12.2.5.3 Probe extensometers. The *multiple-magnet* extensometer uses for reference targets a series of ring magnets that slide along the outside of a plastic guide tube and are fixed to the rock by grout or with mechanical anchors. A probe containing a reed switch is inserted in the guide tube. When it enters the field of a magnet, the reed switch closes, activating a lamp or buzzer in the readout. The depth of the probe and of the magnet target can be measured directly if the probe is suspended by measuring tape. A suspended probe with a single reed switch is used for less precise measurements in vertical holes, such as to monitor soil settlements. Most rock and all inclined hole applications require the greater accuracy of a rod-mounted probe containing two reed switches, which detect a pair of adjacent magnets. Magnet separation is measured by a micrometer at the hole collar. The probe can provide measurements at any number of locations in the hole and in any number of holes. The repeatability of the instrument is in the range ±0.5 to ±5 mm.

Kovari et al. (1979) and Koeppel et al. (1983) describe a portable *sliding micrometer* (Fig. 12.7) that precisely measures the distances between measuring rings at 1-m intervals along a plastic tube. The probe instrument contains a cone that seats on the rings during measurements and can be rotated to pass from one ring to the next. The same sliding micrometer can be incorporated with an inclinometer to determine three-dimensional displacement vectors along the guide tube (Trivec by Solexperts Ltd. of Zurich, Switzerland).

12.2.5.4 Shear strips and time-domain reflectometers. Both of these devices are forms of extensometer in that they can be used to detect the location, but not the magnitude of extension or shear in a drillhole.

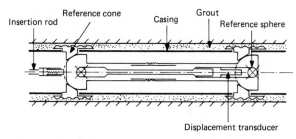

Figure **12.7** Sliding micrometer (Solexperts Ltd., Switzerland). (*After Kovari et al., 1979.*)

The shear strip is a string of electric resistors connected in parallel at regular intervals along the length of a printed circuit conductor, and grouted in the hole. Localized movement breaks the strip. The location of the break is determined from the change in resistance measured at the drillhole collar. Typical applications are to monitor the migration of caving upward from the roof of a mine stope, or the depth of shearing beneath a landslide.

Time-domain reflectometry (TDR), similar in concept to the use of a shear strip but more versatile and less expensive, is an electrical pulse-testing technique originally developed to locate breaks in power transmission cables, which has been adapted for monitoring rock movements. Coaxial cables are installed in drillholes or along tunnels or mine roadways and anchored or grouted to the rock so that any rock movement will pinch or break the cable. Indications are that cable damage occurs at about 2.8 mm of strain per 30-m length of cable between anchors. Cable defects such as crimps, short circuits, or breaks are detected by transmitting voltage pulses from a cable tester. The distance to the cable fault is proportional to the elapsed time between transmission and arrival of a reflected signal. Accuracy is about 2% of the distance between the tester and the cable break. This can be improved by precrimping the cable at 3-m intervals (O'Connor and Dowding, 1984). The crimps distort the signal and act as markers on the arrival waveform, to which any further distortions or breaks may be related.

The cable can continue to be used even though a section of it has been removed. For example, O'Connor and Dowding report one case in which TDR cables were monitored successfully after 20 m of overburden with embedded cables had been removed. For holes deeper than 30 m, TDR cables are less expensive and easier to install than multiple-point extensometers.

12.2.5.5 Inclinometers. Inclinometers are used to measure changes in the inclination or curvature of a drillhole, using either a fixed-in-place or a moving probe system. The readings are interpreted in terms of displacements perpendicular to the drillhole axis.

The most popular version of this instrument, the *probe* inclinometer (e.g., the Slope Indicator by SINCO of Seattle, Wash.; Fig. 12.8), operates by insertion of a probe in the hole each time a set of readings is to be taken. The probe travels along guide grooves in an aluminum or plastic tube grouted to the rock. Inside the probe, the sensing device may be a cantilever pendulum with resistance strain gauges, vibrating wire, or inductive transducers to monitor cantilever deflection, but nowadays will consist more often of a pair of servo accelerometers. Probe inclinometers generally operate in near-vertical holes, although horizontal versions are also available. They can detect differential movements of 0.5–1.0 mm per 10-m run of hole.

A simple form of probe inclinometer, seldom used in rocks because of its limited sensitivity, makes use of a system of go-no-go rods. Close-fit rods are lowered into, or pulled up, a flexible casing until they encounter a bend that impedes their progress. An estimate of the angle of deflection of the casing can be obtained by varying the length and diameter of the rod.

One type of fixed-in-place inclinometer, called a *chain deflectometer,* consists of a chain of pivoted rods. Rotations are measured at the pivots between each pair of rods, by means of resistance strain gauges on flexible steel strips or by inductance or capacitance transducers. Another type uses a series of gravity-sensing transducers (Fig. 12.9).

Horizontal movements can also be measured using *simple-pendulum* or *inverted-pendulum* devices. The simple-pendulum wire is clamped to the top of the structure, and carries at its lower end a heavy weight, which is usually suspended in water or oil to damp oscillations. Horizontal movements are measured at any number of elevations along the wire, using either calibrated scales fixed to plates mounted on the structure, or a vernier microscope sighted onto the wire. The inverted pendulum has the wire anchored at the lower end, usually in stable ground at the base of a vertical drillhole. The upper end is secured to a float in a water tank, which tensions the wire and keeps it vertical. Suspended and inverted pendula are often used to provide a stable datum for geodetic surveying and for routine monitoring of the horizontal movements of large dams.

12.2.6 Monitoring of ground vibrations

12.2.6.1 Overview. Ground movements measured using the methods described develop over hours, days, or weeks. Cyclic variations with periods of seconds or fractions of a second, such as those caused by blasting or by earthquakes, require special dynamic detection and re-

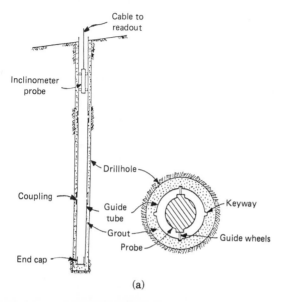

(a)

(b)

(c)

Figure 12.8 Probe inclinometer. (a) Schematic of drillhole installation and probe with its wheels running in guide tube keyways. (b) Monitoring of basement excavation into shale bedrock; Toronto, Ont., Canada (Sinco Digitilt instrument). (c) Monitoring of wall stability at open-pit asbestos mine, Quebec, Canada. (*Bullock et al., 1980.*)

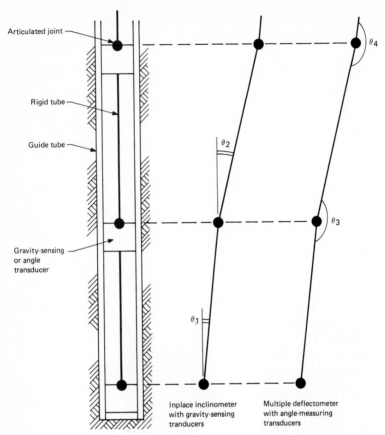

Figure 12.9 Fixed in place inclinometer systems. (*Dunnicliff, 1988.*)

cording devices. Seismic geophone arrays locate the sources of rock noise, monitor landslide movements and rockbursts, and record the waveforms of blast or earthquake tremors, traffic, or machinery vibrations. Strain gauge, piezoelectric, or inductive-type transducers are used depending on the frequency range of interest (Collett and Hope, 1983).

12.2.6.2 Earthquake monitoring. Most large dams are nowadays instrumented to record the magnitude and waveform characteristics of earthquakes that may affect their stability. Automatic triaxial seismographs and very sensitive tiltmeters (i.e., 10^{-9} rad) are available for monitoring ground motions; the recording is triggered by the onset of an earth tremor with amplitude greater than a predetermined minimum. Usually a seismograph consists of a suspended inertial mass that remains stationary with respect to the housing of the instrument. A trace of the oscillation waveform is produced on a chart recorder.

12.2.6.3 Monitoring of machinery vibrations and blasts. Crushing plants and turbines may require monitoring as part of an investigation of foundation damage and stability. Blast vibration monitoring is usually needed as part of a program to optimize blasting sequences, to avoid damage to nearby structures, and to minimize annoyance and inconvenience to local residents. The vibrations are usually of smaller amplitude and higher frequency than those generated by earthquakes. Electromagnetic geophones produce an electric signal proportional to the velocity of ground vibration, and feed this to an oscilloscope or ultraviolet light chart recorder. The criteria and methods are discussed at greater length in Sec. 13.6.2.

12.2.6.4 Rockburst and microseismic monitoring. Vibration monitoring is also used to investigate and predict rockbursts underground and landslides at the surface. Small subaudible noises (called *microseisms* or *acoustic emissions*) are generated early in the development of instability. Listening devices used in an attempt to detect these and to predict locations and times of bursting or collapse comprise, in their simplest form, a detector or geophone, an amplifier, and a counter or chart recorder. A collection of some 10 case studies of microseismic monitoring in Canadian mines together with reviews of the associated instrumentation has been published by CANMET (1984).

Attempts to analyze the characteristics of individual microseismic events and to relate their frequencies and amplitudes to the nature of rock movements have met with only limited success. The Fourier transform is the principal tool employed.

However, a count of the number of events per interval of time (the *microseismic event rate*) has been used successfully as an index to the dynamic activity of the rock mass. Usually the rate increases with the development of rock instability, although it may subside on one or more occasions prior to final collapse. Analysis involves the calculation of such parameters as the event rate, the rms level of the signal, and the distribution of event amplitudes. Continuous data analysis requires fitting an envelope to the peak signals received over an extended time period. An example of the long-term monitoring of an underground natural gas storage reservoir is described in Hardy and Ge (1986).

Source location is the primary goal in most acoustic emission monitoring applications. It allows the delineation of regions of instability within the monitored rock mass, mapping of the progress of unstable regions and areas of stress buildup, objective evaluation of the success of stabilization measures, and investigation of the mechanisms and magnitudes of rockbursts or smaller-magnitude events. Seismic sources can be located in space by an array of geophones distributed around the slope or the underground excavation. The most common method is similar to the trilateration employed for coordinate fixing using electronic distance measuring instruments. Arrival times at three or more geophones are interpreted in terms of stress wave velocities to infer travel distances. In an alternative, P/S-wave, approach the time difference between P- and S-wave arrivals and the polarization of these waves are taken into account (Sec. 9.4.1). Sources can be located with just one three-component geophone (Hardy, 1981).

12.2.7   Monitoring of groundwater pressures

Groundwater is often one of the more important factors controlling the stability of rock structures, and drainage a most effective method for stabilization. Monitoring of water pressures (piezometric monitoring) provides information for design, for control during construction, and for diagnosis and back-analysis in the event of continuing ground movements.

For the sake of completeness, a description of the various types of piezometer instruments for monitoring groundwater pressures was given in Sec. 4.3.1 and will not be repeated here. Piezometric monitoring should not be overlooked when designing the total monitoring program. It often assumes an importance equal to, sometimes even greater than, that of the monitoring of displacements.

12.2.8   Monitoring of stress changes in rock

In Chap. 5 we reviewed methods for measuring the "virgin," or absolute, state of stress in rock, including one-time techniques such as overcoring. Devices that measure virgin stress can also be used to monitor the stress changes that accompany excavation, such as stress redistribution in mine pillars during a room and pillar mining operation, but they are often too delicate or expensive to be left in the ground for prolonged periods and at several locations. They are designed for quite a different, short-term application. For stress change monitoring, simpler, less expensive, but more durable devices are normally used.

12.2.8.1   Drillhole diameter gauges.   This type of gauge permits the calculation of stress changes from the measured changes in the diameter of a drillhole. The drillhole is unstressed, and the rock is usually as-

sumed to behave elastically (i.e., the method is used only in hard or moderately hard rock).

Walton and Worotnicki (1986) compare the conventional CSIRO triaxial stress measurement cell (Sec. 5.4.1.4) with a new *yoke* gauge designed specifically for monitoring stress changes. The drillhole diameter is measured using 1-mm-thick beryllium-copper cantilevers, shaped into the form of a C, or "yoke," with resistance strain gauges mounted on front and back. Three such cantilevers protrude from a thin-walled plastic tube, so that the cantilever tips measure across diameters at 60° intervals around the drillhole circumference. The space surrounding the cantilevers is filled with a silicone moisture-proofing compound. An epoxy cement is extruded between the plastic tube and the wall of the hole to hold the instrument firmly in position and prevent disturbance by blasting.

The yoke system, like the USBM borehole deformation gauge used for stress measurement, has the limitation that stress changes can be calculated only in the plane perpendicular to the drillhole. In contrast, CSIR and CSIRO hollow inclusion cells can be used to measure triaxial states of stress and stress change. However, their durability in long-term applications, for which they were never designed, can be questioned. Over long periods in moist environments, electric insulation is likely to deteriorate and the epoxy body and the cement itself can shrink or expand.

12.2.8.2 Soft inclusion cells. These cells also measure changes in drillhole diameter, but indirectly by the effect of diameter changes on a fluid-filled flat jack or dilatometer. The modulus of rigidity of the "inclusion" is substantially less than that of the rock, hence the term "soft."

One such gauge employs a device similar to the pneumatic/hydraulic cell described earlier. A slot is cut into the rock perpendicular to the direction in which stress changes are to be measured. The flat-jack part of the cell is inserted into the slot and grouted in place, then inflated until the slot is restored to its original width. The cell is then sealed, so that subsequent changes in ground pressure can be recorded as changes in fluid pressure within the cell.

Similar cells can be installed in drillholes, and are the preferred method in soft rocks that tend to squeeze against the cell (Sellers, 1970). Panek and Stock (1964) developed the first soft inclusion cell for monitoring rock stress changes. Similar in concept to the dilatometer described in Sec. 9.2.5, this consisted of a copper flat jack encapsulated before being installed in the drillhole. Later cells such as described by Babcock (1986) use a steel bladder in a concrete cylinder.

This is filled with hydraulic oil connected by tubing to a hydraulic gauge and pressurized by a screw pump. Lu (1984) describes a 37.5-mm-diameter instrument with a combination of one cylindrical and two preencapsulated flat hydraulic pressure cells installed in a single 38-mm-diameter drillhole. Biaxial stresses can be calculated in the plane perpendicular to the hole without knowledge of the elastic modulus of the rock mass. Glycerine or hydraulic oil is forced through steel tubing to expand the copper shell against the drillhole wall.

12.2.8.3 Rigid inclusion cells. Stress change cells designed for drillhole installation in hard rocks are more often of the *rigid inclusion* type, installed in the hole by grouting or wedging. The modulus of the cell is similar to, or greater than, that of the rock (few cells are fully "rigid" in this sense). Changes in rock stress give rise to a proportionate increase in the stress in the inclusion. One type is the photoelastic glass stress meter, consisting of a glass cylinder with a central hole. The cylinder is glued into the drillhole, and stress changes are recorded as a change in colored fringe pattern when the cell is viewed in polarized light.

A vibrating wire stress change meter was developed under contract to the U.S. Bureau of Mines (Hawkes and Bailey, 1973; Hawkes and Hooker, 1974). It consists of a thick-walled stainless-steel cylinder with a vibrating wire across its internal diameter (Fig. 12.10). The cylinder is prestressed against the walls of a 38-mm-diameter drillhole by wedging. Changes in ground stress are measured as changes in the natural frequency of vibration of the wire (Sec. 12.1.2.6). Up to three stress meters can be installed in the same drillhole with the vibrating wires inclined at 45° to each other. Thermistors built into the stress meters allow corrections for temperature change. A similar instrument employing resistance strain gauges and enlarged platens has been modified for use in rock salt (Cook and Ames, 1979; Morgan, 1984), although correct interpretation requires some form of viscoplastic analysis.

12.2.9 Monitoring of loads and pressures on supports

Retaining walls, tunnel liners, anchors, and rockbolts only develop their prescribed working pressures and loads as the rock starts to move against the retaining system. Attempts at predicting these pressures and loads can be unreliable, so monitoring is often specified to check and confirm the actual pressures that are developed.

12.2.9.1 Hydraulic pressure cells. The pneumatic/hydraulic type of pressure cell, such as manufactured by the Glötzl company in Austria,

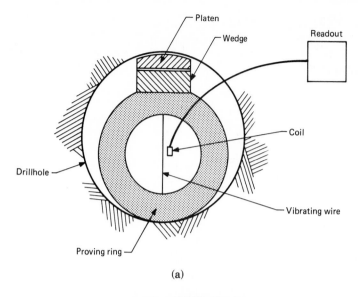

Figure 12.10   Vibrating wire stress change cell. (*a*) Schematic diagram. (*After Dunnicliff, 1988.*) (*b*) Two different sizes of gauge. (*Courtesy of Geokon Inc., Lebanon, N.H.*)

consists of a flat jack connected to a pneumatic or hydraulic transducer (Sec. 12.1.2.4), which in turn is connected by twin flexible plastic tubes to a terminal panel (Fig. 12.11). Various sizes and shapes are available, from small rectangular cells for installation in a thin concrete tunnel lining to large circular flat jacks intended for installation in earth embankments and foundations.

The flat jack consists of two thin sheets of metal, welded around the perimeter and filled with oil or mercury to form a hydraulic pillow. It is connected by a short length of metal tube to the hydraulic trans-

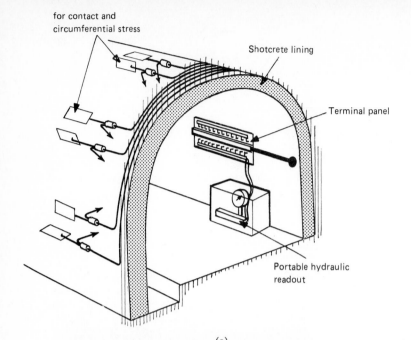

for contact and
circumferential stress

Shotcrete lining

Terminal panel

Portable hydraulic
readout

(a)

(b)

(c)

Figure 12.11 Hydraulic pressure cells. (*a*) Arrangement of a profile of 10 cells in a tunnel, five to measure tangential pressure, and five radial. (*Courtesy of Rock Instruments Ltd., England.*) (*b*) Installing a pair of hydraulic pressure cells in a tunnel crown, Barcelona, Spain. Shotcrete is sprayed to cover the cells and complete the liner. Instruments by Interfels, Austria; project by FH-2, Spain, and BRGM, France. (*c*) Hand-operated readout for pressure cells. (*Courtesy of Rock Instruments Ltd., England.*)

ducer. The complete unit is secured to a rock surface or to concrete reinforcement before pouring or spraying concrete at that location. Care is needed to avoid air pockets and stress raisers, so that the ground pressure perpendicular to the flat jack is equal to the hydraulic pressure in the jack. This pressure is measured in the same way as described for the pneumatic piezometer, except that for higher pressures and over greater lengths of transmission tube, a mixture of oil and kerosene is used as the measuring fluid in place of air or nitrogen.

Compressibility of the cell should be similar to that of the surrounding material, otherwise the stress transfer will be incomplete. Cells for use in soil are usually filled with hydraulic oil, whereas those for use in rock or concrete are filled with mercury. The volume of fluid is kept to a minimum by making the flat jack as thin as possible. When installed entirely in concrete, the cell is provided with a compensating tube to counteract the effects of concrete shrinkage. The concrete initially expands against the cell and then contracts to leave a gap. Crimping of the compensating tube allows the cell to be reinflated by a small amount to reestablish contact with the surrounding concrete.

A further application of this type of cell is in long-wall mining, where bed separation and rock disintegration occur until the roof meets the floor. The amount of load transferred from roof to floor through the broken rock (the *gob*) is of great importance when calculating the stress redistribution to panel boundaries. Maleki et al. (1984) describe attempts to use hydraulic pressure cells to monitor this load transfer. Their cells measured 400 by 400 mm and were protected from collapsing rock by a layer of fine gravel and a piece of old conveyer belt. The hydraulic lines were protected by timber.

12.2.9.2   Electric pressure cells.   These consist of two circular plates of steel, each with a central portion of reduced thickness that forms a flexible diaphragm whose deflection is measured with electric resistance strain gauges or vibrating wires. In the vibrating-wire type of cell, metal pillars are mounted on the internal face at the circle of maximum angular deflection. External pressure causes a small angular rotation of the pillars, between which a vibrating wire is stretched under tension. The two halves are bolted together and waterproofed. Electric leads connect the cell to the remote readout.

12.2.9.3   Load cells.   Load cells (also known as *force transducers* or *dynamometers*) are used to monitor the tension in cable anchors and rockbolts, and the loads in steel ribs, columns, and arches. "Hollow" (center-hole) load cells are used in rockbolting and cable anchoring applications. The bolt or cable passes through the center of the cell, so that the tensile force in the cable is transmitted as a compressive load through the cell to the surface of the rock or concrete (Fig. 12.12).

A simple mechanical cell can be used where access to take readings is available, and in rockbolting applications when large numbers of cells are needed at minimum cost and only moderate accuracy. The sandwich type contains a rubber disk between two metal plates; compression of the rubber is measured with a vernier micrometer. A more reliable but more expensive mechanical cell (accuracy about 0.5%) contains a domed plate between two rigid steel platens. Compression of the dome caused by tension in the bolt or cable, which passes through a central hole, is measured by dial gauge at three locations around the circumference. Cells are manufactured in load capacities of between 100 and 500 kN.

The typical electric load cell consists of one or more metal columns or cylinders, arranged to carry the full load to be measured. Strains are monitored using an electric resistance strain gauge bridge, or a vibrating-wire or an inductive transducer to give a reading proportional to the applied load. Hydraulic load cells measure internal fluid pressure in a sealed hydraulic capsule.

For most designs of load cell to perform reliably, the loads to be measured must act along the axis of the cell. Spherical seatings are needed to prevent the transfer of bending moments that cause incorrect readings. Some designs, however, are more sensitive than others to eccentricity, and some can automatically compensate for at least moderate amounts of bending or eccentricity. Special load cells are available to measure loads simultaneously in two perpendicular directions. These can be installed, for example, beneath the leg of an arch to measure vertical and horizontal components of the transmitted force.

12.2.10   Temperature sensors

Measurements of temperature are required in geomechanics either to permit temperature compensations to be applied to other types of measurements, hence improving their precision, or for more direct reasons. The response of rock to elevated temperatures is being studied in geothermal and radioactive waste storage applications. Measurement is by thermocouple, thermistor, or resistance temperature device (RTD).

The thermocouple operates on the principle that an electric current flows in a closed circuit of two dissimilar metals when one of the junctions is heated to a temperature higher than the other. The measuring junction is referred to as the "hot" junction. High-accuracy thermocouples can measure to ±0.1°C.

(a)

Figure 12.12 Load cells. (*a*), (*b*) Electric resistance strain gauge cell. Different sizes of center-hole load cells. (*Irad Gage, Lebanon, N.H., and Dunnicliff, 1988.*) (*c*) Hydraulic disk load cell. (*d*) Measurement of load beneath a steel rib. (*e*) Measurement of tension in a rockbolt or cable anchor.

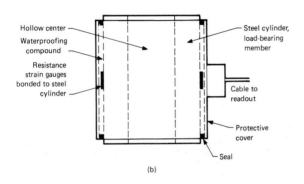

(b)

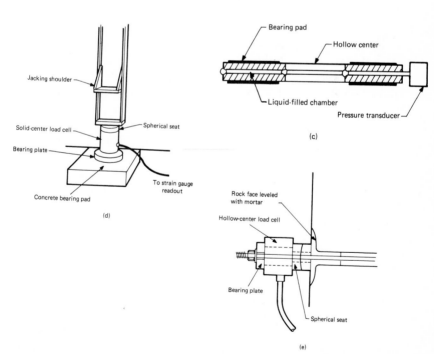

(d)

(c)

(e)

The thermistor is a semiconducting material in the form of a fused bead connected to two leads. It also can detect small changes in temperature.

The RTD makes use of a conducting material whose resistance varies linearly with temperature. Platinum resistance thermometers are protected by stainless-steel pipes with the sensing platinum tip protruding from the end.

12.3 Planning, Installation, and Operation of Monitoring Systems

12.3.1 Planning and design

12.3.1.1 Predicting ground behavior. The requirements for instrumentation location, types, and precision can be very different, depending on whether the anticipated movements, stress, and pressure changes are expected to be small or large, slow or rapid, shallow or deep (Franklin, 1977a). The *locations of instrumented drillholes*, and of instruments within them, are selected taking into account the groundwater, soil, and rock characteristics of the site. The proposed works are drawn up in plan and cross section to show the geology, topography, and geometry of openings. The likely *magnitudes and directions of movements* are evaluated, as are the locations of potential slides or collapses, and the nature of failure, whether it is likely to be by translational movement, bursting, or gradual creep.

These predictions are quantified, if possible, by parametric numerical modeling (Sec. 7.1.3) to cover a range of values for uncertain input data, and to give both optimistic and pessimistic predictions for the movements and the changes in loads, stresses, and water pressures likely to accompany excavation or construction. The calculations show which quantities are the most critical and the most in need of monitoring. Safety considerations will nearly always dictate the measurement of movements, which give the clearest warning of inherent instability. Water pressures, loads, and stress changes may be measured as a further check.

12.3.1.2 Hazard warning levels and contingency plans. *Hazard warning levels* are defined as the magnitudes of movement, water pressure, anchor load, or other observed quantities that constitute a threat to stability or safety. *Contingency plans* are courses of action to be taken in the event that these warning levels are exceeded. When a structure is to be monitored for reasons of safety, the hazard warning levels and contingency plans must be selected and planned in advance. There will hardly ever be sufficient time for discussion when the critical instrument readings become available.

Hazard warning levels can be selected, for example, as some multiple of the elastic displacements predicted by design, or in terms of the movements experienced on a similar earlier project, or as fractions of the tolerances of affected structures. Absolute values of movement are not usually sufficient: velocities or accelerations may be more relevant, or warning levels may be specified in terms of visible structural distress, such as the development of cracking in supports or liners.

For example, displacement measurements have been instrumental in averting major failures in several large underground rock caverns (Cording et al., 1971). Displacements that have exceeded the predicted elastic displacements by a factor of 5 or 10, typically amounting to 10–75 mm, have usually resulted in decisions to modify the methods of support and excavation.

Several different warning levels may be needed, each linked to a contingency plan. For example, the engineer might specify that in a tunnel, a convergence rate of less than 1 mm per week would require the results to be recorded and reported at weekly intervals with no further action. Movements of 1–5 mm per week might call for inspection with a view to increasing the primary support being installed. Accelerations might dictate immediate installation of supplementary rockbolts and shotcrete in the affected areas, and more frequent readings. Tunnel evacuation might be required if cracking and water inflows are observed. From one project to the next, actual hazard warning levels and actions will vary. The principle of establishing hazard warning levels and contingency plans in advance is nevertheless applicable to all projects.

If the data are being recorded automatically, these decision-making levels and other "warning" or "alarm" criteria can be incorporated directly into a computer program which notifies the site engineers with an alarm bell (Bullock et al., 1980; Dusseault, 1986), or even with telephone calls to preprogrammed numbers.

12.3.1.3 Extent of coverage. It is usually a mistake to concentrate on just a few areas to the exclusion of others where unexpected movements may occur. Following the rule of "working from the general to the particular," there should be an extensive area of low-precision and low-frequency monitoring, with more precise and frequent monitoring at the more critical locations. Additional and more frequent measurements can be implemented when and where movement is detected.

12.3.1.4 Calibrations and cross-checking. At the extremities of the monitoring pattern, in extent and in depth, there need to be stable benchmarks and reference instruments. For example, if a landslide is being monitored using probe inclinometers, the inclinometer guide tubing should be taken to a depth of at least several meters beneath the anticipated base of the slide, and in addition, survey observations at the ground surface are referred to datum points sufficiently far removed from the immediate vicinity of the landslide to ensure that they are on stable ground.

When the instruments are to provide a check on design calculations, a series of typical locations or "representative profiles" are selected for monitoring different ground conditions or support types. Thus a tunnel lining may be checked at four or five typical locations along its length.

Redundancy should be built into the system. In deciding the number of instruments, allowance needs to be made for inevitable malfunctions caused by faulty equipment or installation, or by the ground movements themselves. When possible, two or more types of instruments are used to check each other. For example, measurements by conventional surveying can be checked against others by EDM and inclinometer.

12.3.1.5 Selection of instrument features. The range and precision of measurement need to be sufficient but not excessive. The precision of an instrument tends to be inversely proportional to its measuring range, such that precise and sensitive instruments may not be capable of measuring the full range of anticipated movements or water pressures. Costs can be unreasonably high if the specified precision is greater than needed. When the range is exceeded, the resetting of instruments can prove to be difficult and in some cases impossible.

The engineer must decide on whether there is a need for remote or continuous recording and alarm systems, and must choose between instruments that are fixed in the ground and the alternative portable types which can be read at many locations, are less vulnerable to damage, and are easier to calibrate. The fixed-in-place types are usually required only if precision requirements are high, and at locations of difficult access or when there is a need for continuous rather than intermittent surveillance. For reasons of durability, mechanical systems are often preferred to pneumatic ones, and pneumatic to electric systems.

Signal cables and tubes are located where they are least likely to become damaged by ground movements, site traffic, or the effects of rain, snow, and ice.

12.3.1.6 Acquisition of instruments. Lowest cost of an instrument should never be allowed to dominate the selection, and the least expensive instrument is not likely to result in minimum total cost (Dunnicliff, 1988).

To estimate requirements and costs, an instrumentation layout diagram is first drawn up to show locations and depths of transducers, transmitting systems, and readout units. A preliminary schedule of quantities, a cost estimate, and specifications are then prepared, taking into consideration not only the instruments themselves, but also the personnel and materials needed for installation, calibration, and maintenance, and for readings, data processing, interpretation, and reporting.

Comparison of the estimate with the available budget will often require cuts to be made. In spite of budget limitations and cuts, the monitoring system must continue to give results in which the engineer can place confidence, and must meet the specified objectives.

12.3.2 Installation

12.3.2.1 Installation safeguards.
Even the best instrument design can give misleading information if it is incorrectly installed. For example, on one project a roof fall occurred in a tunnel crown, without any preceding movement having been detected by the extensometer at that location. The fall exposed the lower anchor of the extensometer: steel drill casing was seen to surround the anchor and rods, and had prevented the recording of ground movements that would have warned of the impending roof collapse. The drillers had failed to report their inability to remove this casing, perhaps in the hope that it would pass unnoticed.

Drillhole instrumentation needs to be positioned after a careful study of the core, or of the drillhole using a television camera, with a view to selecting optimum locations for ground anchors, piezometer tips, and other components. Correct installation also requires a detailed knowledge of the working of the individual instrument. Printed instructions are seldom adequate, and installation, particularly of complex instruments, should be supervised by an experienced instrumentation engineer. Readings and interpretation, on the other hand, must continue in the long term and are often done by appropriately qualified construction and mine personnel rather than by outside contractors. Specialist monitoring consultants can be retained to advise on the planning and execution of the work and on the interpretation of the results. Instrument manufacturers can provide further assistance.

12.3.2.2 When to install.
The instruments should, if possible, be installed well in advance of construction activities and in periods when instability is least likely to occur, so that checks on drift and background noise levels can be made, and to allow adequate time for a thorough checking and calibration of each instrument before and immediately after its installation.

Instrument locations must be recorded reliably in plan and in section, together with a clear and unambiguous instrument numbering system. Data sheets are designed and drafted to provide space for all necessary observations and calculations, and in particular to allow comparison of one set of readings with the previous set at the time the observations are made. This comparison will ensure that erroneous readings are corrected immediately, and that readings are repeated without delay if anomalies occur.

12.3.2.3 Procedure. Instrument locations are checked on site to ensure that they are the best available and offer the maximum protection for the equipment. Site preparation may be needed to remove loose material or to bury instruments below the depth disturbed by construction equipment and frost. The instrumentation holes are carefully logged. When instruments are to be fixed in the ground, their correct assembly and performance needs to be checked before backfilling or grouting so that adjustments or repairs can be made before it is too late.

Once installed, the instruments are protected against damage by installing cover plates, protective fences, or large boulders. Locations and elevations are surveyed, and each instrument is numbered and marked on a plan. On-site calibration and maintenance facilities are established to allow routine checking of each instrument when readings are taken.

Completion of installation and testing is followed by a *commissioning report*, which will give the instrument specifications, locations, and the numbering system, describe the methods used for installation and calibration, and give details of how to take readings. The commissioning report also provides the necessary data sheets and graphs, complete with initial readings.

12.3.3 Reading, reporting, and interpretation

12.3.3.1 Frequency of observations. An initial period of relatively frequent readings is needed to establish trends in ground behavior. The readings will become less frequent as soon as confidence is established in the system. Thereafter the frequency of observations needs to be adjusted from time to time, so that active locations are monitored more frequently than stable ones. For example, profiles of instruments in the immediate vicinity of an advancing tunnel face may require reading at daily intervals, then weekly or monthly intervals when the heading has advanced several tunnel diameters past the instrument profile. The frequency is adjusted to a level just sufficient for the plotting of meaningful graphs, which are updated to show the new data points as soon as obtained.

12.3.3.2 Graphs and tabulations. A clear graphic presentation of the results is the key to prompt and appropriate interpretation and action. Graphs are much better than tabulations because they show much more clearly the trends and anomalous readings. Axis scales need to be chosen to suit the range and precision of the instrument, the frequency and duration of the readings, and the range and significance of the expected results. Graphs should be annotated to show significant events, such as blasting and excavation, and also incidents of instrument damage and repairs which could otherwise lead to misinterpretation.

The initial commissioning report is followed by *monitoring reports* at intervals, often weekly. However, unusual readings should be reported verbally at once, even before calculations are completed. A typical monitoring report includes field monitoring result tabulations covering all observations since the preceding report, an up-to-date graphic representation of ground behavior, and a brief commentary drawing attention to the significance of observations and to any instrument malfunctions that may have occurred.

12.3.3.3 Interpretation of results. Interpretation of the data obtained by monitoring is a matter of judgment and comparison with the predictions made by numerical modeling and experience gained on previous projects. No hard and fast rules are available. Interpretation is, of course, easier when the results are fresh, and on no account should a backlog of unplotted and uninterpreted data be allowed to accumulate.

Ease of interpretation depends greatly on the care taken in designing the monitoring system and program of readings. Figure 12.13 shows how inclinometer data, superimposed for a single instrument location over a period of time, can reveal shearing in a slope and bulging in the walls of a powerhouse cavern. Readings are even easier to interpret when considered globally along with all other available information. Hence the graphs for several instruments at the same locality are best plotted on a single sheet, one alongside the other, so that a spatial as well as a time pattern can be seen.

Much has been published regarding the anticipated behavior of various types of structure in rock. During the excavation of a tunnel, for example, if convergence is plotted against time, an apparent time effect will be recorded even in elastic rock (e.g., Niwa et al., 1979; Lo et al., 1985). This, however, merely represents the transition from three-dimensional to plane strain conditions as the tunnel face advances. Only when movements continue after the face has advanced to distances beyond about three tunnel radii, is true time-dependent creep or "squeeze" occurring.

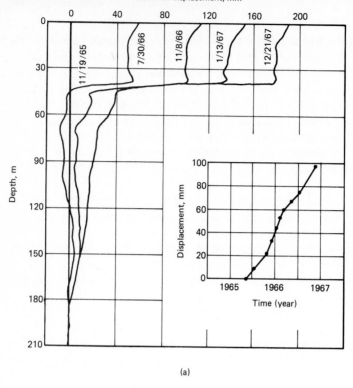

(a)

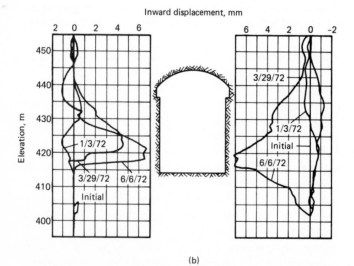

(b)

Figure 12.13 Clear graphic display of results is an essential aid to interpretation of monitoring data. (*a*) Displacement-depth profile for a rock slope, showing inset displacement-time graph for shear zone at 39–46-m depth. Measurements by probe inclinometer. (*b*) Horizontal displacements in walls of a large underground powerhouse chamber. Measurements by probe inclinometer. (*ISRM, 1981.*)

References

Allen, R. V., D. M. Wood, and C. E. Mortensen: "Some Instruments and Techniques for Measurements of Tidal Tilt," *Phil. Trans. R. Soc. London,* ser. A, **274**, 219–222 (1973).

Babcock, C. O.: "Equations for the Analysis of Borehole Pressure Cell Data," *Proc. 27th U.S. Symp. Rock Mech.* (Tuscaloosa, Ala., 1986), pp. 233–240.

Bauer, E. R.: *Ground Control Instrumentation; A Manual for the Mining Industry,* U.S. Bureau of Mines Information Circular 9053, 68 pp. (1985).

Bullock, W. D., A. M. Underwood, and J. A. Franklin: "Rehabilitation of a Failed Haul Road in an Asbestos Open Pit," *Proc. Am. Inst. Min. Eng. Ann. Mtg.* (Las Vegas, Nev., 1980).

CANMET: "Microseismic Monitoring in Canadian Mines," *Proc. Workshop, Sudbury, Ont., Can. Centre for Minerals & Energy Technol.,* Div. Rep. MRP/MRL 85-23, Mining Res. Labs., 160 pp. (1984).

Collett, C. V., and A. D. Hope: *Engineering Measurements,* 2d ed. (Pitmans, London, 1983).

Cook, C. W., and E. S. Ames: "Borehole Inclusion Stressmeter Measurements in Bedded Salt," *Proc. 20th U.S. Symp. Rock Mech.* (Austin, Tex., 1979), pp. 481–485.

Cording, E. J., A. J. Hendron, and D. U. Deere: "Rock Engineering for Underground Caverns," *Proc. Symp. Underground Rock Chambers* (Phoenix, Ariz., 1971), Am. Soc. Civ. Eng., New York, pp. 567–600.

Davis, R. E., F. S. Foote, J. M. Anderson, and E. M. Mikhail: *Surveying: Theory and Practice,* 6th ed. (McGraw-Hill, New York, 1981).

Dunnicliff, J.: "Geotechnical Instrumentation for Monitoring Field Performance," National Cooperative Highway Research Program, Synthesis of Highway Practice 89, Transport. Res. Board, Washington, D.C., 46 pp. (1982).

———: *Geotechnical Instrumentation for Monitoring Field Performance* (Wiley, New York, 1988), 577 pp.

Dusseault, M. B.: "Monitoring in situ Processes," *Proc. Am. Conf. Petroleum Soc. of CIM* (Calgary, Alta., 1986), paper 86-37-63, pp. 351–365.

Franklin, J. A.: "Some Practical Considerations in the Planning of Field Instrumentation," *Proc. Int. Symp. Field Measurements in Rock Mech.* (Zurich, Switzerland, 1977a), pp. 3–13.

———: "The Monitoring of Structures in Rock," *Int. J. Rock Mech. Min. Sci. Geomech. Abstr.,* **14**, 163–192 (1977b).

———: *Mine Monitoring Manual* (Can. Inst. Min. Metall., Montreal, Que., Canada, in press).

Hanna, T. H.: "Field Instrumentation in Geotechnical Engineering," *Series on Rock and Soil Mechanics,* vol. 10 (Trans Tech Publ., Cleveland, Ohio, 1985).

Hardy, H. R., Jr.: "Applications of Acoustic Emission Techniques to Rock and Rock Structure: A State of the Art Review," *Proc. ASTM Symp. Acoustic Emission in Geotech. Eng.,* STP 750 (Am. Soc. Test. Materials, Philadelphia, Pa., 1981), pp. 4–92.

——— and M. Ge: "A Computer Software Package for Acoustic Emission/Microseismic Field Applications," *Proc. 27th U.S. Symp. Rock Mech.* (Tuscaloosa, Ala., 1986), pp. 134–140.

Hawkes, I., and W. V. Bailey: "Design, Develop, Fabricate, Test and Demonstrate Permissible Low Cost Cylindrical Stress Gages and Associated Components Capable of Measuring Change of Stress as a Function of Time in Underground Coal Mines," Contract Rep. H0220050, U.S. Bureau of Mines (1973).

——— and V. E. Hooker: "The Vibrating Wire Stressmeter," *Proc. 3d Cong. Int. Soc. Rock Mech.* (Denver, Colo., 1974), vol. 2, pp. 439–444.

Hedley, D. G. F.: "Triangulation and Trilateration Methods of Measuring Slope Movement," Int. Rep. 72/69, Dept. Energy, Mines and Res., Mines Branch, Ottawa, Ont. (1972).

ISRM: "Suggested Methods for Rock Characterization, Testing, and Monitoring," ISRM Commission on Testing Methods, E. T. Brown, Ed. (Pergamon, Oxford, 1981), 211 pp.

Koeppel, J., C. Amstad, and K. Kovari: "The Measurement of Displacement Vectors with the 'Trivec' Borehole Probe," *Proc. Int. Symp. Field Measurements and Geomechanics* (Zurich, Switzerland, 1983), pp. 209–218.

Kovari, K., C. Amstad, and J. Koeppel: "New Developments in the Instrumentation of Underground Openings," *Proc. 4th Rapid Excavation and Tunneling Conf.* (Atlanta, Ga., 1979), vol. 1, pp. 817–837.

Lo, K. Y., B. Lukajic, and T. Ogawa: "Interpretation of Field Measurements of Stresses and Displacements around Excavations in Rocks," *Can. Tunneling* (Tunneling Assoc. Can.), 107–128 (1985).

Lu, P. H.: "Mining-Induced Stress Measurement with Hydraulic Borehole Pressure Cells," *Proc. 25th U.S. Symp. Rock Mech.* (Evanston, Ill., 1984), pp. 204–211.

Maleki, H., W. Hustrulid, and D. Johnson: "Pressure Measurements in the Gob," *Proc. 25th U.S. Symp. Rock Mech.* (Evanston, Ill., 1984), pp. 533–545.

Morgan, H. S.: "Analysis of Borehole Inclusion Stress Measurement Concepts Proposed for Use in the Waste Isolation Pilot Plant (W.I.P.P.)," *Proc. 25th U.S. Symp. Rock Mech.* (Evanston, Ill., 1984), pp. 212–219.

Morice, P. B., and G. D. Base: "The Design and Use of a Demountable Mechanical Strain Gauge for Concrete Structures," *Mag. Concrete Res.*, 5 (13), 37–42 (1953).

Niwa, Y., S. Kobayashi, and T. Fukui: "Stresses and Displacements around an Advancing Face of a Tunnel," *Proc. 4th Int. Cong. Rock Mech.* (Montreux, Switzerland, 1979), vol. 1, pp. 703–710.

O'Connor, K. M., and C. H. Dowding: "Application of Time Domain Reflectometry to Mining," *Proc. 25th U.S. Symp. Rock Mech.* (Evanston, Ill., 1984), pp. 737–746.

Panek, L. A., and L. A. Stock: "Development of a Rock Stress Monitoring Station Based on the Flat Slot Method of Measuring Existing Rock Stresses," Rep. Invest. RI 6537, U.S. Bureau of Mines, 61 pp. (1964).

——— and W. J. Tesch: "Monitoring Ground Movements near Caving Stopes—Methods and Measurements," Rep. Invest. RI 8585, U.S. Bureau of Mines, Denver, Colo., 108 pp. (1981).

Peck, R. B.: "Advantages and Limitations of the Observational Method in Applied Soil Mechanics," *Geotechnique*, 19 (2), 171–187 (1969); reprinted in J. Dunnicliff and D. U. Deere, Eds., *Judgement in Geotechnical Engineering: The Professional Legacy of Ralph B. Peck* (Wiley, New York, 1984), pp. 122–127.

Sellers, J. B.: "The Measurement of Rock Stress Changes Using Hydraulic Borehole Gages," *Int. J. Rock Mech. Min. Sci.*, 7, 423–435 (1970).

Walton, R. J., and G. Worotnicki: "A Comparison of Three Borehole Instruments for Monitoring the Change of Rock Stress with Time," *Proc. Int. Symp. Rock Stress and Rock Stress Measurements* (Stockholm, Sweden, 1986), pp. 479–488.

Rock Excavation and Stabilization

The last part of this book describes the practical tools of rock construction: the machines and explosives used to create mines and caverns in rock, and the reinforcement, drainage, and grouting measures that maintain these stable and safe.

P3.1 Excavating Methods

Although most rocks with compressive strengths greater than 100 MPa still require blasting, the development of increasingly powerful and more versatile mechanical excavating equipment is leading to its more widespread use. Mechanical methods of excavation, including the use of tunnel-boring machines, are particularly appropriate in urban areas where they produce much lower levels of vibration, cause less damage to the rock, and leave the walls of excavations in a more stable condition, requiring less support.

Blasting methods themselves have advanced considerably since the introduction of gunpowder in the fifteenth and sixteenth centuries. Dynamite, invented by Alfred Nobel in 1867, replaced the slower burning and less efficient black powder except in a few, specialized applications. Dynamite itself is now being superceded by the cheaper and safer AN/FO, a blend of ammonium nitrate fertilizer (AN) with fuel oil (FO), and by emulsion explosives which perform well even in wet conditions. AN/FO was patented in Sweden at about the same time as dynamite, but was not commercially available until much later. Tungsten carbide bits, invented in 1940, have revolutionized the drilling of blastholes, and the cutting of rock by boring machines. The oil-hydraulic percussive drills employed in tunneling projects since about 1971 can now achieve penetration rates of up to 90 m/h even through a granite.

Interesting to note is that in terms of the energy needed to excavate a unit volume of rock, a miner with a pick is more efficient than blasting, and about 100 times more efficient than a tunnel-boring machine. However, energy consumption is neither the best nor the only criterion. More important is how fast can a tunnel be excavated and how cheaply and safely.

Colonel Beaumont's tunnel-boring machine, constructed in 1881, penetrated 1.9 km through chalk beneath the English Channel before being stopped for nontechnical reasons. Advance was sustained at 15.4 m/day for 53 consecutive days. Almost exactly 100 years later, an average advance rate of 20 m/day was reported for six tunnels through sedimentary rock formations in the United States—not much of an improvement, perhaps, although in substantially stronger rock than chalk (see Chap. 15). Modern machines with disk cutters can bore almost as fast even through igneous rock materials.

P3.2   Rock Improvement

The many techniques for improving and stabilizing rock include reinforcement by bolting and anchoring, lining with sprayed concrete (*shotcrete*), and treatment methods such as drainage, grouting, or freezing. Rock that cannot be economically reinforced or held in place can be caught by a berm or fence before it does any damage. The concluding chapters of the book describe these methods and how they are used.

Those who construct using concrete or steel have the singular advantage of being able to choose and adjust their materials to achieve almost any desired combination of properties—and they know what these properties are. Rock engineers, in contrast, have to design structures in rock formations that vary greatly, and whose properties can hardly ever be predicted with confidence. This is why rock improvement measures are so important. Without bolting, shotcreting, drainage, and grouting, construction would be limited to only the best sites. Mining and civil engineering activities would be greatly curtailed.

Rock stabilization technology, like that for excavation, has improved greatly over the years. In the earliest days of tunneling, excavations were supported temporarily with timber until a permanent masonry liner of brick or stone could be installed. Old-fashioned brick and stone were the earliest forms of "precast" construction, and cast-iron segments supported the earlier parts of the London "underground" subway system. Since about 1937, precast concrete segments have replaced cast iron for reasons of cost.

More recently, many open excavations, tunnels, and caverns are lined by spraying directly onto the excavated ground. Gunite (sprayed fine-aggregate concrete) was patented by the Allentown Cement Gun Company in 1909. It was first used underground in the United States in 1914, at the Brucetown Experimental Mine. Spraying of the stronger and less expensive shotcrete (coarse-aggregate concrete) became possible with the invention of the Aliva VS-12 spraying machine in 1942 by a Swiss engineer. Tunneling applications of reinforced and bolted shotcrete date from about 1950 in Europe, with the introduction of the New Austrian Tunneling Method (NATM).

The success of modern ground support stems from the way that rock and support materials work together to achieve stability. A thin and flexible liner is applied as soon as the rock becomes exposed. It forestalls loosening and encourages the rock to "arch" around the excavation. Support costs and quantities are minimized, with also a reduced risk of "loss of ground" and subsidence at the surface.

Blasting

13.1 Introduction

13.1.1 Different objectives

Rock is blasted either to break ore or rock or to create space. The different objectives give rise to different philosophies and techniques.

13.1.1.1 Production. In mining and quarrying, the main objective is to extract the largest possible quantity of valuable resource from the ground at minimum cost. This can include aggregates for construction, gypsum, salt, potash and coal for processing into chemicals or for use as fuels, or metallic sulfides and other ores containing valuable metals. Production rates must be maintained to build adequate stockpiles and meet market demands. At the same time, to maximize profits, the mine or quarry operator must minimize expenditures on drilling and explosives.

13.1.1.2 Fragmentation and throw. In-place rock is broken by blasting and crushing (*fragmentation* and *comminution*) into aggregate sizes or, with additional grinding, into a finer powder suitable for mineral processing.

Large, slabby blocks or the opposite, an excess of "fines," can result from poorly designed blasts or unusual geological conditions. A well-designed blast should produce sizes and shapes that can be accommodated by the available loading and hauling equipment and crushing plant, with little or no need for secondary breakage. While optimizing the fragmentation, it is also important, for safety and ease of loading, to control the throw and scatter of the fragments. Photoanalysis meth-

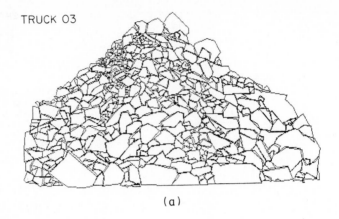

(a)

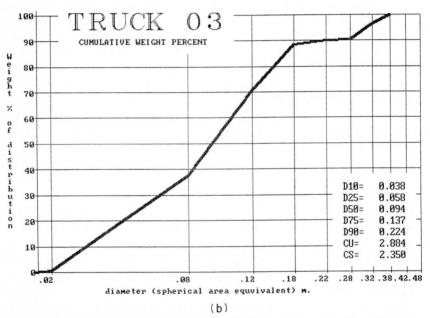

(b)

Figure 13.1 Fragmentation measured by digital photoanalysis. (*a*) Digitized image of broken rock in truck. (*b*) Computer-generated gradation curve giving cumulative weight distribution of fragments. (*Maerz et al., 1987.*)

ods are now capable of measuring grain-size distributions for blasted rock similar to those obtained for sand or gravel by sieving (Fig. 13.1; Maerz et al., 1987). The distributions produced by blasting are typically log-normal.

13.1.1.3 Dilution. An important concern in mining and quarrying is to avoid *dilution* of the mineral by admixtures of waste rock. These have to be processed in the same way and at the same cost as the ore, and are expensive to remove. Dilution is avoided in mine stopes by careful drill and blast design to avoid loosening waste rock in the hanging wall, and by proper stope design to preserve stability at least while the ore is being extracted.

13.1.1.4 Wall control. In civil engineering, rock is removed to create tunnels or caverns, or deep excavations at the ground surface for road cuts, foundations, or basements. The emphasis is not on high rates of production, although the job must be done as quickly and cheaply as possible, but on creating space and leaving behind stable rock walls that are either self-supporting or require little reinforcement and lining. Requirements for smooth walls and long-term stability exist also in mine shafts, and to a lesser extent in mine development drifts that must remain open for moderate periods.

Problems in the blasting of civil engineering works are most often associated with overbreak and underbreak. Special blasting techniques (Sec. 13.4.4) are available that assist in producing smooth-walled excavations. Often only minor and inexpensive adjustments to conventional procedures are needed.

13.1.1.5 Environmental impact. When blasting in near-surface works such as foundations, ditches, and quarries, the goal of space creation or production must be achieved while protecting the environment. Avoidance of blast damage and high levels of noise and vibration requires careful sizing of charges, timing of detonations, and accurate drilling of holes to give burdens that are neither too large nor too small. Blasting mats often are used to limit the projection of fly rock. Side effects of subsidence and underdrainage must also be avoided as part of an overall mining or quarrying plan.

13.1.1.6 Combined objectives. Production and space creation objectives occasionally go hand in hand. In underground aggregate mining, for example, rock is extracted, processed, and sold, and then the underground space can be sold or leased for storage or office accommodation. Mining and quarrying operations have production as a primary goal, but must also take precautions to avoid damaging the rock left behind, at least to an extent sufficient to preserve safe working conditions while the mine remains in production. In terms of mining and quarrying economics, the optimum excavating method is one that maximizes production, fragmentation, and safety; minimizes dilution and excavating costs and environmental impact; and allows the mine or quarry to collapse the day after all the valuable resources have been extracted and equipment removed. Long-term stability needs preserving only if necessary to avoid undesirable subsidence, or if the space is to be converted to other uses.

13.1.2 Mechanisms of explosive fracturing

Pressure in a blasthole can exceed 10 GPa (100,000 atm), sufficient to shatter the rock near the hole, and also to generate a stress wave that travels outward at a velocity of 3–5 km/s. The velocity of detonation (VOD), the speed at which the detonation shock wave travels through a charge of explosives confined in a drillhole, ranges from 1.5 to 7.6 km/s. An explosive reaction that moves through the charge faster than the speed of sound is termed *detonation,* one that propagates more slowly is termed *deflagration.* All high explosives and blasting agents detonate, whereas low explosives such as black powder deflagrate (Clark, 1987).

The leading front of the stress wave is compressive, but it is closely followed by the tensile stresses that are mainly responsible for rock fragmentation. A compressive wave reflects when it reaches a nearby exposed rock surface, and on reflection, becomes a tensile strain pulse. Rock breaks much more easily in tension than in compression, and cratering progresses backward from the free surface. Most researchers consider the stress wave to be the main cause of fracturing, with gas pressures acting to widen and extend stress-generated cracks or natural rock joints.

Explosive processes can be investigated by dynamic numerical modeling. Hong et al. (1986) describe a finite-difference model to investigate the propagation of explosively generated seismic waves and their scattering by joints in the rock mass. Explosives manufacturing companies routinely use programs for the design of blasts.

13.2   Blasting Materials

Blasting materials and methods are described in several excellent reviews, textbooks, manuals, and handbooks (Gregory, 1973; Bauer and Calder, 1974; Dick et al., 1983; Rosenthal and Morlock, 1987; C.I.L., 1973; DuPont, 1977).

13.2.1   Types of explosives

13.2.1.1   Black powder.   Black powder (gunpowder) is a mixture of potassium or sodium nitrate with sulfur and finely ground charcoal. It is still used in conventional munitions, sometimes in coal mining where lump size is to be retained, and in the quarrying of building stones such as granite, where the blocks must be separated along preexisting

joints or planes of weakness rather than by shattering along newly created fracture surfaces.

13.2.1.2 High explosives. In high-explosive dynamites used for rock blasting, the reaction front propagates supersonically, rather than subsonically as in black powder. Most of them have greater energy per unit weight of explosive. They are manufactured in the form of cartridges or "sticks" for use in small-diameter holes (down to 19 mm) (Fig. 13.2a), and have fair to excellent resistance to water. Nitroglycerine (NG), a major component of the dynamites, when in pure form, is relatively unsafe and sensitive to impact.

Straight dynamite contains 20–60% nitroglycerine together with sodium nitrate and a lesser amount of carbonaceous fuel. *Ammonia* dynamites, which have largely replaced straight dynamites, contain ammonium nitrate and generally have a lower velocity of detonation, lower density, higher shock resistance, and better fume characteristics. *Gelatin* dynamites contain liquid nitroglycerine with nitrocellulose; 20% and 60% gelatin dynamites (the percentage refers to the content of NG) have very good fume characteristics and are often used underground; 75–90% gelatin dynamites and 100% gelatin and straight dynamite have very poor fume characteristics, are not used underground, and are mainly employed in underwater and geophysical blasting.

13.2.1.3 Initiating explosives. Initiating explosives can detonate to produce an intense shock even when in the form of very small charges. They are used in small quantities to initiate detonation in larger and less sensitive high-explosive charges. They are supplied in copper or aluminum tubes to form detonators, or in the core of a detonating cord.

13.2.1.4 AN/FO and other nitrocarbonitrate blasting agents. Although most rock is still excavated by dynamites, there has been a trend toward use of nitrocarbonitrate (NCN) "blasting agents" such as AN/FO, which consist of an oxidizer and a fuel. The relatively inactive ammonium nitrate particles, coated with an inert absorbent material or treated with a surfactant to promote thorough mixing, are sensitized with fuel oil in the proportion of 94:6 by weight. The mixture detonates at a velocity of 3–4 km/s.

AN/FO is the least expensive explosive available, giving an explosives cost one-third to one-quarter that of nitroglycerine-based high explosives. When correctly used, AN/FO performs as well or better than the dynamites and is safer to handle. However, it is soluble in water, so cannot be employed in wet conditions. Larger-diameter

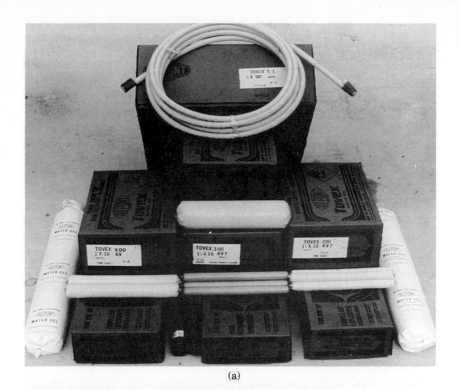

(a)

(b)

Figure 13.2 Typical explosives. (*a*) High explosive; various grades and sizes of prepackaged water gel explosive. (*b*) Bulk loading of AN/FO. (*Courtesy of Du Pont-ETI.*)

blastholes are drilled for economy and because more AN/FO is needed to generate similar amounts of energy.

AN/FO is supplied either in bulk or in waterproof polyethylene bags (Fig. 13.2b). The separate components, delivered for bulk mixing on site, are not classified as explosives and so can be shipped without incurring the extra costs of transportation and storage precautions. For relatively dry holes, bulk free-running AN/FO is best because superior coupling is obtained.

13.2.1.5 Slurries. Specially designed ammonium nitrate emulsions and slurry blasting agents (SBA) have been developed for blasting in wet holes. A combustible fuel is mixed with granular AN dispersed with a sensitizer and thickener and with just enough of an aqueous solution of AN to give a semifluid mixture containing up to 20% water. The combustible fuel component can be a material such as molasses, sugar, sawdust, sulfur, a heat-producing metal such as magnesium or aluminum, or even TNT. Thickeners include starch, water-soluble vegetable gum, or oil with an emulsifier. Sensitizers include microscopic glass bubbles, TNT, pigment-grade aluminum, and water-soluble organic nitrates. The mixture can be sold in final thickened form, or a gelling agent is added before or during loading so that it forms a thick gel after charging into the hole. This shields the slurry blasting agent from external moisture. Sometimes slurries, and particularly emulsions, are blended with prilled AN or AN/FO to increase their energy density.

Slurries, gels, and emulsions have a higher velocity of detonation than AN/FO, ranging from 3.3 to 5.5 km/s. They also yield much higher detonation pressures, so they can be used with larger burdens and spacings to reduce drilling costs. Metallized slurries generate considerable heat on detonation and are used for blasting extremely hard rock.

Slurries are denser than AN/FO, and so sink more readily in water-filled holes. They may be pumped, making loading quicker and easier, thereby avoiding some of the problems of underwater blasting. Slurry costs are comparable to those of dynamite and 2–3 times those of AN/FO.

13.2.2 Initiation of blasts

13.2.2.1 Safety fuse. The safety fuse consists of a central core of black powder around a strand of cotton and enclosed in a textile wrapping. The fuse is cut to a length chosen to provide ample time for the shot firer to withdraw to a place of safety. When lit with a hot flame, the powder burns slowly until burning reaches the end that has been crimped to the detonator. The North American safety fuse burns at a standard rate of 7.6 mm/s (120 s/yd).

13.2.2.2 Igniter cord. The igniter cord (IC) is a cordlike fuse that burns with an intense flame. It is used to ignite the safety fuses leading to individual blastholes to which it is connected at intervals along its length. The igniter cord is color-coded according to slow, medium, and fast burning rates of about 18, 36, and 71 mm/s. As the flame reaches each connector, it automatically lights each safety fuse in turn. All fuses in the round must be alight and burning before the first hole detonates. Therefore the length of the igniter cord between the first and last fuses is chosen to give a time interval shorter than that for the complete burning of the first safety fuse.

13.2.2.3 Detonating cord. The detonating cord, known also by trade names such as Primacord and Detaline, consists of a core of high explosive in a plastic sheath and protective wrapping. Unlike the safety fuse and the igniter cord, it has a high velocity of detonation (6.5 km/s), and when initiated by a single plain or electric detonator to which it is taped, it generates the pressure of a detonator almost simultaneously at all points along its length. Heavy-grade cord can be used for the detonation of high explosives lying alongside it in a blasthole, for simultaneous firing of widely spaced charges, and for the mass initiation of very large charges. More often the detonating cord is attached to a primer or booster, which in turn initiates the charge.

The detonating cord is much safer to handle than a detonator, is extremely water-resistant, and is safer to use because it reduces the hazards of lightning. It is particularly useful for underwater blasting applications. In quarry blasting, a trunk line of detonating cord usually extends along each line of holes, connected to branch or downlines in each hole.

13.2.2.4 Detonators. A powerful localized shock is required to initiate detonation in commercial explosives. This is generated either by a detonating cord as described, or by a self-contained detonator or *blasting cap* inserted in a cartridge of the high explosive to form a *primer* or *booster*.

The detonator consists of a metal tube of copper or aluminum with a base charge of a secondary explosive and a priming charge of a primary explosive. Plain detonators (blasting caps) are ignited using a safety fuse by crimping the mouth of the detonator to the fuse. Plain detonators #6 are in general use, and plain detonators #8 are used when a more powerful detonating effect is needed.

13.2.2.5 Primers and boosters. Primers can be made on site by burying a detonator in a cartridge of cap-sensitive explosive, usually a water gel explosive. An opening about 50 mm deep is formed in the end of the cartridge, using a wooden, brass, or copper skewer. The detonator end of a capped safety fuse is inserted into this hole.

Boosters are factory-made primers with a high density and velocity of detonation. They are initiated by a blasting cap or detonating cord and are used when making up primers on site is inconvenient.

13.2.2.6 Electric shot firing. Blasts are most often initiated electrically (Fig. 13.3a), particularly in wet conditions. Like plain detonators, electric detonators (electric blasting caps) contain an initiating charge of primary explosive and a base charge of secondary explosive, but they also contain a bridge wire and an ignition charge that ignites when an electric current is passed through the wire.

The electric primer is prepared in the same manner as a safety fuse primer, except that the electric wires are half-hitched twice around the cartridge. Wires are kept shunted until just before firing, as a safety precaution. The electric resistance of the circuit must be low to ensure a sufficient current for a satisfactory blast. Either a series or a parallel circuit can be used for a small round, whereas a parallel-series circuit is needed when the round includes a large number of detonators.

Either power mains, or portable *exploders* (blasting machines) equipped with removable firing keys provide the electric impulse. Before firing a series-connected round, the complete electric circuit is tested for continuity. Parallel circuits cannot be so tested, although individual detonators can be checked with an approved resistance

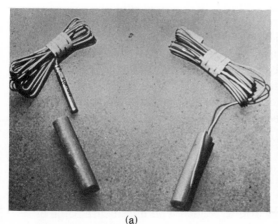

(a)   (b)

Figure 13.3 Typical detonating systems. (a) Electric blasting caps; method of fitting to a booster of flexible explosive. (b) Nonelectric detonating system, including detonating cord and surface and in-hole delay elements. (*Courtesy of Du Pont-ETI.*)

meter. Ordinary meters for measuring resistance can initiate electric detonators. Blasting circuits must be checked with meters designed for this purpose. Precautions are needed to prevent premature blasts caused by lightning strikes, induced currents, and static electricity.

13.2.2.7 Sequential firing and delay blasting. Sequential or delay blasting, a common alternative to instantaneous blasting, tends to give a lower muckpile and a longer throw, increased fragmentation, reduced peak vibration (at the expense of increasing the duration of the round), and reduced overbreak and backbreak. With very short delays, the superimposition of each new detonation wave gives more breakage for the same weight of charge than is given by either long delay or instantaneous firing; or a smaller charge can be used to achieve the same degree of breakage. Most of these advantages result from first removing rock nearest to the exposed free faces to allow easier release of rock at greater depth.

The original method of sequential blasting, hardly ever used nowadays, made use of the safety fuse to blast one or two holes at a time. Fuses can be cut to different lengths, or safety fuse downlines ignited in the desired order using igniter cord.

Electric delay detonator caps give a controlled time gap between pressing the plunger and initiating the charge. The fuse heads in the entire circuit ignite as soon as the current is applied, but the primers do not detonate until the delay elements have burned through. Hence breaking of the circuit by the first detonation does not affect the firing of later charges. An almost unlimited array of charges can be fired safely and in a controlled sequence. Regular-delay (LPV) electric detonators are manufactured with nominal half-second time intervals between each sequence number. The delay number is tagged to the leg wires. Short-delay electric detonators are manufactured with intervals varying from 8 to 1000 ms.

Sequential firing can also be achieved using nonelectric millisecond delays connected into a detonating cord trunk line by crimping. These are copper tubes about 75 mm long with an explosive charge at each end, separated by a delay mechanism (Fig. 13.3b).

13.3 Blasting Methods

13.3.1 Blasthole drilling

Blastholes drilled at erratic locations, angles, and depths are some of the most common causes of overbreak, underbreak, poor fragmentation, air blasts, fly rock, and high vibration levels. To minimize these problems, an upper limit for blasthole deviations is often specified in a blasting contract. Hole locations are carefully marked out on the rock face, and during drilling, their inclinations are measured and adjusted.

In tunnel blasting, a common and to some extent unavoidable tendency is to fan the perimeter holes outward instead of drilling them parallel to the tunnel walls. The resulting sawtooth pattern can be kept to an acceptable minimum by an experienced drilling crew using templates and mechanical guides to improve accuracy. Wandering drillholes limit the length of round. Deviations, which increase as the hole goes deeper, are caused by inaccurate collaring and alignment of holes and by deflection of the drill steel where it crosses joints and seams in the rock.

Hand-held air leg pneumatic drills and stoppers are used in small-scale operations and in small underground drifts, whereas the more powerful vehicle-mounted air track drills and jumbos are employed in larger-scale production blasting. Hydraulically operated drills are becoming increasingly popular. In production blasting, such as in mining and the bulk of civil engineering work, except perimeter holes, the trend is toward the use of larger-diameter blastholes approaching 1 m diameter in some mining applications. The greater development and capital costs of large-hole blasting are offset by an approximately two-fold increase in production rates and a halving of drilling costs per ton. Drilling methods are discussed further in Chap. 14.

13.3.2 Loading and handling

13.3.2.1 Loading with dynamite. After cleaning the blastholes with compressed air from a blowpipe, dynamite sticks are loaded, usually by hand. In horizontal holes, the sticks are firmly squeezed into position with a wooden tamping rod to avoid sparks that might result in premature ignition. An alternative hand-pneumatic loader uses 3 atm of air pressure to force the cartridge into the blasthole through a polyethylene tube. A high packing density is achieved. The method is safe, can load up to one cartridge per second, and is very useful for long-hole and up-hole work.

If the charge is to be fired electrically, the electric primer is the first cartridge to be inserted in the hole. The leg wires are held taut to one side of the hole while the ordinary cartridges are tamped into position. The primer cartridge itself should not be tamped. Bottom priming can also be done safely and effectively using a detonating cord.

13.3.2.2 Charging with AN/FO. In most quarrying work, the holes are loaded by pouring in the bulk material manually from bags or mechanically from an auger feeder. Some operators load the AN/FO into a plastic sleeve if the holes tend to have water running down their walls. In underground work, bulk AN/FO is loaded pneumatically. The pressure type of pneumatic loader applies compressed air above the AN/FO in a pressure vessel, which forces AN/FO from the bottom of the vessel into the placement hose. The alternative ejector type, which draws the AN/FO into an air stream in the placement hose, can load holes to a higher density, and is more suitable for charging long upholes. Loading rates of 2–20 kg/min are achieved in long 50-mm-diameter holes. The loading density is typically about 1 g/cm^3. Bulk AN/FO loaders often take the form of self-propelled rubber-tired vehicles with a 450-kg hopper and extendable boom.

AN/FO is usually detonated by a 150–450-g (⅓–1 lb) primer or several at intervals down a longhole. When detonating cord is used, one or more high-explosive booster units of similar weight are attached to the downline.

13.3.2.3 Charging with slurry. Slurry may be loaded with a hand-pneumatic cartridge loader, a bulk pumper, or in free fall, either in cartridges, slit cartridges, or shucked from cartridges, depending on water conditions and the required loading density. Bulk slurry loaders are sometimes homemade and employ a conventional grout pump, which charges the hole at a rate of about 40 kg/min. Pourable gels bypass the need for a mobile mixing truck. They can be loaded easily into a hole as small as 75 mm in diameter, but may become desensitized if they are poured through too great a depth of water.

Slurries are primed with a #8 cap at room temperature, or with a 10–20-g cast primer at colder temperatures and in longholes. To improve fragmentation, they can be alternated with AN/FO in a single blasthole ("slurry boosting") or placed at the bottom of the hole with a top load of AN/FO.

13.3.2.4 Loading density, coupling, and decoupling. One of the variables of a blasting pattern is the way in which the holes are loaded. Cartridges can be placed end to end, or spaced with cardboard or wooden separators to give any required column loading. Better practice for reliable detonation is to avoid spacing and to maintain a continuous column for each detonator used. Loading density is also controlled by tamping. A dry untamped loaded hole leaves the charge decoupled from the drillhole and reduces the effective detonation pressure. It can result in charge failure caused by pressure desensitization from a shock wave driven up the annulus at a velocity greater than that of detonation. With tight tamping there is less "rifling." The density of AN/FO and slurries is reduced by the inclusion of inert fillers.

Explosive coupling is the efficiency with which the chemical energy of the explosive is transferred to the surrounding rock of the blasthole. Good coupling gives good fragmentation but poor wall control—there is little energy lost, and nearby rock tends to shatter. Perimeter charges are often *decoupled* on purpose by reducing the diameter of the sticks to minimize shatter around the hole and to encourage the propagation of a single clean crack between holes. Decoupled charges are normally used in perimeter holes for controlled-perimeter and smooth wall blasting (Sec. 13.4.4). The special cartridges are generally 15–22 mm in diameter.

13.3.2.5 Stemming. Charges in blastholes must be sealed or "stemmed" to prevent the gases from escaping before they contribute to breaking the rock, and to contain and direct the explosive energy into the rock where it is needed. Unstemmed or poorly stemmed holes can result in "rifling," the shooting of rock and explosives from the hole, with very little or no rock damage and the generation of excessive airblast. Stemming materials are packed into the section of drillhole closest to the free rock face. They consist of sand or sandy clay formed into cartridges or, in large quarry holes, drill cuttings or, preferably, clean crushed stone. Various commercially produced liquid bag and polystyrene plug stemming materials are also available.

13.3.2.6 Treating misfires. The advice of the explosives manufacturer should be obtained. Generally, after a safe interval, broken rock is removed from the face, taking extreme care to identify unexploded cartridges or detonators and to remove them. The face is examined for remaining "butts" of holes which, if they contain explosive remnants, are carefully cleaned out before drilling is restarted. Holes are never drilled into old butts. Close control of drilling angles and depths, together with accurate recording of drillhole positions, can greatly reduce the chance of drilling into an unexploded charge.

AN/FO misfires are much less dangerous to resolve because the ammonium nitrate, which is soluble, can be sluiced out of the hole with a water hose. However, if a primer is present, the misfired hole has to be treated as if it had been loaded with high explosives.

13.3.3 Explosives consumption

Explosives consumption is expressed as the *powder factor* or *specific charge q,* defined as the total weight in kilograms of dynamite (or equivalent weight of other explosives) to excavate one cubic meter of rock.

In quarry and open-pit blasting, the specific charge typically varies from about 0.4 to 0.65 kg/m³, including the explosives required for both loosening and fragmentation; the amounts required for just loosening are even smaller. More explosives are required when tunneling. The specific charge varies from about 0.7 to 3.2 kg/m³. The main reason is that blast energy travels radially outward from a hole, and holes are drilled parallel to the face when bench blasting, but mostly perpendicular to the face when blasting a tunnel.

Explosives consumption also tends to be greater when dynamites are used to excavate weaker and more jointed rocks, which waste energy by crushing and plastic deformation. AN/FO is more efficient than dynamite in weaker rock, because it generates a lower explosion pressure that is better matched to the strength of the rock.

13.3.4 Influence of geology

Problems of poor fragmentation are most often experienced when a strong rock has well-developed planes of bedding or cleavage. Rock containing these geological features tends to break in large slabs that are difficult to handle and to crush. Poor fragmentation can be remedied by reducing the spacing and burden of blastholes and also by a better distribution of the explosive charge over the length of the hole. Millisecond-delay detonators assist in breakage. Turning the face to a more favorable direction relative to the joint planes can also help.

Geological reasons for deviation of the actual blasted perimeter from the intended perimeter include pronounced rock anisotropy or a substantial anisotropy of rock stresses in the plane perpendicular to the direction of drilling. When the rock has well-developed jointing or schistosity, drillholes and also blast-generated fractures tend to follow planes of weakness. The joints act as waveguides; blast energy travels preferentially along the direction of jointing. When the rock is highly and anisotropically stressed, the fractures tend to follow a direction perpendicular to the least principal stress. A further problem develops when the ambient stresses are isotropic but high: a very high blasthole pressure may be needed to overcome the stresses, sufficient to shatter the surrounding rock (Coursen, 1979).

The blasting pattern must be designed to suit the nature of the rock and the stress field, and not borrowed, without modification or checking, from previous projects where success was achieved, possibly under quite different sets of conditions. Changes to the blast patterns are often needed from place to place, even within the boundaries of a single site or within a single mine.

13.4 Open-Cut Blasting

13.4.1 Blast design

13.4.1.1 Elements of the blasting pattern.
Rock is easier to blast toward free faces: the broken rock occupies a greater volume and must therefore have room to expand. When a bench face does not exist, a release cut (called simply a *cut*) is made by drilling, cratering, or cutting.

Variables of the blast pattern include those that relate to the loading of each hole (Fig. 13.4), to the pattern of holes (see also Figs. 13.5, 13.9, and 13.11), and to the firing sequence. By making adjustments, the rock engineer can obtain better fragmentation, reduced damage to the remaining rock walls, reduced costs, reduced vibration and airblast levels, and reduced throw.

Blasts can be designed and results predicted using empirical formulas such as given below, with the aid of a microcomputer. Computer blasting models by the major explosives manufacturers simulate the performance of explosives and predict muckpile profile, throw, backbreak, and drilling and blasting costs.

13.4.1.2 Single-hole variables.
Single-hole variables include drillhole diameter and depth, the explosive used, the diameter of the stick in relation to that of the hole (called *decoupling*), and the method and

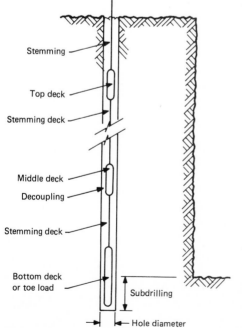

Stemming

Top deck

Stemming deck

Middle deck

Decoupling

Stemming deck

Bottom deck
or toe load

Subdrilling

Hole diameter

Figure 13.4 Variables and terminology of hole loading.

pattern of loading the explosives and stemming materials. This includes the use of decked charges, if any, and the sequence, nominal timing, and precision in timing of firing them. Vertical benches must be subdrilled to avoid leaving a "toe," which subsequently has to be "squared up" at considerable inconvenience and cost. Vibration levels may increase as a result of either excessive toe or excessive charge below floor level.

Ash (1963) defines five ratios that characterize bench blasting at the surface, three of which pertain to individual holes. The *hole depth ratio* K_H, the hole depth divided by the burden, averages 2.6 and ranges from about 1.5 to 4.0. The *subdrilling ratio* K_J, the ratio of depth drilled beneath the proposed floor of the blast to the burden distance, should be at least 0.3, except that no subgrade drilling is needed when joints run parallel to the floor and give a clean floor break. Calder (1977) recommends that subgrade drilling depths be 7–10 times the charge diameter.

Ash reports that the *stemming ratio* K_T, the length of stemming or "collar" divided by the burden, averages 0.7 and ranges from about 0.5 to 1.0. For perimeter blasting, Calder recommends that the length of stemming in the perimeter round be approximately 12 charge diameters in hard rock, 22 diameters in medium-strength rock, and 30 charge diameters in incompetent rock.

The lower parts of blastholes are sometimes (rarely nowadays) expanded to accommodate extra explosives near the toe of the blast, by a process known as *bulling* or *chambering*. A stick of dynamite with an ignited safety fuse is dropped down the hole.

Drillhole pressure can be reduced by decoupling or decking charges. In decoupling, a space is left between the explosive charge and the hole. Decoupling is measured as the ratio of charge radius to hole radius. If blastholes are water-filled, the effectiveness of decoupling is greatly reduced. To deck charges in small-diameter holes, short charges are taped to a detonating cord line. In large-diameter holes, spacers, usually of stemming sand, are alternated with explosives to produce a discontinuous explosive column.

13.4.1.3 Pattern variables. Pattern variables include hole inclination, depth, and spacing; burden, the distance between the exposed rock face and the nearest line of blastholes; orientation of the pattern to any jointing or anisotropic rock fabric; and proximity to surrounding buildings. In bench blasting, if the burden were to exceed the length of the hole, radial fractures and heavy vibrations would be produced, but the rock would remain in place. To make better use of blast waves reflected from the face, drillholes can be inclined at 45–60° to the horizontal. Inclined bench blasting tends to be more efficient in the use of explosives and give better fragmentation, but any fly rock tends to be thrown further. The face has a tendency to tilt back even in supposedly vertical quarry blasting, so a purposely inclined hole tends to give a more uniform burden and a cleaner toe.

The *burden distance* is given approximately by the Anderson formula

$$B = \sqrt{\frac{dL}{12}}$$

where d and L are the diameter and the length of the hole (Ash, 1963). Burden distances calculated by this formula agree closely with those used in practice for a wide range of rock types, from shales to hard igneous rocks. The *burden ratio* K_B, the burden distance divided by the diameter of the explosive cartridge (in the same units), averages 30 and ranges from about 20 to 40.

The *spacing ratio* K_S, the ratio of hole spacing divided by the burden, assumes values of between 1 and 2. Jointing of the rock is the most important factor in determining the spacing between holes in a row. When the spacing is too great, the rock face is left in a ragged condition, and interaction between holes contributes less to fragmentation. By convention, in quarry blasting, the geometry of blasting patterns is expressed as the product of burden and spacing distances, $B \times S$.

In calculating the charge, the volume of rock to be displaced is obtained by multiplying the depth of the hole by the product of burden and spacing. This is then related to the *powder factor*.

13.4.1.4 Sequence. Another variable is the sequence of firing, which includes the number of blastholes detonated in any one round and the time delay between successive rounds. The sequence can be varied using delayed charges, to minimize vibration levels and unwanted rock damage and to give a more efficient pattern of rock removal. The precision with which the desired timing is achieved is a most important consideration.

13.4.2 Quarry and open-pit blasting

13.4.2.1 Bench blasting. Quarry bench blasting operations are usually accomplished by parallel rows of drillholes, detonating first the row nearest to the exposed face to give better release for successive rounds (Figs. 13.5 and 13.6). Small-diameter holes loaded with high explosive can be used or larger holes loaded with AN/FO.

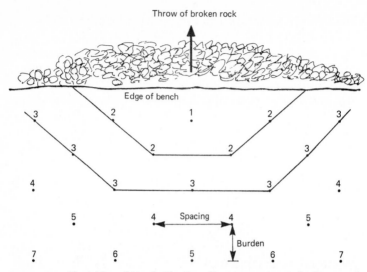

Figure 13.5 Variables of bench blasting. For aggregate production, good fragmentation is commonly achieved by a staggered pattern of holes with spacing 1.5–2.0 times the burden and with relief provided by millisecond delay detonation. (*After U.S. Department of the Army, 1972.*)

(a)

Figure 13.6 Quarry bench blast. Sequence taken by high-speed photography. (*Courtesy of Du Pont-ETI.*)

(b)

(c)

Figure 13.6   *(Continued)*

13.4.2.2 Secondary blasting. Secondary blasting of large blocks is achieved by either *blockholing* or *blistering (mudcapping)*. In blockholing, about a quarter of a cartridge of high explosive is detonated in a shallow drillhole to crack the boulder. In mudcapping, drilling is avoided but more explosive, usually a plaster charge, is required. The primed charges are sealed (*mudcapped*) to the surface of the boulder using clay. Ammonia dynamite or cap-sensitive grades of water gel explosives are employed.

13.4.2.3 Coyote blasting. Quarrying of large quantities of stone for riprap or coastal protection, where the blocks are required in large sizes, is sometimes accomplished by *tunnel* or *coyote* blasting, a modern version of *sapping* (the undermining of military fortifications). The bench is undermined by a drift that is packed with explosives, then detonated. Charges are usually fired with a detonating cord.

Coyote blasting has also been used in removing obstructions to navigation (such as the 1250-t Ripple Rock blast in the Seymour Narrows, B.C., Canada, in 1958), and in preparing copper ore bodies for leaching in place (the 1976 Old Reliable blast near Mammoth, Ariz., in which 6.4 million tons of copper ore was blasted with an 1800-t charge).

13.4.3 Blasting for open-cut engineering works

13.4.3.1 Precautions. In engineering projects, which are often in densely populated areas, special attention has to be given to preventing damage to nearby buildings and services and to safeguarding the public. Blasting mats, constructed of old rubber tires and steel cable, are placed over the area being blasted to limit the throw of fly rock. Vibration and airblast levels are closely monitored (Chap. 12 and Sec. 13.6). Special smooth-wall blast designs may be used to reduce damage to the surrounding rock, as discussed below.

13.4.3.2 Road cuts. In forming road cuts, blasting is made more difficult by the continually varying height of bench; heights of more than 10 m are usually blasted in more than one lift. Deep cuts such as those for cut-and-cover tunnels can be blasted continuously without mucking, using multiple rows of vertical holes. This reduces cost and also the throw and scatter of debris.

13.4.3.3 Trenches and ditches. Trenches in rock are blasted by a variety of methods, depending on the depth and width of the trench, the nature of the rock, the proximity of properties and services, the production rates required, and the available equipment. Blasting is often close to houses, with unusual problems of fly rock, vibrations, and air blast unless done correctly. Pressure desensitization of charges from the action of previously detonated charges can also cause problems unless pressure-resistant detonators and grades of explosive are used, together with a well-designed timing sequence.

There are two basic methods: blasting with the overburden still in place and with it removed. The latter method allows the rock to be assessed and the blast design adjusted accordingly. The trench is advanced by blasting toward a free face. For narrow trenches, say up to 600 mm wide, a single row of holes may suffice, but better results are usually obtained by staggered or paired holes in two rows. Usually 300–600 mm of subdrilling is needed to eliminate the need for secondary breakage at the floor level. In wider and deeper trenches, the holes are drilled in rows and tend to be larger in diameter (Ball, 1976).

Pipeline crossings often require trenching through rock in a river bed. The simplest method is to "plaster" charges directly on the rock under water, calling for at least 4.5 m of hydrostatic head to provide the necessary confinement. Otherwise holes may be drilled in the river bed by divers, or from the surface using barges or crawler rigs. Water-resistant explosives are of course needed.

13.4.3.4 Foundation blasting.

In foundation blasting in particular, throw and vibration levels must be controlled. Short-delay multiple-row blasting with small-diameter drillholes and reduced charges are generally used.

Overbreak is particularly difficult to control in the floors of open-cut excavations, which are usually excavated by vertical blastholes. The depth of drilling has to be monitored carefully to ensure that all holes terminate close to the required elevation. Even with this precaution, unless there is a well-defined geological discontinuity along the required excavation floor, pyramids of underbreak remain between the blastholes, and the floor will be irregular. Floor irregularity can be reduced by using more closely spaced blastholes, but this makes the blasting more expensive. Overbreak is usually preferred to underbreak: a small amount of overbreak can be backfilled with compacted broken rock or with concrete, whereas underbreak hummocks can only be removed by time-consuming secondary blasting or drilling.

The preparation of load-bearing foundations by blasting requires particular care to avoid fragmentation and loosening of the rock. Where the rock is closely jointed or weak, it is best to blast to an elevation somewhat higher than the final foundation level, and then remove the remaining rock with pneumatic tools.

13.4.3.5 Explosive mining. In remote areas, blasting can be used not only to break rock, but also to move rock and earth in large quantities from the outcrop to the point where they are to be used in construction. A dam across the Vakhsh River in the U.S.S.R. was constructed in this manner, by the excavation of 1.55 million m³ of hard limestone. The rocks were ejected from the river bank using 1800 t of explosives to create a 60-m-high dam across the river. A further 30,000 m³ of soft earth was thrown into the river bed to form an impervious upstream blanket (Ariel, 1968).

In open-cast coal mining, *cast blasting* is sometimes used to strip overburden. Here increased throw of the burden can reduce dragline operating costs, but at the expense of an increase in the costs for explosives and drilling. Computer models have had some success in designing cast blasts so as to minimize the overall costs of stripping.

13.4.4 Controlled-perimeter blasting

13.4.4.1 Wall damage, overbreak, backbreak, and underbreak. *Wall damage* is defined as fracturing of wall rock by blasting, including crushing and radial cracking around the blasthole. It is caused by excessive explosion pressures, excessive burden, inadequate time between rows in multirow blasting, or an unfavorable orientation of the blasting pattern relative to the jointing or principal stress directions. The blasthole pressures in most production blasting applications are in the range of 2–5 times the compressive strength of the rock. Perimeter holes should be more lightly loaded, to give pressures similar to, or slightly lower than, the compressive strength. Special explosives are often used for controlled-perimeter blasting, such as Xactex, which consists of an explosive in a 16-mm-diameter cardboard tube loaded into 32–38-mm holes. The charges are thus decoupled with a uniform air space around them, which allows gases to expand and reduces the shattering effect on the rock wall.

Overbreak is defined as unwanted removal of rock beyond a specified perimeter. It can result from overblasting, from inaccurate blasthole drilling (Fig. 13.7), or from the action of gravity combined with geological conditions.

Backbreak is the removal of rock beyond the line of blastholes. In production blasting it is often highly desired, and the operator relies upon it to increase the amount of rock brought down per blast.

Underbreak is rock remaining within a specified excavation perimeter that should have been removed by the blast. The protrusions of rock, called *tights,* interfere with construction or long-term use of the excavation, and must be removed by secondary blasting or breakage. Underbreak can be caused by inaccurate blasthole surveying or drill-

(a)

(b)

Figure 13.7 Extreme example of blasthole drilling deviation. (*Courtesy of Du Pont-ETI.*)

ing, by a charge that is insufficient or poorly distributed within the hole, by misfires or low-order detonations, or by an inappropriate choice of hole spacing or burden.

13.4.4.2 Methods for reducing wall damage and backbreak. Blasthole pressure, and consequently wall damage and backbreak, can be reduced either by reducing the density of the explosive (for example, by adding an inert filler to ANFO, which produces a fracture radius one-half to one-quarter that of dynamite), by reducing the charge diameter in the blasthole (decoupling), or by decking. The zone of crushing can also be eliminated and fracturing minimized by using different strengths of igniter cord instead of dynamite in the presplit holes (Solymar, 1983).

13.4.4.3 Overview of controlled-perimeter methods. When used in open-cut excavations (Fig. 13.8), the special methods for producing a clean perimeter break are termed *controlled-perimeter blasting*. These include, in increasing order of cost, cushion (buffer) blasting, slashing, presplitting, and line drilling. Similar procedures used to produce smooth perimeters in underground excavations are collectively termed *smooth-wall blasting* (Hendron and Oriard, 1972; Calder and Bauer, 1983).

Benefits from controlled-perimeter and smooth-wall blasting include enhanced stability and the possibility of steeper faces when rock quality is poor, reduced requirements and costs for scaling and secondary blasting and for rockbolting and concrete to replace overbreak behind liners, reduced risk of damaging adjacent structures, and improved safety.

13.4.4.4 Buffer blasting. In *buffer,* or *cushion,* blasting, rock is first broken by conventional blasting methods in a *production blast* that stops short of the perimeter holes. The buffer or cushion of unbroken rock protects the perimeter against shatter, and is blasted last in the main blast pattern, but while broken rock from the production blast is still in place.

The *buffer distance,* the width of the buffer zone between the perimeter holes and the nearest row of the production round, is usually about 1 m, about one-half of the regular spacing in the production round.

The burden, spacing, and explosive loads of the perimeter holes are substantially less than those of the production blast. Burden and spacing for the buffer row are typically 0.5–0.8 times those for the adjacent production row. Hole spacing for the buffer row is about 1.25 times the burden, and the powder factor is about 0.6 times that used in the pro-

Figure 13.8 Road cut excavated by controlled-perimeter blasting; La Paz–Cota Pata highway, Bolivia. (*Courtesy of DELCANDA International, Toronto.*)

duction holes. A single row of perimeter holes of 50–125-mm diameter and at 0.5–1.0-m spacing is loaded with light decoupled charges about 30 mm in diameter, usually assembled on detonator cord downlines which are lowered into the holes. Long polyethylene tubes filled with a sensitive grade of water gel explosive are also sometimes used. All air spaces are stemmed. Perimeter charges are fired after the main charge, usually on a delay system.

13.4.4.5 Slashing. *Slashing* is trimming of a final layer of rock from the excavation walls after rock from the production blast has been removed. The perimeter holes are loaded with light, distributed charges, which are fired simultaneously so that the detonation tends to produce a smooth crack between holes. As in buffer blasting, the perimeter holes are usually string-loaded with either dynamite cartridges or small-diameter tubes filled with a sensitive grade of water gel explosive, on Primacord downlines. The loading is 2–3 times as great in the bottom of the hole as in the upper part, to ensure shearing and clean breakage of the toe. The top approximately 1 m of the hole is stemmed. Millisecond delays can be used if necessary to control levels of ground vibration, but the holes should be detonated in sequence around the perimeter, and the delays should be minimal.

13.4.4.6 Presplitting. In *presplitting,* a fracture is created in solid rock along the desired line of break before any production blasting, in some cases even before production drilling. The presplit line protects the rock walls by inhibiting the propagation of cracks across the presplit boundary, by acting as a vent for gases generated during detonation of the main blast, and perhaps to some extent by acting as a reflector for vibrations transmitted from the main blast.

The perimeter (*contour*) holes are drilled at diameters of 50–100 mm and at a spacing of typically 10–20 times the hole diameter. Usually all holes are loaded with a charge about one-tenth the normal. Decoupled charges are often employed. Perimeter holes are preferably detonated simultaneously, at least 50 ms before the main round is fired.

A smooth perimeter crack can sometimes be produced with prenotched blastholes in which the notches act as stress raisers and direct a single crack toward adjacent holes. Holloway et al. (1986) describe blasthole notching with a mechanical broaching tool, with a water jet, and with shaped charges. Their testing showed that at hole spacings 38% greater than normally used, notched holes with half the normal loading density produced results equivalent to those of presplit blasting. Nevertheless, when there is no nearby free face to reorient a high tectonic stress that is in an unfavorable direction, such stresses can control the direction of the fractures beyond several hole radii, regardless of the orientation of notches in the drillhole wall or the position of adjacent holes.

13.4.4.7 Line drilling. The *line drilling* method is used as a last resort in fragile rock, and to protect critical sections of the rock face that must be left intact. A line of holes about 40–75 mm in diameter and 100–150 mm apart are drilled along the required excavation perimeter before detonation of the blast. The perimeter holes are not loaded; their presence and close spacing protect the surrounding rock from blast damage. With appropriate drillhole spacing and design of the remainder of the blast pattern, the rock breaks back to the line of drillholes and no further. Line drilling is used together with a buffer zone about 200–300 mm wide, or about 50–75% of the normal spacing between rows of holes in the production blast.

Costs are high because of the considerable amount of drilling and its required precision. Holes should not deviate out of the general plane of drilling by more than 150 mm. Drilling inaccuracies usually limit the depth that can be drilled.

13.5   Underground Blasting

13.5.1   Features of underground blasting

13.5.1.1   Release.   Adequate release is particularly important within the confines of underground rock excavations. The quantity of explosive required per unit volume of rock excavated (the *specific charge*), because of the confinement of the blast, is much greater than in the case of open-cut excavations.

13.5.1.2   Control of fumes.   The gases produced by blasting are of concern in underground applications. The U.S. Bureau of Mines classifies explosives according to the volume of toxic gases emitted when a 32-by 200-mm 200-g cartridge is exploded in a heavy steel container. Only explosives emitting less than 4.5 liters of toxic gases per cartridge (fume class 1) are permitted underground. In addition, underground openings excavated by blasting need to be be well ventilated, and sufficient time must be allowed for the fumes from a round to clear before workers can reenter the heading to scale the roof and walls and to start mucking.

13.5.2   Blast design

13.5.2.1   Cut, square-up, and perimeter holes.   When blasting a rock tunnel or mine drift underground, the first holes to be detonated (those with the shortest delays) create a *cut,* an opening toward which the rest of the rock is successively blasted (Fig. 13.9). Subsequent delays, called *square-up, helper,* or *reliever* holes, blast rock into the cut, spreading outward from the cut in a pattern of rings of increasing diameter until they reach the *perimeter,* which is the outside line of holes, often with reduced charge density. *Lifters* are those holes in the lower part of the round that remove rock down to the floor or invert level. For best positioning and throw of the muck, the cut should be centrally placed. Cuts low in the face tend to produce larger muck and much less throw. There are two main types of cut, an angle cut in

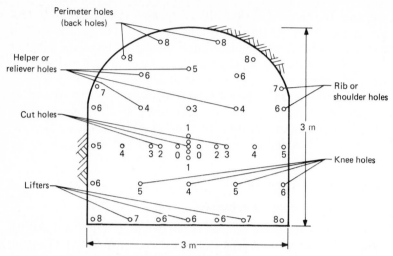

Figure 13.9 Typical 3- by 3-m mine heading with delay sequence. Also shown are some common terms for blastholes in the pattern.

which the blastholes are inclined inward to meet each other, and a parallel cut in which the holes are drilled parallel to each other in the direction of advance (Fig. 13.10).

13.5.2.2 Angle cuts. Angle cuts include *wedge* or *V cuts, fan cuts,* and *pyramid cuts.* A *draw cut* is like one side of a wedge cut, the other side breaking along a bedding plane or a convenient through-going joint. Angle cuts require fewer holes and less explosive, but are more difficult than parallel cuts to drill accurately. They tend to give a more erratic throw of the broken muck. Because advance is limited to about 65% of the heading width, angle cuts are used most often in wide headings, such as mine excavations for iron ore and rock salt.

13.5.2.3 Parallel cuts. Parallel cuts, which include *burn cuts* and *cylinder cuts,* have some of the holes left empty to provide release. Burn cuts have burdens greater than one diameter of the empty hole, and therefore have to be more heavily loaded than cylinder cuts with a closer spacing.

Large empty holes drilled by jumbo typically have diameters of 57, 76, or 125 mm. The smaller charged holes have diameters in the range of 30–50 mm and are loaded with a charge concentration of between 0.25 and 0.55 kg of dynamite per meter. When hand-held drills are used, the empty holes are the same size as the production holes.

13.5.2.4 Smooth wall blasting. The technique is very similar to the buffer blasting used in open-cut excavations (Sec. 13.4.4.4) in that the

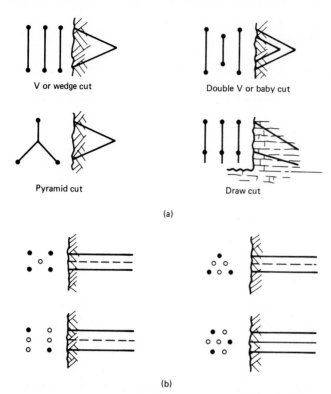

V or wedge cut

Double V or baby cut

Pyramid cut

Draw cut

(a)

(b)

Figure 13.10 Angle and parallel cut configurations. (*a*) Various angle cuts. (*b*) Various parallel (cylinder or burn) cuts. (*From Du Pont, 1980.*)

perimeter holes are closely spaced, lightly loaded, and fired simultaneously to remove the final burden in the tunnel round. Alternatively in underground presplitting, the perimeter holes are detonated immediately after the holes in the cut.

For smooth wall blasting the spacing of perimeter holes should be about 15 times the drillhole diameter, and the hole should be loaded with light distributed charges of 0.18–0.37 kg per meter of hole. Small-diameter cartridges are available that will give these loadings. The burden on the perimeter holes at the time of firing should always be greater than the spacing; a burden of 1.5 times the spacing is common.

13.5.3 Tunnel blasting

13.5.3.1 Full and partial face.
Tunnels and mine drifts of a diameter smaller than about 8 m, and larger headings in good-quality rock, are excavated *full face* by drilling and blasting the entire face in a single

round. As rock conditions deteriorate or the heading becomes larger, an upper section (*top heading*) is often removed first, followed by installation of crown support and then removal of the remaining bench. This procedure is less expensive and also facilitates both mucking out and the installation of support. In really bad ground, a smaller pilot drift is advanced and stabilized in the crown, followed by removal of the remaining rock in several stages.

13.5.3.2 Advance per round. Typically, a tunnel is advanced by blasting one to three rounds per day. The length of advance per round is limited by the quality of the rock and by the diameter of the excavation. The greater the tunnel diameter, the greater the release, and therefore, in general, the greater the advance or *pull* that can be achieved in any given quality of rock. Pull varies from about 0.5 m in very broken ground that requires immediate support, to as much as 3 m in massive and self-supporting rock in a large-diameter excavation.

The tunnel advance as a percentage of depth drilled varies from 50 to 95%. There is no point in drilling much deeper than can be pulled. An appropriate depth is determined from experience and by trial and error, and is quite sensitive to variations in rock conditions (Berg-Christensen and Selmer-Olsen, 1970). An important advantage of parallel cuts is that the advance per round can be adjusted to suit changing ground or support conditions or the degree of experience of the blaster, without modifying the pattern of drilling.

13.5.4 Shafts and raises

Two alternatives for shaft blasting are available, benching and full face. In benching (Fig. 13.11a) the two halves of the shaft bottom are blasted alternately. No cut is required. The method is good for wet conditions because a sump is automatically provided. In the full-face method (Fig. 13.11b), angle cuts are most often employed, although parallel cuts may be needed to limit fly rock and damage to timber supports. Angle cuts permit deeper rounds, but usually with more overbreak. Increasing the depth of a round means a more rapid advance, but overbreak means more mucking, greater difficulty with accurate alignment, and a greater volume of concrete to be poured if the shaft is to be lined.

AN/FO is almost never used because shafts are usually wet; dynamite or, more often, slurries are used instead. The average powder factor for a 3.0- by 4.5-m shaft is 3.25 kg of explosive per cubic meter of rock, varying from about 2 to 6 kg according to shaft size and rock strength.

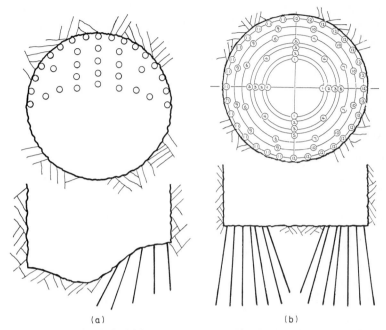

Figure 13.11   Typical shaft blasting patterns. (*a*) Bench method with hand-held drills. (*b*) Full-face method using jumbo-mounted drills.

Raises can be driven blind or can follow a pilot hole. All blasting in raises is electric because the safety fuse may not allow the miner to get out of the raise before the round goes off, nor does it permit the precise timing demanded by modern blasting techniques.

13.5.5   Mine blasting

13.5.5.1   Blasting methods.   The initial development openings (drifts and ramps) are driven in much the same manner as small tunnels, with mobile drilling equipment, small holes, and short rounds. Subsequent production blasting of the mine stopes, the openings from which the bulk of the ore is extracted, calls for much bigger rounds and often larger and longer blastholes. A free face is prepared, and the ore is broken so that it falls downward, letting gravity assist in the flow and fragmentation.

The main considerations are safety, economy, and selectivity. Dilution by waste lowers the grade of the ore and increases the unit cost, whereas leaving ore behind lowers the recovery and shortens the life of the mine.

Traditional methods, now used mainly in less competent ground, employ shorter rounds and smaller holes. Increasingly popular, particularly for mining of massive ore bodies in relatively stable ground conditions, are longer and larger blastholes (diameter 100–200 mm) drilled in a fan pattern 30° to either side of the vertical.

13.5.5.2 Short-hole blasting. In traditional short-hole blasting, one of three standard types of round is employed (Fig. 13.12). *Breasting* involves slashing a layer from the roof of the stope, using horizontal holes drilled 2–5 m into a vertical face. In *overhand stoping*, blastholes are drilled upward using a stoper drill, so that the rock breaks horizontally to a vertical or near-vertical free face. *Underhand benching*, similar to bench blasting in a quarry, uses blastholes drilled downward.

Dynamite is now often replaced by small-diameter slurries or by AN/FO when the holes are dry. These blasting agents give a better charge distribution in the hole, and so better fragmentation. Initiation is usually by safety fuse and igniter cord when the blastholes are short. However, electric blasting with short-period delay caps is best in wet conditions, and when sequencing may be a problem, particularly when the blast is extensive. Hand-held drills are the most common, although jumbo drilling can be used to increase production where there is sufficient headroom, particularly in room and pillar mining operations.

13.5.5.3 Long-hole blasting. Benching and crater retreat methods of mining make use of large-diameter (44–63 mm) longholes. The production blast is detonated one or two rows at a time, using either full column loading, decking (where individual charges are separated by about 1.5 m of sand and individually primed to ensure complete detonation), delay decking using detonating cord to initiate the charge sequentially upward, or bench slicing in which only the bottom of the slice is loaded. The greater development and capital costs of blasting

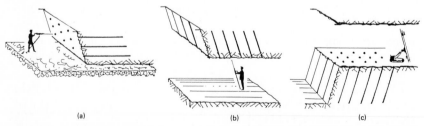

Figure 13.12 Short-hole methods for mine blasting. (*a*) Frontal stoping (breasting) with horizontal holes. (*b*) Overhand stoping. (*c*) Underhand stoping (benching). (*Hustrulid, 1982.*)

using long and large-diameter holes can be offset by an approximately twofold increase in production rates and a halving of drilling costs.

13.6 Control of Blasting Operations

13.6.1 Blasting specifications

Blasting patterns in mines are optimized over months or even years with the help of trial blasting, particularly in the early stages of mine development. The blasters soon gain an intimate knowledge of the ground conditions that pertain at the mine, although outside specialists are often retained to assist in the early stages, and to investigate problems that develop from time to time.

Blasting on civil engineering projects must be optimized much more quickly. Specifications are formalized. They often require the contractor to engage the services of a blasting specialist for designing and directing each blast, and to submit blast designs for the engineer's approval before the start of drilling and loading. Approval does not relieve contractors of their responsibility for safety, for producing the specified results, and for obtaining breakage sufficient for the handling of broken rock. Provisions are included for vibration control, as discussed below, and for the various types of controlled-perimeter blasting that may be needed at different locations (Hendron and Oriard, 1972).

Often an upper limit of about 100 mm is placed on the blasthole diameter to remove the temptation of bidding the job on the assumption that larger holes and large spacings can be used to give maximum production, at the expense of increased vibrations and damage to the perimeter. Maximum depth of lift and depth of subdrilling should be specified, and frequently the lift depth should be linked to the installation of rock support before blasting of a subsequent bench.

Hendron and Oriard (1972) recommend that rather than giving maximum powder factors, the specification should state maximum vibration levels together with locations and methods of vibration monitoring. This gives the contractor greater flexibility, allowing him to come up with low-cost, high-production blast designs while demonstrating that these make adequate provision for control of vibrations and damage.

Specifications for controlled-perimeter blasting should further give details of stemming materials and hole alignment tolerances, and should state the use of either simultaneous firing or millisecond delays to reduce ground vibrations. Those for presplit and cushion blasting should give the diameter, depth, and spacing of perimeter holes, and an approximate range for the charge per meter of hole. Unloaded perimeter holes are specified for line drilling. For smooth-wall blasting in tunnels, the specifications should require that all perimeter holes except the lifters be fired on the last delay period of the round.

The engineer has to choose between specifying the required end result (in terms of lines of excavation, backbreak, vibration levels, etc.) and the methods whereby these results are to be achieved. Specifications containing reference to both the methods and the results will often be contradictory, frustrating the contractor, and leading to poor results and disputes. Given the choice and a blasting contractor who knows the job, an end-result specification is usually the better alternative.

13.6.2 Control of blast vibrations

13.6.2.1 Vibration problems. In civil engineering work, blast vibration levels must be limited to minimize environmental impact, damage to nearby structures, and damage to the rock walls of the perimeter (Dowding, 1985). In mining, high levels of vibration can damage the open-pit slopes or underground pillars, and lead to substantial problems of safety and of subsequent recovery of ore from the blast-affected areas (Oriard, 1972). In stopes, damage to the hanging wall can lead to slabbing, high dilution, and the need for secondary blasting to relieve ore passes that become blocked by large fragments of ore and waste rock.

13.6.2.2 Remedies. Empirical predictions of blast vibration levels are often in error because of unknown and varying ground conditions. This leads to a requirement for vibration monitoring, and a program of trial blasting in which the blast pattern is progressively modified to achieve the required results. Seismographs can be used to record either the vibration pattern generated by a single round of blasting or, over a longer time period of weeks or months, the peak particle velocities of successive rounds (Bollinger, 1971; Dowding, 1985; see also Chap. 12).

Vibration levels can be reduced by limiting the charge weight per delay to an amount sufficient to achieve the required degree of fragmentation and no more. However, drilling costs tend to increase because more holes are needed. Vibration levels are also slightly reduced if there is no stemming, but in most civil engineering projects at least, problems of air blast and fly rock are more than sufficient to offset this advantage.

Another effective and usually supplementary method for reducing vibrations is to use delayed detonation and sequential blasting so that the vibrations from one line of drillholes can dissipate before the next row is detonated. Alternatively, short delays can be timed such that vibrations are out of phase with those from adjacent decks, holes, and rows. Ground vibration levels are related more to the weight of the explosive in any one short-delay period than to the total explosive in the round. In general, because of the timing errors of most initiation systems, some of the charges not separated by at least a nominal delay of 8 ms (or 15 ms, allowing a margin for more imprecise delays) may act together in causing vibration, whereas when delay blasting is used, the maximum level of vibrations is usually limited to about twice that from a single delay. At the Kidd Creek Mine in northern Ontario, Canada, for example, delays and decking were used together to reduce the vibration levels to 130 mm/s, in comparison with values of 600 mm/s that were recorded 30 m from the initial pattern of large-diameter blastholes.

13.6.2.3 Damage criteria for ground-transmitted vibrations. Permissible vibration levels are specified in relation to the levels that can be tolerated by the wall rock, by different types of building or manufacturing operations, and by people. Table 13.1 summarizes the typical levels of damage and human response experienced in relation to the peak particle velocity of ground-transmitted vibrations produced by the blast.

The most common criterion for the prevention of structural damage at the surface, for example, as recommended by the U.S. Bureau of Mines (Nichols et al., 1971), is that the peak particle velocity should

TABLE 13.1   Levels of Vibration in Relation to Blast Damage and Human Response

Peak particle velocity threshold, mm/s	Effect
600	New cracks form in rock
300	Falls of rock in unlined tunnels
190	Falls of plaster and serious cracking in buildings
140	Minor new cracks, opening of old cracks
100	Safe limit for lined tunnels, reinforced concrete
50	Safe limit for residential buildings
30	Feels severe
10	Disturbing to people
5	Some complaints likely
1	Vibrations are noticeable
<1	Barely perceptible vibrations

SOURCE: Hendron and Oriard, 1972.

not exceed 50 mm/s (2 in/s). Chae (1978) proposes instead a range of safe peak particle velocities as a function of the class of structure:

100 mm/s   Class A: structures of substantial construction

50 mm/s   Class B: relatively new structures in sound condition

25 mm/s   Class C: relatively old residential structures in poor condition

13 mm/s   Class D: old residential structures in very poor condition

He also presents graphically a relationship between maximum charge weight and distance for each of the four structural classes (Fig. 13.13). This permits estimating the safe charge per delay or the minimum safe distance. Note that for a distance of 30 m between the blast point and a class D structure, 1.8 kg of explosive per delay could be safely used, whereas the safe amount is increased to about 45.3 kg per delay for a class A structure at the same distance. Test blasts are recommended at the project site to establish a more site-specific scale distance if at all possible.

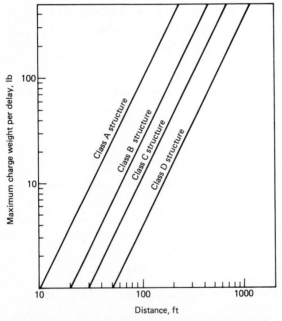

Figure 13.13   Maximum charge weight versus safe distance for various classes of structure. *A*—structures of substantial construction; *B*—relatively new structures in sound condition; *C*—relatively old residential structures in poor condition; *D*—old residential structures in very poor condition. (*Chae, 1978.*)

The frequency spectrum of the transmitted vibrations also plays a role in determining which blasts cause the most complaints. Frequencies in the range of 5–20 Hz are apparently the most annoying.

13.6.2.4   Damage criteria for air blasts.   Tolerances must sometimes also be placed on air-transmitted vibrations (*airblast* effects). Only exceptionally are airwaves from downhole blasting sufficient to endanger structures. Damage can occur, however, during above-ground demolition, during unstemmed blasting of tunnels and shafts, and where large quantities of detonating cord are exposed at the surface. Airblast overpressures greater than 0.7 kPa (0.1 lb/in^2) can break windows, and pressures greater than 7 kPa (1 lb/in^2) will almost certainly break all windows (Hendron and Oriard, 1972). Even in the absence of damage, complaints and legal action resulting from annoying levels of noise and vibration can close an operation down.

13.6.2.5   Predicting vibration levels.   Having specified an acceptable upper limit for vibration levels, the engineer has to predict the intensity of ground vibration as a function of charge weight, distance from the detonation, and properties of the ground (Northwood and Crawford, 1965; Hendron and Oriard, 1972).

The peak particle velocity v, which gives the best indication of whether a particular blast is likely to cause structural damage, is given by the empirical equation

$$v = H \left(\frac{D}{W^b} \right)^a$$

where D is the distance from the point of detonation, W is the maximum charge weight per delay interval, and H and a are constants determined by trial blasting. As can be seen, the peak particle velocity is proportional to the quantity D/W^b, which is called the *scaled distance*. The U.S. Bureau of Mines employs square-root scaling where $b = \frac{1}{2}$, which gives more conservative predictions than the alternative of cube-root scaling ($b = \frac{1}{3}$) used by some others.

The safe charge weight per delay W_s can be determined, using square-root scaling and for any given blast, from the formula

$$W_s = \left(\frac{d}{D} \right)^2$$

where d is the distance between the blast site and the structure, and D is the minimum safe scaled distance allowable.

References

Ariel, R. S.: "Explosively Constructed Dam of the Baipazinsk Hydro-Complex on the Vakhsh River," transl. 789, U.S. Bureau of Reclamation, Denver, Colo. (1968).

Ash, R. L.: "The Mechanics of Rock Breakage," *Pit and Quarry,* 98–112, 118–123, 126–131 (1963).

Ball, M. J.: "The Use of Explosives in Pipeline Construction Work," *Pipes Pipelines Internat.,* 29–33, 40 (Aug. 1976).

Bauer, A., and P. N. Calder: "Trends in Explosives, Drilling and Blasting," *Bull. Can. Inst. Min. Metall.,* **67** (742) (1974).

Berg-Christensen, J., and R. Selmer-Olsen: "On the Resistance to Blasting in Tunneling," *Proc. 2d Int. Cong. Rock Mech.* (Belgrade, Yugoslavia, 1970), vol. 3, paper 5-7.

Bollinger, G. A.: *Blast Vibration Analysis* (Southern Illinois Univ. Press, 1971), 132 pp.

Calder, P.: "Perimeter blasting," *Pit Slope Manual,* Chap. 7, Rep. 77-14, CANMET, Can. Centre for Minerals and Energy Technol., 82 pp. (1977).

—— and A Bauer: "Pre-split Blast Design for Open Pit and Underground Mines," *Proc. 5th Int. Cong. Rock Mech.* (Melbourne, Australia, 1983), vol. E, pp. 185–190.

Chae, Y. S.: "Effects of Blasting Vibrations on Structures and People," *Proc. 19th U.S. Symp. Rock Mech.* (Reno, Nev., 1978), pp. 312–318.

C.I.L.: *Blasters' Handbook,* 6th ed. (Canadian Industries, Montreal, Que., 1973), 545 pp.

Clark, G. B.: *Principles of Rock Fragmentation* (Wiley, New York, 1987), 610 pp.

Coursen, D. L.: "Cavities and Gas Penetrations from Blasts in Stressed Rock with Flooded Joints," *Acta Astronaut.,* **6**, 341–363 (1979).

Dick, R. A., L. R. Fletcher, and D. V. D'Andrea: *Explosives and Blasting Procedures Manual.* U.S. Bureau of Mines Information Circular IC8925 (1983).

Dowding, C. H.: *Blast Vibration Monitoring and Control* (Prentice-Hall, Englewood Cliffs, N. J., 1985).

Du Pont, Inc.: *Blasters' Handbook,* 175th anniv. ed. (E.I. du Pont de Nemours, Wilmington, Dela., 1977), 494 pp.

——: *Blasters' Handbook,* 16th ed. (E.I. du Pont de Nemours, Wilmington, Dela., 1980).

Gregory, C. E.: "Explosives for North American Engineers," *Series on Rock and Soil Mechanics,* vol. 1, no. 4 (Trans Tech Publ., Cleveland, Ohio, 1973), 276 pp.

Hendron, A. J., Jr., and L. L. Oriard: "Specifications for Controlled Blasting in Civil Engineering Projects," *Proc. North Am. Rapid Excavation and Tunneling Conf.* (Chicago, Ill., 1972), pp. 1585–1609.

Holloway, D. C., G. Bjarnholt, and W. H. Wilson: "A Field Study of Fracture Control Techniques for Smooth Wall Blasting," *Proc. 27th U.S. Symp. Rock Mech.* (Tuscaloosa, Ala., 1986), pp. 456–463.

Hong, M., L. J. Bond, and S. A. F. Murrell: "Computer Modelling of Seismic Wave Propagation and Interaction with Discontinuities in Rock Masses," *Proc. 27th U.S. Symp. Rock Mech.* (Tuscaloosa, Ala., 1986), pp. 488–495.

Hustrulid, W. A., Ed.: *Underground Mining Methods Handbook* (Soc. Min. Eng., Am. Inst. Min. Metall. Petrol. Eng., New York, 1982), 1603 pp.

Maerz, N. H., J. A. Franklin, and D. L. Coursen: "Measurement of Rock Fragmentation by Digital Photoanalysis," *Proc. U.S. Soc. Explosives Eng. 3rd Mini Symp. Blasting Research* (Miami, Fla., 1987).

Nichols, H. R., et al.: "Blasting Vibrations and Their Effects on Structures," Bull. 656, U.S. Bureau of Mines (1971).

Northwood, T. D., and R. Crawford: "Blasting and Building Damage," Canada National Research Council, Ottawa, CBD 63, 4 pp. (1965).

Oriard, L. L.: "Blasting Effects and Their Control in Open Pit Mining," in C. O. Brawner and V. Milligan, Eds., *Geotech. Practice for Stability in Open Pit Mining, Proc. 2d Int. Conf. Stability in Open Pit Mining·*(Vancouver, B.C., 1972), chap. 13, pp. 197–222.

Rosenthal, M. F., and G. L. Morlock: *Blasting Guidance Manual* (Office Surf. Mining Reclamation & Enforcement, U.S. Dept. Interior, 1987).

Solymar, Z. V.: "Blasting and Slope Stability," *Proc. 5th Int. Cong. Rock Mech.* (Melbourne, Australia, 1983), vol. C, pp. 123–128.

U.S. Department of the Army: "Systematic Drilling and Blasting for Surface Excavations," *Engineer Manual* EM 1110-2-3800, 140 pp. (1972).

Drilling, Breaking, and Cutting

14.1 Drilling

14.1.1 Drilling methods and objectives

14.1.1.1 Overview. Components of a drilling operation include a means of breaking or cutting the rock or soil; a means of cooling the drill bit and bringing the cuttings to the surface in a process known as *flushing;* and sometimes also a means for taking samples and for supporting the walls of the hole.

In exploratory drilling, water is the usual flushing medium, whereas compressed air is the main method in percussive drilling and also in rotary coring. Viscous drilling mud is used when the uncased drillhole walls need supporting to prevent collapse, and high-density mud prevents blowouts of high-pressure oil and gas in oil-well drilling operations.

In *open-hole* drilling, such as for blastholes, which accounts for the great majority of total rock length drilled in quarrying, mining, and tunneling, the two main objectives are to achieve maximum penetration rates and to minimize drill-bit wear. In other applications, such as rock exploration, the further objectives of producing high-quality core and stable drillhole walls take precedence. In oil-field drilling, minimizing damage to the formation takes a high priority.

14.1.1.2 Percussive and rotary drilling processes. Drilling processes can be broadly classified as percussive or rotary. Percussive drills break rock by a reciprocating, hammering action. In the earliest blasthole drilling, the "drill steel" was driven by a sledgehammer, and penetration rates of about 1 m/h were the best that could be achieved. Pneumatic-percussive rock drills were first employed in the latter part of the nineteenth century for blasthole drilling in the Fréjus tunnel in France and the Hoosac tunnel in Massachusetts (Ottosson and Cameron, 1976). Tungsten carbide drill bits, introduced in 1940, gave much improved penetration rates and increased bit life (Fig. 14.1). The oil-hydraulic percussive drills used in tunneling projects since about 1971 can achieve penetration rates of up to 90 m/h in hard rock.

Present-day soil exploration *churn* or *cable tool* drills still operate by percussion, using a heavy chisel which is repeatedly dropped and retrieved. Soil samples are taken, when required, using a falling weight to drive a tube sampler into the ground. Churn drills can only penetrate satisfactorily through soils and very soft rock. Drilling is relatively slow and produces a hole of between 10 and 30 cm in diameter.

Soils are usually bored with an auger rather than drilled using a coring bit. Bulk samples are taken from the auger flights, or the auger is withdrawn intermittently and a *split spoon* sampler is hammered or a thin-walled tube is pushed into the base of the auger hole. Soil boreholes usually require casing to prevent collapse, particularly when boring is to be followed by rock drilling in the same hole. The casing is advanced closely behind the auger, then drilled or driven for a short distance into the top of the rock to provide a positive seal.

Figure 14.1   Percussive drill bits; various designs with button and blade inserts of tungsten carbide. (*Courtesy of Boart Canada, Inc.*)

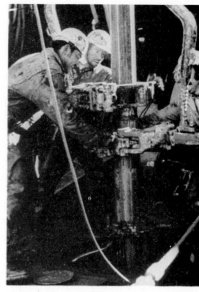

(a)   (b)

Figure 14.2   Rotary oil well drilling. (a) Drill rig. (*Courtesy of Hughes Tool Company.*) (b) Adding sections of pipe to the drill string.

The simplest form of rotary drilling is auger boring, in which a solid or hollow-stem auger is rotated into the ground without mud. *Continuous-flight* augers convey the cuttings continuously to surface. *Single-flight* augers are used to bore large-diameter holes, such as for foundation caissons. A short length of auger is screwed into the ground and withdrawn to surface, and the cuttings on the flights are "unloaded" by spinning the auger.

Rotary drills of various types give the rapid and deep penetration required for the exploration and development of oil, gas, and geothermal energy resources (Fig. 14.2). Major advances in the twentieth century have permitted holes as deep as 10 km to be drilled in extremely difficult rock.

Core bits used for rock exploration (Fig 14.3) cut an annular groove, leaving the central core sample in a relatively undisturbed condition. Rotary drag bits and tricones equipped with tungsten carbide or diamond-impregnated studs or milled teeth (Fig. 14.4) bore large-diameter open holes through the harder rock formations.

Tricone bits were developed in the 1930s. More recently, solid rotary bits impregnated with diamonds and new synthetic materials have increased penetration rates in the harder rocks, but the basic tricone bit still dominates rotary drilling (Fig. 14.4). High-pressure, high-

Figure 14.3 Diamond surface-set coring bits, A- to N-sized, with face and side discharge, stepped wireline, and straight types. A casing shoe and a noncoring bull-nosed bit are shown top and bottom left. (*Courtesy of JKS Boyles International, Inc., Toronto.*)

capacity hydraulic pumps developed in the first two decades of this century remove cuttings from the drillhole and provide superior bottom hole cleaning when combined with small nozzle jets on the bits themselves. The engineered drilling fluids introduced in the 1930s, usually of expanding clay (bentonite) with chemical additives to control properties, give improved drillhole stability and allow the containment of high formation fluid pressures.

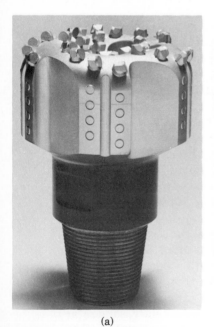

(a)	(b)	(c)

Figure 14.4 Rotary drill bits. (*a*) Drag bit (Blue Chip polycrystalline diamond bit). (*b*) Tricone bit (tungsten carbide, journal-bearing bit). (*c*) Bull-nosed bit (Hughes thermally stable polycrystalline diamond bit P18). (*Courtesy of Hughes Tool Company.*)

Much rock mechanics research has been directed toward understanding the mechanisms of fragmentation during drilling, in attempts to improve the design of conventional mechanical bits (diamond bits, percussive bits, or rotary drag bits) and to develop new and improved methods of drilling (Marshall and Sowden, 1971; Mouraz Miranda and Mello Mendes, 1983; Chugh, 1985; Bourgoyne et al., 1986). Various novel techniques have been investigated, including abrasion by waterborne pellets, breakage by projectiles fired against the rock surface, thermal cutting using flame jets and lasers, and erosion of rock by high-pressure water jets. Of these, flame and water jet cutting are the most developed and are used commercially in certain specialized operations, not only for drilling but also to assist in tunnel boring.

14.1.2 Percussive drilling

14.1.2.1 Pneumatic-percussive drills.
Pneumatic-percussive drilling tools employ an air-driven hammer. The conventional pneumatic drill is powered by a 100-mm-diameter air supply at a pressure of 690 kPa (100 lb/in^2). The kinetic energy of a pneumatically driven piston is transformed into a strain pulse traveling along the full length of the drill rod, which acts as a wave guide. Rock is broken by the drill bit, usually a tungsten carbide insert blade or button design (Fig. 14.1). This traditional method, with the hammer inside the drill and out of the hole, can produce a 150-mm-diameter hole to depths of up to 30 m, and smaller holes to depths of about 120 m. Energy losses occur at the hammer/drill steel contact, at rod couplings, and at the interface between drill bit and rock. Deep holes are drilled with considerable energy losses and slow rates of penetration. Hustrulid and Fairhurst (1971, 1972) have derived a formula for predicting the penetration rate knowing such operational parameters as blow energy and blow frequency. Various other researchers report on the efficiency of the drilling process (e.g., Paone et al., 1969; Wu, 1986).

Drills using down-the-hole (DTH) air-driven percussion hammers can penetrate much deeper because they avoid losses of energy during transmission of the stress wave down the drill rod. Hole sizes vary, although most drilling with downhole hammers is at 100–150-mm diameter, sometimes larger. Capital and operating costs are usually reduced, although higher air pressures are needed, and DTH hammers also give less hole deviation.

14.1.2.2 Hydraulic rock drills.
Ottosson and Cameron (1976) describe the hydraulic-percussive drills that are increasingly replacing pneumatic drills. Large compressors and air lines are replaced by hydraulic

pumps powered by electric or diesel motors right on the rig. Pressure losses and air leaks that lead to inefficient drilling are eliminated, and the self-contained hydraulic drills are smaller and lighter. Capital costs of air compression equipment are saved, operating costs are lower, and holes can be more precisely placed, reducing overbreak.

Hydraulic drills give greater penetration rates without increasing drill rod stresses. By adjusting pressure and flow rate using variable displacement pumps, the impact energy and impact rate can be varied to obtain maximum performance. Net penetration rates are usually 50% higher than those obtainable with pneumatic rock drills, resulting from an impact power output about 50% higher than the output of the equivalent pneumatic rock drills.

Whereas water is injected down the center of a pneumatic drill steel, in hydraulic drilling a more powerful flushing is achieved at pressures of up to 1 MPa by a flushing head mounted at the front of the drill. Hydraulic rock drills also make less noise because there is no discharge of compressed air. The absence of air exhaust and associated water mist leads to an improved working environment and visibility.

Electric power means electronic and automatic control and optimization of drilling operations; an automatic feed control monitors advance and prevents the steel from becoming stuck. When the correct depth is reached, the impact mechanism is turned off and the drill retracted. Automation and electronic control relieve the driller of much of the physical labor traditionally associated with hard-rock drilling, but require new skills and additional training.

14.1.2.3 Drill mounting systems. The weight and the method of mounting the drill limit the available thrust, and the depth and diameter that can be drilled. By 1963 pneumatic drills supported by *air legs* (pneumatically operated pistons) were penetrating rock at rates of 40 m/h. Pivoted air legs give the thrust needed when drilling horizontal or inclined holes (Fig. 14.5). *Stoper* drills have an air leg in line with the drill, and are used for drilling upward only, such as in roof bolting.

For drilling blastholes, track-mounted bench drills and mechanical *jumbos* for tunneling and drifting underground have largely replaced hand-held drills (Fig. 14.6).

Most quarries use a pneumatic drill mounted on a wheeled or tracked undercarriage, termed *wagon drill* or *air track*, respectively. Drills producing holes with diameters of typically 50–75 mm are most often used for long-hole production blasting in hard-rock quarries. They will comfortably penetrate 6–9 m in average ground. Self-propelled crawler drills using jointed steel (Fig. 14.6a) can drill 100-mm-diameter holes up to 30 m deep. They are often equipped with a

Figure 14.5   Hand-held air leg drilling. Tunnel in Riyadh, Saudi Arabia.

separate flushing attachment through which high-pressure air or water is forced down the steel in a continuous stream to remove dust and rock chips.

A jumbo (Fig. 14.6b) is a self-propelled machine with several drills mounted on hydraulically operated booms. It gives a higher rate of production and lower operating cost. Development drilling is more rapid, which is particularly important in long-hole mining. Jumbos, however, require several headings if they are to be used effectively. Capital costs and maintenance costs are high.

Drilling jumbos equipped with heavy pneumatic drills can now penetrate at rates of 70 m/h through granite, and are standard in tunneling. Drill steels employed in tunneling are typically 37 mm in diameter, with a forged thread at the shank end. For the central cut holes in the tunnel face, 76- or 102-mm button bits are employed. The remaining holes of the round are completed using conventional drill bits with diameters of 43–48 mm. The latest drill jumbos being delivered to mines can be programmed to drill an entire face without human intervention; hence they fall into the category of robots. Roof bolters operated remotely with the operator standing in a safe location represent the latest approach to high productivity and safety in bolting operations (see Fig. 16.6).

14.1.2.4 Aids to drilling precision. Precise drilling is important if smooth, undamaged walls are to be produced by blasting. Angle-measuring devices such as the Swedish Ilmeg system are available

(a)

(b)

Figure 14.6   Drilling vehicles. (*a*) Hydraulic crawler for benching (Tamrock Zoomtrack). (*b*) Hydraulic jumbo for drilling blastholes underground (Tamrock Maximatic).

which, when fitted to a drilling jumbo, allow the drill operator to monitor both vertical and horizontal drilling angles to the nearest 0.1°.

14.1.3 Rotary drilling

14.1.3.1 Applications. Rotary methods are employed in practically all oil-field operations and for most exploration drilling and large-diameter blasthole drilling in mining. Most water-well drilling is rotary except in harder rock, where DTH hammers (Sec. 14.1.2.1) are used to advantage. Rotary drills break apart the rock by either a crushing or a milling action.

In mining, the trend is toward large-diameter rotary blasthole drilling with tungsten carbide toothed tricone bits, which produce holes of 171–270 mm diameter for general-purpose blasting, and up to 380 mm diameter for stripping overburden in open-cast coal mining. Holes are typically 18–27 m deep, and are drilled at the rate of 60–120 m per shift for harder rocks. Output is high, but the machine is large and slow to move from hole to hole.

14.1.3.2 The drill string. The rotary drill bit penetrates rock by transmitting torque and normal load to the base of the hole through the drill string (Fig. 14.2b). In deep oil-well drilling, the entire drill string rotates, but only the special lower portion, consisting of extremely heavy pipe sections, is in compression; the remainder is in a state of tension. When a bit wears out, the drill pipe is removed, the bit is changed, and the pipe is lowered down the hole to resume drilling.

Turbo drilling, where torque is transmitted to the bit by a downhole turbine activated by the drilling fluid, is used extensively in the U.S.S.R. and is also becoming popular in North America and elsewhere, especially when used for *directional* drilling. The drill string is not required to rotate, so it can be oriented in a given direction and the bit made to "sidetrack" in that direction by means of either a short bent pipe section, a bent drill housing, or a deflection shoe on the turbo drill.

Drilling successfully to the depths of most oil and gas wells requires the placement of casing (a tubular steel drillhole liner) to a depth of typically 10% that of the drillhole, to isolate the unstable, permeable, and environmentally sensitive near-surface deposits and to provide a safe means of shutting in high fluid pressures encountered in some strata. A smaller drill bit is then used to deepen the hole, and other smaller casings are installed as needed to isolate zones of high pressure or unstable ground. Bit diameter is reduced by about 65–80 mm each time a new casing string is inserted. Therefore deep holes often start off with bits of 600-mm diameter or larger, and end up with bits as small as 125 mm, drilling a hole into which a production casing of 100-mm diameter is lowered.

14.1.3.3 Optimization of drilling variables. The major controllable factors affecting the penetration rate are the bit type for a particular rock, the thrust on the bit (also called *bit weight*), the speed of rotation, the properties of the drilling fluid, and the hydraulics of the fluid system. The drilling engineer's responsibility is to optimize these factors.

Softer rocks are usually drilled with faster rotation (60–100 r/min) than the harder limestones and sandstones (40–70 r/min). Harder rocks also require more thrust. Rotary speed and thrust need to be optimized together, for as one is increased, the other must usually be decreased. Solid *plug* bits without bearings (Fig. 14.4c) can be rotated at higher speeds because there are no bearings to overheat or wear out.

14.1.3.4 Bit types. Toothed bits are used for drilling soft to medium-hard rock (usually less than 200-MPa compressive strength), and button bits with tungsten carbide inserts for the harder rocks. Bits can come with regular bearings, sealed bearings, or journal bearings. The former are by far the least expensive and are generally used for quarry and mining applications. Deeper oil-field applications warrant the use of the latter types.

The tricone bit (Fig. 14.4b) works mainly by subjecting rock beneath the teeth to a cyclical compressive stress, which generates spalling and tensile fractures. The cone axes are also offset from center to impart a gouging, scraping action. In oil-well drilling, soft rocks such as clay shales, chalk, calcareous shales, and sands and gravels with little cohesion are drilled with bits having long, widely spaced teeth and a large cone offset. Hard strata such as dense limestones, anhydrite, dolomite, chert beds, dense quartzitic sands, and volcanic rocks, for example, basalt, are most efficiently drilled with short, closely spaced teeth, usually of tungsten carbide, with a minimum of cone offset.

Recent trends are toward increased use of tungsten carbide teeth, even for long teeth because of added bit life, and of synthetic industrial diamonds and other new materials. Diamond plug bits are employed to drill very hard and abrasive rock or at depths below about 3 km, when the time consumed in changing bits exceeds the life of the toothed or button bits. New materials, such as Stratapax developed by General Electric, in solid bits without rotating cones are gradually replacing tricone bits because of much longer bit life (they have no bearings to wear out) and superior scraping action under high rotation speeds and bit weights.

14.1.3.5 Bit selection. Proper bit selection is a major factor in reducing the number of bit changes, increasing penetration rates, and reducing drilling costs. The correct bit choice is critical, particularly for offshore and deep-hole drilling where bit costs are minimal in relation to the huge costs of even an hour of delay.

Improvement of penetration rates is the most active area of research in drilling engineering. Most oil companies and bit suppliers have developed elaborate computer programs to guide bit selection. Criteria are developed from thousands of borehole records, and by laboratory tests followed by field proving. In several laboratories, full-scale (100–170-mm) rock drilling is carried out in high-pressure chambers. Measurements and analyses are so comprehensive that drilling parameters for a given rock type, drill bit, and fluid can be optimized in a few minutes, although it may take several weeks and considerable expense to set up the test. The success of the bit choice and the suitability of new bit designs in various rock types are evaluated by keeping careful drilling records.

Variables that influence bit choice and affect the wear and breakage include rock properties (compressive strength, hardness, abrasiveness, porosity, permeability, the presence of dispersible and swelling clays), rock temperature and stress, capabilities of the hydraulics and drill rig, and properties of the drilling fluid being used. The choices are most difficult when closely spaced alternating beds of very different rocks must be drilled, such as beds of dense limestones and swelling clay shales.

14.1.4 Hydraulics of rotary drilling

14.1.4.1 Role of a drilling fluid. A drilling fluid cleans cuttings from the base of the hole and carries them to the surface, exposes fresh rock for cutting, and also cools and lubricates the bit. Drilling *muds* create a low-permeability filter cake on the drillhole wall adjacent to porous or water-sensitive strata, reducing friction and abrasion on the drill string.

Another important mud function in drilling is to preserve hole stability and, when drilling for hydrocarbons, to prevent oil and gas blowouts. Extremely high fluid pressures, sometimes approaching the magnitude of the entire overburden stress (approximately 23 MPa/km in most sedimentary basins) can exist at depth and must be contained. Drillhole stability in uncased sections is provided by the pressure of a dense drilling mud and the formation of the filter cake which slows down liquid penetration into the formations. Pressure control requires a mud weight (i.e., density) sufficient to counteract the fluid pressure in the formation.

The specialized science of drilling hydraulics involves designing the drilling fluid and selecting pumping rate, velocity, and pressure to give optimum hole stabilizing and cleaning characteristics.

14.1.4.2 Jet ports. The drilling fluid is discharged through nozzles in the bit to strike the base of the hole with high velocity, removing loose rock chips and penetrating pores and fissures in the intact rock just ahead of the bit. The higher the pore pressure, the lower the effective stress, therefore the lower the resistance to drilling. Chips are popped off the bottom, and turbulence keeps the bit teeth clean of balled, remolded clay and packed cuttings.

Jet ports, up to three per tricone bit, are as small as 5 mm under some conditions. For soft rocks, bits with extended nozzles bring the jet close to the hole bottom. In many weak mudrocks, such as the shallower Cretaceous and Tertiary clay shales found from the Gulf Coast to northern Alberta, Canada, the drill string can virtually be "pumped" down because the jets are very effective in removing soft shale. However, straight jetting is usually avoided because uncontrolled washouts need to be backfilled by large volumes of cement after casing has been installed.

14.1.4.3 Types of fluid. Water as a drilling fluid is inexpensive and often adequate when drilling at shallow depth (less than 1 km) in nonpetroliferous rocks. Usually the most rapid penetration rates are achieved with the minimum solids content in the mud. Sometimes potassium chloride or a surfactant such as sodium polyphosphate are added to inhibit the swelling of clays.

Greater viscosity is required when drilling deeper. Before the development of scientifically controlled muds, a natural mud was often allowed to develop by letting clayey cuttings disperse into the drill water, or using overburden clays mixed in a mud pit. The remarkable colloidal properties of bentonite (smectite) were recognized in the 1920s, and by the 1970s, virtually all fresh-water muds used in rotary drilling were slurries of bentonite.

Common additives include barium sulfate (barite, specific gravity = 4.25), a dense low-abrasion material abundantly available in nature. Stable drilling fluids with a density approaching 25.5 kN/m^3 can be made with water, bentonite, barite, and chemical stabilizers. Dispersion of clay particles is controlled by adjusting the pH and the surface activity of the clays using organic compounds such as lignins, tannins, lignosulfonates, or polyphosphates. To improve filter cake behavior, polymers such as sodium carboxymethylcellulose can be added. In the case of a sodium chloride mud, used to drill through halite, attapulgite clays provide viscosity and gel strength, with natural pregelatinized starch added to improve the filter cake.

Polymer drilling muds are fresh-water based and use macromolecular organic agents such as guar gum, treated cellulose products, and polyacrylamides. Potassium chloride may be added to make the filtrate slightly saline, so as to prevent the sloughing of shales, but polymers rapidly lose their effectiveness as the level of salinity rises. Polymer muds achieve low filtration and medium viscosity without use of solids such as clays, which tend to reduce drilling rates. Clay and silt contaminants are removed by a hydrocyclone at the surface. Because polymer is removed along with impurities, small quantities need to be added continuously to maintain drilling fluid properties.

Oil-based muds consist of oil, an emulsifier, and a water phase present as stabilized colloidal droplets. The desired rheological and filtration properties are achieved by adding colloidal-sized asphalt, soaps of high molecular weight, and inert minerals to control density. They are used when drilling through water-sensitive shales and soluble formations and through reservoirs likely to be damaged by water infiltration. For example, they have been employed when drilling perimeter freeze holes for shaft sinking in the potash rocks of Saskatchewan, Canada (Sec. 18.3).

Lightweight fluids are often used when there is no requirement to prevent blowouts. With a mud weight less than the formation pressure, flow toward the hole bottom assists in chip removal. This has resulted in air-drilling technology, foams, and lightweight oil muds, all giving higher penetration rates than conventional mud.

14.1.4.4 Selecting mud density. In a deep drillhole passing through several formations, a mud column dense enough to control rock squeeze or blowout at the base of the hole can cause hydrofracture higher up, leading to uncontrolled loss of fluid. To avoid this, a string of casing is set to a depth just above the zone of elevated pressure, and then a mud of the desired density is placed before drilling deeper.

The producing horizon must not be damaged by the introduction of incompatible fluids or solid particles. The fluid pressure must be controlled to prevent hydraulic fracture. Usually it must be close to the formation pressure, otherwise high seepage gradients keep the rock "plastered down" and give reduced rates of penetration.

14.1.4.5 Fluid circulation and monitoring. The fluid circulation system includes surface settlement and treatment tanks, and vibrating screens to remove cuttings. Tank volume also provides a surge capacity to handle emergencies, and special application muds can be prepared while active circulation continues. An area of one tank, usually just before the entry to the major circulating pump, has a hopper and mixer to add and mix chemicals, bentonite, or barite weighting material.

In most modern drilling operations, detailed records are maintained of all mud additions, properties, pressures, and rates of circulation, as well as of drilling parameters such as torque, rotation rate, bit weight, drill string assembly details, and penetration rate. These records are critical in developing drilling expertise with different subsurface conditions.

The major relevant properties of water-based drilling muds are the density, used to control pressures; the dynamic viscosity, for chip-carrying capability; the static gel strength, to keep particulate matter in suspension during static periods; and filter cake development, to protect formations from damage and prevent the drill pipe from sticking in the hole. Other relevant properties are corrosion potential, the emulsified oil content which aids lubrication, the nonbentonitic solids content which reduces penetration rates, the liquid salinity and chemistry, and the pH of the liquid phase.

14.1.5 Directional drilling

14.1.5.1 Applications and methods. Directional methods are needed for multiple-hole drilling from an offshore platform, to intersect mineral ore bodies with certain configurations and alignments, and for the long horizontal holes that are starting to gain popularity for geotechnical exploration. Holes can be maintained straight or deflected by wedging or by oriented turbo drilling. They can be surveyed during drilling or on completion after having been allowed to follow a path dictated by the rock and the flexibility of the drill string.

Directional drilling is particularly useful from offshore platforms, and allows holes to reach many targets without having to move the drill rig and set up a new site. Holes can be drilled so that they are almost horizontal when they enter the target reservoir, which has particular application in thermal recovery methods.

Deviated holes are more difficult to drill than conventional ones, a major factor being the high friction of the rotating drill pipe lying against the drillhole wall. Nevertheless, in 1955 horizontal exploratory holes 75 mm in diameter and more than 500 m long were drilled from each end of the planned Pennsylvania Turnpike tunnel. A hole 305 m long was bored for the Straight Creek highway tunnel. In 1972 horizontal exploratory holes as long as 1.6 km were driven for the Seikan tunnel in Japan through soft to medium-hard volcanic rocks, using conventional rotary drilling techniques and a 170-mm-diameter tricone bit. Special techniques were used to deviate the hole in the vertical and horizontal directions. The hole was surveyed every 10 m with either a Sperry-Sun EC inclinometer or a magnetic multishot survey instrument.

The U.S. Bureau of Mines has investigated the use of long small-diameter horizontal holes drilled in coal seams for purposes of methane relief. Drilling can be guided accurately to its target with cableless telemetry survey systems.

14.1.5.2 Drillhole surveying. Hole surveying equipment can measure various combinations of inclination, horizontal bearing angle, and axial orientation of the drilling assembly. *Single-shot* devices only produce one reading per survey. *Multishot* or continuous-readout survey instruments record many hundreds of readings photographically or electronically.

The hydrofluoric acid etch tube consists of a small glass tube partly filled with hydrofluoric acid. It is run into the hole and brought to rest, and after 15–20 min is withdrawn. The vertical angle of the hole is determined directly from the etch line created during the period that the tube remained at rest. In the melting gelatine method, a small glass tube filled with liquid gelatine and containing a small open floating magnetic compass is run into the hole and brought to rest; the gelatine congeals and fixes the compass needle in position.

In a plumb-bob single-shot device, the survey instrument is loaded with a paper bull's-eye and run down the center of the drill string to the hole bottom. After a certain elapsed time, a needle plumb bob is pressed against the paper, leaving a needle hole that shows the inclination. The device comes with a compass option as well. In another method, radioactive paint on the tip of a compass needle is recorded on film. Photographic inclinometers take pictures of a ball bearing or pendulum.

Probably the most common drillhole survey instrument used today is the gimbal-mounted magnetic compass, which is positioned in a section of nonmagnetic drill pipe near the bit. In some instruments, survey data are sent to the surface through the drill pipe using a telemetry signal. One of the more widely used survey tools is produced by the Sperry-Sun well surveying company. A steering survey tool is pumped down the inside of the drill pipe and retrieved on a wire line. Survey data from the tool are transmitted to the surface through an electric conductor. Data include the bearing angle, vertical angle, axial orientation angle, and bottom hole temperature. Gyroscopic devices are available from borehole geophysical logging companies which give accurate total hole surveys, and the hole bottom can be located within 1 m at distances in excess of 2 km.

14.1.6 Coring

Rock coring methods were introduced in Chap. 6. In the present chapter, the technology of obtaining good-quality core samples is reviewed in greater detail (Acker, 1974; Cumming, 1975; Heinz, 1985).

14.1.6.1 Core barrels and bits. Conventional rock coring drills make use of sectional *drill rods* in 2- or 4-m lengths (more often, even nowadays, in 5- or 10-ft lengths), with threaded connections and a central hole to carry the water or air used in flushing the ground-up rock and cooling the bit. A core barrel to hold the core and bring it to the surface is attached at the cutting end.

The barrel is armed with a drill bit that can either be diamond impregnated or diamond or tungsten carbide studded, depending on the rock to be cut. The following standard sizes and designations of coring bits are in common use:

Designation	Hole diameter (OD), mm	Core diameter (ID), mm
EX	37.3	21.5
AX	47.6	30.1
BX	59.6	42.0
NX	75.3	54.7
HX	98.8	76.2

The correct choice of drill bit is important. The larger diameters tend to give better recovery of core. The flushing method and pressure also have to be selected to suit the rock conditions. Water is more commonly used than air as a flushing medium, and tends to be less damaging to the core. Water is stored and recirculated when in short supply, or when it would otherwise have to be pumped from a remote site. Good water circulation must be maintained, particularly in clay-bearing rocks, otherwise mud may cake inside the core barrel and cause overheating, excessive friction, and breakage of the core.

Older designs of rock drill have a mechanical gear drive that rotates the rods at constant speed, irrespective of resistance at the bit or in the barrel. Only the thrust is hydraulic. Rotational speed can be varied only by changing gears. More recent designs are equipped with hydraulic motors for both thrust and rotation. Thus they have a continuously variable rotational speed that is responsive to blockages and tends to be less damaging to the core.

14.1.6.2 Double- and triple-tube methods.

The most common method of coring is the *double-barrel* method, where a nonrotating inner barrel is suspended on an axial thrust bearing, and the bit protrudes below the entry to the inner barrel. As the rock is drilled, the core enters the inner barrel and is retained by a core catcher. Upon completion of a single coring run, usually 1.5–3 m long in geotechnical studies (Chap. 6), but 6–20 m long in oil-well exploration, the entire drill string is brought to surface. The core is removed, and the barrel and the string are reassembled.

If the formations are fragile, the inner barrel can be fitted with a PVC or split steel liner and a one-way check valve at the top to protect the core even more. The combination of a double barrel with an inner liner is called *triple-tube* coring. This inner liner is then typically pumped out of the inner tube when the barrel is brought to surface. The core can be preserved in a much less disturbed state than after dumping it from the end of the inner tube in a double-tube barrel.

Exploratory drilling can be difficult in weathered rocks, where slow progress and poor recovery can result from the contrast between weak residual soils and strong unweathered corestones. In the weathered granitic and volcanic rocks of Hong Kong, more than 500 m of hole are drilled each day. Exploratory drilling in "soil" (rock with grade IV to VI weathering; see Chap. 6) makes use of rotary wash boring. If in an urban area, the drill casing is often surged down using water from a fire hydrant. On refusal to wash boring, drilling is continued with a double-tube core barrel to confirm usually 5 m of less weathered rock (Brand and Phillipson, 1984).

Triple-tube core barrels are employed in Hong Kong for good-quality sampling, using nonretractable barrels for rock (grade I to III weathering), but retractable ones for coring weak or variable materials (grade IV to VI weathering). The *retractable barrels* have a cutting shoe that projects ahead of the drill bit when "soft" material is being cored, to protect the sample from the drilling fluid. The inner tube and shoe retract against a spring when hard material is encountered. The Mazier triple-tube sampler is commonly used but cannot core rock. A Triefus triple-tube barrel which produces a 63-mm-diameter core is frequently employed both as a retractor barrel for soils and as a nonretractor barrel for rocks. Triefus triple-tube barrels producing 102-mm-diameter cores have also been used successfully with air foam as the flushing medium to obtain excellent core quality even in residual soil and colluvium.

14.1.6.3 Wire-line drilling. Conventional diamond coring requires making a round-trip in and out of the hole to recover the core once the barrel is full, or a core block occurs. A tall derrick or drilling mast is desirable to increase "stand" lengths and thus reduce turnaround time, but even then the drill rods must be uncoupled and reconnected at each stand, which becomes slower and more costly as the hole penetrates deeper.

Wire-line diamond coring is faster, particularly for drilling deep holes, because the inner tube is retrieved on the end of a wire rope without the drill rods or outer core barrel. This eliminates the considerable time and cost of withdrawing and reinserting the complete drill string. After removal of the core, the inner tube is dropped, lowered, or pumped back to the bottom of the hole. The retrievable inner split barrel used for oil-well drilling is usually 3.5 m long. Geotechnical and mineral exploration applications more often use 1.5- or 3-m split inserts (Bridwell, 1967; Varlamoff, 1979).

14.1.6.4 Special precautions in coring sensitive rocks. The most difficult materials to core successfully are oil sands, highly compressible rock, heavily jointed rock, sensitive shales and swelling mudstones, and carbonates with large vugs. When sampling loose materials, special core catchers are required to prevent the sample from dropping or flowing out of the core tube.

If double-tube or triple-tube technology is still insufficient to obtain high-quality rock cores, other methods can be employed. In one method, a latex or neoprene rubber sheath is extended to enclose the core (*rubber-sleeve* coring). It gives some confinement, helping to keep the core intact. With the *pressure-core barrel,* sometimes used in the oil industry, a 3-m segment of core is drilled and drawn up into a chamber which has been prefilled with a special noninvasive fluid. The chamber is sealed while still at the base of the hole. Then the core is brought to surface and frozen to the temperature of dry ice before the barrel is disassembled.

14.1.6.5 Continuous coring. Continuous coring is a method in which core is broken off in 150-mm or shorter lengths behind the bit, and pumped to the surface inside the drill pipe by reverse circulation of the drilling fluid. It can be used only if the core is competent enough to withstand the trip up the entire drill string while being flushed by the moving fluid column. Core quality suffers, but progress is faster than with wire-line core retrieval since penetration is not halted to retrieve the core-laden inner tube. Some systems make use of dual-walled drill pipe, where fluid is pumped down the annulus between concentric pipes. Others rely on pumping the fluid down the annulus between the hole and the drill pipe, requiring that the rock be tight enough to ensure that most of the fluid returns to surface rather than being lost in the formation.

14.1.6.6 Calyx coring. Steel shot or calyx coring is a method, now seldom employed, for drilling large-diameter vertical holes and shafts to relatively shallow depths. It used small steel balls trapped in the kerf of, and rotated by, a heavy walled barrel. Tungsten carbide cutters welded to the kerf of the core barrel more typically replace this technique today.

14.1.7 Water jet drilling

High-pressure water jets are being used commercially for certain specialized rock drilling and slot cutting applications. For example, a portable water jet can be purchased for drilling rockbolt holes, and has the advantage of producing a rough hole for better bonding with grout. Water jets have also been tried as a means of assisting mechanical breakage in tunnel-boring operations (Chap. 15). Cooley (1972) reports that water jets can break any type of rock, and do not require high axial forces to be applied to the rock face. Potential disadvantages are a high noise level unless the jets are properly muffled, and a water disposal problem.

Continuous jets are produced by pumps or pressure intensifiers, and *pulsed* jets by "water cannons," in which the impact of a free piston extrudes water through a nozzle. When a continuous jet is directed against rock, the erosion rate decreases as the eroded hole deepens. Water present in the hole absorbs energy and decreases the efficiency of cutting. Experiments have mostly been with jets smaller than 4 mm and at pressures of around 100 MPa. Jet pressures of up to 1200 MPa have been tried with nozzle diameters of 1 mm.

A self-propelled water cannon on tracks, with a pulsed water jet at a pressure of 620 MPa, has been used in the U.S.S.R. to break shale and sandstone with compressive strengths of up to 117 MPa, and also to break coal. Water cannons appear well suited for coal mining because of the reduced spark hazard and control of coal dust. Soviet researchers state that the jet pressure of a water cannon should be at least 10 times the compressive strength of the rock.

14.1.8 Thermal drilling

Taconite ore, presently the chief source of iron in the United States, is extremely tough and difficult to drill by conventional means. To overcome this problem in the late 1940s, Union Carbide developed a jet-piercing tool that burned fuel oil with oxygen to spall the rock. The combustion chamber is water-cooled, and mechanical cutters are used to remove rock not amenable to spalling (Williams, 1986). Hole diameters can be increased just by reducing the advance rate of the burners, and holes can be enlarged by making another pass with the same burner. These jet-piercing machines have since drilled more than 10 million meters of shallow blastholes in the taconite mines (Carstens, 1972). Increasingly, however, rotary drills are taking over since they produce a more consistent hole diameter.

Research by Williams (1986) indicates that spallation could reduce drilling costs in granite by 45–49% because of the rapid advance rate, low wear on mechanical components, and reduced trip time (the time spent in removing and replacing bits or other downhole equipment). The ratio of fuel volume consumed to rock volume excavated works out to about 0.27 (2 U.S. gal/ft^3), which might be improved by reducing the amount of water needed to quench the exhaust gases and protect hoses. Others taking a more pessimistic view have commented that only under extremely difficult drilling conditions can the high fuel consumption be justified.

Whereas spalling relies on differential thermal expansion to crack the rock and works only in certain rock types, fusion (melting) is effective in all rocks if a temperature of between 1100 and 2200°C can be attained. Electric fusion drilling has been tried experimentally at the Los Alamos Scientific Laboratory in New Mexico. The rock-melting probe can make a 50-mm hole 15–18 m horizontally into rock at a rate of about 1.5 m/h and leaves a lining of solidified glass on the walls of the hole, which could be helpful in stabilization.

14.2 Excavating

14.2.1 Mechanical excavators

Tractor-mounted excavating equipment allows economic excavation without the help of explosives not only through soils, but also into many of the softer or more broken rock formations. Some 30% of the earth's crust is shale or mudstone, much of which can be excavated using a backhoe, front-end loader, or heavy mechanical excavator. For example, most of the deep foundation excavations that penetrate into the Ordovician shales of Toronto, Ont., Canada, are routinely excavated using a heavy mechanical excavator with a bucket armed with pointed teeth. Weak and closely jointed schists and weathered igneous intrusives in sections of the Rubira tunnel in Barcelona, Spain, were excavated by conventional front-end loader.

Crawler-mounted or rubber-tired shovels, bulldozers, and scrapers are available, depending on the nature of the material to be moved or loaded. A face shovel is ideal for working a rock stockpile because, compared with a dragline, for example, it gives positive penetration of the rockpile. In an open-pit mining, quarrying, or large-scale excavating operation, the choice of excavating machine depends on output requirements, the type of material to be worked, capital and operating costs, and repair and maintenance requirements (Atkinson, 1971). Bucket capacities are selected by calculation from the volume of the in situ material with an additional *swell* factor, which can vary from 12% for sand or gravel to 40% for clay and 67% for limestone (Lakin, 1971).

The largest rock excavation works are those of large open-pit mines, where bucket-wheel excavators continuously mine huge quantities of soft rock materials such as lignite, coal, or poorly cemented overburden (Fig. 14.7). The method is ideally suited for continuous mining

Figure 14.7 Krupp bucket-wheel excavator mines 240,000 m³ of coal per day. The machine weighs 13,000 t. (*Courtesy of Krupp Industrietechnik, Duisburg, Federal Republic of Germany.*)

of soft materials. For example, bucket-wheel excavators are in use at the Suncor operation in the Athabasca Oil Sands, north of Fort McMurray, Alta., Canada, where a 50-m ore body is mined in two benches. At the Morwell project in Australia (Brawner, 1968) 20 million tonnes of brown coal is being mined each year for the generation of thermal power. The coal stratum varies in thickness between 105 and 137 m and is sufficiently soft to be mined by bucket-wheel excavators and ladder dredgers, many separate levels being mined simultaneously. The brown coal mines of Germany are famous; they are the deepest and largest open pits in the world and use many bucket-wheel excavators on numerous levels with tens of kilometers of conveyer belts.

14.2.2 Ripping

14.2.2.1 Equipment. A ripper consists of one or more teeth, called *shanks,* pulled through soil or rock to loosen it for excavation (Fig. 14.8). To build the Appian Way, the Romans used a ripper mounted on

Figure 14.8 D10 tractor with single-shank ripper. (*Courtesy of Caterpillar, Inc.*)

wheels and pulled by oxen. Rippers helped build railways in the United States from 1860 to 1880.

The early rippers were drawn by, rather than mounted on, tractors, and extra tractors were added until as many as three were pulling a single ripper tooth (Caterpillar, 1983). Tractor-drawn rippers were employed in 1931 on the Hoover Dam project. The advent of tractor-mounted rippers in the mid 1950s made possible the excavation of stronger materials because of increased weight on the tooth, improvements to ripper design, and a greater traction.

Normally, in moderate to difficult materials, single-shank ripping gives optimum production and is easier on the operator and the machine. One shank centers the load and mounting assembly and allows full force to be exerted at a single point. Twin shanks, however, can be more effective in softer, more easily fractured materials that are going to be loaded by scraper. Three teeth should be used only in very easy-to-rip materials, such as clay shale.

Rippers of the hinge or parallelogram type have varying degrees of adjustment. The shanks can be either straight or curved. Straight shanks provide a lifting action and the tensile stress needed in tight, laminated materials, whereas curved shanks work well in soils and porous rocks. Caterpillar introduced a ripper with a percussive action which, they claim, increases ripping rates and enables the ripping of harder rocks.

14.2.2.2 Ripping technique. Ripping directions are dictated by the extent, shape, and depth of the area to be ripped and by the dip directions of rock jointing. A scraper cut is always best ripped in the same direction that the scrapers will load. This assists in scraper loading and also permits traffic to flow in the same direction.

Downhill ripping is best, whenever possible. Gravity helps the tractor take maximum advantage of its weight and horsepower. Uphill ripping is occasionally used to obtain more rear-end pressure or to get beneath and lift slabby horizontal material. When ripping bedded rock with a slight bedding dip, the direction of ripping should if possible coincide with that of the dip to facilitate the lifting of beds.

Cross ripping (ripping the same area in two different directions) is occasionally used in particularly difficult conditions, but requires more passes and directional changes. Preblasting is sometimes used to prepare the ground for ripping. In such cases, the ripping serves mainly to improve fragmentation.

14.2.2.3 Assessing rippability. Even harder rocks such as limestones and sandstones, when closely jointed or bedded, can be removed by heavy rippers, at least down to the limit of weathering and surficial stress relief. Sedimentary rocks are usually the easiest ripped; metamorphic rocks such as gneisses, quartzites, schists, and slates vary depending on their degree of lamination and mica content. Igneous rocks are often impossible to rip, unless very thinly laminated, as in some volcanic lava flows.

Ripping is easiest in open excavations. An experienced operator can use the ripper to full advantage, lifting rather than breaking or gouging the slabs. Thinly bedded rocks, even when hard, can be excavated in this manner to almost any depth, given sufficient room for the vehicle to maneuver. In the confines of a narrow trench, however, the same rock often requires blasting.

Contractors preparing tenders for excavating work have to decide whether, and to what depth, the rock formations on site can be excavated by ripping rather than blasting. Cost estimates, competitiveness, and profits depend on selecting the right equipment and methods, and on accurate predictions of the quantities to be excavated by heavy equipment or by explosives. The best guide is experience or, in the absence of experience in a particular rock, a full-scale trial on site. Other, more quantitative methods and indexes to rippability come into their own when a trial excavation is impractical because of insufficient time or site remoteness or because the rocks are covered by thick soil.

When the rock can be observed directly in outcrops, quarries, or core, rock mass quality classification systems can be used, which then are related to ease of excavation and other measures of rock mass performance. Kirsten (1982) proposes a classification system to predict how easily rock can be excavated using digging, ripping, and blasting. Abdullatif and Cruden (1983) compared three other systems: the Franklin (1974), the Norwegian Q, and the South African RMR systems, all based primarily on block size and strength (Secs. 3.3.3.4 and 3.3.3.5). They conducted excavation trials with rock mass quality measurements in limestone, sandstone, shale, and igneous rocks at 23 sites in Great Britain. The RMR system gave the best prediction. RMR values of up to 30 could be dug, and of up to 60 could be ripped, whereas rock masses having an RMR greater than 60 required blasting.

The indirect methods all are based on correlations with various rock mass parameters. Uniaxial compressive strength is a useful indicator in the more massive and/or weaker materials, whereas sonic velocity, investigated as a rippability index in 1958 by the Caterpillar Tractor Company, gives a good indication of rippability in closely jointed yet strong rock masses (Fig. 14.9). The ratio of sonic velocity measured in the laboratory to that determined for the same, but jointed, rock in the field reflects the degree (intensity and openness) of jointing (Chap. 3),

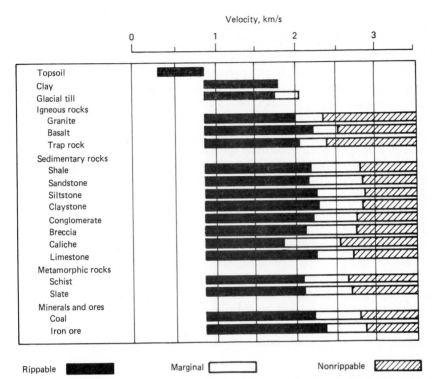

Figure 14.9 Rippability as a function of sonic velocity. (*After Smith, 1986; Caterpillar, 1983.*)

which is the main rock mass parameter governing whether or not ripping will be effective. Light portable seismographs can measure to a depth of 10–15 m under favorable field conditions and to deeper than 30 m under ideal conditions. Seismic velocity may be as fast as 6 km/s for a strong, dense, and unweathered rock, or as slow as 300 m/s for loose unsaturated soil.

Smith (1986) compares predictions with measurements of the time spent ripping a given test area, and the weight and volume of excavated materials. Such predictions make sense only if related to a particular model and design of ripper. The example in Fig. 14.9 relates to the Caterpillar D8L ripper no. 8. The D8L ripper can excavate rock to an upper velocity limit of approximately 2 km/s, compared with limits of approximately 2.3 and 2.6 km/s for D9L and D10 ripper types.

A suggested rippability rating chart, which relates rock index properties to the required ripping equipment, has been proposed in Singh et al. (1986). Tensile strength is obtained by Brazilian disk or point-load testing (Secs. 2.3.1.3, 2.3.1.4, 8.2.5.4, and 8.2.5.5). An overall rating is obtained by adding the individual ratings for each parameter.

14.2.3 Impact breakers

14.2.3.1 Pneumatic rock breakers. When the phrase "excavation by hand" appears in a specification, it usually means excavation with the aid of splitters or hand-held pneumatic tools. The same pneumatic jackhammer used to break up highway pavements can also remove rock, although slowly. Hand excavation is often specified beneath sensitive structures where blast vibrations cannot be tolerated, and in the final stages of trimming delicate rock foundations or blocks for in situ testing. It may be preferred even for routine excavating in parts of the world where there is a problem of underemployment, where machines, spare parts, servicing, and skilled operators are in short supply, and where explosives are under strict military control. Even substantial rock cuts have been completed, under such circumstances, by an army of workers equipped only with pneumatic tools.

Boom-mounted pneumatic picks (Fig. 14.10) are becoming more popular, versatile, and powerful, and can now cut trenches and tunnels into even quite strong rock materials. Beus and Phillips (1981) conducted trials using a pneumatic impact breaker with attached mucking boom and bucket to excavate 7.3-m-diameter test shafts in various igneous rock types, with uniaxial compressive strengths as high as 196 MPa. The average rate of advance was 0.4 m/h, including

Figure 14.10 Boom-mounted pneumatic impact breaker; Hecla Mines Ltd., Utah (Model TM-16/700, Teledyne Mining Products).

breaking and mucking the rock. Advantages in comparison with blasting included no loss of time waiting for fumes to clear, reduced overbreak, and more stable shaft walls. Cost, for a 610-m-deep shaft, was estimated as 30% less than using drill and blast methods, and there was a 16–34% increase in the average rate of sinking.

14.3   Splitting and Cutting

14.3.1   Splitting of stone

Before the age of gunpowder, rock excavation was done by hand with the help of picks, levers, and wedges made of bone, flint, bronze, and subsequently iron. The pyramid builders 4700 years ago found that large blocks could be quarried by taking advantage of jointing and preferred planes of splitting. Holes were bored, wooden sticks were driven into them, and these, swelling when soaked in water, cracked apart the rock. The pyramids and the monuments of ancient Greece and the cities of the Incas have been built from blocks dressed with only the simplest of hand tools, so perfectly dimensioned and assembled that scarcely a place can be found to insert even a hair between them.

Dimension stone is still quarried using the *plug and feather* method. Two or three shallow holes are drilled along a weakness plane if available, or along the required line of splitting in solid rock. A wedge-shaped plug is inserted and driven between "feathers" (wedge-shaped steel strips). By hammering on successive plugs, considerable force can be developed to split and separate the block from the mass.

The work is nowadays made easier by *hydraulic rock splitters*, employed in North American quarries since about 1968 for producing dimension stone and for secondary breaking of boulders (Duncan and Langfield, 1972). A hole is driven with a rock drill, and the splitter, a hydraulically driven plug and feather assembly, is inserted and activated. Expansion forces of up to 400 t are developed. Outside the quarry, splitters can be used to break rock in trenches and underground works when blasting is undesirable because of the need to limit vibrations or rock damage. As in rock blasting, the sequence of splitting should be arranged so as to break rock toward a free face.

The Okumura Corporation of Japan has developed a jumbo-mounted slot drilling and splitting method for driving tunnels that has given excellent results. In a 9-m-diameter highway tunnel in Kobe, Japan, a rate of advance of 1 m per 24 h was achieved in a high-strength and almost joint-free granite (Fig. 14.11). Slots 250 mm long were cut in the form of five overlapping 52-mm-diameter percussive-

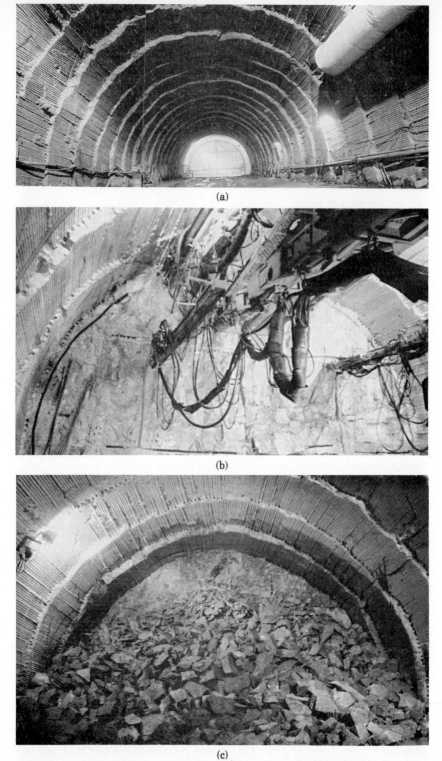

Figure 14.11 Tunneling by a slot drilling and splitting method; second Shin Kobe tunnel, Kobe, Japan. (*a*) View along tunnel, showing the ribbed appearance produced by line drilling the perimeter. (*b*) Jumbo line drilling to cut slots in the face. (*c*) Slotted rock is removed by impact breaker. (*Courtesy of Okumura Corp.*)

drilled holes, driven simultaneously from a single boom. The tunnel perimeter was fully slotted, and further slots were cut to subdivide the face into blocks of about 2–4-m dimension. The blocks were further drilled with isolated holes, into which a Hydrocracker device was inserted and expanded. The cracked rock was broken out and removed using a crawler-mounted hydraulic impact breaker.

14.3.2 Slot cutting

Wire rope sawing is used extensively in quarrying (Moebs et al., 1986) and, on a smaller scale, can be a convenient method for taking large block samples for testing. At the Pretty Boy dam site in Maryland, a wire rope saw excavated a cutoff trench into highly foliated schists that would have been severely loosened had blasting been employed. Calyx core holes provided access for exploration and carried the wire rope used to cut the trench.

Chain-driven slot cutting machines, similar in concept to a chain saw but armed with conical or chisel teeth, have simplified and greatly speeded up the task of cutting narrow trenches for underground electric power lines, water and gas pipes, and other services. They are used routinely in dimension and facing stone quarries for cutting limestone and marble into slabs and in salt, potash, and coal mines.

A hand-held version of the spallation burner discussed in the context of thermal drilling (Sec. 14.1.8) has been used for the past 25 years to cut slots in granite. Developed by the Browning Engineering Company of Hanover, N.H., this device, which resembles a small jet engine with its exhaust pointing downward, is now standard equipment for quarrying granite throughout the world (Williams, 1986). More recent versions have been used to spall experimental holes in granite at rates in excess of 30 m/h. Exhaust gases, along with steam produced from the cooling water, blow the fragmented rock from the slot.

References

Abdullatif, O. M., and D. M. Cruden: "The Relationship between Rock Mass Quality and Ease of Excavation," *Bull. Int. Assoc. Eng. Geol.* 28, 183–187 (1983).

Acker, W. L., III: "Basic Procedures for Soil Sampling and Core Drilling," Acker Drill Co., Scranton, Pa., 246 pp. (1974).

Atkinson, T.: "Selection of Open Pit Excavating and Loading Equipment," *Trans. Inst. Min. Metall.* (U.K.), **80**, A101–A129 (1971).

Beus, M. J., and H. E. Phillips: "Impact Breakers for Shaft Sinking," *Proc. Rapid Excavation and Tunneling Conf.* (San Francisco, Calif., 1981), vol. 1, pp. 815–830.

Bourgoyne, A. T., M. E. Chenevert, K. K. Millhelm, and F. S. Young: *Applied Drilling Engineering* (Soc. Petrol. Eng., Richardson, Tex., 1986), 502 pp.

Brand, E. W., and H. B. Phillipson: "Site Investigation and Geotechnical Engineering Practice in Hong Kong," *Geotech. Eng.* (J. Southeast Asian Geotech. Soc.), **15** (2), 97–153 (1984).

Brawner, C. O.: "The Influence and Control of Groundwater in Open Pit Mining," *Proc. 5th Can. Symp. Rock Mech.* (Toronto, Ont., Dec. 1968); also *Western Miner,* **42** (4), 42–55 (1969).

Bridwell, H. C.: "Development of Wireline Retractable Core Bits: A Progress Report," *Proc. 12th Symp. Explor. Drilling* (Minneapolis, Minn., 1967), pp. 113–133.

Carstens, J. P.: "Thermal Fracture of Rock—A Review of Experimental Results," *Proc. 1st North Am. Rapid Excavation and Tunneling Conf.* (Chicago, Ill., 1972), pp. 1363–1392.

Caterpillar: *Handbook of Ripping,* 7th ed. (Caterpillar Tractor Co., Peoria, Ill., 1983), 29 pp.

Chugh, C. P.: *Manual of Drilling Technology,* 1st ed. (A. A. Balkema, Rotterdam and Boston, 1985), 567 pp.

Cooley, W. C.: "Water Jets and Rock Hammers for Tunneling in the U.S. and the U.S.S.R.," *Proc. 1st North Am. Rapid Excavation Tunneling Conf.* (Chicago, Ill., 1972), vol. 2, pp. 1325–1360.

Cumming, J. D.: *Diamond Drill Handbook,* 3d ed. (J. K. Smit and Sons Diamond Products Ltd., Toronto, Ont., 1975), 547 pp.

Duncan, N. J., and E. R. Langfield: "Rock Excavation by Hydraulic Splitter," *Proc. 1st North Am. Rapid Excavation and Tunneling Conf.* (Chicago, Ill., 1972), vol. 1, pp. 785–791.

Franklin, J. A.: "Rock Quality in Relation to the Quarrying and Performance of Rock Construction Materials," *Proc. 2d Int. Cong., Int. Assoc. Eng. Geol.* (Sao Paulo, Brazil, 1974), paper IV-PC-2, 11 pp.

Heinz, W. F.: *Diamond Drilling Handbook,* 1st ed. [South Afr. Diamond Drilling Assoc. (SADA), Johannesburg, 1985], 517 pp.

Hustrulid, W. A., and C. Fairhurst: "A Theoretical and Experimental Study of the Percussive Drilling of Rock," *Int. J. Rock Mech. Min. Sci.,* 8, 311–356 (1971); 9, 417–431 (1972).

Kirsten, H. A. D.: "A Classification System for Excavation in Natural Materials," *Siviele Ingenieur in Suid Afrika,* 293–308 (July 1982).

Lakin, A. E.: "Excavation Techniques," *Quarry Manager's J.,* 155–168 (May 1971).

Marshall, D. R., and J. E. Sowden: "Drilling in the Seventies," *Quarry Manager's J.,* 96–104 (Mar. 1971).

Moebs, N. N., G. P. Sames, and T. E. Marshall: "Geotechnology in Slate Quarry Operations," Rep. Invest. RI9009, U.S. Bureau of Mines (1986), 38 pp.

Mouraz Miranda, A., and F. Mello Mendez: "Drillability and Drilling Methods," *Proc. 5th Int. Cong. Rock Mech.* (Melbourne, Australia, 1983), vol. E, pp. 195–200.

Ottosson, L., and T. I. Cameron: "Hydraulic Percussive Rock Drills—A Proved Concept in Tunneling," in M. J. Jones, Ed., *Tunneling '76* (Inst. Min. Metall., London, 1976), pp. 277–285.

Paone, J., R. Madsow, and W. E. Bruce: "Drillability Studies. Laboratory Percussive Drilling," Rep. Invest. 7300, U.S. Bureau of Mines (1969).

Singh, R. N., B. Denby, I. Egretli, and A. G. Pathan: "Assessment of Ground Rippability in Opencast Mining Operations," *Univ. Nottingham (U.K.) Min. Dept. Mag.,* **38**, 21–34 (1986).

Smith, H. J.: "Estimating Rippability by Rock Mass Classification," *Proc. 27th U.S. Symp. Rock Mech.* (Tuscaloosa, Ala., 1986), pp. 443–448.

Varlamoff, I.: "New Developments in Drilling Machine, Wireline Core Barrel and Synthetic Diamond Bit Systems," *Proc. 36th Ann. Mtg. Conv., Can. Diamond Drill. Assn.* (Victoria, B.C., 1979), paper 7, 19 pp.

Williams, R. E.: "Thermal Spallation Excavation of Rock," *Proc. 27th U.S. Symp. Rock Mech.* (Tuscaloosa, Ala., 1986), pp. 816–820.

Wu, W. Z.: "Microcomputer Application for the Efficiency Evaluation of Percussive Drilling," *Proc. 27th U.S. Symp. Rock Mech.* (Tuscaloosa, Ala., 1986), pp. 647–652.

Tunnel-Boring Machines

15.1 Types and Designs of Boring Machines

15.1.1 History of boring machines

The present-day tunnel-boring machine (TBM) competes with drill and blast methods, and is often faster and more efficient than explosives for excavating rock.

Probably the earliest TBM was one used in 1856 to cut a 33-cm-wide groove around the portal of the Hoosac tunnel. After advancing just 3 m it was considered a failure and was scrapped (Brunton and Davis, 1922). In contrast, a boring machine designed by Colonel Beaumont, patented in 1864 and constructed in 1881, achieved remarkable success through chalk beneath the English Channel. The head, armed with pick cutters, was driven forward on a rotating shaft under the force of compressed air. Two similar machines, advancing from the French and English sides of the English Channel, drove 2.1-m-diameter tunnels for a distance of 1.9 km. On the English side the maximum advance was 24.7 m in a single day, and 15.4 m/day was sustained for 53 consecutive days. In 1882 work was stopped for political, military, and economic reasons. When the project was reactivated in 1958, the unsupported tunnel was still standing.

A 60-year period elapsed before the first commercial use of rock excavating machines in the mines of North America. A coal borer was developed by James S. Robbins in 1947, while similar machines were being developed by European manufacturers. Some were used also in mining soft sandstone, rock salt, and potash, and by the early 1950s, machine mining of soft rock was commonplace in the United States and Germany. By the mid 1950s these machines were cutting medium to hard sandstone and limestone.

The first successful civil engineering application of a Robbins machine was at the Oahe dam in South Dakota where 6.9 km of an 8-m-diameter tunnel were driven through shale (Williamson, 1970). Robbins experimented with different types and combinations of cutter, and in 1956 used circular disk cutters effectively for the first time to excavate the Humber River sewer tunnel in Toronto, Ont., Canada. The disks were able to penetrate sedimentary rocks including limestone with a strength of 140 MPa. Since then, increased thrust and an improved layout of disk or roller tools on the cutting head of the machine have enabled satisfactory rates of advance through rocks with uniaxial compressive strengths approaching 300 MPa.

15.1.2 Modern TBM designs

Successful tunnel boring calls for fast, economical excavation of materials, while maintaining the heading and tunnel perimeter in a stable and safe condition. These two essentials, rapid excavation and effective ground support, govern machine design.

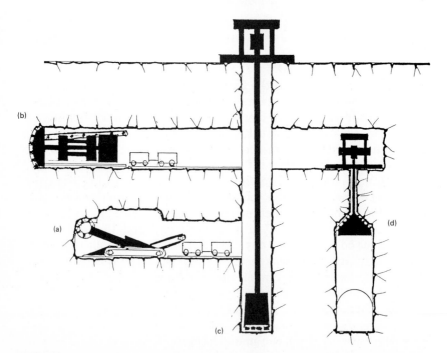

Figure 15.1 Types of boring machine. (*a*) Partial-face (roadheader) machine. (*b*) Full-face machine. (*c*) Blind shaft borer. (*d*) Raise borer.

TBMs and shaft borers fall into one of four main categories according to their cutting action (Fig. 15.1). *Partial-face (roadheader)* machines (Fig. 15.2) attack only a limited area of the tunnel face at any one time using a cutter head mounted on the end of a hydraulically operated boom, whereas *full-face* machines (Fig. 15.3) excavate the complete heading. *Blind shaft* borers are similar to full-face TBMs but bore downward and have special facilities for removing cuttings. *Raise* borers (Fig. 15.4) make use of a pilot hole to pull a rotating cutting head upward, with cuttings falling downward.

Most full-face TBMs excavate a circular cross section and use a rotating circular cutter head of the same diameter as the tunnel. A few models use one or more cutter heads smaller than the tunnel, covering the whole face by planetary rotation or up-down oscillation. These can be used to excavate an oval or rounded rectangular cross section, which is convenient for small-diameter tunnels for sewers, pipes, or cables.

Figure 15.2 Alpine Miner AM75 roadheader working at American Borate's Billie Mine in Death Valley, Calif. (*Courtesy of Voest-Alpine International Corp., New York, and Wajax Industries Ltd., Ottawa.*)

(a)

(b)

Figure 15.3 Full-face tunnel-boring machines. (*a*) 9.83-m-diameter Atlas Copco—Jarva hard rock TBM armed with disk cutters. This machine, weighing 900 t, bored a 7.7-km section of the TARP project through the Chicago limestone bedrock. Advance rate averaged 28.4 m/day in the last 2.5 km. (*b*) Typical soft rock boring machine, fully shielded and armed with chisel and pick cutters. (*Courtesy of Lovat Tunnel Equipment Inc., Toronto.*)

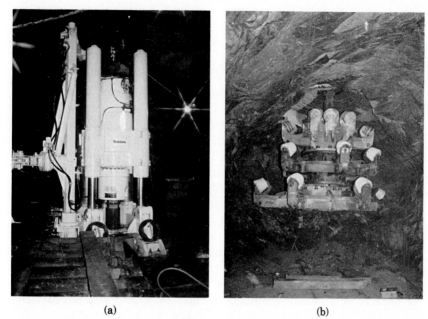

(a)   (b)

Figure 15.4 Raise boring a 150-m-long 3.65-m (12-ft)-diameter raise at Westmin's H-W Mine, Campbell River, B.C., Canada. (*a*) Robbins 82R raise boring machine. (*b*) Tamrock 8-, 10-, and 12-ft staged reamer collaring the raise. (*Courtesy of J. S. Redpath Ltd., Mining Contractors, North Bay, Ont.*)

Irrespective of the category of TBM or shaft borer, all machines have certain essential components. With reference to the most common full-face type of machine (Fig. 15.5), these are the cutter head armed with cutting tools; a system of propulsion and reaction to thrust the cutter head and rotate it against the rock face; a system for muck removal; and, when needed, facilities for erecting permanent support

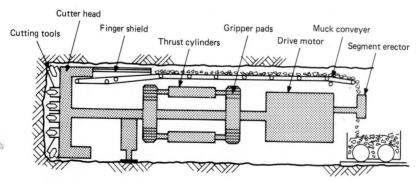

Figure 15.5 Components of a typical full-face TBM.

and a shield to provide temporary ground support and protection for the operating crew.

15.1.3 Boring or blasting?

Given an appropriate choice of machine, TBM excavation gives a smoother and more stable profile than can be achieved by blasting. Support is reduced and often eliminated entirely. Rates of advance can be much greater than with drill and blast methods.

The purchase price of a full-face TBM increases in proportion to the tunnel diameter, from about $1.5 million for a machine of 4-m diameter to about $7 million for one of 10 m (1988 prices). In addition to the capital outlay, boring machines are expensive to mobilize, so a full-face TBM usually is only justified for tunnels longer than 1 km. The costs of maintenance and cutting tool replacement also are high.

These costs are just part of the reason why the choice of whether to bore or blast a tunnel is a crucial one. Once committed in the tunnel, a full-face TBM cannot be removed easily. The engineer and contractor have to be sure that ground conditions have been well forecast and will permit uninterrupted boring and support.

Variable ground conditions are the main reason why more tunnels are not excavated by TBM. Boring machines work best in relatively uniform ground, and often no one machine will do the job in ground that in places is strong and massive, and in others soft and closely jointed. Unlike a drill bit, a tunneling machine cannot be changed at will to suit rock conditions. Completely versatile machines have not yet been developed, such as would allow the spacing, size, and type of cutting disk or pick to be varied, as well as the thrust and speed of cut. TBMs are usually designed with a combination or compromise of cutting tools. Quick removal and replacement of tools is provided, for example, by the Greenside-McAlpine TBM, which uses a drag bit with tungsten carbide insert picks mounted on steel shanks and fitted into a box holder with a quick-release mechanism (Pirrie, 1969).

Mixed-face conditions present even more severe problems than those caused by conditions that change along the route. Sometimes the face itself is composed of combinations of hard and soft rock layers, or exposes the boundary between rock and soil. Attempts to cut these materials of extreme contrasts can lead to severe problems of vibration, cutter wear and loss, and deviations from alignment and grade.

Another critical consideration is the possibility of unexpected inrushes of groundwater. A TBM can be immobilized by a collapse or water inflow before defensive measures can be taken. Access for preventive or remedial measures is much more limited when using a full-face TBM than when drilling and blasting.

15.1.4 Choice between full- and partial-face machines

Many of the limitations of a full-face TBM do not apply to the roadheader, which is much more versatile. However, it is less powerful, and currently available roadheaders can make only modest progress in rock of low to medium strength.

Full-face machines are usually the choice when a circular bore is acceptable, and are ideal when a segmental liner is to be the means of support. They achieve much greater thrusts and therefore usually twice the rate of advance possible with a roadheader. For the same reason, they can penetrate most types of rock, whereas partial-face machines are at present limited to relatively weak materials such as shales and thinly bedded sandstones. Roadheaders are most often used for mining in soft rocks, such as coal or potash, and particularly for benching, undercutting, or overcutting of rectangularly shaped openings (McFeat-Smith, 1978).

The turning radius of the full-face TBM, because of the length and diameter of the machine and its trailing gear, is at best about 30 m, and more often about 60 m. Some full-face machines can change direction while they bore whereas others reset direction at the end of each stroke. Maximum operating grade is about 45° when boring uphill (inclines), and 18° when boring downhill (declines).

Partial-face machines, on the other hand, can be used to cut right-angle corners and can cope with any tunnel cross sections, grade, and radius of curvature. They are more versatile in their ability to excavate any required tunnel or mine geometry, and can be used to excavate underground chambers of irregular shape.

There is room to work around a partial-face machine so that support can be installed close to the working face. The partial-face machine can cope more readily with changing rock conditions (by selective mining) and can be easily withdrawn to allow blasting of difficult sections. An Alpine Miner, for example, was selected to excavate variable cross sections at the Nevada test site of the U.S. Atomic Energy Commission because of its versatility in maneuvering.

Partial-face machines have long been used in soft-rock mines because of their versatility, which adapts well to an irregular pattern of stoping. Full-face TBMs are a more recent innovation, capable of rapidly driving long access roads and other forms of development heading, even through hard rock (Handewith, 1980).

Athorn and Snowdon (1986) compare the performances of a Robbins full-face TBM and a Tyssen Titan roadheader used to drive a pair of roadways during development of the Selby coalfield in England, where most coal mine roadways are driven by either roadheaders or drill and blast. The full-face machine was very efficient in competent ground conditions, where it produced a smooth, well-profiled tunnel at a faster rate, and hence an earlier return on the high capital investment. However, only the roadheader was able to excavate engine houses, substations, junctions, and cross-cut roadways. It continued its work through badly faulted ground that temporarily stopped the full-face TBM. Machine utilization for the roadheader was considerably higher, and total excavation costs 16.4% lower than for the full-face TBM. A conclusion of this study was that if full-face tunneling machines are to make a significant impact on the mining market, their performance under poor ground conditions will need to be improved significantly.

15.1.5 Machines for shaft sinking and raise boring

The type of machine used for blind shaft boring is similar to a full-face TBM except for provisions to remove muck against the action of gravity. Drilling mud supports the shaft walls, cools the cutter head, and removes the cuttings. Because of the large diameter of the shafts, cuttings cannot be removed by conventional means, and double-walled pipe with reverse circulation is required (Fig. 15.6). At the ground surface, wire mesh screens separate cuttings from the drilling fluid, which is then recycled.

Moss et al. (1986) describe the blind hole drilling of drainage shafts 200 m deep and 1.68 m in diameter through closely jointed andesite at the Bougainville open-pit copper mine in Papua New Guinea. An attempt using a raise bore had been abandoned because of difficult ground conditions. Blind hole shaft sinking gave better results, aided by a drilling mud of specific gravity 1.06 to counterbalance the rock and groundwater pressures.

The drill rig had a 150-t lift capacity and supplied 100 kN·m of torque. A 310-mm pilot bore was driven to stabilize the full-sized bit. This allowed checking of hole alignment before reaming, and gave warning of zones of poor ground. A Robbins cutter head with 17 disk tools evenly spaced across its face was used for reaming. Welded tubular steel liners were floated into place and back-grouted by concrete placed through pipes attached to the outside of the liner tube.

In raise boring, a large circular boring head armed with cutting tools is advanced by pulling upward through a pilot hole (Fig. 15.4). The pilot hole is bored at about 2 m/h and a bit lasts typically from 90 to 370 m. If the pilot hole fails to break through, its location has to be found and opened by drifting. The cutting head is then assembled at the lower level, and the raise is reamed from the bottom up. Reaming usually proceeds at a rate of between 1.5 m per shift (hard rock) and 12 m per shift (very soft rock). Usually the last 3 m is blasted, or the raise borer is elevated to allow the reamer to complete its break-

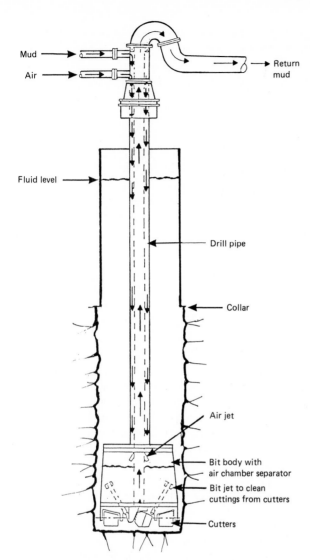

Mud

Air

Return mud

Fluid level

Drill pipe

Collar

Air jet

Bit body with air chamber separator

Bit jet to clean cuttings from cutters

Cutters

Figure 15.6 Schematic drawing of blind shaft boring procedure.

through. The smooth bore is often left unlined. In ground that tends to ravel, a steel liner can be advanced with the reamer, or the completed raise can be rock-bolted (Harrison et al., 1972).

15.2 Full-Face Machines

By far the majority of TBMs are of the rotating-drum type with an axis of rotation coincident with the axis of the tunnel. Others have several contrarotating cutter heads, or a single pivoting and rotating head which "undercuts" the rock.

15.2.1 Undercutting machines

One of the best known undercutting machines is the Atlas Copco Mini Fullfacer, introduced in 1970–1971, which uses a pivoting cutter head with chisel-shaped *drag bits* to bore minitunnels 1.5 m wide and 2.4 m high, with a relatively flat invert and circular crown (Barensden and Cadden, 1976). These oval, human-sized tunnels, which have a cross-sectional area of only 3.3 m^2, offer advantages in terms of speed, stability, and cost, and are especially suitable for carrying urban utilities or as exploration adits to precede larger tunneling works. Undercutting starts at the invert, and as the cutter head is swung upward, cutting increases from nil to a maximum at mid-swing; only a small part of the total rock face is attacked at any one time. The swing speed and the length of advance can be adjusted to give the required amount of undercut. Because of their dragging action, these undercutting machines tend to have a short cutter tool life in harder rocks.

15.2.2 Multiple-rotor mining machines

In the Saskatchewan, Canada, potash industry, multiple-rotor Marietta mining machines cut cross sections with rounded or almost square sides using a chain cutter. The primary cutters are picks mounted on two or three rotating three-arm cutter heads. About 250 mm behind the primary cutter heads, picks on a chain travel along a series of rollers placed to conform to the desired cross section. The Marietta mining machine travels on dozer tracks and, when equipped with a continuous-loading device and an extensible belt conveyer, can achieve very high rates of advance in potash and halite.

15.2.3 Rotating-drum machines

15.2.3.1 Diameter and weight. Historically most TBMs have been of the rotating-drum type, with diameters in the range of 2.3–4.1 m. Attempts to standardize diameters have met with only limited success. Weights vary in proportion to the diameter, from about 20 t at 2-m to about 130 t at 5-m diameter.

15.2.3.2 Thrust and reaction. Effective rock cutting requires a total thrust of about 3–5 MN for machine diameters in the range of 3–6 m (Williamson, 1970). Mellor and Hawkes (1972) give the required thrust as about 0.43 MN for every square meter of cross section. Average thrust per individual disk cutter (in kilonewtons) is about 50 times the tunnel diameter (in meters). The average rolling force per cutter is typically one-fifth of the thrust per cutter in relatively high-strength rock such as crystalline limestone or indurated sandstone, and one-quarter of the thrust in low- to medium-strength rock such as shale or porous sandstone (Nelson and O'Rourke, 1983).

Hydraulic cylinders apply thrust to the cutting head through the main bearing, which consists of a sealed system of tapered roller bearings to allow rotation of the head. When the ground is strong enough to provide the necessary reaction, as is usually the case when boring rock, the rear part of the TBM is held by gripper pads that press outward against the tunnel walls, mounted above and below or to either side of the TBM body (Fig. 15.5). Cutting stops when the thrust rams are fully extended, at the end of their stroke. The machine then advances by retracting the rear set of gripper pads and drawing the back part of the TBM forward hydraulically, in an inch-worm type of movement.

In other designs the TBM is pulled rather than pushed forward, using a single gripper inserted and expanded in a pilot hole drilled ahead of the face. In weaker rocks and soils, thrust is obtained by jacking against a segmental liner or a temporary lining of steel sets which must be strongly strutted and braced.

15.2.3.3 Torque, power, and speed of rotation. Substantial torque is needed to rotate the cutter head, typically 1.2–3.5 MN·m for machine diameters in the range of 3–6 m. The rotary drive of most Robbins machines consists of multiple electric motors acting through a ring gear around the perimeter of the heavy domed cutter head. Other machines use hydraulic motors. Few nowadays employ a central drive shaft, mainly because this tends to impede access to the working face in small-bore machines. However, a central drive shaft can be an advantage when excavating larger tunnels in that it permits easier access to the tunnel crown for early installation of primary support.

Interrelationships between thrust, power, torque, and machine diameter have been given in Mellor and Hawkes (1972) for a variety of boring machines. Rated head power is typically 0.2–0.6 MW for tunnel diameters of between 3 and 6 m, increasing in proportion to the square of the diameter.

The cutter head rotation rate is usually in the range of 8–18 r/min, most often toward the lower end of this range, and is in practice limited by the capability of the cutting tools, and particularly their bearings, to withstand friction and impact.

15.2.4 Cutting tools

15.2.4.1 Selection to suit rock type. The cutter head is armed with tools whose shape and mode of action depend on the rock to be re-

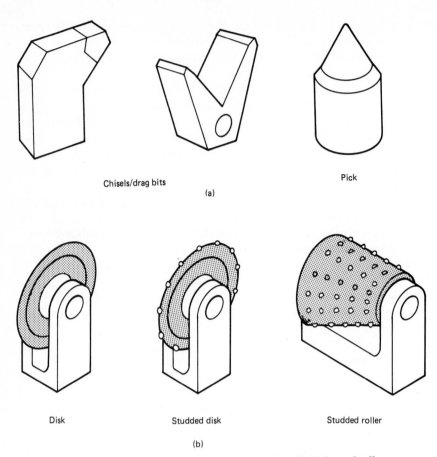

Chisels/drag bits

Pick

(a)

Disk

Studded disk

Studded roller

(b)

Figure 15.7   Cutting tools. (*a*) Chisels, picks, and drag bits. (*b*) Disks and rollers.

moved (Fig. 15.7). Cohesive soils are best cut by a blade, and soft rocks by a chisel, a pick, or a drag bit that breaks and separates the rock along preexisting joints (Fig. 15.7*a*). Picks used by roadheaders are typically operated at cutting speeds greater than 50 m/min and cutting depths of only a few millimeters. Under these conditions they cannot efficiently penetrate rock of more than 80 MPa compressive strength. Barensden and Cadden (1976) report that the same picks used at slower cutting speeds, of less than 15 m/min, and depths of 10–15 mm can cut rocks of strengths of up to 300 MPa.

Harder rocks are easier to break by the compressive spalling action of single- or multiple-disk cutters made from nickel molybdenum steels. Gear rollers that spall rock between the teeth are also used, and extremely hard rocks can be bored using edge-studded narrow disk cutters with tungsten carbide inserts (Fig. 15.7b). Tungsten carbide studded broad roller cutters are often employed in rocks with compressive strengths in the range of 190–280 MPa.

Disks and rollers apply a high thrust to a thin line or point of contact, which generates tension and results in spalling or cratering (Claasons and Pike, 1970; Gaye, 1972). Sharp-edged disks, with cutter angles of between 55 and 60°, are best for rapid penetration through hard rock, but need frequent replacement. Disks at blunt angles, for example, 100–110°, give a slower but steadier and more continuous rate of penetration.

Appropriate choice of cutters is essential. Picks and chisels are nearly always inefficient when excavating harder rock materials, producing only a scraping and polishing action. Disks and rollers can become clogged and less efficient when they encounter softer rock or clay.

15.2.4.2 Arrangement of tools on the cutter head.

The performance of a boring machine cannot be judged by that of a single cutter. The tools are arranged geometrically on the cutting head (e.g., Fig. 15.3) so that in combination they produce the maximum rate of penetration with the minimum wear, maintenance, and tool replacement, and the largest possible size of fragment that can be handled by the mucking system.

The number of cutters typically increases in proportion to the diameter from about 25 at 4 m to about 50 at 8 m. The *center cutters* follow a very tight, circular path and travel slowly. The perimeter or *gauge cutters* travel at higher speeds and have to be tilted outward to aid in steering and to provide an opening big enough for the machine to move forward and through. The remaining tools, called *face cutters,* are set normal to the rock face to make the best use of available thrust.

Efficient design of the cutter head requires an optimization of the depth and separation of the grooves or *kerfs* so as to excavate the maximum of rock for the minimum of energy. Efficient excavation has been shown by many investigators to correlate with the formation of large chips as the rock between cutter disk rolling paths is removed. In the very hardest rock, the largest kerf spacing currently used is about 90 mm. The number of cutters in each kerf should be increased as the kerf spacing is increased (Gaye, 1972).

Nelson and O'Rourke (1983) report that the muck produced by disk-armed TBMs while cutting through sedimentary rock formations typically has a log-linear (well-graded) size distribution in the range of 2–70 mm, with about 25% of muck finer than 2 mm. Muck gradation is a consideration in the design of mucking systems.

15.2.5 Tunneling shields

Unless the ground is completely self-supporting, the tunneler must be provided with some measure of protection, and the tunnel walls must be held stable until more permanent supports can be placed. Early tunneling in soft ground was accomplished by people working under the cover of a *tunneling shield*. They erected *temporary supports* of timber, later to be replaced by a *permanent liner* of masonry. With the advent of mechanized boring, the same shield can now be carried, similarly to the shell of a tortoise, as part of a self-propelled boring machine.

Different types of shielding are employed to support different types of ground. Boring machines designed to operate in hard rock often require little or no shielding, at least along much of the tunnel route, while passing through self-supporting ground. However, many tunnels pass through both rock and soil, and must be equipped with the shielding appropriate for the worst of the anticipated ground conditions.

Among the worst are the running ground conditions associated with saturated sands or silts with little or no cementation. Then not only must the tunnel be shielded by a steel tube, but also the face has to be supported using a closed shield with narrow or adjustable slots that allow the soil materials to pass into the shield at a controlled rate. Ground that is very fluid can be supported by compressed air, either in the entire tunnel or in the direct vicinity of the working face. Earth support and bentonite shield alternatives prevent collapse of the face by providing support behind a bulkhead, either with a dense slurry of the excavated materials or with a bentonite mud.

Heavy shields are not without disadvantages. They impede ground inspections and monitoring, and they interfere with or prevent work on the rock such as drilling of holes for probing ahead of the face, for grouting and drainage and for the installation of rockbolts and other forms of more permanent support. In highly stressed ground, converging rock walls have been known to grip a fully shielded machine, making advance or even retraction difficult. Fully shielded machines are usually tapered to aid withdrawal in the event of a squeeze problem.

A shield is therefore best avoided when the rock is likely to be either self-supporting or stable once primary support is applied close to the face. One compromise is to use a *finger shield* of parallel steel strips separated by gaps through which rockbolts can be installed. This gives adequate protection while permitting inspection and installation of support, and lessens or eliminates the risk of jamming caused by squeeze.

15.2.6 Ground treatment and primary support

The tunneling machine and its trailing gear, even when unshielded, interfere with installing ground support close to the advancing face. Rockbolts can be placed just behind the cutter head, but often at angles other than the optimum, and in small numbers during the short period while the TBM is retracting ready for the next cycle of thrust and cut. Most support work is done either in the confined space between the TBM and its trailing gear, or further back down the tunnel. In less stable conditions, a segmental liner can be erected in the tail of the TBM, with the help of a mechanical erecting device attached to the machine. Problems can occur in zones of poor rock when conditions elsewhere in the tunnel are too good to warrant such a liner. Tunnel support systems are discussed further in Chap. 17.

15.2.7 Mucking

Muck (also called *tunnel spoil*) is the excavated rock, and *mucking* is the process of removing it from the tunnel. The muck-handling capacity limits the rate of advance that can be achieved, regardless of the power and effectiveness of the TBM itself. Delays in muck handling often account for up to 40% of the available advance time, as a result of waiting for muck cars; hence the need to increase the speed of in-tunnel traffic and to improve the design of handling equipment.

Mucking systems include the conveyers and other *short-range* feeding and loading equipment within the TBM, used to move the spoil from the immediate vicinity of the cutter head and load it. Also, *long-range* transportation equipment is needed to take the spoil from the TBM to the shaft, portal, or disposal area.

Full-face TBMs have an auger or screw that directs rock fragments from the base of the cutter head onto a conveyer that runs through the body of the machine and dumps into a rail car or free-wheeled vehicle. The roadheaders that operate in confined mining areas, such as those used to excavate thin coal seams, are equipped with a chain or grab mechanism. This feeds broken rock or minerals from the cutter head to a conveyer belt, which in turn feeds into a small truck or rail car.

Long-range muck hauling uses unitized systems such as free-wheeled vehicles or rail cars, or continuous systems such as conveyers and pipelines. Wheeled vehicles are more often used in large-diameter excavations where there is room for them to turn, whereas rail-mounted muck trains are best in tunnels of small diameter. Trucks and rail cars, in contrast to continuous systems, have the advantage that they not only remove muck, but also carry workers and support materials to the working face. However, automatically extensible conveyers fixed to the TBM tail permit continuous operation without the need to lay more track and shunt rail cars.

Continuous systems carry the spoil on a conveyer or by pipeline in an air suspension. Pneumatic handling systems were introduced in the nineteenth century, and under ideal conditions they provide a very rapid and cheap way of removing tunnel spoil. However, they are rarely used. A single mechanical failure or blockage brings mucking to a standstill, and restarting is difficult. Pipeline systems also require that the muck have certain characteristics of moisture content, grain size, and plasticity. The pipeline and pumping equipment must be designed to suit the particular muck, or the spoil itself needs to be modified by crushing or screening (Golder and MacLaren, 1976).

For any given tunneling operation, the required muck-handling capacity can be calculated from the tunnel diameter and the overall rate of advance. The volume of excavated muck must be increased by a *bulking factor* of about 40%, and the estimates of advance rate must take into account stoppages for repairs, installation of support, and other inactive periods. Excavation proceeds more rapidly in soft than in hard rock, and soft-ground boring places the greatest demand on the handling equipment. Typically a capacity of 30–70 m^3/h may be required when boring in shale, compared with only 5–10 m^3/h when excavating through hard rock.

15.3   TBM Performance

15.3.1   Rates of penetration and advance

15.3.1.1   Terminology. The various parameters used as measures of TBM performance reflect the rapid rates of penetration during continuous boring, and the slower, average advance rates caused by stoppages for maintenance, rock support, and other activities in the tunnel.

Penetration rate (mm/rev or m/h) is the rate at which the TBM progresses while actually cutting rock.

Advance rate (m/month, m/shift, m/h, etc.) is the average rate at which the tunnel heading advances over a longer period that includes cutting and other work, and is substantially less than the penetration rate because of limited TBM utilization.

Availability (%) is the percentage of tunneling time during whichthe TBM is available for tunneling, excluding periods of breakdown, maintenance, and so on.

Utilization (%) is the percentage of shift time during which the TBM is cutting rock, which may be less than the availability because of delays while installing support, moving conveyers, and so on.

Advance rate is perhaps the most meaningful of these indexes, at least when assessing the costs of tunneling work (Fourmaintraux, 1985). High penetration rates are of little value if they are achieved only locally, and are interspersed by long intervals of slow progress. Robbins (1976) suggests that "...it is better to advance at a moderate but uniform rate throughout the job...than to move at high rates through good ground and struggle painfully and expensively through bad ground."

15.3.1.2 Penetration rate. The rate of penetration increases in proportion to the speed of rotation of the cutter head, which is limited by the capabilities of the cutters and bearings. Penetration is more rapid through weaker rocks (Fig. 15.8), typically about 7.3 m/h through shale with a uniaxial compressive strength of 30 MPa, compared with 1.3 m/h through dolomite of strength 180 MPa. For tunnels through sedimentary rocks in the United States, Nelson and O'Rourke (1983) report penetration rates of between 2.1 and 2.8 m/h, or 7.3 mm/r on the average.

Nelson and Fong (1986) recommend that disk performance be measured in terms of a field penetration index R_f, defined as the ratio of the average force per cutter to the penetration rate. The required force F increases in an approximately linear manner as either the spacing of disks s increases or the penetration p decreases (corresponding to an increase in the s/p ratio).

Traditionally, uniaxial compressive strength has been the sole index for predicting the rate of penetration (e.g., Williamson, 1970; Hibbard and Pietrzak, 1973). More recently a better correlation has been found between penetration rate and the critical energy release rate of the rock $G_{Ic'}$, determined by a fracture toughness test as described in Chap. 8 (Nelson and O'Rourke, 1983; Nelson and Fong,

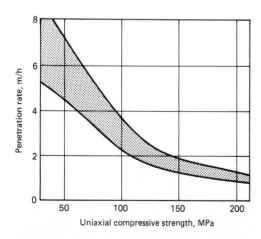

Figure 15.8 Penetration rate for typical full-face TBM as a function of rock strength.

1986). *Fracture toughness* depends not only on rock strength but also on the modulus of elasticity and is a measure of the force needed to drive a crack through the rock material. Sanio (1985) was able to predict disk cutter performance using measurements of point-load strength index and strength anisotropy.

Abrasion hardness is another important rock property related to the penetration rate. Nelson and Fong (1986) found that linear regression of R_f against total rock hardness (Taber test, Sec. 2.3.2.4) or a combination of abrasion and rebound hardnesses gave a high degree of statistical correlation, whereas strength tests gave very low degrees of correlation.

15.3.1.3　Advance rate. Typical average rates for the 9.85-m-diameter Robbins machine used on the TARP project in Chicago, Ill., in 1970 were 914 m/month, with infrequent high record rates exceeding 1830 m/month (Williamson, 1970; Anon., 1980). Nelson and O'Rourke (1983) quote an advance rate of 0.75–1.0 m/h, average 0.82 m/h, during the excavation of six tunnels for a total distance of 22.9 km through sedimentary rock formations. Much other information is available, and the value of high-quality information is critical for evaluation (Fourmaintraux, 1983).

15.3.1.4　Availability and utilization. A TBM becomes unavailable for tunneling during periods of downtime for maintenance, which include stoppages for inspection, cutter changes, and repairs to the machine, conveyer, and backup systems. Even when available in a state of good repair, the TBM often must stop boring during ground stabilization operations, which include tunnel dewatering and installation of steel sets, rockbolts, and other forms of support. A machine availability in the range of 60–90% (average 86.7%) was observed in the projects studied by Nelson and O'Rourke. Utilization rates were 30–36% (average 33.2%).

15.3.2　Cutter wear and replacement

15.3.2.1　Tool replacement costs. The cost of replacing cutting tools can amount to a significant proportion of the total cost of tunneling by machine. For roadheader machines, high wear rates such as 1.5 tools per cubic meter of excavated rock are equivalent to a tool cost alone of $500 for every meter of a typical sewer tunnel or mine access roadway. This does not take into account the downtime needed to actually replace the tools. Failure to replace tools can lead to even more expensive downtime to replace major components, and to a greatly reduced cutting performance (Johnson and Fowell, 1986).

Under normal operating conditions, cutter wear has no significant influence on the penetration rate until about 38 mm of the cutter disk diameter has been worn away, and even then the penetration rate can be maintained constant with an approximately 10% increase in thrust (Nelson and O'Rourke, 1983).

In 1970 it cost between $40,000 and $70,000 to replace a set of disk cutters on a 5-m-diameter full-face machine in the sedimentary rock tunneling operations studied by Nelson and O'Rourke. Cutter changes were required about every 100 h and for every 300 m of tunnel advance. The gauge, face, and center cutters had to be replaced at different intervals:

	Gauge	Face	Center
Disk life, hours cutting	148	281	189
Distance rolled, km	100	500	50

Gauge cutters are mounted at an angle to the tunnel axis and travel at relatively high velocities, so their bearings are subjected to higher temperatures, which decrease the bearing life. Also, they travel through muck accumulations in the invert, and their mountings are subjected to more abrasion than those of the centrally located cutters.

15.3.2.2 Mechanisms of wear. Johnson and Fowell (1986) identify four main wear mechanisms: abrasion wear, a function of the distance traveled by the tool in contact with the rock; impact wear, including microspalling and gross brittle failure; thermal fatigue, which produces deep cracks in the tungsten carbide structure; and vibration damage. Abrasion wear appears to predominate.

The body of a cutter disk is seldom replaced, just the cutter ring, which wears according to the abrasive nature of the rock, particularly its quartz content. Tarkoy and Hendron (1975) indicate that the consumption of cutter rings varies from near zero in rocks weaker than 70 MPa to a value of about 1.5 rings per meter of tunnel at rock strengths approaching 200 MPa. Bearings fail as a result of the forces and temperatures generated in the hub. In contrast, the picks used in softer rocks have no bearings to wear out or keep cool. When they require replacement, this is usually the result of abrasion and severe impact loading.

15.3.2.3 Methods for predicting wear and loss of tools. Uniaxial compressive strength gives a rough and often sufficient guide to tool consumption, although rates of wear and replacement can perhaps be predicted more reliably with the help of other indexes. West (1981) discusses the various testing methods for determining rock abrasiveness, and recommends measurements of quartz content and uniaxial compressive strength. Nelson and O'Rourke found Taber abrasion hardness (Sec. 2.3.2.4) to correlate better than other rock indexes with the rate of disk wear. For assessing rates of pick wear, Johnson and Fowell suggest the Cerchar scratch abrasivity test (Bougard et al., 1974), which is simple and can be performed on small hand specimens. They also developed an *instrumented cutting test* (McFeat-Smith and Fowell, 1979), which provides a wear flat value on a standardized tungsten carbide insert used to cut a core sample of rock of specified length at a standardized depth of cut and tool speed.

15.3.3 Developments in tunnel-boring technology

TBM manufacturers are constantly researching new and improved methods of tunnel boring, with the object of achieving faster rates of penetration through a greater variety of harder and more abrasive rocks, and with reduced wear on cutters and other components. The greatest advances in recent years have been through relatively unspectacular but steady improvements to the design of conventional machines and components, such as bearings, mucking equipment, and lining segment erectors. Research has continued into completely new methods of boring, so far with limited success. Some of these areas of research are outlined below.

15.3.3.1 Water jet assisted cutting. Experiments into the use of high-pressure water jets to assist mechanical cutting tools have been performed in the United States, Great Britain, France, Germany, and South Africa. Water jet assisted roadheaders are now being manufactured commercially, and kits are also available that allow water jets to be added to existing machines (Plumpton and Tomlin, 1982; Ip et al., 1986).

At the University of Newcastle in England, cutting trials have made use of jet pressures in the range of 70–207 MPa. The water jet assists drag bit cutting by flushing out the fine debris and chips, and by cooling the interface between tool and rock. A jet penetration of 20–30% of the mechanical depth of cut is reported as ideal, being sufficient to relieve the tool tip without removing the crushed zone necessary for crack initiation. Water jet assistance reduces the rate of development of a wear flat on the drag bit, and good performance is maintained for much longer. Water jets used in combination with disk cutters produce the initial kerf so that the disk travels along a preformed jet-cut slot. The disk becomes more efficient at breaking the rock between kerfs.

The South African Chamber of Mines reported a fivefold improvement in the performance of a mine boring machine by using 50-MPa jets to assist drag bits. Henneke and Knickmeyer (1979) describe water jet cutting tests in which a 2.6-m-diameter Wirth TBM was adapted for trials in a sandstone quarry with rock strengths of about 100 MPa. Water was delivered at 120 L/min to 94 nozzles at a maximum pressure of 360 MPa. A penetration rate of up to 4 m/min could be achieved at approximately half the thrust required using rollers only, when the water jets were turned off. Water jet tunneling has also been used to excavate sandstone bedrock in the Minneapolis area (Nelson and O'Rourke, 1983). The Robbins Company reports a substantial increase in advance rate using water jet assisted cutting through 280-MPa granite.

Water jets are not without their problems. About 2 MW of power is needed to produce the 2–4 L/s of high-pressure water necessary to excavate a tunnel of 6-m diameter. The high energy consumption raises rock temperatures substantially, and makes ventilation more difficult and expensive. Further problems arise because the muck is wet and must be removed from the heading as a slurry.

15.3.3.2 Thermally assisted tunneling. As early as in the third century B.C., miners used fires to heat and crack rocks (Carstens, 1972). Thermal techniques have been investigated in recent times and occasionally put to use. Infrared heating is used, for example, at the White Pine copper mine to break and weaken shale.

Rock can be induced to spall as a result of intense local heating and the accompanying differential thermal expansion (Williams, 1986; Sec. 14.1.8). The main application has been in drilling blastholes and cutting slots in extremely tough rock such as taconite, for which purposes an oxygen-hydrogen torch is employed. High quartz content favors spalling, whereas micas and ferromagnesian minerals make the rock more resistant. Carbonate rocks are not easily spalled unless they contain dolomite in percentages greater than 30%.

Rock can also be excavated by melting, if the temperatures are sufficiently high (Chap. 14). To date, thermally assisted rock tunneling remains more a concept than a reality.

15.3.3.3 Rock breakage with projectiles. Dardick (1981) describes a system for firing projectiles to break a tunnel into hard rock, such as granite, claimed to be much faster and cheaper than conventional drilling and blasting. A 4-m-diameter tunnel, 17 m long, was excavated into granodiorite for the U.S. Bureau of Mines, using 90- and 105-mm army guns. The ammunition was conventional, but the projectiles were of concrete with plastic covers, and were able to shatter 300 times their own weight in rock from the tunnel face. The report suggests that a 6-m-diameter hard-rock tunnel could be excavated at

167 m/day and \$713 per linear meter, as compared with 22 m/day and \$1221 per linear meter using conventional drilling and blasting.

The ultrarapid-fire Tround gun has three barrels that fire concrete or cast-iron projectiles at a muzzle velocity of 1600 m/s. This is estimated to be capable of removing over 2 t of hard rock with each salvo. Evidently the projectile method has important potential for a rapid excavation of protective tunnels for the military, although its peacetime uses might be more restricted.

References

Anon.: "Chicago's TARP Project," *Mining Mag.*, **142**(3), 4 pp. (Mar. 1980).

Athorn, M.-L., and R. A. Snowdon: "Performance of a TBM and a Roadheader in the Coal Measures," *Proc. 27th U.S. Symp. Rock Mech.* (Tuscaloosa, Ala., 1986), pp. 771–774.

Barensden, P., and R. G. Cadden: "Machine-Bored Small-Size Tunnels in Rock with Some Case Studies," in M. J. Jones, Ed., *Tunnelling '76* (Inst. Min. Metall., London, 1976), pp. 423–433.

Bougard, J.-F., et al.: "Proposals Relating to Measurements and Tests for a Mechanical Excavation Project" (in French), *Tunnels et Ouvrages Souterrains,* 5 (Sept./Oct. 1974).

Brunton, D. W., and J. A. Davis: *Modern Tunneling,* 2d ed. (Wiley, New York, 1922).

Carstens, J. P.: "Thermal Fracture of Rock—A Review of Experimental Results," *Proc. 1st North Am. Rapid Excavation and Tunneling Conf.* (Chicago, Ill., 1972), pp. 1363–1392.

Claasons, G. C. D., and D. R. Pike: "Machine Tunneling in Hard Rock," *Proc. South Afr. Tunneling Conf.* (Johannesburg, S. Africa, 1970), vol. 1, pp. 209–216.

Dardick, D.: "Tround Rapid Excavation and Tunneling System," *Proc. Rapid Excavation and Tunneling Conf.* (San Francisco, Calif., 1981), pp. 859–873.

Fourmaintraux, D.: "Performances of TBM in Rocks: Current Results, Forecasting and Control" (in French), *Proc. 5th Int. Cong. Rock Mech.* (Melbourne, Australia, 1983), vol. D, pp. 211–216.

Fourmaintraux, D.: "Utilisation des tunneliers en terrains difficiles," *Rev. l'Industrie Minérale—Les Techniques,* 441–446 (Nov. 1985).

Gaye, F.: "Efficient Excavation with Particular Reference to Cutting Head Design of Hard Rock Tunneling Machines," *Tunnels and Tunneling,* 39–48 (Jan. 1972); 135–143 (Mar. 1972).

Golder Associates and James F. MacLaren Limited: "Tunneling Technology," Ont. Ministry Transport. and Commun., Toronto, 166 pp. (1976).

Handewith, H. J.: "Mining Applications of Tunnel Boring Machines," *Bull. Can. Inst. Min. Metall.,* 133–136 (Nov. 1980).

Harrison, G. P., N. E. Green, and W. E. Bennett: "Some Aspects of the Art of Raise Boring," *Proc 1st North Am. Rapid Excavation and Tunneling Conf.* (Chicago, Ill., 1972), vol. 2, pp. 1161–1183.

Henneke, J., and W. Knickmeyer: "Possibilities and Limitations of Waterjet-Assisted Tunnel Boring in German Coal Mines," *Proc. Rapid Excavation and Tunneling Conf.* (Atlanta, Ga., 1979), vol. 2, pp. 1012–1031.

Hibbard, R. R., and L. M. Pietrzak: "Computer Simulation of Hard Rock Tunneling," Rep. AD-763 567, General Research Corp., Arlington, Va. (May 1973), 192 pp.

Ip, C. K., S. T. Johnson, and R. J. Fowell: "Water Jet Rock Cutting and Its Application to Tunnelling Machine Performance," *Proc. 27th U.S. Symp. Rock Mech.* (Tuscaloosa, Ala., 1986), pp. 883–890.

Johnson, S. T., and R. J. Fowell: "Compressive Strength Is not Enough (Assessing Pick

Wear Rates for Drag Tool-Equipped Machines)," *Proc. 27th U.S. Symp. Rock Mech.* (Tuscaloosa, Ala., 1986), pp. 840–845.

McFeat-Smith, E.: "Effective and Economic Excavation by Roadheaders," *Tunnels and Tunneling,* **10** (1), 43–44 (1978).

—— and R. J. Fowell: "The Selection and Application of Roadheaders for Rock Tunnelling," *Proc. Rapid Excavation and Tunneling Conf.* (Atlanta, Ga., 1979), pp. 261–279.

Mellor, M., and I. Hawkes: "Hard Rock Tunneling Machine Characteristics," *Proc. 1st North Am. Rapid Excavation and Tunneling Conf.* (Chicago, Ill., 1972), vol. 2, pp. 1149–1158.

Moss, A. S., J. Zeni, and J. Linkous: "Shaft Development Using Blind Hole Drilling Methods: A Case History," *Proc. 27th U.S. Symp. Rock Mech.* (Tuscaloosa, Ala., 1986), pp. 304–308.

Nelson, P. P., and F. L. C. Fong: "Characterization of Rock for Boreability Evaluation Using Fracture Material Properties," *Proc. 27th U.S. Symp. Rock Mech.* (Tuscaloosa, Ala., 1986), pp. 846–852.

—— and T. D. O'Rourke: "Tunnel Boring Machine Performance in Sedimentary Rock," Geotech. Eng. Rep. 83-3, Cornell Univ. School of Civ. and Environm. Eng., Ithaca, N.Y., 438 pp. (1983).

Pirrie, N. D.: "Hinkley Tunnels Prove Economics of Machine for Short Distance Tunnels," *Tunnels and Tunneling,* **1** (3) (1969).

Plumpton, N. A., and M. G. Tomlin: "The Development of a Water Jet System to Improve the Performance of a Boom-Type Roadheader," *Proc. 6th Int. Symp. Jet Cutting Technology* (Surrey, U.K., 1982).

Robbins, R. J.: "Mechanical Tunnelling—Progress and Expectations," in M. J. Jones, Ed., *Tunnelling '76* (Inst. Min. Metall., London, 1976), pp. xi–xx.

Sanio, H. P.: "Prediction of the Performance of Disc Cutters in Anisotropic Rock," *Int. J. Rock Mech. Min. Sci. Geomech. Abstr.,* **22** (3), 153–161 (1985).

Tarkoy, P. J., and A. J. Hendron, Jr.: "Rock Hardness Index Properties and Geotechnical Parameters for Predicting Tunnel Machine Performance," Rep. for National Science Foundation Grant GI-36468, Dept. Civ. Eng., Univ. of Illinois, Urbana-Champagne, 325 pp. (1975).

West, G.: "A Review of Rock Abrasiveness Testing for Tunneling," *Proc. Int. Symp. Weak Rock* (Tokyo, Japan, 1981), vol. 1, pp. 585–594.

Williams, R. E.: "Thermal Spallation Excavation of Rock," *Proc. 27th U.S. Symp. Rock Mech.* (Tuscaloosa, Ala., 1986), pp. 816–820.

Williamson, T. N.: "Tunneling Machines of Today and Tomorrow," *Highway Research Record* 339, pp. 19–25 (1970).

16

Rock Reinforcement

16.1 Rockbolting Methods and Materials

16.1.1 Introduction

Reinforced rock becomes a competent structural entity either on its own or as part of a composite rock-steel or rock-concrete structure. Rockbolts perform their task by one or a combination of several mechanisms. The simplest, that of suspension when a loose block is secured in the crown of a tunnel, is quite uncommon. Much more often, bolts act to increase the stress and the frictional strength across joints, encouraging loose blocks or thinly stratified beds to bind together and act as a composite beam (Lang, 1972; Underwood and Distefano, 1964; U.S. Army Corps of Engineers, 1980).

The components that make up rock reinforcing systems include rockbolts of various types and their accessories; high-capacity grouted ground anchors; and the mesh, straps, and cable nets that, when used with bolts and anchors, help form a composite system.

Hobst and Zajic (1983) report the use of roof bolting as early as in 1918 in the mines of upper Silesia (now in Poland). By the 1920s and 1930s roof bolts were being used in several underground mines in the United States where, by 1952, annual consumption had reached 25 million bolts. Rockbolting very largely replaced timber, made the excavations safer, released space previously obstructed by timber, and gave improved ventilation. Today about 120 million rockbolts are installed annually in U.S. mines, and over 90% of underground coal production is mined under a bolted roof (Bieniawski, 1984).

Rockbolts anchored by mechanical expanding shells are the most common type, being used in about 60% of applications. Resin-bonded bolts are now employed in about 30% of applications. The remaining 10% include patented systems such as the Swellex and Split Set steel tubular types discussed in this chapter.

16.1.2 Mechanically point-anchored bolts

16.1.2.1 Slot and wedge anchors. Slot and wedge anchored rockbolts were the earliest and, although they are not the best, continue to be used in some temporary support applications. They consist of a steel bar with a slot cut at one end, which contains a steel wedge (Fig. 16.1*a*). The wedge end is placed in the hole first, and the wedge is driven into the slot by hammering the exposed end of the bolt. Expanding halves grip the hole, allowing the bolt to be tensioned and to carry load. Bolts of this type are easily loosened by blast vibrations and ground movements.

16.1.2.2 Expanding shell anchors. The expanding shell anchor (Fig. 16.1*b*) contains toothed blades of malleable cast iron with a conical wedge at one or both ends. One or both of the cones are internally threaded onto the rockbolt so that when the bolt is rotated by a wrench, the cones are forced into the blades to press them against the walls of the drillhole. Unlike anchors of the slot and wedge type, the expansion shell increases its grip on the rock as tension on the bolt increases.

Expanding shell anchors are the least expensive in current use and the most widely used, particularly for short-term rock support in underground mines. They are most effective in harder rock types, and in softer rocks they tend to slip and loosen. Loosening can be avoided by

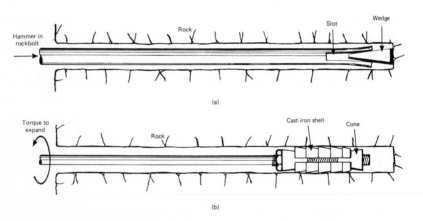

Figure 16.1 Mechanical rockbolt anchors. (*a*) Slot and wedge anchor. (*b*) Expanding shell anchor. Different designs of shell have either a single or a double wedging action.

introducing a cement grout through a plastic tube running alongside the bolt. Grouting of point-anchored rockbolts not only reinforces the anchor, but also protects against corrosion. Grout is injected at the lowest point in the drillhole, using a bleeder tube to allow escape of air from the upper end.

The Williams type of rockbolt is provided with a central hole running the length of the bolt, through which air can be bled. Hollow-center rockbolts offer a simpler and more efficient means of introducing grout, in return for a somewhat higher price.

16.1.3 Grouted dowels

16.1.3.1 Definition. A *dowel* is a fully grouted rockbolt without a mechanical anchor, usually consisting of a ribbed reinforcing bar, installed in a drillhole and bonded to the rock over its full length (Fig. 16.2a). Pender et al. (1963) describe the important part dowels have played in supporting major underground works such as in the Snowy Mountains hydroelectric project in Australia.

Dowels are self-tensioning when the rock starts to move and dilate. They should therefore be installed as soon as possible after excavation, before the rock has started to move, and before it has lost its interlocking and shear strength.

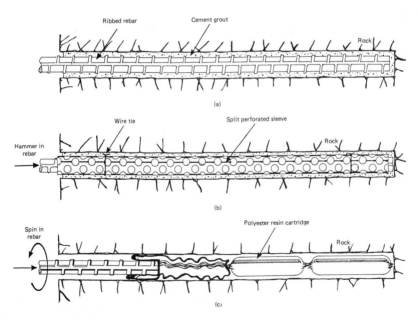

Figure 16.2 Grouted dowels. (*a*) Cement grouted dowel. (*b*) Perfobolt. (*c*) Resin-bonded bolt.

16.1.3.2 Cement-grouted rebar. Although several grout types are available, in many applications where the rock has a measure of short-term stability, simple portland cement grouted reinforcing dowels are sufficient. They can be installed by filling the drillhole with lean, quick-set mortar into which the bar is driven. The dowel is retained in upholes either by a cheap form of end anchor, or by packing the drillhole collar with cotton waste, steel wool, or wooden wedges.

The holding capacity depends on the grouting materials and methods. Unpublished research by Ontario Hydro gave typical 28-day steel-to-grout bond strengths at failure in the range of 2.0–5.5 MPa for deformed (ribbed) bolts, and in the range of 0.7–2.0 MPa for plain (smooth) bolts.

Cement-grouted dowels cannot be used for immediate support because of the time needed for the cement to set and harden. Also, cement grout can be damaged by blasting because it is brittle. Some tender documents in the United States prohibit grouting of bolts with cement until the tunnel heading has advanced by about 40–50 m beyond the point of bolt installation, thereby preventing blast damage.

The drillhole needs to be thoroughly flushed with water or air to remove dust, soft cuttings, and mud cake from the walls which, if left in place, reduce the bond strength. The normal grout mix consists of sand and cement in ratios between 50/50 and 60/40. The sand should be well graded, with a maximum grain size of 2 mm (Schach et al., 1979). The water-to-cement ratio should be no greater than 0.4 by weight; too much water greatly reduces the long-term strength. To obtain a plastic grout, bentonite clay can be added in a proportion of up to 2% of the cement weight. Other additives can accelerate the setting time, improve the grout fluidity allowing injection at lower water-to-cement ratios, and make the grout expand and pressurize the drillhole. Additives, if used at all, should be used with caution and in the correct quantities to avoid harmful side effects such as weakening and corrosion.

Grout injection, particularly in upholes, requires care to ensure that the complete drillhole is filled. Pockets of air and half-filled holes are difficult to detect and reduce the holding capacity of the dowel, sometimes to zero. The Bergjet grout pump (Schach et al., 1979) uses a container in the form of an upturned cone with a hemispherical lid. Compressed air is supplied at the top to force grout into an injection tube attached to the lower end of the cone. The hose is pushed fully into the drillhole and then, while injecting, is pulled out in small jerks to keep it in contact with the injected grout. When the hole is full, air pressure is turned off, the hose withdrawn, and the bolt pushed home.

16.1.3.3 Perfobolts. The Perfobolt system was introduced to simplify and speed up grouted doweling. The grout, instead of being pumped into the hole, is troweled into the two halves of a split perforated sleeve. The halves are placed together and bound at the ends with soft iron wire, and the tube full of cement is inserted in the hole (Fig. 16.2b). A reinforcing dowel is then driven by sledgehammer into the sleeve, forcing grout out through the perforations and into contact with the rock. Perfobolting has been used extensively in Scandinavia, where it supports many large underground excavations in hard igneous rock. It is no less effective in weaker rock types.

16.1.3.4 Resin-grouted dowels. Dowels grouted with polyester resin are the strongest form of rockbolt and also the most expensive (Franklin and Woodfield, 1971; Dunham, 1973; Snyder et al., 1979). The expense can be justified, however, where high-strength and rapid-strength bolting is needed.

A ribbed reinforcing rod is cemented into the drillhole by a polyester resin, which in a few minutes changes from a thick liquid to a high-strength solid by a process of catalyst-initiated polymerization. Resin and catalyst can be injected as a two-component fluid. This method is preferred for the grouting of long high-capacity anchors. In more routine rockbolting applications, the resin is supplied in sausagelike cartridges, each containing a strip of catalyst (Fig. 16.2c).

One or several cartridges are inserted into the hole, followed by the rockbolt which is pressed and rotated through the cartridges using a pneumatically or electrically operated drill. Rotation combined with thrust from the jackleg of the drill mixes the resin and catalyst, and forces the bolt fully home. As is the case with all grouted systems, the drillhole must be initially clean for good bonding. It should also be drilled accurately to the required length, because any overdrilling results in resin being wasted in the open end of the hole rather than being used to bond the bolt. Best results are obtained with a hole diameter 8–12 mm larger than the bolt diameter. A small annulus is necessary both for high strength and to avoid wastage.

In comparison with cement grouts, resin has the advantage of setting and reaching its full strength very quickly, in a period of from 2 min to half an hour using fast, medium, or slow setting cartridges according to requirements. The bond strength is also much greater than that of cement grout. One or two fast setting cartridges are inserted to provide an immediate point anchor, or a greater number are inserted for full-column grouting.

Complete grouting combined with tensioning can be achieved by inserting several slow-setting cartridges behind the fast ones. The bolts can be tensioned after the fast cartridges have set, and while the slower ones are still fluid. Subsequent setting of the slower "column grout" cartridges locks tension in the bolt and gives an ideal pretensioned, postgrouted system. To achieve this in practice, and without waiting and loss of production, requires a skilled bolting crew and careful timing.

Advantages of resin-bonded bolting include high early strength for immediate support, which is very important close to the face of an advancing tunnel. In average rock conditions, resin-bonded bolts are as much as 10 times stronger than their mechanical or cement-grouted equivalents, with a bond strength of approximately 4 kN/cm (1 t/in) of bonded length up to the limit of the strength of the steel bolt itself. Disadvantages include higher cost, nearly twice that of a bolt with a mechanical expanding shell anchor, and the slightly longer time needed for installation. A skilled crew, however, can drill and install a resin-grouted rockbolt in 2 or 3 min.

The maximum length of a fully grouted bolt is limited by the thrust capacity of the drill. Bolts longer than about 3 m cannot be installed by jackleg or stoper drills, and require a heavier drill to continue spinning through the viscous resin. If the resin starts to set before installation is complete, the bolt is left sticking out of the hole and is partially or completely ineffective.

Fully resin-bonded dowels made of wood in place of steel have been used to restrain floor heave in coal mines. They are ideal in this application because they can be cut without damaging the mechanical excavators used to trim the floor. Wood is also cheaper than steel, and surprisingly strong in relation to its weight.

16.1.3.5 Fiberglass and other systems. The Spokane Mining Research Center of the U.S. Bureau of Mines has developed a method in which a rope of fiberglass threads is fed from a spool into the hole while injecting a polyester grout. The fiberglass rope is cut off outside the hole, after the grout has set. The hole can be of any length and need not be straight. Holding capacities of 5 t/m of hole have been achieved in wet holes, and of up to 37 t/m in dry holes.

Because of the high cost of polyester and other resin adhesives and the slow curing time of portland cement, there is a demand for alternative grouts which combine rapid strength with low cost. Cartridges of inorganic gypsum-based grouts are now commercially available, which offer much faster curing rates (Fabjanczyk, 1982).

16.1.4 Tubular steel rockbolts

These patented rockbolt systems act as ungrouted dowels, relying on the friction developed between an expanded steel tube and the rough walls of the drillhole. They eliminate grouting and are very quick to install, although somewhat weaker than cement-grouted dowels. Corrosion of a thin, tubular steel bolt might limit its use in long-term applications.

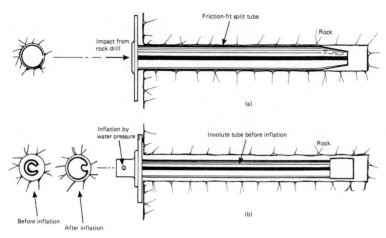

Figure 16.3 Steel tube bolts. (*a*) Split Set (Ingersoll-Rand). (*b*) Swellex before inflation (Atlas Copco; see also Fig. 16.4).

16.1.4.1 Split Set system. The Split Set rockbolt marketed by Ingersoll-Rand (Fig. 16.3*a*) consists of a steel tube split along the full length of one side to give a C-shaped cross section. When driven by sledgehammer or pneumatic drill into a slightly undersized drillhole, the C-shaped tube compresses and develops friction between the steel tube and the rock. The bolts typically have a 38-mm outside diameter for installation in 35-mm-diameter holes. The drillhole diameter must be controlled within close tolerances for the bolt to be effective. Drillhole roughnesses and deviations from linearity add to the frictional resistance and give the bolt its strength. Split Set bolting is intermediate in cost and has been found effective in harder rock types in particular, in which an anchorage of between 22 and 44 kN per meter length of bolt is obtained.

16.1.4.2 Swellex system. The Swellex rockbolt, marketed by Atlas Copco (Figs. 16.3*b* and 16.4) operates on a similar principle, relying on friction between a steel tube and the walls of the drillhole. In this case, however, the 2-mm-thick steel tube, instead of being split up the side, has an involute along its length (see cross section of tubing, Fig. 16.3*b*), which allows it to be expanded by hydraulic pressure. A portable pump, operated by compressed air, supplies about 27 MPa of water pressure to a small hole at the collar of the bolt. The bolt expands from its initially small diameter to much larger diameters to fill irregularities and cavities in the hole. Its holding capacity is therefore little affected by drillhole diameter and is usually greater than that of the Split Set type. Swellex tubes are lightweight and easy to transport and install. They are supplied in lengths of up to 3.6 m.

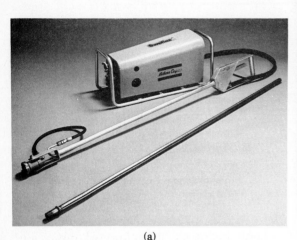

(a)   (b)

Figure 16.4   Swellex bolt and lance system. (*a*) Swellex portable high-pressure pump, installation lance, and an uninflated Swellex bolt. (*b*) Installing a roofbolt with the lance. (*Courtesy of Atlas Copco.*)

16.1.5   Relative costs

Costs vary widely, depending on the required quantities and on the variable costs of transportation and installation. An idea of relative costs can be gained from the following results of a survey carried out in 1980 by a Canadian mine. The costs cover supply, drilling, and installation of 2.1-m-long bolts.

	Cost per bolt type, $Cdn			
	Shell	Split Set	Swellex	Resin
Material	4.60	7.47	9.55	11.95
Transportation and handling	0.42	0.42	0.43	0.45
Drilling	7.00	7.00	7.00	7.00
Installation	1.96	2.28	1.78	3.42
Total	13.98	17.17	18.76	22.42

16.1.6   Rockbolting accessories

A steel bearing plate or *faceplate* prevents rock from raveling around the bolt head by converting the bolt tension into a compressive load applied to the rock face. Raveling, if permitted, rapidly undermines the bolts and leaves them protruding and ineffective. Plates have the further purpose of supporting steel mesh, strapping, or ribs.

Bearing plates can be circular, square, or triangular. Commonly the plates are square and measure 10 by 10 cm for use in hard massive rock, and 15 by 15 cm for use in softer or more closely jointed rock. They may be flat or domed so that when used together with washers, they can rotate to accommodate angles other than perpendicular between the bolt and the rock face (Fig. 16.5). Holes in the plates must be large enough to allow this rotation. Plates with slotted rather than circular holes are sometimes used to facilitate rotation and rock contact without bending. Domed plates also have greater flexibility, and by their compression can give a useful indication of the level of bolt tension and developing rock pressure.

Spherical washers or a pair of beveled, wedge-shaped washers are installed above the faceplate and behind the nut. They are needed because bolts are seldom installed exactly perpendicular to the rock face. High-capacity bolts in particular must make use of some type of washer to protect them against bending, which severely limits the tension that can be applied and can result in premature bolt failure if the ground is highly stressed. Special nuts with hemispherical heads are available to combine the functions of nut and hemispherical washer.

An ungrouted (point-anchored) bolt can be tensioned if threaded at either end, whereas a grouted bolt (dowel) can only be tensioned if

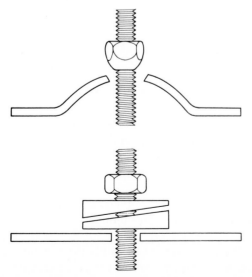

Figure 16.5 Drilling angles other than normal to the rock face are accommodated either by a domed plate and hemispherical nut or by a pair of beveled washers.

threaded at the free end. Relatively short (e.g., 2–3-m) point-anchored bolts are therefore often supplied with integral forged or "upset" square heads, which are cheaper and less vulnerable to damage than threads and nuts. Longer bolts and those of higher capacity are usually supplied with both ends threaded.

The thread at the free end should not be cut deeply or the bolt will be weakened, particularly if the steel cross section must first be reduced to remove a ribbed surface. A rolled thread is preferred to a cut one because the cross section is maximized and because the notching effect is less pronounced.

Bolts with rolled threads yield in a ductile manner, which is a particular advantage in highly stressed, rockburst-prone ground. Ortlepp (1970) reports a full-scale field test in which these "yielding" rockbolts were used to secure wire mesh to one side of a tunnel, the other side being identically supported with conventional rockbolts. The support was loaded by the detonation of explosives in drillholes around the periphery of the tunnel. The conventionally bolted side was completely destroyed, whereas the yielding bolts and mesh remained intact and prevented any appreciable damage.

16.1.7 Installation, grouting, tensioning, and testing

16.1.7.1 Mechanized installation of bolts. Installation of large numbers of rockbolts underground is nowadays greatly assisted by the use of mechanized bolting machines as shown in Fig. 16.6. The machine drills the hole and performs the complete installing operation. A supply of bolts is stored in a magazine on the rig. The operation is much faster than conventional manual installation, and also safer because the operator no longer stands beneath the rock while it is being reinforced.

16.1.7.2 Whether to grout. Ungrouted, mechanically anchored bolts are used in many mines and in some tunnels for short-term support, mainly because they are much less expensive and much quicker to install. They perform well in medium- to high-strength rock, provided that the anchors are well designed, that the bolts are effectively tensioned, and that the tensioning can be checked and reinstated from time to time.

However, full-column grouting is always an advantage because it adds considerable strength and stiffness to a mechanically anchored rockbolt. Grouting is usually considered essential for long-term applications, and for reinforcement of soft or highly stressed rock. It inhibits rock loosening and dilation along joints. A short, stiff length of

Figure 16.6 Mechanized bolting machine. This Swellbolter H 311 is specifically designed for drilling and installing Swellex bolts, but other models can install cement- and resin-grouted dowels.

grouted bolt responds directly to resist dilation, whereas an ungrouted bolt, which can stretch along its entire length, is much more flexible. Grouting also eliminates stress concentrations around a point anchor, and hence reduces slippage of anchors in soft rock, or as a result of blast vibrations.

Grouting also protects against corrosion of the bolt and anchor, which is important in many engineering applications. The bolts used in dam abutments and road cuts, for example, must have a life comparable with other permanent parts of the works. Bolts that are attacked and corroded by aggressive groundwater are likely to lose their tension and holding capacity within a matter of months or a few years. Protection can be provided either by full column grouting or, before the rockbolts are installed, by galvanizing or painting the steel with a coating of rust-resistant material. Epoxy resin has been used as a coating in this application. Stainless-steel bolts were used on one recent project for long-term support of a tunnel to carry aggressive waste waters from a pulp mill.

If a point-anchored bolt is to be grouted, it is doubtful whether a mechanical shell is needed in the first place. Once set and hardened, the grout bond is usually stronger than the mechanical anchorage, and a grouted reinforcing rod with no anchorage has a similar holding capacity to one with an expanding shell at the end. Probably the only useful functions of a shell in this application are to hold the bolt in an uphole while it is being installed, and to provide short-term support during the month or so while the grout is gaining strength.

16.1.7.3 Whether to tension. Point-anchored, ungrouted rockbolts must of course be tensioned and remain tensioned if they are to work at all. Usually a point-anchored bolt is tensioned to about one-half of its anchor-holding capacity, leaving a "reserve" between the applied tension and the strength of the rockbolt system. The action of opposing forces on the anchor and bearing plate tightens the superficial zone of loose rock, allowing it to make a significant contribution to the support of rock at greater depth.

Grouted dowels, on the other hand, are self-tensioning as a result of small rock movements and dilation along joints. While undergoing these movements, the rock tends to bind together and "arch" and, with the aid of the dowels, to support itself. Tension applied to a dowel after the cement grout has hardened serves little purpose except the important one of pulling mesh, straps, and other support items tight to the rock face. Tensioning of a grouted dowel, if sufficient, will shear the bond between grout and steel or rock. The shearing migrates along the bar over a distance proportional to the applied force, and damages rather than improves the support system.

The ideal system, at least in theory, is one in which the bolt is first tensioned and then grouted. It combines the benefits of applying immediate compression to the rock with those of increased holding capacity and corrosion resistance. Simple dowels are often preferred, however, in view of the cost and time required for two-stage installation and grouting.

16.1.7.4 Methods of tensioning. Point-anchored rockbolts for long-term support are tensioned to about 50% of their yield strength, and those for short-term applications, to about 70% of yield. Tensioning with an ordinary wrench is the simplest but also the least satisfactory method. Even with a piece of pipe over the wrench handle, the leverage, applied torque, and resulting tension are seldom enough and vary from bolt to bolt. Better to use a torque wrench, which allows the torque to be measured and standardized. Bolt threads should also be greased to reduce friction and to give a more repeatable and measurable tension.

A compressed air impact wrench is best for torquing large numbers of bolts very quickly and to a much higher preset level. An even more reliable but slower method is to use a hydraulic tensioner, a small hydraulic jack activated by a screw piston, to apply a measured direct pull to the bolt.

16.1.7.5 Testing. Simple pull tests are used at the start of a project to evaluate the suitability of one or several alternative rockbolt systems in a particular suite of rocks, and later for proof testing (quality control) of the selected system, once installed (Fig. 16.7).

Initial evaluation typically includes the pull testing of some 10–20 bolts in rocks of the various qualities expected on the project. Each is tested to failure, recording pull force and displacement measured with a dial gauge (Barry et al., 1956; ISRM, 1981). For grouted or resin-bonded systems the *bond factor* (tonnes holding capacity per meter of bonded length) is measured using bolts grouted over a range of bond lengths (Franklin and Woodfield, 1971). The maximum length of bond should be selected as sufficiently short to allow testing of the bond without breaking the steel bar.

Proof testing is done periodically on a specified percentage of bolts installed. The bolts are usually tested to a given percentage of their specified ultimate holding capacity and not necessarily to failure. As a separate operation, bolt tension may be monitored from time to time with a torque wrench and adjusted if tension has been lost by "bleed-off."

A seismic device called the Boltometer has been developed for nondestructive testing of grouted rockbolts and dowels (Bergman et al., 1983). Compression and flexural elastic waves are transmitted into the bolt and are reflected from its buried end. The amplitudes of the transmitted and reflected waves are compared. On the basis of tests on 271 cement-grouted bolts and 21 resin-grouted bolts, the manufacturers claim that the instrument can detect low-quality grouting, air pockets, bolt breaks, and bond failure; also the length of the bolt can be determined from the time interval between transmitted and received signals.

16.2 High-Capacity Rock Anchors

16.2.1 Applications

16.2.1.1 Civil engineering applications. Applications of cable anchors include rock reinforcement in large caverns, the securing of tied-back

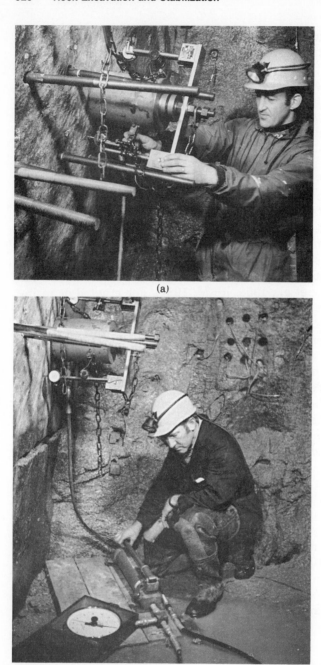

(a)

(b)

Figure 16.7 Testing rockbolts in granite; Cornwall, England. (*Franklin and Woodfield, 1971.*)

retaining walls, and the stabilization of landslides. Marine anchors provide uplift resistance for mooring ships and for anchoring structures to the sea bed. Anchors in combination with pressure-relief drain holes prevent uplift by hydrostatic pressure in foundation excavations. They also are needed to secure the cables of suspension bridges and the guy wires of tall radio and radar masts.

At the Morrow Point power plant (Brown et al., 1971) prestressed anchors were effective in stabilizing shear zones intersecting the cavern walls, which contained up to 1.5-m thickness of gouge material and fractured rock. The 80,000 kN of thrust calculated as sufficient to resist sliding was provided by 35- and 43-mm-diameter bars, 23 m long, installed through the shear zones. A drainage tunnel driven beneath the potential slide wedge and into the shear zones provided access for installing a further 25 prestressed anchors between the cavern wall and the tunnel.

Tensioned anchors are often used to augment shearing resistance along the base of a dam and along rock joints in the foundation, so as to increase the factors of safety against sliding and overturning. The Cheurfas dam in Algeria, a 30-m-high gravity masonry dam, was built during the 1880s on sandstone bedrock. In 1934 the dam was raised a further 3 m with the help of 37 prestressed anchors to provide the necessary additional stability. Each anchor carried a working load of 10 MN (1000 t) and was formed from 630 high-tensile-steel wires. After a period of 20 years, only 3% of the prestress had been lost (Mohamed et al., 1969).

Concrete can be saved by anchoring a dam to its foundation. A metric tonne (10 kN) of anchoring force can replace up to two metric tonnes of concrete deadweight, depending on the angle of installation and the friction angle at the foundation contact (Hobst and Zajic, 1983). Thus anchored dams can be made thinner. The cost of installing anchors is typically about one-third to one-half of the same force established using the weight of concrete. With suitable design, up to one-half of the mass of the structure can be replaced by anchoring forces. At the St. Michel dam in France, this cost reduction amounted to 20%.

16.2.1.2 Mining applications. Grouted cables, called *cable bolts* in this application, can be installed in mine stopes either upward to control caving of the stope back (roof), or horizontally to stabilize the hangingwall. Old hoist ropes can be used if they are degreased by washing with solvent. However, for routine cable bolting, the materials usually are purchased specifically for the application.

Fabjanczyk (1982) describes applications in Australian underground mines, where cables up to 18 m long, usually untensioned, provide extensive support in large openings, extraction horizons, critical drawpoints, and access drives. The most common tendon is a 15.2-mm, seven-wire, stress-relieved strand with a minimum breaking load of 250 kN (25 t). Usually two of these tendons are grouted into a hole of about 60 mm in diameter, using neat cement mixtures with a low water content. This two-strand cable bolt is sufficiently flexible for coiling up and handling, yet rigid enough for installing into 30-m upholes. Neat cement gives a relatively high strength, yet can still be pumped into longholes.

Upward cable bolting of open-stope crown pillars and backs can give improved support, an increase in stope span, and a reduction of pillar dimensions, and can eliminate the need for timber and other forms of support within the stope. In the cut-and-fill mining method, the cable lengths exposed during each successive mining lift are cut off and discarded. Cables have been used both in low-stress, near-surface applications where raveling of blocks is the main problem, and in highly stressed ground where pillars are overstressed. Fabjanczyk (1982) reports a significant number of failures of cable-supported crown pillars in the mines he visited. These occurred either through partial raveling of the pillar, stripping or shearing the cables, or through massive collapse following raveling of the stope walls. The number of cables actually broken was apparently very low. To minimize this type of raveling failure, cable bolts should be "stitched" to each other or anchored into a continuous and integral system using steel straps.

For stope hangingwall support, cable bolt holes are drilled either from the stope into the hangingwall, or from development drifts behind the hangingwall, back toward the wall. However, the excavation of special development headings is usually too expensive to be considered, so most cable bolting is done from within the stope, where it is limited to sublevels usually separated by 30-m vertical intervals. This restriction means that most support patterns offer little or no support to the large areas of wall between sublevels. The supported zones themselves are supposed to behave like reinforced beams along the length of the wall, but collapses between levels often leave the cables stripped and hanging like spaghetti. As mentioned, hangingwall support can be made more effective if the cables are tensioned and connected into a coherent system by stitching or strapping and mesh. All such systems have to be sufficiently robust to resist blasting.

16.2.2 Types of rock anchor

High-capacity reinforcing bars and cables are generally termed *rock anchors* rather than bolts, although their principle of operation is similar to that of a rockbolt. Conventional rockbolting is by definition limited to maximum working loads on the order of 200 kN, achieved using bolts of up to about 25-mm diameter and with a maximum length of 2 or 3 m. The working loads of high-capacity anchors are typically in the range of 200 kN to 2 MN, depending on the application. The anchors are constructed from multistrand cable or single or multiple rods.

The terminology for the various anchor components is defined in Fig. 16.8. In the most common, *tension* type of anchor (Fig. 16.9a), the pretensioning load is transmitted to the end of the grout plug nearest to the hole collar. Some shearing of bond between the grout and the rock, or the grout and the steel, is inevitable. The grout plug is placed under tension by the shear stresses that develop along the grout interfaces and tends to crack, exposing the bar or cable to potentially corrosive groundwater. In the alternative *compression* anchor (Fig. 16.9b), the entire length of anchor rod or cable is sheathed to prevent bonding. An end plate attached to the tendon at the base of the drillhole transfers a compressive load to the grout plug, which therefore does not crack. Compression anchors are often preferred for longterm applications to avoid corrosion problems.

16.2.3 Anchor tendons

The tendons for anchors can be formed from bar, wire, or strand. Wire and strand are more flexible and easier to store, whereas for smaller loads, a bar is sometimes easier to use and cheaper to install.

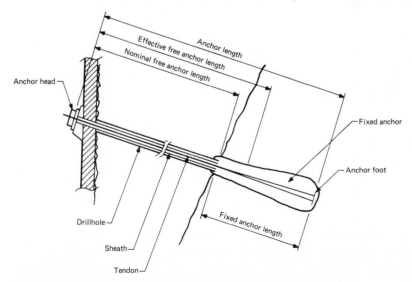

Figure 16.8 Terminology for high-capacity anchors. (*After ISRM, 1985.*)

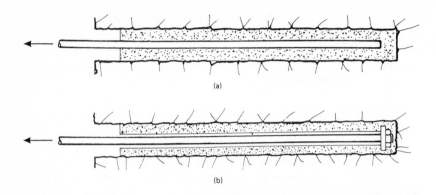

Figure 16.9 Types of high-capacity anchor. (*a*) Tension anchor. (*b*) Compression anchor. (*c*) Typical tieback cable anchors.

Bar anchors made of normal reinforcement steel typically have a yield strength of up to 400 MPa and an ultimate strength of 600 MPa. High-strength steels are often used, with a yield strength of 1100 MPa and an ultimate strength of 1300 MPa. High-strength alloy steel bars vary in diameter from 12 to 40 mm and can be either plain (smooth) or deformed (ribbed). Smooth bars are provided with threads at the anchorage end. Clusters of two, three, or four bars are sometimes used.

Wires vary in diameter from about 2 to 8 mm, and can be supplied either "as drawn" or "annealed." Annealing is a stress-relieving heat treatment that improves the elastic properties. Strand consists of a group of wires spun in helical form around a straight wire. Seven-wire and 19-wire strands are common (Hanna, 1982). Strand is usually also preannealed at a low temperature.

16.2.4   Anchor grouting

16.2.4.1   Single- and double-stage grouting.   The tendon is grouted over part of its length (the *bonded,* or *anchor,* length) after which the remaining *free* length is tensioned to about 80% of the yield strength of the steel.

Anchors can be grouted in either one or two stages. In the single-stage method the entire drillhole is grouted at the time of installing the anchor. The free length is protected by grease or by a plastic sheath to prevent bonding, and to allow free extension when tension is applied two to four weeks later, by which time the grout has gained sufficient strength. In the double-stage method the anchor length alone is grouted in the first stage. The free length is grouted in a second stage, after applying tension.

16.2.4.2   Design of mix and anchor.   High-quality grout and grouting are particularly important when anchors are to operate at high stress levels for extended periods. A neat portland cement grout with a water-to-cement ratio of 0.35–0.40 is nearly always used. Farah and Aref (1986) compare the bond strengths of cables grouted with sand-cement mortar and neat cement paste. Those grouted with mortar tend to be stronger, and the cost of the grout is reduced because part of the cement is replaced by sand. However, sand also makes a grout less pumpable. Additional water is therefore needed to achieve the same flow characteristics, which reduces the long-term strength. The anchor holes need to be larger and are more expensive to drill.

The required bond length is calculated knowing the shear strength between grout and rock and between grout and steel. Littlejohn and Bruce (1975, 1976) review grout-to-rock bond values assumed for design purposes on a wide range of projects. Working bond stress usually lies in the range of 0.35–1.40 MPa, the lower values being relevant for anchors in weaker rocks. Design values are best determined, or at least checked, by pull testing on site. Further design checks are needed to compare the proposed working load with the strength and yield limit of the steel itself, taking into account the sometimes lower strengths of threads, nuts, and faceplates.

The design anchor length should incorporate a safety factor, defined as the ratio of the ultimate holding capacity of the anchor to the working load. Littlejohn recommends minimum safety factors of 1.3 for temporary anchors with a service life of less than 6 months and few serious consequences of failure; 1.6 for temporary anchors with a service life of up to 2 years and quite serious consequences of failure; and 2.0 for all permanent anchors, high-risk situations, and temporary anchors in a highly corrosive environment.

16.2.4.3 Anchor Installation.

Individual anchors should be inclined at an angle slightly updip from the normal to the plane of potential sliding. This increases the safety factor in two ways: by increasing the normal force across the plane and by reducing the activating shear force along the plane. For a simple slab-slide situation, the angle between the drillhole and the normal to the slide plane is given by the equation

$$\tan \theta = \frac{1}{\tan \phi}$$

where θ is the required angle of drilling upslope of the normal, and ϕ is the friction angle (Fig. 16.10).

Holes for rock anchors are usually drilled either with percussive air-track equipment or with a rotary drill equipped with a tricone bit. The lower end of a soil anchor hole can be expanded ("belled") using an underreaming drill or by the pressure of grouting to increase the holding capacity of the anchor. The same effect can be achieved in rock by undercutting with a water jet or flame drill (Sec. 14.1.7), or by "bulling" (expanding the base of the hole) with explosives. Bulling is not often done because of the risk of damaging the ground or adjacent anchors.

In fractured rock, particularly with flowing groundwater, the open joints intersecting the anchor hole must be thoroughly sealed to eliminate the risks of groundwater diluting or washing out sections of the grout column, leaving pockets of air. The hole is drilled, filled with a thick grout, and then redrilled after the grout has set. It is water-tested and the cementing operation repeated as often as necessary to obtain a complete seal.

Cement mixes are injected using pneumatic grouting equipment as for rockbolting applications, or more efficiently by mechanical pumps. Pressure grouting vessels are usually equipped with a pneumatic turbulent mixer and operate at pressures of up to 0.5 MPa. Positive displacement pumps of various types are used for longer holes and higher grouting pressures of up to 12 MPa.

Complete air bleeding is easy to achieve in downward holes, but those inclined upward require an air bleed tube running to the upper limit of the drillhole. Grout is injected from below, at the hole collar,

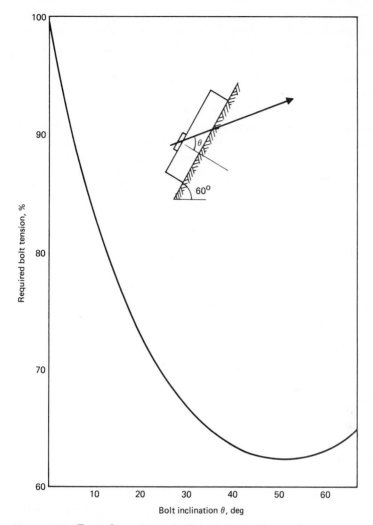

Figure 16.10 Example: optimum inclination for anchoring loose blocks and slabs on a 60° slope.

and injection is continued until good-quality grout emerges down the air bleed pipe. Because of unavoidable segregation of the grout and sometimes its dilution by groundwater, the first grout to emerge is of little value as a bonding material. Good practice is to form test cubes from the grout emerging from each anchor hole (not from the mixer) for crushing at ages of 7 and 28 days. Values of 28-day compressive strength specified by different authorities vary from 17 to 28 MPa (Hanna, 1982).

Grouting pressures must be limited to avoid hydraulic fracturing, heave, and disturbance to adjacent structures and anchors. A rule of thumb suggested by Hanna is 0.2 MPa/m depth of overburden. Permissible grouting pressures are further discussed in Sec. 18.2.4.5.

16.2.4.4 Corrosion protection. Because most anchors are to remain under tension for many years, they need to be protected against corrosion. At the Joux dam in France, several 1300-t-capacity anchors failed after a few months in use, even though stress levels in the tendons were only about 67% of full strength. After many tests, the phenomenon of corrosion under tension was blamed for the failures (Portier, 1974). As a result, the stress levels in permanent anchors are nowadays usually restricted to about 50% of tendon strength (Hanna, 1982).

The grout protects the bonded section of the bar or cable. The alkalinity (pH 9.8–12.3) of cement mortar prevents corrosion for as long as the grout column remains unfissured and impermeable. Compression-type anchors (Fig. 16.9b) are best in this application, because the compressed plug of grout is unlikely to crack and give access to corrosive groundwater. Sections that remain ungrouted, and the exposed anchor plates and other components, need to be galvanized or painted with a protective coating. Grease and bituminous and epoxy paints are often used.

16.2.5 Quality control

16.2.5.1 Testing. Anchor quality is ensured by specifying the appropriate grout mix and components, by monitoring grouting procedures, and by proof testing the installed anchors at the time they are tensioned. Further long-term monitoring of tension and movement is often needed.

High-capacity rock anchors are put through rigorous testing. *Design* testing is carried out to demonstrate or investigate, in advance of the installation of working anchorages, the quality and adequacy of the design in relation to the ground conditions and materials used. *Proof* testing is carried out on each installed rock anchor to demonstrate the short-term ability of the anchorage to support a load greater than the design working load.

Testing of multiple-wire or multiple-tendon anchors is complicated by the possibility of the different strands stretching differently and therefore carrying different levels of load. Usually the tendon is pulled as a single unit. The individual wires or cables that make up a compound tendon can be pull-tested one by one, but the results are difficult to interpret in terms of behavior of the anchor as a whole.

Anchor testing can be by either *coaxial* loading or *remote* loading methods (ISRM, 1985). Coaxial loading does not permit the surrounding rock mass to rupture, so is used only where an anchorage is deeply embedded (rock mass rupture unlikely) or solely for the purpose of rock reinforcement. Remote loading, which permits failure not only of the tendon and grout plug, but also of the surrounding rock, is applied to evaluate the strength of shallow anchors designed to resist uplift.

In the coaxial testing method, a hollow-center jack is installed over the top of the bar or tendon and bears against a hard rock surface or a bearing plate when the surface is softer (Fig. 16.11a). Remote loading is obtained by using a beam to span the anchor and transmit reaction

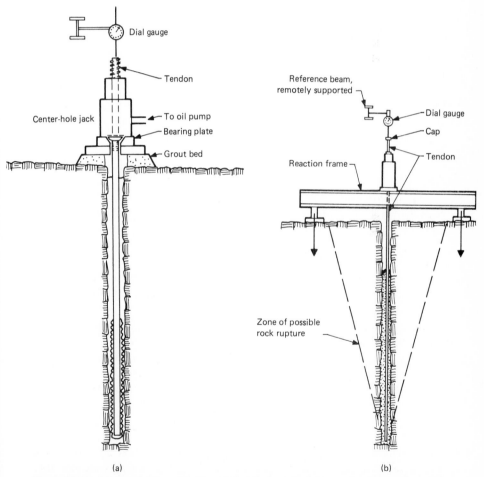

Figure 16.11 Anchor testing. (a) Coaxial loading method for evaluating tendon and bond. (b) Remote loading method for evaluating tendon, bond, and also the holding capacity of the rock. (*ISRM, 1985.*)

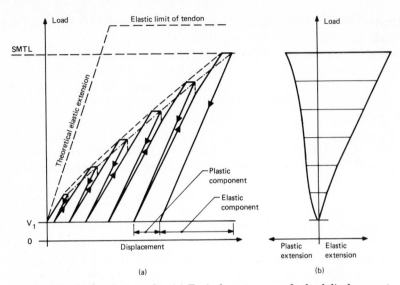

Figure 16.12   Anchor test results. (*a*) Typical appearance of a load-displacement curve, showing the effects of load cycling. V_1 is the specified bedding-in load, SMTL the specified maximum test load. (*b*) Diagram of plastic and elastic extensions. (*ISRM, 1985.*)

to remote bearing pads. It allows a conical zone of failure to develop (Fig. 16.11*b*).

Results are plotted in the form of a load-displacement curve for several cycles of loading and unloading. Superimposed on the graph are the theoretical elastic extension of the tendon or bar and its elastic limit (Fig. 16.12). From the data one can calculate the effective free length of the anchor, which should be greater than 90% of the nominal free anchor length and less than the nominal free anchor length plus 50% of the nominal fixed anchor length. This ensures that at least half of the theoretical anchor bond remains unsheared.

Load is applied to a level greater than the required working load and maintained at this level for several hours. Loss of load and increases of extension during the period of constant-load maintenance are symptomatic of anchor creep. If the anchor continues to creep at its prescribed test load, it may, depending on the specifications, be rejected or downgraded to a lower operating capacity.

16.2.5.2 Monitoring.   Various kinds of load cell are used to monitor long-term relaxation (loss of load) in ground anchors, so that when necessary, further tensioning can be applied. Surveying methods and also inclinometers, multiple-position extensometers, and other types of instruments are installed to monitor the movements of anchored retaining structures and of the rock mass itself, to confirm the effectiveness of the anchorage system. The instruments and the techniques are discussed in Chap. 12.

16.3 Bolted and Anchored Support Systems

16.3.1 Rockbolted systems

16.3.1.1 Spot bolting. *Spot bolting* is the placing of bolts at individually selected and marked locations and angles to stabilize isolated blocks that have been identified as potentially unstable. Bolt lengths are selected to pass through the potentially loose block and into solid, stable rock for a further distance of 1–2 m. This distance may need to be calculated from measurements of block dimensions and angles of the key joint faces. Blocks that are evidently likely to fall, and which cannot be safely drilled or scaled, are secured by strapping or mesh that spans between bolts on either side.

16.3.1.2 Pattern bolting. *Pattern bolting* is the systematic installation of an array of rockbolts to stabilize an area of a rock face or roof. The bolts are usually installed on a regular grid and at angles perpendicular to the rock face, or to a specified set of joints. Pattern bolting is usually best when there is insufficient time or information to determine which "key blocks" may be the unstable ones. Pattern bolting is much more common, and the spot bolting alternative is only used when systematic support is evidently unnecessary. More than 90% of underground coal in the United States is mined beneath a pattern-bolted roof. A square grid pattern at 1.22- by 1.22-m (4- by 4-ft) centers is the most common.

Maleki et al. (1986) report a study of the performance of 13,000 mechanically anchored roof bolts in a U.S. coal mine. The variables of significance in determining roof sag were shown to be the consistency of bolt tension, bolt anchorage capacity, roof rock type, roof shape, and the dip of bedding in the roof.

16.3.1.3 Wire mesh. The wire mesh, also called *screen,* used in ground support applications can be made from either woven or welded wire. Its main purpose is to support the rock between bolts, which is particularly necessary when the rock is closely jointed and the bolts are moderately to widely spaced. The wire mesh can also serve as reinforcement for shotcrete (Chap. 17).

Woven mesh as used for chain-link fencing is easy to handle and to install because of its flexibility, but this same flexibility can result in greater sag when the mesh is used to support broken rock. Welded mesh tends to be springy and difficult to install unless supplied in the annealed condition. Annealing, a form of heat treatment, removes much of the springiness and makes the mesh easy to bend and press into close contact with the rock.

A number of different wire gauge thicknesses and mesh apertures are available. A square aperture of about 150 mm is usually the most satisfactory in ground support applications. An aperture this large is essential when shotcrete is to be sprayed over the mesh. Anything smaller encourages the rebound of sprayed coarse aggregate particles and air pockets between the mesh and the rock face.

16.3.1.4 Straps, crown plates, roof ties, and truss bolts. Various methods can be used to lend support to the intervening rock between bolts, and can give a much more positive protection than small faceplates against raveling and falls of blocks and wedges. Straps, crown plates, roof ties, and truss bolts are available for use in combination with mesh or shotcrete or both. Straps and ties are particularly useful to span major discontinuities such as clay-filled faults or zones of intense jointing.

Straps (Fig. 16.13a) are strips of steel plate or channel connecting two or more rockbolts. They are perforated with drilled or flame-cut holes to fit over the bolt and under the faceplate, and may be corrugated to give extra stiffness. Straps are commonly fabricated from 6-mm steel about 2 m long and 100 mm wide and are held in place by three bolts or dowels.

Crown plates for installation in the crown of a tunnel are similar, but are supplied in a prebent configuration to fit the curved tunnel perimeter (Fig. 16.13b). Crown plates are very effective in supporting machine-bored tunnels through shale. In Toronto tunnels, they have been used together with a pattern of four rockbolts, two installed to the left and two to the right of the tunnel centerline in the excavation crown. Resin-bonded rockbolts can be quickly installed by an operator working from a platform near the TBM cutter head. Four bolts followed by a crown plate can be installed in just a few minutes while the TBM gripper pads are being relocated for the next cutting cycle (Chap. 15).

Roof ties serve the same purpose as straps and crown plates but are lighter, more flexible, and less expensive. They are manufactured from a pair of steel-reinforcing rods separated and held parallel by short lengths of welded reinforcing bar or by specially manufactured clamp-on spacers (Schach et al., 1979).

Truss bolts consist of two angled rockbolts connected by a central turnbuckle for tensioning. Developed by the Birmingham Bolt Com-

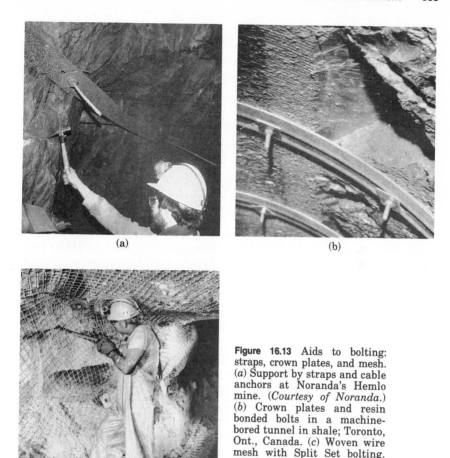

(a)

(b)

(c)

Figure 16.13 Aids to bolting: straps, crown plates, and mesh. (a) Support by straps and cable anchors at Noranda's Hemlo mine. (*Courtesy of Noranda.*) (b) Crown plates and resin bonded bolts in a machine-bored tunnel in shale; Toronto, Ont., Canada. (c) Woven wire mesh with Split Set bolting. (*Courtesy of Ingersoll-Rand, Split Set Division.*)

pany in England, their main use is for the support of flat, horizontal roofs such as those in coal mines.

Roof trusses are skeleton beams fabricated from welded reinforcing bar and bolted to the rock face. When shotcreted they form a strong reinforced concrete rib. Their light weight makes them easy to install.

16.3.2 Design of pattern-bolted systems

The development of design methods for pattern bolting originated in coal mining, where the bedded nature of coal deposits suggested a "reinforced beam" model. Design is often largely empirical, although built around a simple beam and column theory (Panek, 1956, 1984; Coats and Cochrane, 1970; Tang and Peng, 1984). Bolt loads are assumed to be a pair of opposing point forces acting at each end of the bolt. Their action augments the normal stress levels along bedding planes, and "welds" the thin strata into a single thick stratum that behaves more rigidly. The more recent design methods take into account not only friction between beds, but also the shearing resistance of the bolt itself.

Bieniawski (1984) gives an empirical design method for pattern bolting in coal mines, based on the rock mass rating parameter RMR of his geomechanics classification system (Sec. 3.3.3.4),

$$h_t = \frac{B(100 - \text{RMR})}{100}$$

where h_t is the *rock load height, B* is the roof span, and RMR is the rock mass rating. The bolt length is then given as one-half of the rock load height or one-third the span of the opening, whichever is greater.

The bolt spacing S is calculated from the bolt capacity C_b established by pull-out tests in the field:

$$S = \sqrt{\frac{C_b}{1.5} \, Gh_t}$$

or from the criterion that the ratio of bolt length to spacing should be between 1.2 and 1.5. Design charts are also given for entries and intersections.

Whether or not the rock is stratified, Lang (1972) shows how a compression zone is developed behind the exposed face of an excavation provided that the ratio of bolt length to spacing is less than 2.0. For systematic bolting in tunnels, Schach et al. (1979) recommend the selection of bolt length L in relation to tunnel span B (in meters) according to the following formulas:

$$L = 1.4 + 0.184B \qquad \text{untensioned dowels}$$

$$L = 1.6 + \sqrt{1 + 0.012B^2} \qquad \text{tensioned bolts}$$

Gerrard (1983) reviews recent numerical models that try to simulate more realistically the behavior of reinforced rock. Only limited success has been achieved. Such methods have potentially two applications, first to permit a rational and economic design of a rockbolted support system, and second to model the behavior of works in rock containing individual anchors and pattern bolting. Individual bolts are difficult to model in detail. An alternative approach, not fully satisfactory, is to make use of equivalent continuum elements with properties that account for both rock reinforcement and jointing.

16.3.3 Cable net and mesh systems for slopes

16.3.3.1 Overview. When stabilization of rockfalls by reinforcement is impractical, for example, along the extensive rock slopes that border transportation routes, preventive reinforcing systems can be replaced by protective rock-catch systems. Cable nets and anchored mesh can be used to collect and hold rock as it loosens from the face. As an alternative, falling rocks can be intercepted before they cause damage by a variety of catch systems, including ditches, benches, berms, fences, nets, walls, rock sheds, and tunnels.

16.3.3.2 Cable nets. Anchored cable nets restrain masses of small loose rock or individual blocks as large as 2–3 m. Cables that form the net are slung around the rock to be supported and are gathered together at each side by main cables leading to rock anchors. Cable lashing is a variation in which large rock blocks are supported by one or several individual horizontal cables anchored to the more stable parts of the rock face. Ribs of concrete or steel can be used to bridge between the cables. The required anchorage capacity is calculated from estimates of the weight of loosened rock that could conceivably need supporting.

16.3.3.3 Draped wire mesh blankets. In contrast to anchored mesh, which is designed to prevent rockfalls, the purpose of draped mesh is merely to reduce fall velocities and bounce heights, and to guide falling rock into a toe ditch, so only a narrow ditch is required. It is used when the bounce heights and roll distances of blocks would otherwise carry them beyond a normally dimensioned ditch and fence system. It works best when no individual blocks are larger than 1 m and when the slope is uniform enough for continuous contact between the mesh and the slope face.

The blanket is anchored at the crest by grouted rock dowels attached to a cable threaded through the upper edge of the mesh. It is rolled over the crest and then the vertical seams are wired together. The bottom end is usually left open, 1 m or so above ditch level. The mesh usually is 9 or 11 gauge galvanized, standard chain-link or gabion wire. The gabion mesh alternative has a double-twist hexagonal weave that does not unravel when broken.

16.3.4 Anchors for wedge stabilization

The method of limiting equilibrium, introduced in Chap. 7, is usually the best way of designing an anchorage system to stabilize large rock wedges or potential slab slides in both open cuts and large underground chambers.

Bolts or anchors are installed at angles selected so that anchor tension augments the stress normal to the potential slide plane, and at the same time contributes to resisting the downslope weight component. The tendon or bolt needs to be sufficiently long to ensure that the anchor is entirely beneath the surface of potential sliding. Anchorage capacity and pretension are calculated to be sufficient to augment the safety factor to the required level. Methods for designing anchored rock slope stabilization systems are extensively reviewed in Barron et al. (1970).

References

Barron, K., D. F. Coates, and M. Gyenge: "Artificial Support of Rock Slopes," Res. Rep. R228, Can. Dept. Energy, Mines & Resources, Mining Res. Centre (1970).

Barry, A. J., L. A. Panek, and J. A. McCormick: "Anchorage Testing of Mine Roof Bolts, Parts 1 and 2," Rep. Invest. 5040, 12 pp. (1954), 5194, 19 pp. (1956), U.S. Bureau of Mines.

Bergman, G. A., N. Krauland, J. Martna, and T. Paganus: "Non-destructive Field Test of Cement-Grouted Bolts with the Boltometer," *Proc. 5th Int. Cong. Rock Mech.* (Melbourne, Australia, 1983), vol. A, pp. 177–181.

Bieniawski, Z. T.: "Tunnelling in Coal Mines—Designing Development Entries for Stability," *Proc. 2d Int. Conf. Stability in Underground Mining* (Lexington, Ky., 1984), Szwilski, A. B., and C. O. Brawner, Eds., Am. Inst. Min. Eng., New York, pp. 3–22.

Brown, G. L., E. D. Morgan, and J. S. Dodd: "Rock Stabilization at Morrow Point Power Plant," *Proc. Am. Soc. Civ. Eng.,* **97** (SM1), 119–140 (1971).

Coates, D. F., and T. S. Cochrane: "Development of Design Specifications for Rockbolting from Research in Canadian Mines," *Proc. 6th Int. Min. Cong.* (Madrid, Spain), 12 pp.; also *Can. Min. J.,* 37–45 (Mar. 1971).

Dunham, R. K.: "Some Aspects of Resin Anchored Rockbolting," *Tunnels and Tunneling,* 376–385 (July 1973).

Fabjanczyk, M.: "Review of Ground Support Practice in Australian Underground Metalliferous Mines," *Proc. Austr. Inst. Min. Metal. Conf.* (Melbourne, Australia, 1982), pp. 337–349.

Farah, A., and K. Aref: "An Investigation of the Dynamic Characteristics of Cable Bolts," *Proc. 27th U.S. Symp. Rock Mech.* (Tuscaloosa, Ala., 1986), pp. 423–428.

Franklin, J. A., and P. F. Woodfield: "Comparison of a Polyester Resin and a Mechanical Rockbolt Anchor," *Trans. Inst. Min. Metal.* (London), Sec. A, **80**, A91–A100 (1971).

Gerrard, C.: "Rockbolting in Theory: A Keynote Lecture," *Proc. Int. Symp. Rockbolting* (Abisko, Sweden, 1983), pp. 3–32.

Hanna, T. H.: *Foundations in Tension: Ground Anchors,* 1st ed. (Trans Tech Publ. and McGraw-Hill, New York, 1982), 573 pp.

Hobst, L., and J. Zajic: *Developments in Geotechnical Engineering,* vol. 13, *Anchoring in Rock* (Elsevier, Amsterdam, 1983), 390 pp.

ISRM: "Suggested Methods for Rock Characterization, Testing, and Monitoring," ISRM Commission on Testing Methods, E. T. Brown, Ed. (Pergamon, Oxford, 1981), 211 pp.

———: "Suggested Method for Rock Anchorage Testing," ISRM Commission on Testing Methods, *Int. J. Rock Mech. Min. Sci. Geomech. Abstr.,* **22** (2), 71–83 (1985).

Lang, T. A.: "Rock Reinforcement," *Bull. Assoc. Eng. Geol.,* **9** (3), 215–239 (1972).

Littlejohn, G. S., and D. A. Bruce: "Rock Anchors: State of the Art," *Ground Eng.,* **8** (3), 25–32 (May 1975); **8** (4), 41–48 (July 1975); **8** (5), 34–35 (Sept. 1975); **8** (6), 36–45 (Nov. 1975); **9** (2), 20–29 (Mar. 1976); **9** (3), 55–60 (May 1976); **9** (4), 33–44 (July 1976).

Maleki, H. N., M. P. Hardy, and C. J. H. Brest van Kempen: "Impact of Mechanical Bolt

Installation Parameters on Roof Stability," *Proc. 27th U.S. Symp. Rock Mech.* (Tuscaloosa, Ala., 1986), pp. 526–535.

Mohamed, K., B. Montel, A. Civard, and R. Luga: "Cheurfas Dam Anchorages: 30 Years of Control and Recent Reinforcement," *Proc. 7th Int. Conf. Soil Mechanics and Foundation Eng.* (Mexico City, Mexico, 1969), Speciality Session 15, pp. 167–171.

Ortlepp, W. D.: "An Empirical Determination of the Effectiveness of Rockbolt Support under Impulse Loading," *Proc. Int. Symp. Large Permanent Underground Openings* (Scandinavian Univ. Books, Oslo, Norway, 1970), pp. 197–205.

Panek, L. A.: "Principles of Reinforcing Bedded Mine Roof with Bolts," Rep. Invest. RI 6138, U.S. Bureau of Mines, 59 pp. (1956).

———: "Design for Bolting Stratified Roof," *Trans. U.S. Soc. Min. Eng.,* 113–119 (June 1984).

Pender, E. B., A. E. D. Hosking, and R. H. Mattner: "Grouted Rockbolts for Permanent Support of Major Underground Works," *J. Inst. Eng.* (Australia), **35,** 129–150 (1963).

Portier, J.: "Protection of Tie-Backs against Corrosion," *Prestressed Concrete Foundations and Ground Anchors, 7th FIP Cong.* (New York, 1974), pp. 39–53.

Schach, R., K. Garshol, and A. M. Heltzen: *Rock Bolting: A Practical Handbook* (Pergamon, Oxford, 1979), 84 pp.

Snyder, W. V., J. C. Gerdeen, and G. L. Vigelahn: "Factors Governing the Effectiveness of Roof Bolting Systems Using Fully Resin-Grouted Non-tensioned Bolts," *Proc. 20th U.S. Symp. Rock Mech.* (Univ. of Texas, Austin, 1979), pp. 607–614.

Tang, D. H. Y., and S. S. Peng: "Methods of Designing Mechanical Roof Bolting in Horizontally Bedded Strata," *Proc. 25th U.S. Symp. Rock Mech.* (Evanston, Ill., 1984), pp. 615–626.

Underwood, L. B., and C. J. Distefano: "Development of a Rockbolt System for Permanent Support at Norad," *Proc. 6th U.S. Symp. Rock Mech.* (Rolla, Mo., 1964), pp. 43–86.

U.S. Army Corps of Engineers: *Engineering and Design Rock Reinforcement Engineer Manual,* EM1110-1-2907 (1980).

Chapter 7

Support and Lining Systems

17.1 Support Using Rock

17.1.1 Self-support by in-place rock

An important principle when designing a support system is to use rock to support rock wherever possible. Nothing is to be gained by replacing rock with concrete that may even be weaker than the material removed. However, if the rock is to support itself effectively, it must not be allowed to loosen. This requires careful blasting and the installation of rockbolts and cables to keep joints tight and to encourage friction and arching between blocks.

To further encourage self-support, excavated openings are shaped to the best configuration taking into account the stress field and the pattern of jointing. The openings are kept as small as their function will permit. Often the final size has to be larger than can be maintained stable without some form of artificial support, because of the need to extract an ore body or to house a specific size and shape of facility. Even then, the excavations should be kept as small as possible for as long as possible.

Open-pit mining is one example of this important concept of "phasing" the work of excavation. The pit walls are steepened to final grade only as a last phase of the mining cycle. Rock buttresses are often left behind to span between the walls of a narrow, elongated pit. A deep basement excavation in poor rock can be excavated in two parts with an intervening buttress, and opened to full size only after the initial excavations have been completed and stabilized. In many underground mines, the pillars of ore are at first quite large, and then wholly or partially recovered by retreat mining as a final phase. In a large tunnel passing through poor-quality ground, a small top heading is excavated first and stabilized before removing the remaining "bench."

17.1.2 Support by broken rock

17.1.2.1 Support by rock rubble, broken ore, or backfill. Often an economic alternative to leaving solid rock behind is to use broken rock for stabilization. Natural slopes that have been oversteepened by geological erosion can be restabilized by replacing rock and earth materials at the toe. In the shrinkage stoping method of mining, broken ore is used to support the stope walls, and in cut and fill mining, openings are stabilized using a backfill composed of waste rock or tailings with or without cement to bind these materials together.

17.1.2.2 Gabions. Gabions (Fig. 17.1) are an alternative for toe protection where slopes are undercut by fast flowing water, along river banks and the tailrace channels of generating stations. They are constructed by filling wire baskets with durable crushed rock or cobbles. The baskets, made of hexagonal woven-steel galvanized wire mesh, are wired together and filled in groups. They are placed to overlap each other, forming a buttress wall that is rapidly and inexpensively

Figure 17.1 Filling gabions with crushed limestone blocks for river-bank reconstruction in Ontario, Canada.

constructed and has the further advantage of being able to tolerate substantial settlements and deformation.

17.1.2.3 Dressed masonry. Rock masonry consists of dressed blocks, that is, blocks shaped by sawing or by splitting and chiseling. Masonry arches were in use for permanent support in tunneling before the advent of brick, steel, and concrete. Masonry is still used widely for the construction of retaining walls (Fig. 17.2).

Requirements for stone masonry include durability and the local availability of suitably sized and, if possible, suitably shaped blocks. When natural materials are close to hand, they are often preferred to expensive artificial alternatives. They are attractive from an architectural standpoint also.

17.1.2.4 Dental masonry. Patches of *dental masonry* serve well to repair undercuts and sections of loose or deteriorated rock. Loose rock is removed and replaced either by concrete or by the same rock blocks with mortar-filled joints. The latter method preserves the natural ap-

Figure 17.2 Masonry retaining wall, Hong Kong. Masonry blocks are bedded on a minimum 75-mm-thick layer of free-draining crushed stone or gravel, and are mortared and provided with weep holes. (*Geotechnical Control Office, 1984.*)

pearance of the rock face, so it is often used for the renovation of architectural landmarks and historic sites. Dental masonry was chosen to stabilize highway cuts through shaly limestone in England. The same technique was employed to restore the steep faces of a gorge at the Balls Falls grist mill in Ontario, Canada. Undercut sections of shaly limestone were scaled and backfilled with blocks of the more resistant local dolostone.

17.2   Artificial Supports and Liners

17.2.1   Timber

In the earliest days of tunneling, excavations were supported with timber at least until a permanent liner could be installed. Some tunneling specifications still call for provision of standby timber for use in emergencies, and timber props and cribs (*timber packs*) are often installed when mining through highly stressed ground. Timber has the advantage of flexibility and can accommodate large strains before it breaks. Crushing of timber becomes apparent long before failure, allowing support to be added or, in the extreme, the unsafe area to be evacuated. More up-to-date and brittle forms of support, with the notable exception of flexible shotcrete and rockbolted systems, often give much less warning of excessive straining.

A new development is the use of timber-filled pipe props, called *pipe sticks*, in which the timber protrudes from each end of a steel tube. The timber acts as the axial load-bearing member, and the steel pipe provides lateral constraint, thereby greatly improving the yieldability of the prop (South African Chamber of Mines, 1977).

One limitation is that timber supports, and particularly thin timber lagging between steel supports, tend to rot (Fig. 17.3). Temporary timber should not be incorporated into a final concrete wall or liner, because it weakens the support system. A timber "temporary" support can be difficult and dangerous to dismantle and to replace by a more permanent system.

17.2.2   Steel

17.2.2.1   Anchored beams. Anchored beams are a compromise between anchors with localized bearing pads and tied-back retaining walls that cover the entire rock face. The beams bridge between anchors and prevent the intervening rock from loosening. The anchors then act in combination so that fewer are needed than when installed individually. The beams can run horizontally or in any direction to suit the geometry of the rock face and jointing system. They can be made of concrete, reinforced shotcrete, or steel, and of any weight from

(a)   (b)

Figure 17.3 Rotten timber lagging in a pilot heading for the Rubira tunnel, Barcelona, Spain. The timber collapsed a few days after the photograph was taken, leading to a roof fall. Rock debris flowed from the pilot tunnel and filled the main tunnel. The caved region vacated by the debris stopped just short of the road and buildings above the tunnel line. The debris and cave were grouted, and tunneling was continued without further incident.

lightweight "strapping" and corrugated "pans" (Sec. 16.3.1.4) to heavy H or I beams.

17.2.2.2 Steel piles. Heavy-steel H piles are used for supporting open-cut excavations. The piles are usually cantilevered by cementing them in augered or percussion-drilled toe-in sockets in the rock, and are supported at higher elevations by walers (horizontal beams) that are themselves retained by tieback anchors. Lagging boards of timber or precast concrete are inserted between the flanges of the H piles as excavation proceeds downward, to span the gaps and support the ground between adjacent piles. Often a wall of H piles and lagging, socketed into rock, is employed to support overlying soil.

Interlocking steel sheet piling avoids the need for lagging and can be driven through soil overburden and into soft underlying rock such as a weak shale. However, the driving tends to break up the rock, making toe resistance unreliable except in the more ductile and very soft types of shale. Driving problems occur when the rock turns out to be stronger than expected, or contains hard and thick interbeds.

17.2.2.3 Steel sets and lagging. Steel sets and lagging followed timber as the main means of temporary support in mining and tunneling work; they themselves are now largely superseded by shotcrete and rockbolts. The roadways of many mines are still supported by traditional H-section steel sets that are bolted together. Connections usually consist of fish plates bolted to the webs only.

Steel sets and lagging are the underground equivalent of the H piles and lagging used in open excavations. The only difference is that the sets are curved to fit the underground opening, and come in shorter lengths for ease of handling and assembly. Lagging boards of timber or precast concrete slide between the flanges of the H to span the gaps between adjacent sets.

Advantages include speed and ease of erection, a relatively high and reliable load-carrying capacity, and little if any maintenance. If the ground is moderately stable, the sets can be quite easily recovered and reused. Disadvantages include a high cost compared with rockbolted and shotcreted systems. Steel sets also take up more space, restrict air flow, and require additional rock excavation. They are difficult to adapt to varying rock conditions, so usually more steel is used than needed. The supports are not as strong as they look, and the bolted H sections are susceptible to sudden buckling or torsional failure when ground pressures are high. Also, the rigid support fits imperfectly in an irregularly blasted tunnel. Gaps between rock and lagging allow the rock to move and loosen and to develop greater pressures on the support members than if movement were more confined.

17.2.2.4 Steel ribs. Steel ribs usually have a U-shaped rather than an H-shaped cross section. They are much lighter than sets, therefore less expensive and more flexible. Lightweight steel ribs usually have rolled sections with depths of 100–150 mm. They are not so convenient for use with lagging, which must pass behind the rib in the absence of H flanges, and are more often used to provide added reinforcement in combination with rockbolts and shotcrete.

The ribs are coupled together in sections with sliding telescopic connectors (usually consisting of U bolts at the overlaps between rib segments) so they can accommodate rock squeeze without becoming overstressed. This allows them to be pulled tight against the rock surface by rockbolts. The U section should be pointing inward, away from the rock face, if it is to be later filled with shotcrete. Otherwise a pocket of air is trapped with inevitable weakening of the system. Ribs can be widely spaced to provide only nominal support, or as close as 0.5–1.0 m where heavy ground squeeze is expected.

17.2.2.5 Bin walls. Whereas the cantilever-supported steel sheet pile wall commonly used to support soil excavations is usually not possible

Figure 17.4 Steel bin wall for slope stabilization; Hamilton, Ont., Canada. (*Courtesy of Proctor and Redfern Ltd., Toronto.*)

in rock, a similar wall can be supported by anchored tiebacks. A corrugated and galvanized steel sheeting *bin wall* is often used for rock cut stabilization (Piteau and Peckover, 1978). Figure 17.4 shows a bin wall anchored by shallow rockbolts to stabilize a cut in limestone and shale at Hamilton, Ont., Canada. The void behind the wall was backfilled with free-draining material, and water was collected and conducted to a storm sewer.

17.2.2.6 Steel liner plate. The *steel liner plate* is the underground equivalent of the bin wall. However, the plates are usually flat rather than corrugated, welded in situ rather than bolted, and they are thicker. Many small-diameter tunnels are supported by liner plate erected and welded on site. The method is expensive but simple and reliable. Steel tube liners are also used in tunnels that carry high-pressure water, such as the penstock tunnels of hydroelectric generating stations. They guarantee freedom from leaks and give a strength that is reliable and can be calculated.

Tunnels lined with steel plate are usually back-grouted with a portland cement grout, or back-packed with pea gravel, to fill the void between steel and rock, and to lend some rigidity to an otherwise very flexible structure. Plates can be transported flat and rolled on site, so take up very little space. Billington and Jacomb-Hood (1976) present a design method to predict factors of safety against elastic buckling and plastic collapse mechanisms.

Steel tube liners are also used with the *pipe-jacking* method of soft ground tunneling, in which the tube is equipped with a cutting edge and is forced through the soil by the action of hydraulic jacks on the trailing edge. This method can evidently not be used even in the softest of rocks.

17.2.2.7 Tunneling shields. Tunneling shields are discussed in Chap. 15. A similar concept, the *trench box,* is employed in open excavations. In contrast with stationary supports, shields move at the same pace as the excavation. Support is provided where it is most needed, at the working face of the tunnel. The shield gives protection both for excavation work and for the installation of primary or secondary liner systems. It can be self-propelled as part of a tunnel-boring machine, or independently propelled for use with a partial-face TBM or other form of mechanical miner. It may cover the entire rock surface, including the face, or just the curved surface, or it may give partial shielding in the crown only.

17.2.3 Cast concrete

17.2.3.1 Concrete walls. The gravity and cantilever walls so often used to retain soils are hardly ever used to retain rock. They require too much excavation and too much space. The ability of rock to hold high-capacity anchors makes the alternative of an anchored or tieback wall attractive. Rockbolts, mesh, and shotcrete are more often used when complete lining is needed. Concrete may be economic if there is easy access for ready-mix trucks, such as along highway cuts through weathered rock. Cast in situ retaining walls, beams, and slabs are employed only where there is ample room for concreting work. Precast members are more convenient at less accessible locations.

Caisson walls are often used as support in deep excavations passing through soil and into soft rock, particularly when the rock is erodible, permeable, and unstable. They are constructed by augering a row of slightly overlapping caisson holes, typically of 1–2 m diameter, which are then backfilled with concrete. The excavation proceeds downward under the protection afforded by surrounding concrete, if necessary installing walers and tiebacks through the wall at intermediate levels.

17.2.3.2 Concrete buttresses. Concrete buttresses are localized sections of anchored or gravity retaining wall, by definition in the form of vertical support members that transfer part of the weight of the rock to the ground. They are most often used to support overhangs, such as in road cuts at highway level and where river erosion has taken place. They are often expensive but may be less costly than the removal of overhangs, especially in steep terrain.

17.2.3.3 Cast concrete tunnel liners. Cast in situ tunnel liners are most suitable when the tunnel has an irregular cross section such as produced by blasting. A formwork is erected behind which concrete is pumped to fill the gap. Traveling forms can be collapsed and reerected. Telescopic forms are available that can be stripped, collapsed, moved through the center of previously erected forms, and then reerected. A further type is the nontelescopic form that collapses enough only to break away from the concrete surface. Lengths of up to 30 m are now common in tunnels in the United States.

The most common sequence is first to pour the invert and then to complete the concrete arch. This provides a firm base for the arch formwork and a smooth working roadway. The alternative of pouring the arch first and then the invert has the merit of eliminating relaying of track and final cleanup of the invert. A full-section pour requires that the forms be carried on needle beams or trusses, so the method is used only in small tunnels where the weight of formwork and its limited maneuverability are less of a problem.

17.2.3.4 Bernold liner system. In the patented Bernold system of tunnel lining, a pressed and perforated steel sheet acts as both formwork and reinforcement (Fig. 17.5). The steel is supplied in curved segments that are assembled in the tunnel and connected together by pins. Concrete is pumped behind the steel, flows around the annular gap, and partially emerges from the openings in the mesh. The result is a very high strength and moderately flexible reinforced concrete structural liner that can carry substantial rock loads.

17.2.3.5 Extruded (slipform) concrete liners. Horizontal slipforming is common practice for canal lining. The concrete leaves the slipform compacted to such a degree that it can support itself, so the advance

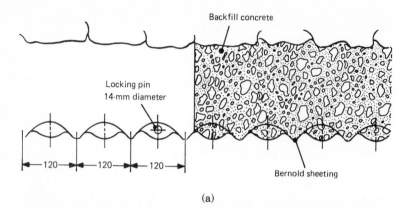

(a)

(b)

Figure 17.5 Bernold liner system. (*a*) Bernold sheeting, overlapping and retained by pins. (*b*) Completed Bernold liner, military cavern, Europe. (*Courtesy of Bernold A.G., Switzerland.*)

rate is limited only by how fast the concrete can be delivered and placed by the machine. Advance rates as high as 10 m/min have been achieved. Vertical slipforming has been used widely in recent years for constructing tall buildings with relatively uniform cross section.

Underground applications of slipforming have been reported in the U.S.S.R., where concrete has been successfully placed continuously behind the mechanized shield of a TBM (Ounanian et al., 1981). Using this extruded tunnel lining system (ETLS), in which the moist concrete was compacted under 1.4 MPa of pressure, 10 km of tunnel had been lined by late 1976 at rates of up to 10 m/day. The method has undergone feasibility testing in the United States. In a prototype system, very high early-strength concrete was continuously delivered to a slipform and bulkhead system that supported the concrete liner until strong enough to support itself. The combined TBM and slipforming machine advanced at rates of 0.6–3.7 m/h.

17.2.4 Precast support systems

17.2.4.1 Precast retaining systems. Old-fashioned brick and stone were the earliest forms of "precast" construction, but for reasons of cost, they have largely been replaced by precast concrete. Rock excavations at surface can be supported by concrete panels with anchored ribs or waling. Cribs can be constructed of crisscrossing concrete beams backfilled with stone or earth. Rock-filled timber cribs are used underground to resist very high roof-to-floor pressures (Fig. 17.6). Cribs constructed of crisscrossing alternating concrete and wooden beams, called *sandwich packs*, are used in deep South African mines (South African Chamber of Mines, 1977). Their high resistance to crushing permits a controlled rate of energy release.

Among the advantages of precast over cast-in-place concrete are that quality control and uniformity can be maintained to a higher standard and much more easily in a manufacturing plant than at the construction site. This makes possible lighter sections and less concrete. Erection is quicker and can be carried out during excavation. These advantages are offset to some extent by the increased costs of shipping, handling, and losses by breakage.

17.2.4.2 Precast tunnel liners. Cast iron, which is both strong and resistant to corrosion, was first used in the form of segments for the London subway system in England. Precast concrete segments, first introduced in 1937, have become the preferred method for reasons of cost. They were used extensively during World War II for air raid shelters. By 1970–1977 the demand for cast-iron segments had fallen to between 1 and 70 km per annum. During this period, most tunnels

Figure 17.6 Timber crib crushed by rock squeeze; Central Canada Potash mine, Colonsay, Sask. (*Courtesy of Noranda Minerals, Inc.*)

were lined with bolted segmental concrete liners (48–80 km/yr). Expanded concrete liners accounted for between 3 and 30 km/yr, with a similar demand for in situ concrete linings or unlined tunnel excavations (McBean and Harries, 1970; Halcrow and Partners, 1978).

When first introduced, both iron and concrete segmental tunnel linings were flanged and bolted together to form a rigid ring designed to resist bending moments. Boltless expanded concrete tunnel linings were patented in 1942. Expansion of the liner against the surrounding ground eliminated the annular void and the need for backfill. Other jacked-in-place designs soon followed. Most of the currently used concrete segmental systems are articulated or flexible and incorporate knuckle joints so as not to transmit bending moments around the ring (Fig. 17.7). The more uniform pressure distribution permits use of lighter-weight and therefore less expensive segments.

Precast segments are best used in circular tunnels such as those excavated by a full-face boring machine, where they can be fitted tightly against the rock or soil. Concrete segments for soft-ground tunneling are usually reinforced to resist TBM thrust, whereas those for rock tunneling are reinforced only to prevent damage during handling, if at all. Their interlocking joints have a caulking groove for water sealing after erection. Alignment and sealing of segments can be a real problem if the tunnel bore becomes nonuniform. Segments also are easily damaged and must be han-

(a)

(b)

Figure 17.7   Segmental concrete tunnel liner. (*From slide set courtesy of U.S. National Committee on Tunneling Technology.*)

dled and installed with care, usually by special erecting arms built into the trailing gear of the boring machine.

17.2.5 Concrete and mortar coatings

17.2.5.1 Definitions. *Shotcrete* is a concrete applied by spraying (Fig. 17.8). Just like concrete, it contains fine- and coarse-grained aggregate (sand and gravel), cement, water, and sometimes additives to accelerate setting or to improve flow (Lorman, 1968; Poad and Serbousek, 1972; Reading, 1973).

Its sister and predecessor material Gunite is a cement mortar applied by spraying, which contains no coarse aggregate and is less widely used because of its higher cost and somewhat inferior performance. Either material can be applied directly to rock with no other form of support, or can be sprayed over mesh, rockbolts, and ribs to form part of an integrated support system. Steel fibers can be added to give fiber-reinforced shotcrete, a particularly tough liner material (Kaden, 1974).

A further kind of coating, Chunam (Fig. 17.9), is a portland cement or lime mortar applied by plastering with a trowel. Where used for slope stabilization in Hong Kong, the aggregate consists of a coarse-grained sand derived from the weathered granite detritus at the place where the Chunam is to be applied (Geotechnical Control Office, 1984).

17.2.5.2 Uses. Sprayed concrete is used mainly for stabilizing exposed rock faces that otherwise would ravel because of their close jointing, or would deteriorate under conditions of wetting and drying. It also works well when applied to soils. The uncemented sands and gravels of the Milan subway and the unconsolidated slide debris and

Figure 17.8 Shotcrete application. (*From slide set courtesy of U.S. National Committee on Tunneling Technology.*)

Figure 17.9 Chunam plaster used for stabilization of weathered rock and residual soils, Hong Kong. (*Geotechnical Control Office, 1984.*)

wet clays of the Serra Ripoli highway tunnel in northern Italy are "...two of the most outstanding examples of shotcreting in the worst imaginable tunneling conditions" (Sutcliffe and McClure, 1969; Curzio et al., 1973).

Shotcrete is used extensively for stabilizing slopes and the walls of deep excavations (Piteau and Peckover, 1978). Deere and Patton (1971) report the stabilization of a slope in colluvium, residual soil, and weathered rock above a tunnel portal in the western United States. Weak and friable ground was cut at slopes of 1:1.5 and covered with steel mesh, anchored to untensioned grouted dowels 2.5 m long on 2.5-m centers. The slope was then treated with 70–100 mm of Gunite. Erosion, raveling, and sloughing were prevented. Shotcrete is also employed to repair damaged or deteriorated concrete, and for lining canals, reservoirs, and swimming pools.

Shotcrete is particularly effective for stabilizing underground excavations, most often in combination with mesh, bolts, and, when necessary, lightweight steel ribs (Fig. 17.10), which are the basic materials of the New Austrian Tunneling Method (NATM). Shotcrete support systems maintain close contact with the rock and so reduce rock movements to a minimum. Faulted or altered rock is often excavated and replaced with shotcrete, which can be reinforced to cope with any anticipated loads. In a tunnel, for example, a reinforced concrete ring beam can be constructed in this manner, and anchored to sound rock using rockbolts.

Because it is durable, a shotcrete liner can be incorporated into a final cast concrete liner, or augmented with further layers of shotcrete, mesh, and bolts. Shotcrete is nowadays used both as primary support and as a finished liner, and gives satisfactory long-term stability provided that there has been close control over materials and methods of installation (Gullan, 1975). A flexible liner of shotcrete, mesh, and rockbolts tends to suffer less from blasting, rockbursting, or even close-proximity bomb blasts than does a more rigid concrete liner (Lorman, 1968; Selmer-Olsen, 1970; Kendorski et al., 1973).

17.2.5.3 History of sprayed concrete. Gunite (sprayed fine aggregate concrete) started with the patenting of *cement gun* and *Gunite* by the Allentown Cement Gun Company in 1909. Gunite was first used underground in the United States in 1914, at the Brucetown Experimental Mine (Verity, 1971).

Coarse aggregate shotcrete only became possible with the invention of the Aliva VS-12 spraying machine in 1942 by a Swiss engineer (Kobler, 1976). Tunneling applications date from about 1950 in

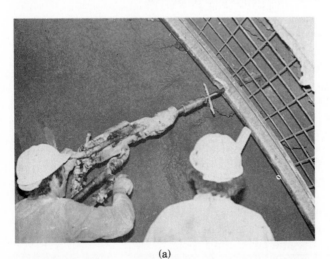

(a)

Figure 17.10 Support components of the New Austrian Tunneling Method (NATM); Rubira tunnel, Barcelona, Spain. (*a*) Installation of resin-bonded bolts and steel ribs. The ribs and steel mesh reinforce the shotcrete. (*b*) Drilling from the jumbo to install further rockbolts through a previously shotcreted and ribbed layer.

Europe with the introduction of the very successful NATM technique (Rabcewicz, 1964, 1965, 1969, 1972; Cecil, 1970; Golser, 1976). This had spread by 1960 to Venezuela, by 1963 to Japan (Nakahara, 1976), and by 1966 to Canada and the United States (Mason and Mason, 1972; McClure, 1973). The Canadian National Railway's Burnaby-Vancouver railway tunnel (1966–1968) was the first in North America to be driven with a support system of shotcrete, making a saving of approximately $3.25 million on the contract. In the United States, shotcrete was first used in 1968 by the Hecla mining company, and in 1968–1969 it was the sole lining of the Balboa tunnel near Los Angeles, Calif., where the ground was poorly consolidated sandstone with running gravel, silt, clay, and considerable water (Verity, 1971).

17.2.5.4 Shotcreting processes. There are two processes for applying shotcrete, termed *dry mix* and *wet mix*. In the dry-mix method (Fig.

(b)

Figure 17.10 *(Continued)*

17.11*a*), a mixture of coarse and fine aggregates, cement, and some-times accelerator powder is pumped by compressed air in a dry or semidry condition down a delivery hose. Water is added at the deliv-ery nozzle just before the shotcrete emerges. The operator at the noz-zle controls the amount by adjusting a valve in the water delivery line. This contrasts with the wet-mix process (Fig. 17.11*b*), in which the shotcrete is mixed like concrete, with wet and dry components to-gether, before being pumped down the delivery line.

Dry-mix shotcrete has been claimed to give better results and greater long-term strength, probably because of the higher velocity of projection that can be achieved. The water-to-cement ratio can be kept low, whereas in the wet-mix process the water content must be high enough for pumping. The dry-mix method also permits the introduc-tion of an accelerator at the nozzle.

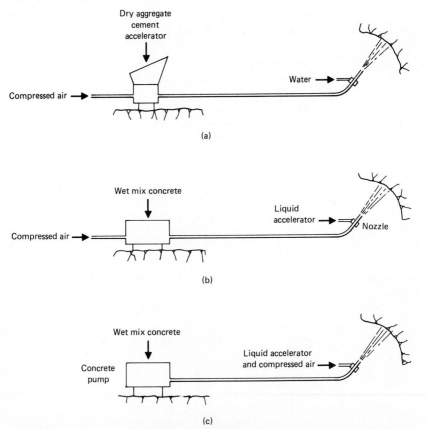

Figure 17.11 Shotcreting processes. (*a*) Dry-mix method. (*b*) Wet-mix method with pneumatic feed. (*c*) Wet-mix method with positive feed (concrete pump).

The major advantage of the wet-mix process is that ready mix can be brought to the site in concrete trucks. The water-to-cement ratio is under full control, whereas dry-mix shotcreting requires a more experienced operator who is fully responsible for the water content and therefore the long-term strength. Dry-mix shotcreting also tends to be a more dusty operation and there is often a greater percentage of "rebound," that is, shotcrete that bounces off the surface of application and so is wasted.

17.2.5.5 Mix design. A correct water-to-cement ratio is most important. With too much water, the shotcrete has a reduced long-term strength and may not even adhere. With too little, it will not pump (wet-mix process) or will give excessive rebound (dry mix). The optimum water-to-cement ratio ranges from 0.35 to 0.50, increasing with the decreasing maximum size of aggregate.

Figure 17.12 shows a typical range of gradations for aggregate in shotcrete. Between 290 and 300 kg of cement is normally mixed with 1470–1500 kg of aggregate. Type 1 portland cement is usual, but type

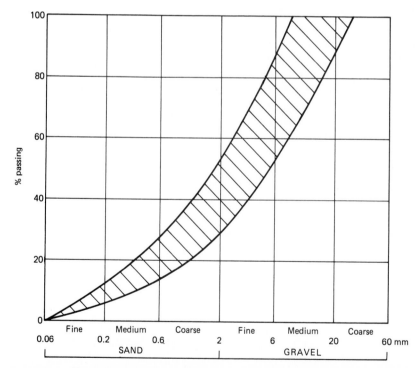

Figure 17.12 Typical size gradation of shotcrete aggregate. Dry-mix shotcrete with rounded, coarse aggregate tends toward the lower curve, whereas wet-mix shotcrete with crushed rock aggregate tends toward the upper (finer) gradation.

V sulfate-resistant portland cement can be employed when the groundwater contains potentially aggressive amounts of dissolved sulfates. High early strength is important in tunneling applications, so a type III portland cement might be more appropriate. However, accelerators are normally used instead.

Powder accelerators are added to the cement-aggregate mix in the amount of 3–6% of the weight of cement. However, the moisture in the aggregate can be sufficient to initiate reactions before the mix reaches the nozzle. An alternative is to add a fluid accelerator to the water entering at the nozzle. Whichever alternative is selected, the compatibility of cement and accelerator must be tested in advance.

ACI standard 506-66 recommends that strengths higher than 4000 lb/in^2 (27.6 MPa) be specified only for the most carefully engineered and executed shotcrete jobs (ACI, 1966). Brekke (1972) suggests that for most routine work, the required minimum 28-day strength should be set much lower, for example, at 2500 lb/in^2 (17.2 MPa) if a substantial amount of accelerator is used, and at 3000 lb/in^2 (20.7 MPa) with little or no accelerator.

17.2.5.6 Application. No pretreatment of the rock surface is usually necessary because the shotcrete is abrasive and does its own cleaning. Wire mesh, if included as reinforcement, should have at least 100-mm and preferably 150-mm openings to allow free entry of shotcrete; otherwise rebound will be excessive. Mesh and any geomechanics instruments to be embedded in the shotcrete need to be very firmly secured or they will move when hit by the high-pressure spray. The correct pressure in a dry-mix shotcreting process is normally around 480–550 kPa. Water is added at the nozzle at a pressure of about 410 kPa. The sand used normally will contain enough moisture to supply up to half of the total water needed.

Only with skill can a uniform coating of the required thickness be obtained, free from pockets of air between the rock, ribs, and mesh. The spraying should be perpendicular to the rock surface to minimize rebound. The maximum thickness that can be applied in a single layer is about 100 mm. Thicker layers are likely to separate from the rock substrate, leaving a gap between shotcrete and rock, or to break away entirely, leaving bare rock areas to which new shotcrete must be applied.

17.2.5.7 Advances in shotcreting. For work in large tunnels a robot shotcreting machine ensures better protection for the nozzle operator and makes it easier to spray perpendicular to the rock surface. Perpendicular spraying by hand application is uncomfortable because of the rebound, and can be dangerous because of the risk of blocks becoming loosened.

Lately the trend is away from hand-applied dry-mix shotcreting toward a remotely controlled wet-mix process with microsilica and steel-fiber additives. Wet-mix shotcreting improves productivity, reduces rebound, and gives safer and cleaner working conditions (Overlie and Rippentropp, 1987).

The addition of 8–10% microsilica by weight of cement gives easier mixing and improved pumping and spraying. Microsilica, a waste by-product of steel manufacture, consists of silicate particles of less than 1 μm in diameter. These react with calcium hydroxide in the cement to form calcium-silicate-hydrate gel (CSH), which increases the strength and reduces the permeability of the hardened shotcrete.

Steel fibers are often used to reinforce a shotcrete mix. This eliminates the time-wasting installation of wire mesh and costly rebound from the mesh, and it avoids the extra thickness of shotcrete needed to protect wire mesh against corrosion. The fibers are usually deformed for better bond. Typically they are 15–20 mm long and comprise about 1% by weight of the total mix.

17.2.5.8 Inspection and testing. Quality control provisions ensure that the mix is within specifications, and that application techniques are satisfactory. In practice, shotcreting is to some extent self-regulatory. Really poor shotcrete does not adhere, it falls from the rock; 10% rebound must be expected from walls and 25% from roofs. Initially mainly the coarser aggregate particles rebound, leaving a mortar pad at the shotcrete-rock contact.

Inspection includes checking that the specified thickness of shotcrete has been achieved. The finished surface feels warm if the cement is setting and hardening. It should be "sounded" to check for hollow areas which, if found, should immediately be broken out and replaced.

Inspection is supplemented by a program of quality control testing designed to give early results. Long-term tests such as for the 28-day strength of shotcrete cores can be useful to confirm the quality of shotcrete known to be good, but are useless for the much more important objective of promptly correcting any deficiencies.

Of the long-term tests, *core tests* are the most reliable. The cores are taken from the in-place shotcrete by diamond drilling, and are tested for uniaxial compressive strength in the laboratory at intervals of typically 4, 7, and 28 days after shotcrete application.

Less definitive but quicker are such tests as the *panel test*, for which shotcrete is sprayed into special boxes or panels on site at the same location and time as the rock surfaces themselves are coated. The slabs of shotcrete taken from the boxes are cut into cubes using a circular saw and then tested for compressive strength. The results correlate with, but are not identical to, those from core testing, because the sizes and shapes of specimens are different, as are the densities of application because of the greater resilience of the plywood panel.

Another alternative is a *pull test*. Miniature anchors are inserted in holes drilled into the shotcrete, to which a pulling device is attached. A small cone of shotcrete is pulled from the surface (Fig. 17.13). Short-term cone pull strength correlates closely with long-term compressive strength and so can be used as a quality control index at the time of construction. An alternative shotcrete penetration test was developed by Sallstrom (1970).

The *Schmidt rebound hardness test* described earlier in a rock-testing role (Sec. 2.3.2.3) can be applied also to shotcrete. However, the roughness of the shotcrete surface leads to considerable scatter of results.

Any stresses that develop within a shotcrete liner system are revealed by convergences and by a pattern of shotcrete cracking that can readily be observed. Shotcrete liners therefore have their own built-in monitoring system, and unlike rigid liners, they can easily be strengthened when necessary by installing more bolts or further shotcrete layers.

Figure 17.13 Pull tests on shotcrete. The Lok-Test portable tester manufactured by Germamm of Denmark provides a convenient method for measuring the short-term strength of in situ shotcrete. (*Courtesy of Trow Inc., Consulting Engineers, Toronto.*)

References

ACI: "Standard Recommended Practice for Shotcreting," ACI 506-66, Am. Concrete Inst. Committee 506, SP-14, pp. 193–217 (1966).

Billington, C. J., and E. W. Jacomb-Hood: "New Approaches to Steel Supports for Tunnels and Mines," in M. J. Jones, Ed., *Tunnelling '76* (Inst. Min. Metall., London, 1976), pp. 349–368.

Brekke, T. L.: "Shotcrete in Hard Rock Tunneling," *Symp. Rock Support Sys.* (Oregon, Oct. 1971); also *Bull. Assoc. Eng. Geol.*, **9** (3), 241–264 (1972).

Cecil, O. S.: "Shotcrete Support in Rock Tunnels in Scandinavia," *Civ. Eng., ASCE*, **40**(1), 74–79 (1970).

Curzio, P. Q., G. Barazzoni, F. Nobili, and A. Anselmi: "Use of Shotcrete for Tunneling in Difficult Grounds," *Proc. Eng. Foundation Conf. Use of Shotcrete for Underground Structural Support* (South Berwick, Me., 1973), pp. 79–95.

Deere, D. U., and F. D. Patton: "Slope Stability in Residual Soils," *Proc. 4th Pan-Am. Conf. Soil Mechanics and Foundation Eng.* (San Juan, Puerto Rico, 1971), pp. 87–170.

Geotechnical Control Office: *Geotechnical Manual for Slopes* (Geotech. Control Office, Eng. Develop. Dept., Hong Kong Gvt., 1984), 295 pp.

Golser, J.: "The New Austrian Tunneling Method, NATM," *Proc. Eng. Foundation Conf.* (Easton, Md., Oct. 4–8, 1976), Pub SP-54, Am. Concrete Inst., pp. 323–347.

Gullan, G. T.: "Shotcrete for Tunnel Linings," *Tunnels and Tunneling*, **7** (5), 37–47 (1975).

Sir William Halcrow & Partners: "The Annual Length and Volume of Tunnels Constructed in the United Kingdom (1977)," Rept., Pt. 2, U.K. Dept. of the Environment (1978).

Kaden, R. A.: "Slope Stabilized with Steel Fibrous Shotcrete," *Western Construct. Mag.* (Apr. 1974).

Kendorski, F. S., C. V. Jude, and W. M. Duncan: "Effect of Blasting on Shotcrete Drift Linings," *Mining Eng.* (Am. Soc. Min. Eng.), 38–41 (Dec. 1973).

Kobler, H. G.: "Review of Dry-Mix Coarse-Aggregate Shotcrete as Underground Support," *Proc. Eng. Foundation Conf.* (Easton, Md., Oct. 4–8, 1976), Pub. SP-54, Am. Concrete Inst., pp. 188–200.

Lorman, W. R.: "Engineering Properties of Shotcrete," Publ. SP-14A, Am. Concrete Inst. (1968).

Mason, E. E., and R. E. Mason: "Shotcrete Support with Special Reference to Mexico City Drainage Tunnels," *Rock Mech.*, **4** (2), 115–128 (1972).

McBean, R. J., and D. A. Harries: "Development of High-Speed Soft Ground Tunneling Using Pre-cast Concrete Segments and Tunneling Machines," *Proc. South Afr. Tunneling Conf.* (Johannesburg, 1970), vol. 1, pp. 129–134.

McClure, C. R.: "Use of Shotcrete from the Standpoint of the Designer," *Proc. Eng. Foundation Conf. Use of Shotcrete for Underground Structural Support* (South Berwick, Me., 1973), pp. 18–21.

Nakahara, A.: "Shotcrete Application for the Seikan Tunnel," *Proc. Eng. Foundation Conf.* (Easton, Md., Oct. 4–8, 1976), Publ. SP-54, Am. Concrete Inst., pp. 460–474.

Ounanian, D. W., J. S. Boyce, and K. Maser: "Development of an Extruded Tunnel Lining System," *Proc. Rapid Excavation and Tunneling Conf.* (San Francisco, Calif., 1981), vol. 2, pp. 1333–1351.

Overlie, F. E., and G. Rippentropp: "Steel Fiber Microsilica Shotcrete with Remote Controlled Equipment." *Proc. Rapid Eng. Tunneling Conf.* (New Orleans, La., 1987), chap. 23, pp. 351–370.

Piteau, D. R., and F. L. Peckover: "Engineering of Rock Slopes," in *Landslides, Analysis and Control*, Spec. Rep. 176, U.S. Transport. Res. Board, Chap. 9, pp. 192–228 (1978).

Poad, M. E., and M. O. Serbousek: "Engineering Properties of Shotcrete," *Proc. 1st North Am. Rapid Excavation and Tunneling Conf.* (Chicago, Ill., 1972), pp. 573–591.

Rabcewicz, L. V.: "The New Austrian Tunneling Method, Parts I, II, and III," *Water Power* (Nov./Dec. 1964; Jan. 1965).

————: "Stability of Tunnels under Rock Load, Parts I, II, and III," *Water Power* (June, July, Aug. 1969).

————: "Application of the NATM to the Underground Works at Tarbela," *Water Power* (Sept., Oct. 1972).

Reading, T. J.: "Corps of Engineers Study of Shotcrete," *Proc. Eng. Foundation Conf. Use of Shotcrete for Underground Structural Support* (South Berwick, Me., 1973), pp. 263–276.

Sallström, S.: "Improving Initial Compressive Strength of Shotcrete by Accelerating Agents," in T. L. Brekke and F. A. Jorstad, Eds., *Large Permanent Underground Openings* (Universitetsforlaget, Oslo, Norway, 1970), pp. 227–232.

Selmer-Olsen, R.: "Experience with Using Bolts and Shotcrete in Areas with Rock Bursting Phenomena," in T. L. Brekke and F. A. Jorstad, Eds., *Large Permanent Underground Openings* (Universitetsforlaget, Oslo, Norway, 1970), pp. 275–279.

South African Chamber of Mines: "An Industry Guide to the Amelioration of the Hazards of Rockburst and Rockfalls," Publ. PRD216, High-Level Committee on Rockbursts and Rockfalls, Chamber of Mines, 178 pp. (1977).

Sutcliffe, H. F., and C. R. McClure: "Large Aggregate Shotcrete Challenges Steel Ribs as Tunnel Support," *Civ. Eng., ASCE,* 51–55 (Nov. 1969).

Verity, T. W.: "Ground Support with Sprayed Concrete in Canadian Underground Mines," Can. Dept. of Energy, Mines and Resources, Information Circular IC 258, 26 pp (1971).

Drainage, Grouting, and Freezing

Whereas rock reinforcement and surface coatings are more or less superficial treatments, the alternative or supplementary measures of drainage, grouting, and freezing are treatments that improve the mechanical properties and behavior of the bulk rock mass (Lancaster-Jones, 1968; Bell, 1975).

18.1 Drainage

18.1.1 Objectives

Just as water can be the major cause of rock instability (Chap. 4), so drainage and dewatering can be the most direct and effective methods of stabilization (Sabarly et al., 1970; Sowers, 1976). Wet conditions alone, without extraordinary inflows or pressures, are sufficient to slow down construction. Equipment and explosives become difficult and less safe to handle. High water pressures contribute to instability by reducing the shear strength of joints. High inflows cause erosion of weaker materials and are expensive to remove by pumping. Goodman and John (1983) recommend that: "systematic and positive dewatering of rock masses affords the most effective, reliable, and cost-effective rock stabilization. Remedial works by other means such as bolting or anchoring, without fully integrated dewatering, are believed to be poor rock engineering practice."

18.1.2 Design of drainage systems

Four principal methods are used alone or in combination for the control of groundwater: pumping, gravity drainage, freezing, and grouting. When excavations are deep or the rock is very permeable, the cost can be high, although the benefits can be substantial.

Groundwater can be brought under control before, during, or after construction. Pretreatment methods include grouting, the drilling of relief wells and positive dewatering by pumping to lower the water table. Control during construction often uses a collector system of ditches and sumps. High pressures, when suspected, can be relieved from within the excavations before they cause a problem, often at less cost than pretreatment. Probe holes drilled ahead of the advancing face of a tunnel serve such a purpose. Postconstruction treatments sometimes are needed, such as drainage for landslide remedial work or back-grouting behind a tunnel liner to stop undesirable seepages.

Drainage schemes should be designed with the help of the modeling techniques discussed in Chap. 4, in order to estimate drainage and pumping requirements, to predict the performance of the system, and to avoid adverse environmental effects such as consolidation, subsidence, and interference with regional and local water supplies and vegetation.

For example, realignment of the Welland Canal in Ontario, Canada, required excavation of about 13 km of new canal and the permanent depressuring of a regional aquifer to lessen the risk of uplift and slope failure (Farvolden and Nunan, 1970). This aquifer, widely exploited by wells in the region, consisted of jointed dolomite at the surface of the bedrock, beneath 20–30 m of low-permeability glacial clays. Calculations based on piezometer observations and pumping tests showed that, given the high conductivity of the aquifer, the entire construction site could be dewatered from just four pumping wells. Pumping rates of about 100 L/s would provide the necessary 10 m of drawdown. A check on the environmental effects of the proposed dewatering, however, showed that the high conductivity of the aquifer would lead to extensive areal propagation of the drawdown cones. The pumping would affect groundwater levels as far as 12 km from the canal.

Design of a drainage system requires a substantial safety margin: more drainage should generally be provided than the calculations may suggest. This is because drainage may have little effect on flow conditions except where the drain happens to intersect one of the flow channels in the jointing system. A further problem is that drains often clog because of siltation or the precipitation of soluble impurities such as calcium bicarbonate. Piezometers are required to monitor the groundwater conditions during drainage and to demonstrate the effectiveness of the system.

18.1.3 Drainage around underground works

In underground excavations, excessive groundwater inflows often lead to high pumping costs, blasting problems, unsafe conditions in the handling of drills and other equipment, and traffic problems caused by the formation of mud in the invert. Excessive pressures have resulted in blowouts of rock and water sufficient to flood an excavation. Blowouts most often occur when a tunnel heading penetrates high-pressure pockets of water trapped behind clay-bearing and impermeable strata and faults.

Grouting, groundwater lowering, or, in extreme cases, freezing or compressed air, can be applied either before or during construction, depending on which is likely to be the most effective and economic choice (Fig. 18.1). In rock work, grouting combined with drainage from within the excavation is usually best.

Groundwater lowering and compressed-air methods are most often used when tunneling at shallow depth through coarse-grained water-laden soils, although occasionally also through fissured rock. They become competitive when the predicted inflows are too great to be economically pumped from within the excavation. Lowering is achieved by pumping continuously from a network of well-points drilled from the surface, stopping only after the final lining has been installed. Compressed air works by balancing the surrounding water pressure, and also because injection of air into joints and pores reduces the hydraulic conductivity of the ground.

Compressed air can be used only if the head is less than 10 m of water. The method is employed when groundwater lowering is too expensive or unacceptable because of the risk of subsidence or interference with surrounding water wells. Work in compressed air is expensive because of the cost of air locks and continuously operating compressors and, more particularly, because workers can operate only for a short time and must "decompress" slowly before coming to surface.

Much more often, the works are drained from within. Groundwater pressures are explored by probing ahead of the face, monitoring water yields from the probe holes. Rockbolt and blastholes, although too short, themselves act as probes and provide sufficient drainage to remedy most moderate-pressure groundwater situations. Problems of excessive inflow are solved usually by grouting. Further drainage can be provided by a fan pattern of purpose-drilled drain holes. For continued drainage after construction, the internal drains must be connected to a piping and sump system.

18.1.4 Drainage beneath dams

A drainage curtain is nowadays mandatory beneath the downstream toe of a large dam. In combination with grouting, this curtain relieves

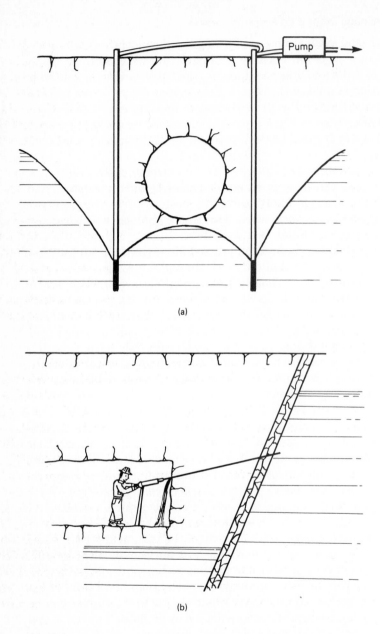

(a)

(b)

Figure 18.1 Drainage alternatives for underground works. (a) Dewatering before construction. (b) Probe drilling and drainage during construction.

high uplift pressures that would present a risk of rupturing the jointed rock and endangering the stability of the dam.

Grout curtains are often unreliable because the cement grout fails to penetrate fine or infilled fissures. Local "windows" are difficult to avoid; the curtain can be improved by jetting to remove fillings, or by a secondary treatment with chemical grout, but these methods are expensive. In such cases, a drainage curtain can be used as a supplementary or alternative method for controlling uplift and works particularly well in finely fissured rocks where grouting is difficult. Drainage achieves a similar pressure reduction, the only difference being that drainage increases the amount of leakage whereas grouting reduces it.

Supplementary grouting and cutoff systems reduce the amount of drainage needed. These treatments are located as far upstream as possible from the point of drainage so that hydraulic gradients are kept low in the vicinity of free rock faces, and the risks of initiating failure during grouting, or of blowouts during later service, are kept to a minimum. Steel sheet piling can be driven into softer rock, but may do more harm than good by fracturing the rock and enhancing its hydraulic conductivity. An alternative is to construct a concrete or bentonite clay slurry cutoff wall in a rock trench. Slurry walls are most effective when the seepage velocity is low; otherwise there is a risk that the cutoff wall will be disrupted before the blockage is complete.

18.1.5   Drainage and protection of slopes

Because of their lack of confinement, rock slopes are more susceptible than underground works to the destabilizing effects of water pressures and flows. Water pressures within the rock joints reduce factors of safety and can induce sliding. Water running over and through the slope face erodes soft or weathered rock materials. Freezing and thawing of water in rock crevices results in loosening and raveling of the slope face.

Various drainage alternatives are available (Fig. 18.2). The aim of subsurface drainage is to lower the water table and hence the water pressure, if possible to a level below that of potential frost action and surfaces of sliding. A drain does not have to produce a large flow of water to be effective, as long as it is keeping any pressures from building up. Furthermore, the absence of damp spots on the rock face does not necessarily mean that groundwater conditions are favorable.

18.1.5.1   Surface water control at the slope crest.   Water flowing over the crest and down the slope face can erode the face, and if allowed to enter the ground by way of tension cracks, will also contribute to a

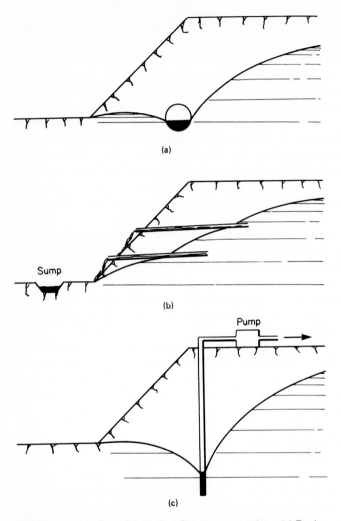

(a)

Sump

(b)

Pump

(c)

Figure 18.2 Drainage alternatives for open excavations. (a) Drainage gallery. (b) Horizontal gravity drains. (c) Vertical well point dewatering.

high water table and joint water pressures and to the risk of deepseated instability.

Provisions for the control of surface water should be included in the design of any rock excavations. Existing and unstable slopes should be inspected to determine whether surface water is flowing toward unstable areas and into the ground. The upper slope should be contoured (graded) to provide surface runoff. Surface ditches and culverts used to divert flows into nonproblem areas should be lined by paving with clay, asphalt, or thick slush grout. Existing vegetative cover should be left undisturbed or should be reestablished by postconstruction planting.

Maintenance is needed to prevent plugging and ponding. Water-filled depressions should be drained and cracks or depressions filled with concrete or asphalt, although attempts to seal cracks often fail because seals are broken by further movements. Grout used in tension cracks should be thick to prevent penetration and the buildup of hydraulic pressures that can trigger failure.

Water that reaches the base of an excavation, either by surface run-off or as groundwater seepage, can usually be controlled by pumping from sumps or drainage ditches.

18.1.5.2 Slope face trenches. Trench drains, often in a chevron or herringbone pattern, are used to lower the water table directly beneath the surface of a slope face in soft or weathered rock or soil. Their main application is to counteract frost deterioration and shallow sliding. Wider trench drains are sometimes used, because they can be excavated and backfilled by bulldozer. Hutchinson (1977) gives a flow-net analysis of trench drain efficiency.

The *counterfort* type of trench drain, widely used in France and England in the early 1800s, is excavated into firm ground beneath the surface of sliding or potential sliding and then backfilled with coarse stone. Thus it provides some mechanical support in addition to reducing groundwater pressures.

18.1.5.3 Horizontal drain holes. Gravitational drainage has the advantage of permanence as long as the drain holes remain open, which cannot often be guaranteed. "Horizontal" drain holes are usually inclined on a rising gradient of about 5° from near the slope toe, so as to discharge by gravity. Such drains have, for example, been used to stabilize 70° slopes in the weathered granite of Hong Kong (Choi, 1974). The drain pipes (open percussion-drilled holes in hard rock) are typically 50–200 mm in diameter and 10–15 m apart. They can be installed in soil by helical auger and in rock by rotary drill. The maximum practical length is about 100 m, but lengths of about 60 m are more common.

The pattern of holes should intersect as many water-conducting joints as possible. An orientation approximately normal to the principal joint set is best. Holes drilled in a fan pattern minimize the time to set up and move the drill. Drain holes should be thoroughly cleaned of cuttings by jetting with high-pressure air, water, or even a detergent. In cold climates the outlets need to be protected against freezing using insulating materials such as straw, gravel, or crushed rock.

18.1.5.4   Blasted toe drains.   An alternative with its own frost protection is to blast the entire lower bench around the toe of a slope and to leave the broken rock in place. The blasted toe acts as a high-permeability drain, and water issuing from it is collected and pumped away.

18.1.5.5   Vertical well points.   The *well-point dewatering* method, common for short-term dewatering during construction in soils, is seldom practical as a means of long-term stabilization. This is because vertical holes are only self-draining if connected to horizontal drains or galleries, and otherwise require continuous pumping or blowing with compressed air.

The method can be effective for a limited period until a more permanent system of stabilization can be installed, and also to help stabilize the slopes of open-pit mines. Stability depends on continuous pumping to maintain the groundwater table at a low elevation. Each well should if possible be pump-tested in advance. The risk of pump breakdown should be countered by installing backup pumps and warning devices (Sowers, 1976).

18.1.5.6   Drainage adits.   Drainage tunnels, adits, or galleries at one or several elevations may be required to drain major slopes. These are typically 2–4 m in diameter and excavated parallel to the slope face. Although expensive to excavate, they are usually effective and reliable. Their effectiveness can be increased by fanning holes outward from the adit. As a refinement, it may be possible to place the entire adit under a partial vacuum (Sharp, 1970).

18.2   Grouting

18.2.1   Objectives and history

Because open joints make a rock mass weak, deformable, and permeable, the characteristics of the mass can be greatly improved by filling and cementing the joints with a suitable grout material (Fig. 18.3). Grouting is used to consolidate loose rock in the foundations and walls of excavations; to reduce leakage beneath dams and into open-cut or underground excavations beneath the water table; and, together with drainage measures, to control uplift and pore pressures in rock slopes, dam foundations, and abutments.

Early in the nineteenth century, French engineers forced clay into voids below lock walls and floors and used cement grouts behind tunnel linings. The first cement grouting in Germany was carried out in 1864 in the Rhein-Preussen shaft at Moerz to a depth of 70 m. Cement

Figure 18.3 Hydraulic grout plant with twin high-speed mixers, feeding a holding tank and 3L8 Moyno grout pump. (*Courtesy of Groundation Inc., Brampton, Ont.*)

grouting was used beneath the foundations of a dam on the New Croton project in New York State in 1893, and tunnels in New York were grouted on a large scale during construction of the Catskill water supply scheme (Glossop, 1961).

A more recent focus has been the development of methods for sealing the shafts and vaults of underground radioactive waste repositories. Treatments under investigation include vault grouting, borehole sealing, buffer packing, and backfilling (Gyenge, 1980). Very high standards of grout quality and durability are required in this application.

18.2.2 Types of treatment

18.2.2.1 Consolidation grouting. Various foundation grouting treatments are shown schematically in Fig. 18.4. *Consolidation* grouting, also known as *blanket* grouting, is used in closely jointed foundations to increase the bearing capacity and stiffness of the rock mass and to reduce settlements. The grout is usually portland cement, which when set has a strength similar to that of intact rock. It is injected through a rectangular grid of short vertical grout holes.

The treatment is most often used in the foundations and abutments of arch dams that impose heavy loads and are sensitive to settlement. It may be extended to less sensitive structures such as power stations and concrete gravity dams, and even to the foundations of earth em-

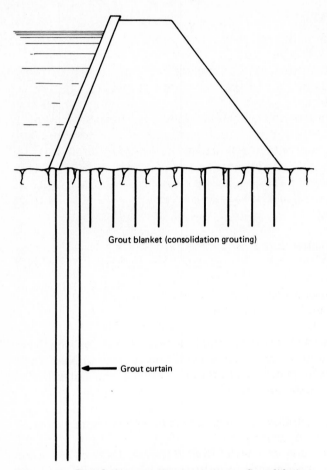

Figure 18.4 Foundation grouting treatments. Consolidation grouting (or blanket grouting) to increase the stiffness of a dam foundation, and curtain grouting to reduce pore-water pressures and uplift.

bankment dams and ordinary buildings if the rock is very closely jointed, and if the alternatives of taking the excavation to a deeper and more competent horizon, or spreading the load over a wider area, prove more expensive than grouting (Seed, 1972).

18.2.2.2 Groundwater control in surface works. *Grout curtains,* which consist of one or two rows of closely spaced grout holes, are used to reduce rock mass permeability and the flow of groundwater. They are employed usually beneath dams, but sometimes around the perimeters of deep excavations and along rock cuts. Their function is threefold. By lengthening the flow path, they reduce joint water pressures that might cause uplift and foundation failure. By reducing the hydraulic conductivity of the rock mass, they minimize leakage from reservoirs and the requirements for pumping from excavations. By reducing hydraulic gradients and flow velocities, they lessen the risk of internal and external erosion.

The object is to seal all open joints within a narrow zone defined by the required thickness of the grout curtain. Portland cement grouts are the least expensive, and are used when the rock joints are quite open. However, strength is not the main consideration. Grouts of much lower strength but much greater penetration capability, such as sodium silicate gels, are used to seal rock masses in which the jointing is relatively tight.

18.2.2.3 Groundwater control in underground works. Grouting techniques applicable to underground works are sketched in Fig. 18.5. Grouting is an expensive way of reducing water inflows during construction and, in addition, delays the work. It is therefore used only when drainage and pumping, which are cheaper, are impractical; where seepage must be reduced substantially in the long term as well as the short term; or where the strength as well as the water tightness of the rock mass need improvement (Johnson, 1982). Grouting also can be beneficial in underground mining situations where water inflows would otherwise lead to difficult and unsafe mining conditions (Dietz, 1982).

Shallow underground excavations can be grouted ahead of construction using vertical grout holes drilled and injected from the surface. Shafts are pregrouted to control water inflows in weathered and jointed rock near the rock-soil contact, using vertical grout holes parallel to the shaft walls. Deeper works have to be grouted during construction by injecting ahead of the advancing face through a fan pattern of drillholes. This slows down the work, particularly where there is no room to blast and grout simultaneously. Sometimes grouting is done from a pilot heading or from special grouting chambers excavated laterally into the tunnel or shaft walls.

Another technique for shaft sinking through wet ground is to inject grout in stages during the sinking operations. The grout drillholes are angled outward from the shaft bottom, and "spun," that is, drilled in an inclined rather than a vertical plane, so that each grouted zone overlaps that of adjacent holes. Typically the holes are 30 m (100 ft) long. After each stage of grouting from the base of the shaft, sinking is continued through about 90% of the grouted depth before repeating the treatment from the new floor.

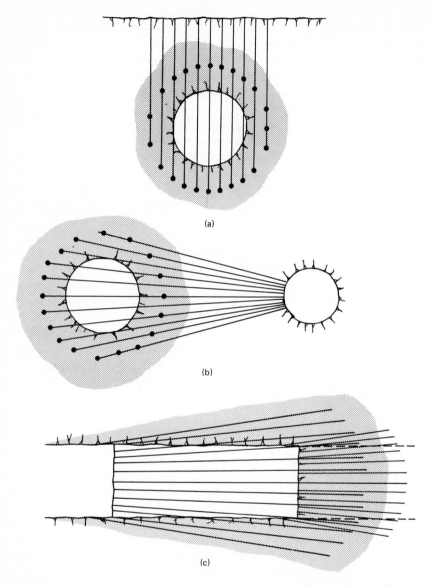

Figure 18.5 Grouting around underground works. (*a*) Pretreatment from surface. (*b*) Grouting from adjacent gallery. (*c*) Grouting ahead of a tunnel face.

18.2.2.4 Contact grouting and repairs. *Contact grouting* or *backgrouting* is the filling of voids behind retaining walls and tunnel and shaft liners. A sand-cement or neat cement grout is best for filling small gaps between concrete and rock, and open joints in the rock itself. Larger voids such as behind segmental tunnel liners are filled either with cement or by blowing in "pea gravel," a small and uniform-sized, well-rounded gravel selected for its ability to flow. Angular gravels tend to arch and leave voids.

18.2.3 Grouting materials

18.2.3.1 Particulate and nonparticulate grouts.
There are two main categories of grouting material; *particulate* grouts are suspensions of particles in water, whereas *nonparticulate* grouts, also termed *chemical* grouts, are completely fluid. Use of one or the other type depends mainly on the aperture of the joint to be filled (Fig. 18.6). The minimum groutable joint aperture for ordinary portland cement grout is about 0.2 mm. Translated in terms of water test results (Sec. 4.3.3.4), grouting with neat portland cement is probably not worthwhile in rock with a permeability of less than 50 lugeon units. Efficient grouting of fissures with finer apertures needs a chemical grout.

Particulate grouts include portland cement, bentonite clay, sand, and fly ash, used individually or blended with each other to give the required characteristics of flow, setting time, and cost. The most widely used chemical grouts are sodium silicates, chrome lignins, resins, and polymers.

Particulate grouts are most often used for grouting gravel soils and rock masses with open jointing or solution cavities. In these applications they are ideal, being able to block and cement the void spaces and being cheaper than nonparticulate materials. In rapidly flowing

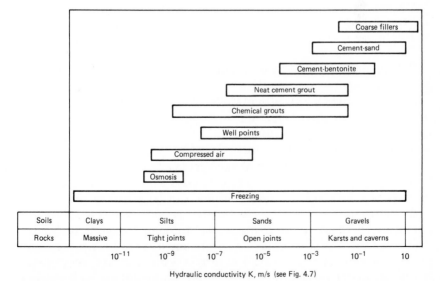

Figure 18.6 Grout materials to suit different classes of ground. (*Johnson, 1958.*)

groundwater, a cement grout is often diluted and carried far from the intended location before it sets. To solve this problem, for example, to plug cavities in karstic limestone beneath dam sites, very coarse "grouts," such as concrete, bales of hay, and mattresses, are sometimes used. Having achieved at least partial blockage, finer grained and more cementitious materials can be injected without being washed away.

Chemical grouts are most often employed to seal sandy and silty soils and finely fissured rocks. Multiphase grouting is sometimes used where an initial phase of cement grout injection is followed by secondary grouting with nonparticulate materials such as silicates. Progressively finer materials are injected until the filling of voids is complete.

18.2.3.2 Clay grouts. Natural grouts include clays, silts, and sands, suspended in water with the addition of cement and other additives (Johnson, 1958; Deere, 1982). Clay grouts are used to control seepage through alluvial materials and also in cavernous bedrock.

18.2.3.3 Portland cement grouts. Portland cement is a finely ground powder manufactured by crushing and milling the clinker produced when limestone and clay are mixed and fired in a furnace. When water is added, the powder sets after a period of about 1 h and continues to strengthen during two or more weeks. Most common is a *neat cement* grout consisting only of portland cement and water. Its consistency, expressed in terms of water-to-cement ratio, nearly always is in the range 10:1 to 1:1 (water to dry cement, by volume). A consistency of 10:1 gives a wet or thin grout, and 1:1 is a relatively thick grout (Albritton, 1982; Littlejohn, 1982).

Extremely finely ground portland cements with a maximum particle size of 5 μm rather than the 20–25 μm of ordinary portland cement are now available for grouting (Moller et al., 1983). The *microfine* cement penetrates tight joints and pores, and is stronger, more chemically resistant, and easier to handle.

Additives such as accelerators, retarders, lubricants, colloids, gas-producing agents, and nonshrinkage agents are also used with cement grouts. Sand, clay, or other inert materials such as fly ash, a waste product of coal-burning power stations, can be added to reduce cost and modify flow and setting characteristics, provided that these can penetrate, without undue segregation, through the sizes of fissure present in the rock. Bentonite clay contributes improved sealing characteristics and fluidity. The *grout take* (volume that can be injected per given length of packed-off grout hole) can sometimes be increased in finely fissured rocks by adding up to 5% by weight of bentonite to the portland cement mix.

Lau and Crawford (1986), from a study of the flow characteristics of portland cement grouts injected between transparent plastic plates, proposed the following equation for grout penetration R:

$$R = \sqrt{\frac{PbD}{\tau_0} + \frac{D^2}{4}} - \frac{D}{2}$$

where D is the diameter of the drillhole, P is the grout pressure, $2b$ is the joint aperture, and τ_0 is the fluid yield strength of the grout mix. Factors affecting grout penetration included the consistency (water-to-cement ratio) of the mix, the grain-size distribution of the cement in relation to the joint aperture, and the type and dosage of admixtures used to improve the flow properties. These factors influence the stability and penetrability of a grout, which in turn determine grout penetration and the strength, permeability, and durability of the hardened grout. According to Deere and Lombardi (1985), fluid yield strength determines how far the grout will penetrate, whereas viscosity determines the rate of penetration. Very thin mixes are undesirable because of their instability.

18.2.3.4 Nonparticulate (chemical) grouts. Chemical grouts usually consist of two or more chemicals that react to form a gel or solid precipitate (Karol, 1982). Their penetration is controlled by reaction rates and setting times rather than by particle sizes. The materials are usually fluid, with a viscosity approaching that of water, so they can be injected into fine cracks and pores. The chemical constituents, accelerators, and additives must be selected and injected to give the required depth of penetration, viscosity, and flow. Chemical grouting is therefore a specialized operation.

Sodium silicates are perhaps the most common and least expensive grouts. They have a viscosity of 4–50 centipoise (cP), only a little greater than that of water (1 cP), so they can be injected into fine fissures and into soils down to a silty fine-grained sand. They set by a chemical reaction to form a gel with little strength, but good sealing characteristics. Sodium silicates are used with calcium chloride, aluminum sulfate, formamide, or sodium bicarbonate to promote gelation. The Joosten process, patented in 1926, used alternating injections of sodium silicate and calcium chloride.

Chrome lignin grouts are derived from waste lignin liquor, a byproduct of paper making, and from a solution of sodium dichromate. Their viscosity varies from 2.5 to 50 cP.

Saltwater inflow, a serious concern in salt and potash mines, can sometimes be sealed and prevented by injecting supersaturated calcium chloride brine. As a result of differing solubilities and the common ion effect, sodium chloride forms as a particulate precipitate, which bridges and blocks pores and fissures. The method becomes inefficient if the inflow is diluted by unsaturated groundwater from other formations.

Epoxy resin and other types of polymer are the most expensive but the strongest grouts. They set by polymerization to form a solid as strong as, or even stronger than the rock itself. Formaldehyde polymerizes with the addition of resorcinol at ambient temperatures in aqueous solution. The viscosity is slightly higher than that of chrome lignin but lower than that of silicate grouts. The proprietary grout AM9, a vinyl polymer with a viscosity of only 1.2 cP, was found to be a neurotoxin and is no longer used. Special safety precautions are required when using most chemical grouts, which can be poisonous, particularly in poorly ventilated environments.

Crow et al. (1971) describe a survey to investigate impregnation of pyroclastic and sedimentary rocks using five different monomers. Compressive strengths were increased by factors of between 2 and 5, and elastic moduli by factors of between 2 and 3.5. The monomers injected included methyl methacrylate and styrene. Polymerization was accomplished by either radiation or chemical means.

Bituminous grouts are injected hot, and they set or at least become highly viscous as a result of cooling. A bituminous emulsion is a suspension of fine bitumen particles in water. Cationic asphalt emulsion with a hydrated lime breaking agent has been used for grouting. The breaking time is controllable by varying the amount of lime. Experimental asphalt grouting was done at the Morrow Point dam in 1969 (Gebhart, 1974). Ten thousand gallons of asphalt reduced leakage by approximately 65% as a preliminary to cement grouting of the foundations.

Molten sulfur at 120–160°C has a viscosity similar to that of water and so has excellent penetrating properties, although it sets rapidly over short distances of travel from the point of injection. When cooled by the rock and groundwater, it sets into a solid nearly as strong as the host rock. Grouting with molten sulfur shows potential for sealing not only fine fissures, but also large cavities carrying fast-flowing water, which are difficult and expensive to seal by other means. Sulfur is inexpensive but requires special grouting equipment, including heated tanks and hoses (Franklin and Hungr, 1981).

18.2.4 Grouting methods

18.2.4.1 Grouting patterns and sequences. *Primary* grout holes are drilled at a spacing of between 5 and 20 m, depending on the jointing in the rock mass. They are then flushed, pressure tested, and injected. *Secondary* holes are drilled and grouted midway between the primary holes, followed by *tertiary* holes if warranted by the results of pressure tests and the monitored grout "take" (Fig. 18.7). Final spacing is determined by the results achieved, and on most projects it is between 1.5 and 3 m.

Individual holes can be grouted from the top down or the bottom up. In the top-down *stage* method, the hole is drilled to a depth of about 1 m, and a grout nipple (a short length of pipe) is fixed in the hole collar by molten sulfur or a fast-setting grout. The hole is extended in stages by drilling through the nipple, stopping at each stage to connect the supply line and grout one stage before drilling the next. Grouting pressures are increased progressively as the hole deepens.

In the *series* grouting method, holes for an upper zone are grouted starting at the maximum spacing. Split spacing is then employed until the zone has been grouted satisfactorily. Another series of holes is then drilled and grouted to seal the next lower zone, and the procedure repeated until the final depth has been grouted. The advantage over stage grouting is that grout holes need not be cleaned. The amount of drilling required is evidently much greater.

Packer grouting is done in the reverse fashion. The hole is drilled to full depth, a packer is inserted and expanded to isolate the lower section of the hole, and grout is then injected. The packer is raised 3–5 m at a time, grouting at successively lower pressures closer to surface.

Stage grouting is simpler, and with no packers there are no leakage problems. Less trouble is experienced with caving holes because of the absence of downhole equipment. Packer grouting, however, is quicker and usually cheaper. The drill rig need not be moved on and off the hole. Better information is obtained regarding where grout take is high and where low, and there is better control over leakage to surface.

18.2.4.2 Drilling, flushing, and jetting. The rock is drilled, flushed, water-tested, grouted, and water-tested again. At the start of grouting it is impossible to predict accurately the number of holes or the exact quantity of grout needed. The program has to be adjusted as the work progresses, based on results. Grouting must therefore be carried out under the continuous field supervision of a specialist.

Grout holes are usually drilled with an EX (37-mm-diameter) diamond-impregnated plug bit, an "open-hole" bit that grinds up the rock without taking a core. This is faster than core drilling because the drill string is left down in the hole until the final depth is reached. A few holes may be core-drilled at a larger diameter to assess the quality of grouting.

Percussive air-track drilling is sometimes used for grout holes shorter than 15 m, but may introduce a risk of drill cuttings being

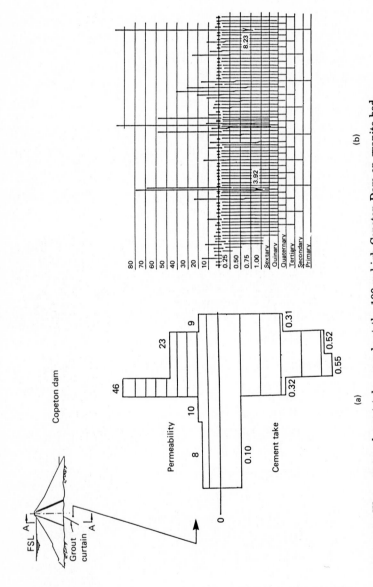

Figure 18.7 Water test and grout take records at the 100-m-high Copeton Dam on granite bedrock. Permeabilities (in lugeon units) were measured in the primary grout holes before grouting. They relate closely to the grout takes measured later (cubic feet of cement per foot along the dam axis). (*a*) Average conditions. (*b*) Individual grout holes. (*Houlsby, 1982.*)

forced into the fissures to be grouted. For the same reasons, air or water flush can be used, but never drilling mud. The rotary methods usually specified for grouting, at least in North America, are 2–5 times more expensive than percussive drilling. Deere (1976) has suggested that percussive drilling be permitted on more contracts.

Before injection, the holes are flushed to remove drill cuttings and, to the extent possible, to remove filling materials from the joints. Cuttings or fillings restrict the flow of grout and may necessitate a more closely spaced and therefore more expensive pattern of grout holes.

Flushing can be carried out either from within a single hole, by injecting water at the base of the hole so that cuttings and fillings flow upward to the collar, or by pressurizing one hole and draining from adjacent ones until the discharge water is running clear. This requires skill and is time-consuming, expensive, and often incomplete. Once a "pipe" of water is formed between adjacent holes, little subsequent washing is achieved. Flushing is more efficient at high pressures and can be assisted by pumping air into the wash water, which causes a boiling action.

Surface jetting is a high-pressure treatment given to joints exposed at rock surfaces such as in galleries. In special cases beneath the foundations of dams, joints with substantial amounts of clay filling can be treated by fine jets of water under pressures as high as 70 MPa, gaining access by a network of exploratory adits. These are expensive to excavate, but the results can be monitored and a high degree of quality control can be achieved.

18.2.4.3 Pressure testing. Grout holes are often pretested with water, section by section, to identify zones of intense jointing likely to need precementing and to help plan the best methods for grouting. Water is pumped into packed-off sections of drillhole and the inflow rates are measured. Pressure testing just before grouting also serves to check the pump, lines, and connections, and to wet the rock to reduce takeup of water from the grout.

In the Lugeon method (Sec. 4.3.3.4), which originates from French grouting practice, the water flow rate is reported in lugeon units, where 1 lugeon is 1 L/min per meter of packed-off hole under a standarized injection pressure of 1 MPa. Grouting is often restricted to those locations where the water loss exceeds 1 lugeon, approximately equivalent to a hydraulic conductivity of 10^{-7} m/s. The standard 1-MPa pressure, however, cannot be applied in jointed rock at shallow depths because it would jack the joints open.

18.2.4.4 Cement grouting. Cement grouting requires a mixer, an agitator, a grout pump, and hoses connected to a downhole packer or a grout nipple in the hole collar (see packer and stage systems, Sec. 18.2.4.1). Expanding rubber or leather "mechanical" packers are robust and inexpensive. Pneumatic inflatable packers 300–600 mm long have been used to assist in the grouting of soft or fractured rock in which the seating of mechanical packers is seldom effective (Gourlay and Carson, 1982).

A lump-free grout, essential for thorough and uninterrupted grouting, is produced by a mixer of the high-speed impeller type. Chemical additives, usually sulfonates, improve the mixing and wetting of the colloid particles. Injection is usually by an air-driven positive displacement pump.

Alternative *circulating* and *single-line* systems are available. In the circulating system, all grout not accepted by the drillhole is returned to the agitator tank. The continuously moving grout has no opportunity to set in the supply hose. In the alternative single-line system, all unused grout is discharged through a blow-off valve. Single-line grouting generates pressure pulses that some believe help to force grout into the rock.

The grout should be fed steadily into the hole at three quarters to full final grout pressure. It is best to start with a thin grout in order to avoid blocking the joints and possibly losing the hole. Grouting is often started at a consistency of between 3:1 and 5:1 (water to cement, by volume) and thinned if injection is slow or thickened if it is so fast that the specified pressure cannot be reached at the maximum pump speed. If the required pressure cannot be obtained even with a thick grout, pumping is stopped, the grout is allowed to set, and the hole is redrilled.

18.2.4.5 Permissible grouting pressures.

The higher the grouting pressure, the greater the penetration, and the fewer the drillholes needed. There is, however, an upper limit to the pressure that can safely be used. If this is exceeded, hydraulic fracturing occurs. Rock joints are forced apart and voids are created where none existed before. Buildings and other structures above and close to the zone of injection may be uplifted and damaged. The strength of the rock mass can actually be reduced by grouting at too high a pressure, because of loss of interlocking and shear strength along joints.

Excessive pressures can, in the extreme, trigger a rock collapse such as the one that destroyed a hydropower plant built at the foot of the Niagara gorge. Grouting work was in progress to seal vertical open joints in the limestone rock. The pressure of grout appears to have been sufficient to trigger a large slab slide that destroyed the power station at the foot of the cliff. It has never been rebuilt.

Hydraulic fracture is actually encouraged in some European grouting practice (Rigny, 1974) on the premise that grout travels further

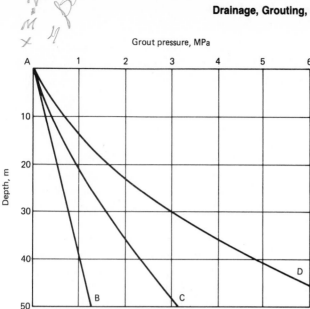

Figure 18.8 Grouting pressure variations according to depth. Line AB defines an overburden pressure of 25.5 kPa/m, and is a conservative lower limit of grouting pressure for closely jointed rocks. Line AC is a commonly accepted limit for massive sedimentary rocks that have been pregrouted above the point of injection. Line AD is a commonly accepted limit for strong and massive rocks.

with fewer holes and that joint closure on release of pressure consolidates and squeezes water from the in-place grout. Whether or not this viewpoint is accepted, hydraulic fracture should evidently be avoided wherever uplift might cause damage or instability.

Guidelines for grouting pressures in dam foundations are given in Fig. 18.8. In horizontally bedded sedimentary rocks it is usually considered unsafe to grout at pressures above the overburden pressure (curve AB in Fig. 18.8), which is calculated as the product of bulk unit weight and depth (Sec. 5.3.1.2). However, in massive hard rocks, grout pressures of several times overburden pressure have been used safely. The only sure way to establish a reliable permissible grouting pressure at a given site is by testing with packers at different elevations. Pressure is increased while carefully monitoring pressure and injected volume. The point of hydraulic fracture is usually marked by a sudden increase in the rate of grout take.

18.2.5 Quality control and monitoring of grouting

18.2.5.1 Grout take and leakage. On typical projects the grout take can vary from practically zero to 1000 kg of cement per lineal meter of

grout hole. Occasionally more than 50,000 sacks of cement (over 1,000,000 kg) have been pumped into a single hole (Houlsby, 1982). When the rate of injection shows no sign of tailing off, grouting should be stopped after several hundred sacks. After letting the grout set, nearby holes or special techniques are used to confine the grout to the zone of rock to be treated. Surface leakage around the grout holes can be stopped by low-pressure grouting of an upper stage, by excavating cutoff trenches and filling them with concrete, by caulking cracks and the holes themselves with wooden wedges and other materials, by applying slush grout or shotcrete to the rock surface, or by grouting the packed-off rock at depth before removing the surcharge of overburden or weathered and broken rock.

Grouting in karstic strata can be a frustrating experience, requiring many cycles of injection before the formations will hold pressure. The Grand Rapids dam in northern Manitoba, Canada, needed 20 times more grout than had been estimated in the call for tenders because of the severity of karstic channels. During injections, a helicopter circled the site to see when and where the grout emerged at surface. Injection was then stopped and the cement allowed to set before repeating the treatment.

18.2.5.2 Monitoring of grouting work. Grout pressures and takes should be systematically recorded (e.g., Fig. 18.7) together with the results of water testing in primary, secondary, and tertiary holes (Huck and Waller, 1982). Areas surrounding the zone of treatment need continual inspection to detect leakages and heaves as soon as they occur. Near-surface grouting, particularly when close to buildings, should be accompanied by displacement monitoring. Precise geodetic leveling or one of the settlement measuring systems described in Chap. 12 may be used.

Sound velocity measurements provide a useful index for assessing the effectiveness of grouting (Sec. 3.3.2). A fully grouted rock mass has a sonic velocity approaching that of unjointed rock in the laboratory, whereas jointed and open rock has a much lower velocity. Measurements during site investigation help in deciding whether and where to grout and the likely grout quantities and costs. A comparison of velocity measurements before, during, and after grouting provides a measure of the improvement achieved.

18.3 Freezing

18.3.1 Uses and limitations

Freezing, which converts the interstitial water in soil or rock into ice, can be used to stop groundwater inflows for the period of construction, to consolidate loose ground temporarily or to provide structural under-pinning (Jessberger, 1979). Its main application is in cofferdams and shaft sinking through silts and weathered and clay-bearing rock strata that cannot be effectively or economically grouted, and partic-ularly for stabilizing excavations in bouldery, soft, or running ground below the water table. The method is also useful for stabilizing tun-nels where they intersect the soil-rock contact. Permanent freezing systems are rare outside the arctic regions, where they are sometimes used to maintain frozen ground beneath heated buildings or pipelines and to seal in-ground storage containers for cryogenic liquids.

Ground freezing has the advantage of being easily confined to the immediate vicinity of the excavation, and, using the cryogenic method, it can be completed quite quickly. However, the methods are expensive and are considered only when there are serious technical problems with alternatives.

18.3.2 Freezing methods

Controlled ground freezing for mining and civil engineering construc-tion work is described in Shuster (1972), who reports that it was first used in Wales in 1862. Frozen conditions are created by circulating a cold medium through a series of pipes so that when the individual col-umns of frozen ground merge, they form a continuous membrane around the volume to be excavated (Fig. 18.9). There are two main al-ternatives, the *brine* method, using a salt solution, and the *cryogenic* method, using liquid carbon dioxide or, more often, nitrogen.

Brine freezing (Fig. 18.10a) is similar to conventional refrigeration. The brine is cooled by a refrigerating unit and then circulated through pipes in drillholes. The brine method most often employed nowadays was developed by F. H. Poetsch in Germany around 1880. The pri-mary source of refrigeration is a one- or two-stage refrigeration plant which compresses either ammonia or freon; two stages are needed for temperatures below $-25°C$. The most common coolant is water with calcium chloride added in sufficient quantities to depress the freezing point so that the brine remains liquid and pumpable. In northern win-ter climates, the cost of freezing can often be reduced by taking ad-vantage of atmospheric cooling through heat exchangers.

Brine freezing is slow, because brine remains liquid only down to temperatures of about $-35°C$. In contrast, the liquid nitrogen most of-ten used for cryogenic freezing evaporates at about $-196°C$, so that cryogenic freezing is much quicker. The cryogenic refrigerant is ex-pendable but expensive. It is applied directly to the freeze tubes, evap-orating at the point where freezing is required, and exhausting to the atmosphere (Fig. 18.10b). The method is most often used for "rescue

(a)

(b)

Figure 18.9 Ground freezing to sink a tunnel ventilation shaft through gravel overburden, Canada. (*Courtesy of Cementation Company Ltd.*)

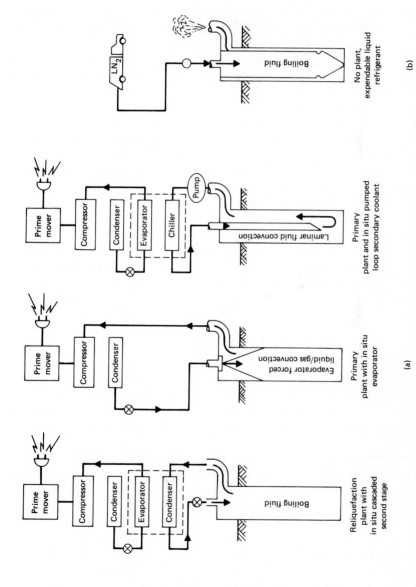

Figure 18.10 Freezing processes. (*a*) Brine freezing alternatives. (*b*) Cryogenic freezing. (*After Shuster, 1972.*)

operations" lasting only a few hours to a day or two, when the cost of delays is high.

Alternatives of freezing with brine or liquid nitrogen were considered on a tunneling project in North Dakota, where a thin shale barrier separated the tunnel crown from water-laden gravels and a lake. The boring machine had been brought to a halt by unexpected lenses of strong and brittle limestone within the shale. Blasting to break up the limestone could crack the shale and lead to a catastrophic inflow. A quick assessment indicated that nitrogen freezing, including drilling of 80-mm-diameter holes, would require a week, compared to 10 weeks using brine circulated through 130-mm holes. No power would be needed for nitrogen freezing, whereas the brine method would require a 275–300-hp supply weighing 35–40 t. With nitrogen freezing the rock might be disrupted by ice lens formation, and thawing might be irregular, placing uneven stresses on the segmental liner. In the end, neither technique was employed. The limestone lenses were broken by the boring machine after making modifications to the cutter head.

Drilling to freeze the ground around a shaft is often accomplished from surface, before shaft sinking, and this requires great accuracy. Oil-well drill rigs are often used, with directional drilling in which the drillhole is surveyed and the drill bit redirected if it wanders off line. Freezing has succeeded in this manner even to depths of 600 m in shaft sinking operations for mining projects in Saskatchewan, Canada.

18.3.3 Design and monitoring of freezing operations

Freezing elements must be located and aligned precisely. Misalignment leaves unfrozen "windows" and allows leakage and rapid thawing of the frozen barrier. Accidental freezing within the zone to be excavated makes the ground tougher and excavation work more difficult.

Most problems and failures on freezing projects have been related to groundwater flow. Salinity and water flow rate should be known in advance, since they determine the required temperature and duration of treatment. If water flowing into the freezing zone supplies energy at a greater rate than it can be removed by the refrigeration plant, the zone will not freeze. For brine systems, the maximum transverse groundwater seepage velocity that can be frozen is about 1–2 m per day, whereas for cryogenic systems, flows as high as 50 m per day have been stopped, although at greater expense.

Ground movements and pressures need to be monitored to determine the effects of freezing. Freezing of saturated ground causes expansion if there is no way for excess water to be expelled, and further expansion can accompany thawing. This can lead to problems of ground heave if not controlled. Any associated rupturing of freezing tubes will result in a leakage of brine into the ground, which thaws the ground and makes it impossible to refreeze by normal brine-freezing methods.

To control the freezing process and ensure completion of a satisfactory frozen zone, the ground temperatures must be monitored at critical locations. Thermocouples are most often employed, but can be supplemented by fluorescein frost-penetration markers consisting of 0.1% fluorescein in water, mixed with clean sand and encased in a clear plastic tube.

Computer-controlled logging is to be used to monitor freezing operations at the U.K. National Coal Board shafts at Asfordby, and a preprogrammed alarm system will warn of any anomalies in ground temperature (Martin, 1986). Two 500-m shafts will be sunk through 100 m of highly permeable sandstone, with the help of 39 freeze holes drilled from a collar just above the sandstone formation. Six weeks will be needed to grout the freeze pipes in place and to connect them to the ring main, after which freezing will take another 15 weeks.

References

Albritton, J. A.: "Cement Grouting Practices, U.S. Army Corps of Engineers," *Proc. Conf. Grouting in Geotech. Eng.* (New Orleans, La., 1982), Am. Soc. Civ. Eng., New York, pp. 264–278.

Bell, F. G., Ed.: *Methods of Treatment of Unstable Ground* (Newnes-Butterworths, London, 1975), 215 pp.

Choi, Y. L.: "Design of Horizontal Drains," *J. Eng. Soc. Hong Kong*, 37–49 (Dec. 1974).

Crow, L. J., D. J. Kelsh, M. Steinberg, and P. Colombo: "Preliminary Survey of Polymer-Impregnated Rock," Rep. Invest. RI7542, U.S. Bureau of Mines, 35 pp. (1971).

Deere, D. U.: "Dams on Rock Foundations, Some Design Questions," in *Rock Engineering for Foundations and Slopes, Proc. Speciality Conf.* (Boulder, Colo., 1976), Am. Soc. Civ. Eng., New York, vol. 2, pp. 55–85.

————: "Cement-Bentonite Grouting for Dams," *Proc. Conf. Grouting in Geotech. Eng.* (New Orleans, La., 1982), Am. Soc. Civ. Eng., New York, pp. 279–300.

———— and G. Lombardi: "Grout Slurries, Thick or Thin?," *Proc. ASCE Conf. Issues in Dam Grouting* (Denver, Colo., 1985), pp. 156–164.

Dietz, H. K. O.: "Grouting Techniques Used in Deep South African Mines," *Proc. Conf. Grouting in Geotech. Eng.* (New Orleans, La., 1982), Am. Soc. Civ. Eng., New York, pp. 606–620.

Farvolden, R. N., and J. P. Nunan: "Hydrogeologic Aspects of Dewatering at Welland," *Can. Geotech. J.*, **7**, 194–204 (1970).

Franklin, J. A., and O. Hungr: "Stabilization of Soils and Rocks Using Molten Sulphur," *Proc. Sulphur 81 Conf.* (Calgary, Alta., 1981), 18 pp.

Gebhart, L. R.: "Foundation Seepage Control Options for Existing Dams," in *Inspection, Maintenance and Rehabilitation of Old Dams, Proc. Eng. Foundation Conf.* (Pacific Grove, Calif., Sept. 1973), pp. 660–676 (1974).

Glossop, R.: "The Invention and Development of Injection Processes," *Geotechnique*, **10**, 91–100, 255–279 (1961).

Goodman, R. E., and K. W. John: "Surface and Near-Surface Excavations," Gen. Rep., Theme B, *Proc. 5th Int. Cong. Rock Mech.* (Melbourne, Australia, 1983), 12 pp.

Gourlay, A. W., and C. S. Carson: "Grouting Plant and Equipment," *Proc. Conf. Grouting in Geotech. Eng.* (New Orleans, La., 1982), Am. Soc. Civ. Eng., New York, pp. 121–135.

Gyenge, M.: "Nuclear Waste Vault Sealing," *Proc. 13th Can. Rock Mech. Symp.* (Toronto, Ont., 1980), pp. 181–192.

Houlsby, A. C.: "A Digest of Typical Cement Grouting Takes," *Proc. Conf. Grouting in Geotech. Eng.* (New Orleans, La., 1982), Am. Soc. Civ. Eng., New York, pp. 1000–1014.

Huck, P. J., and M. J. Waller: "Quality Control for Grouting," *Proc. Conf. Grouting in Geotech. Eng.* (New Orleans, La., 1982), Am. Soc. Civ. Eng., New York, pp. 781–791.

Hutchinson, J. N.: "*Assessment of the Effectiveness of Corrective Measures in Relation to Geological Conditions and Types of Slope Movement,*" Gen. Dep., Theme 3, *Proc. Symp. Landslides and Other Mass Movements*, Prague, Czechoshovakia, 1977), *Bull. Int. Assoc. Eng. Geol.* 16, pp. 131–155.

Jessberger, H. L., Ed.: *Ground Freezing* (Elsevier, Amsterdam, 1979), 550 pp.

Johnson, S. J.: "Cement and Clay Grouting of Foundations; Grouting with Clay-Cement Grouts," *J. Soil Mech. and Found. Div., ASCE,* **84** (SM1), paper 1545, pp. 1–12 (1958).

Johnson, G. D.: "Thorough Grouting Can Reduce Lining Costs in Tunnels," *Proc. Conf. Grouting in Geotech. Eng.* (New Orleans, La., 1982), Am. Soc. Civ. Eng., New York, pp. 892–906.

Karol, R. H.: "Chemical Grouts and Their Properties," *Proc. Conf. Grouting in Geotech. Eng.* (New Orleans, La., 1982), Am. Soc. Civ. Eng., New York, pp. 359–377.

Lancaster-Jones, P. F. F.: "Methods of Improving the Properties of Rock Masses," in K. G. Stagg and O. C. Zienkiewicz, Eds., *Rock Mechanics and Engineering Practice* (Wiley, New York, 1968).

Lau, D., and A. Crawford: "The Cement Grouting of Discontinuities in Rock Masses," *Proc. 27th U.S. Symp. Rock Mech.* (Tuscaloosa, Ala., 1986), pp. 854–861.

Littlejohn, G. S.: "Design of Cement Based Grouts," *Proc. Conf. Grouting in Geotech. Eng.* (New Orleans, La., 1982), Am. Soc. Civ. Eng., New York, pp. 35–48.

Martin, D.: "Ground Freezing Sees Mine Shafts to a Dry Bottom," *Tunnels and Tunnelling,* 67 (Mar. 1986).

Moller, D. W., H. L. Minch, and J. P. Welsh: "Ultrafine Cement Pressure Grouting to Control Groundwater in Fractured Granite Rock," Publ. SP-83-8, Am. Cement Inst., pp. 129–151 (1983).

Rigny, P.: "European versus U.S. Grouting Practices," *Foundations for Dams Conf.* (Asilomar, Calif., 1974), pp. 37–45.

Sabarly, F., A Pautre, and P. Londe: "Quelques réflexions sur la drainabilité des massifs rocheux," *Proc. 2d Int. Cong. Rock Mech.* (Belgrade, Yugoslavia, 1970), Rep. 6–12.

Seed, H. B.: "Foundation and Abutment Treatment for High Embankment Dams on Rock," *J. Soil Mech. and Found. Div., ASCE,* **98** (SM10), 1115–1128, paper 9269 (1972).

Sharp, J. C.: "Drainage Characteristics of Subsurface Galleries," *Proc. 2d Int. Cong. Rock Mech.* (Belgrade, Yugoslavia, 1970), vol. 3, pp. 197–204.

Shuster, J. A.: "Controlled Freezing for Temporary Ground Support," *Proc. 1st North Am. Rapid Excavation and Tunneling Conf.* (Chicago, Ill., 1972), vol. 2, pp. 863–894.

Sowers, G. F.: "Dewatering Rock for Construction," in *Rock Engineering for Foundations and Slopes, Proc. Specialty Conf. ASCE* (Boulder, Colo., 1976), pp. 200–216.

Index

ABOUT THE AUTHORS

JOHN FRANKLIN is a consulting engineer and Research Professor at the University of Waterloo in Ontario, Canada. He is perhaps best known for having introduced, in 1970, the point-load and slake-durability index tests that are now widely used for rock classification, and for his contributions to international standardization of testing methods. Recently, his research group has developed new ways to obtain information on rock jointing patterns and blasting fragmentation using digitized photographs. Dr. Franklin is currently president of the International Society for Rock Mechanics.

MAURICE DUSSEAULT is a consulting engineer and Professor of Geological Engineering in the Earth Sciences Department at the University of Waterloo in Ontario, Canada. Dr. Dusseault and his staff do research mainly in the area of resource development and waste management, specifically the geomechanics of potash, salt, coal, and clay shales, underground and open pit mining, and heavy oil recovery.